Engineering Materials and Processes

Series Editor
Professor Brian Derby, Professor of Materials Science
Manchester Materials Science Centre, Grosvenor Street, Manchester, M1 7HS, UK

Other titles published in this series:

Fusion Bonding of Polymer Composites
C. Ageorges and L. Ye

Composite Materials
D.D.L. Chung

Titanium
G. Lütjering and J.C. Williams

Corrosion of Metals
H. Kaesche

Corrosion and Protection
E. Bardal

Intelligent Macromolecules for Smart Devices
L. Dai

Microstructure of Steels and Cast Irons
M. Durand-Charre

Phase Diagrams and Heterogeneous Equilibria
B. Predel, M. Hoch and M. Pool

Failure in Fibre Polymer Laminates
M. Knops
Publication due January 2005

Materials for Information Technology
E. Zschech, C. Whelan and T. Mikolajick
Publication due March 2005

Gallium Nitride Processing for Electronics, Sensors and Spintronics
S.J. Pearton, C.R. Abernathy and F. Ren
Publication due March 2005

Thermoelectricity
J.P. Heremans, G. Chen and M.S. Dresselhaus
Publication due August 2005

Computer Modelling of Sintering at Different Length Scales
J. Pan
Publication due October 2005

Computational Quantum Mechanics for Materials Engineers
L. Vitos
Publication due January 2006

Fuel Cell Technology
N. Sammes
Publication due January 2006

M.M. Kamiński

Computational Mechanics of Composite Materials

Sensitivity, Randomness and Multiscale Behaviour

 Springer

M.M. Kamiński, MSc, PhD
Division of Mechanics of Materials, Technical University of Łódz,
Al. Politechniki 6, 93 - 590 Łódz, Poland

British Library Cataloguing in Publication Data
Kamiński, M.M.
   Computational mechanics of composite materials :
   Sensitivity, randomness and multiscale behaviour. —
   (Engineering materials and processes)
   1. Composite materials — Mathematical models   2. Mechanics,
   Applied — Data processing
   I. Title
   620′.001518
   ISBN 1852334274

Library of Congress Cataloging-in-Publication Data
Kamiński, M. M. (Marcin M.), 1969–
   Computational mechanics of composite materials: sensitivity, randomness, and
   Multiscale behaviour / M.M. Kamiński
       p. cm.—(Engineering materials and processes, ISSN 1619-0181)
   Includes bibliographical references and index.
   ISBN 1-85233-427-4 (alk. Paper)
       1. Composite materials—Mechanical properties—Mathematical models.   I. Title.   II.
   Series.
   TA418.9.C6   K345   2002
   621.1′892—dc21   2002033327

Engineering Materials and Processes ISSN 1619-0181
ISBN 1-85233-427-4
Springer Science+Business Media
springeronline.com

Typesetting: Electronic text files prepared by author
69/3830-543210   Printed on acid-free paper   SPIN 10791506

Acknowledgements

Chapter 2 includes material, © Civil-Comp Press, 2001 previously published in M. Kamiński, "Homogenization Method in Stochastic Finite Element Analysis of some 1D Composite Structures", in Proc. 8th Int. Conf. on Civil & Structural Engineering Computational Technology, B.H.V. Topping, ed., (paper no 60), Civil-Comp Press, Stirling, United Kingdom, 2001. This material is reprinted with permission from Civil-Comp Press, Stirling, United Kingdom.

Chapter 5 includes material, © Civil-Comp Press, 2001 previously published in Ł. Figiel, M. Kamiński, "Mechanical and Thermal Fatigue of Curved Composite Beams", in Proc. 8th Int. Conf. on Civil & Struct. Engineering Computational Technology, B.H.V. Topping, ed., (paper no 61), Civil-Comp Press, Stirling, United Kingdom, 2001. This material is reprinted with permission from Civil-Comp Press, Stirling, United Kingdom.

Chapter 5 includes material, © Civil-Comp Press, 2002 previously published in Ł. Figiel, M. Kamiński, "Numerical Analysis of Fatigue Damage Evolution in Composite Pipe Joints", in Proc. 6th Int. Conf. on Computational Structures Technology, B.H.V. Topping and Z. Bittnar, eds., (paper no 134), Civil-Comp Press, Stirling, United Kingdom, 2002. This material is reprinted with permission from Civil-Comp Press, Stirling, United Kingdom.

Chapter 7 includes material, © Civil-Comp Press, 2002 previously published in M. Kamiński, "Multiresolutional Homogenization Technique in Transient Heat Transfer for Unidirectional Composites", in Proc. 6th Int. Conf. on Computational Structures Technology, B.H.V. Topping and Z. Bittnar, eds., (paper no 138), Civil-Comp Press, Stirling, United Kingdom, 2002. This material is reprinted with permission from Civil-Comp Press, Stirling, United Kingdom.

Figures 2.38 – 2.41 are reproduced from M. Kamiński, M. Kleiber, Stochastic finite element method in random non-homogeneous media, in *Numerical Methods in Engineering '96*, J.A. Desideri *et al.*, eds. pp. 35 – 41, 1996, © John Wiley & Sons Limited. Reproduced with permission.

Figures 2.98 – 2.119 are reproduced from M. Kamiński, M. Kleiber, Numerical homogenization of n-component composites including stochastic interface defects, *Int. J. Num. Meth. Engrg.*, 47: 1001-1027, 2000, © John Wiley & Sons Limited. Reproduced with permission.

Figures 7.16 – 7.21 and 7.54 – 7.60 are reproduced from M. Kamiński, Stochastic perturbation approach to wavelet-based multiresolutional analysis, *Num. Linear Algebra with Applications*, 11(4): 355-370, 2004, © John Wiley & Sons Limited. Reproduced with permission.

Figures 2.65 and 2.66 reprinted from *International Journal of Engineering Science*, Vol 38, Kamiński, M., Homogenized properties of n-components composites, pp. 405-427, Copyright (2000), with permission from Elsevier.

Figures 2.4 – 2.13 reprinted from *International Journal of Solids and Structures*, Vol 33, Kamiński, M. and Kleiber, M., Stochastic structural interface defects in composite materials, pp. 3035-3056, Copyright (1996), with permission from Elsevier.

Figures 2.1 – 2.3 and 2.30 – 2.40 reprinted from *Computers and Structures*, Vol 78, Kamiński, M. and Kleiber, M., Perturbation-based stochastic finite element method for homogenization of two-component elastic composites, pp. 811-826, Copyright (2000), with permission from Elsevier.

Figures 4.1 – 4.9 reprinted from *International Journal of Engineering Science*, Vol 41, Kamiński, M., Homogenization of transient heat transfer problems for some composite materials, pp. 1-29, Copyright (2003), with permission from Elsevier.
Figures 2.67 – 2.69, 2.77, 2.78, 2.88, 2.89 and 4.17 – 4.52 reprinted from *Computer Methods in Applied Mechanics and Engineering*, Vol 192, Kamiński, M., Sensitivity analysis of homogenized characteristics of some elastic composites, pp. 1973-2005, Copyright (2003), with permission from Elsevier.
Figures 3.1 – 3.12 reprinted from *Computational Materials Science*, Vol 22, Figiel, Ł., Kamiński, M., Effective elastoplastic properties of the periodic composites, pp. 221-239, Copyright (2001), with permission from Elsevier.
Figures 2.129 – 2.140 reprinted from *Computational Materials Science*, Vol 11, Kamiński, M., Probabilistic bounds on effective elastic moduli for the superconducting coils, pp. 252-260, Copyright (1998), with permission from Elsevier.
Figures 5.1 – 5.4 and 5.66 –5.73 reprinted from *International Journal of Fatigue*, Vol 24, Kamiński, M., On probabilistic fatigue models for composite materials, pp. 477-495, Copyright (2002), with permission from Elsevier.
Figures 7.2 – 7.15 are reprinted from *Computational Materials Science*, Vol. 27, Kamiński M., Wavelet-based homogenization of unidirectional multiscale composites, pp. 613-622, Copyright (2001), with permission from Elsevier.
Figures 7.30-7.43 and 7.46-7.53 reprinted from *Computer Methods in Applied, Mechanics and Engineering*, Kamiński, M., Homogenization-based finite element analysis of unidirectional composites by classical and multiresolutional techniques, in press, Copyright (2005), with permission from Elsevier.

Figure 2.49 reprinted from Kamiński, M., Stochastic computational mechanics of composite materials", in *Advances in Composite Materials and Structures VII*, de Wilde, W.P., Blain, W.R. and Brebbia, C.A., eds., pp. 219 – 228, Copyright (2000) WIT Press, Ashurst Lodge, Ashurst, Southampton, UK. Used with permission.

Figures 2.42 and 4.10 – 4.16 reprinted from *Archives of Applied Mechanics*, Material sensitivity analysis in homogenization of the linear elastic composites, Kamiński, M., 71(10): 679 – 694, 2001, copyright Springer-Verlag Heidelberg. Used with permission.

Figure 2.143 is reproduced from Kamiński, M., Stochastic finite element in homogenization of linear elastic composites. *Arch. Civil Engrg.* 3(XLVII): 291-325, 2001. Copyright property of the Polish Academy of Science. Used with permission.

Figures in Chapter 6 are reproduced from Kamiński, M., Stochastic reliability in contact problems for spherical particle reinforced composites. *Journal of Theoretical and Applied Mechanics*, 3(39): 539-562, 2001. Used with permission.

Figure 7.1 appeared in Kamiński, M., Multiresolutional wavelet-based homogenization of random composites in *Proceedings of the European Conference on Computational Mechanics* Cracow, 26 - 29 June 2001.

Preface

Composite materials accompanied the human activity from the beginning of the civilisation. Apart from natural composites, like the wood, applied in various structures people invented many multi−component materials even in ancient times. One of the most famous applications of the old−time composites is the Chinese Wall, whose durability and stability was ensured by contrastively different materials incorporated into a single structure. Next applications worked out and popularised in Central Europe in the Middle Ages was known as the Prussian wall combining the wooden skeleton filled with the bricks. One of the most significant milestones in the history of modern composites was the application of the concrete reinforced with the steel bars in France at the end of the nineteenth century.

Nowadays composites play a very important role in engineering from aerospace technology and nuclear devices to microelectronics or structural engineering applications [37,128,203,286,298,351,367,389]. Considering this fact and the growing role of numerical experiments in the designing of structures and industrial processes, one of the most important purposes of computational mechanics research and direction of progress appeared to be precise numerical modelling of these materials. On the other hand, experimental sciences prove that every structural parameter has a random, in fact stochastic, character. Thus, many probabilistic approaches and methodologies have emerged recently to simulate more accurately the real behaviour of mechanical systems and processes. These methods show that the random character of parameters discussed is very important for the systems simulated [14,121,357]. This conclusion may lead us to the hypothesis, that the random character of the material and physical parameters should play an essential role in multi−component structures [32,34,151,154,275].

Modern computational mechanics of composite materials follows many various ways through different science domains from experimental materials science to advanced computational techniques and applied mathematics. They engage more and more complicated and precise testing methods and devices, stochastic and sensitivity analysis algorithms and multiscale domain theoretical solutions for partial and ordinary differential equations reflecting some practical engineering and physical problems. Commercial computer programs based on the Finite Element Method enable now visualisation of the multifield, multiphase and non-stationary physical and mechanical problems and even introducing uncertainty into computer simulation using random variables (ANSYS, for instance). The growth of computer power obtained from technological progress and advances in parallel numerical techniques practically eliminated the parameter of the cost of computational time in modelling, which resulted in the efficient implementation and use of Monte Carlo simulation.

The basic idea behind this book was to collect relatively up−to−date approaches to the composite materials lying somewhere in between experimental measurements and their opportunities, theoretical advances in applied mathematics and mechanics, numerical algorithms and computers as well as the practical needs of the engineers. The methods are well–documented in the context of computer batch files, scripts and computer programs. It will enable the readers to start from this point and to continue and/or replace the ideas with newer, more accurate and efficient ones. The author believes that this book will appear to be useful for applied mathematicians, specialists in numerical methods and for engineers: civil, mechanical, aerospace and from related branches of industry. Some elements of probabilistic calculus and computation as well as general ideas can also be applied by students, who can incorporate these concepts into new research or into the existing well–documented knowledge dealing with composite materials.

A primary version of the book was completed in Texas, during the author's postdoctoral research at Rice University in Houston in the academic year 1999/2000 under auspices of Prof. P.D. Spanos. The author would like to appreciate the help of many people, whose valuable comments and the time spent enabled finishing of the book. Special thanks are directed to Prof. Michał Kleiber from the Institute of Fundamental Technological Research, Polish Academy of Science in Warsaw, who expressed many precious ideas during a common research in random composites and who promoted this research. Prof. Tran Duong Hien from Technical University of Szczecin influenced the work in the area of stochastic finite elements. The cooperation with Prof. B.A. Schrefler from the University of Padua in Italy concerning numerical analysis of superconducting composites remarkably enhanced the relevant computational illustration included in the book. The help of Mr. Łukasz Figiel, M.Sc. and Mr. Marcin Pawlik, Dr. Eng., two of my younger colleagues, was decisive for finishing of some computations devoted to heat transfer and fracture analysis. The author would like to express his respect to all the colleagues from Chair of Mechanics of Materials at the Technical University of Łódź for their advising voices, too. Last but not least, the role of the unknown reviewers, the editors and the people who commented and criticised this work is also appreciated.

Layout of the Book

Mathematical preliminaries open the book considerations and consist of basic definitions of random events, variables and probabilistic moments as well as description of the Monte Carlo simulation technique with the relevant statistical estimation theory elements. The stochastic perturbation approach (second order second central moment generalised to the nth order and higher moments technique) is explained using two examples: a transient heat transfer equation and the solution of the linear elastodynamic problem. The solution to these problems in terms of expected values and standard deviations as well as spatial and temporal cross-covariances is demonstrated and it illustrates the applicability of the method. An important part of this opening chapter is a probabilistic algebraic description of some transforms of random variables, which is necessary for further formulation and development of the stochastic interface defects model. Some of them are valid for the Gaussian variates only, which essentially bounds the application. However, it leads to the specific formulae implemented further in the computer software attached. An important issue raised in this chapter is to show a difference between Gaussian and quasi–Gaussian random variables defined on some unempty and bounded real subsets.

Elastic problems related to deterministic and probabilistic systems are collected in Chapter 2. They are divided into two essentially different parts – the first shows the linear elastic behaviour of some composite materials and structures in boundary value problems connected with their real microstructure. The other part contains description of the homogenisation technique together with the relevant numerical tests documenting the computational determination of so-called homogenisation functions, *a posteriori* error analysis related to homogenisation problems, probabilistic moments of effective material tensors and their variability with respect to some input parameters.

The first part of this chapter starts from the mathematical model of composite, whose material characteristics are given arbitrarily as constant deterministic values or by using the first two probabilistic moments constant through the given component material region (or volume). Further, the stochastic interface defects concept is presented, which originated from some computational contact mechanics models. The interface defects are introduced as semicircles lying on the interface into a weaker material. The radii and total number of these defects are input cut-off Gaussian random variables defined using their expected values and the variances (or standard deviations) with elastic properties equal to 0. The modeling is performed through the following steps: (i) determination of the interphase – a thin film containing all the defects with thickness determined from defect probabilistic parameters, (ii) probabilistic spatial averaging of the defects over the interphase area, (iii) computational analysis of a new composite with the new extra component. Obviously, it is not possible to approximate the real composite with stochastic interface microdefects very accurately. However it can be and it is done intermediately – by comparison with the composites with the weakened interphase or interface, for instance. Computational experiments validating the model are performed using the system ABAQUS [1] (in the deterministic approach) and the specially adapted academic package POLSAP (for the Stochastic Finite Element Method – SFEM needs) [183]. All the results obtained for various composites and various combinations of interface defect parameters demonstrate a high level of structural uncertainty in the case of their presence as well as a significant increase of the structural state functions stresses and displacements around the interface region. The second part of the chapter concerns the homogenisation method both in deterministic and probabilistic context. Computational experiments dealing with a numerical solution of the homogenisation problem are done thanks to the FEM commercial system ANSYS [2], where most of the databases for these experiments are available from the author to be used in further extensions of mathematical and mechanical homogenisation model.

Interface defects model and probabilistic homogenisation using both Monte Carlo simulation techniques are analysed using the authors FEM implementation called MCCEFF. The results of simulation are compared in terms of expected values and variances with analogous results obtained through the stochastic second order perturbation methodology. The appendix to this chapter consists of necessary fundamental mathematical theorems and definitions for the asymptotic homogenisation theorem.

Elastoplasticity of composites discussed in the next chapter is focused on the alternative homogenisation technique, where instead of periodicity conditions imposed on the external boundaries of the RVE, some combination of the symmetry conditions and strain fields are applied to this element. The application of this method to the homogenisation of a periodic superconducting coil cable is also shown − an effective elastoplastic constitutive law is determined numerically and shown as a function of the homogenising uniform strain applied at the RVE boundary. Analogously to the methods typical for elastostatic problems, the closed-form equations for effective yield stresses are formulated in various ways, which can next be extended on probabilistic analysis. This chapter is completed with the transformation matrices algebraic definition, which is the essence of the computational implementation of the method. Probabilistic moments of the effective elastoplastic constitutive law can be obtained as a conjunction of this method with the Monte Carlo simulation technique discussed in the previous chapter. The fundamental issue is however experimental determination of higher order probabilistic moments for the superconductor material characteristics; otherwise the analysis is useful in the context of the sensitivity of the homogenised characteristics with respect to the adopted level of input randomness only.

Sensitivity analysis presented in Chapter 4 is entirely devoted to a relatively new research area − determination of the sensitivity gradients for homogenised material characteristics. For this purpose two essentially different homogenisation methods are used − algebraic approximation and asymptotic methodology. Starting from a traditional description of the effective parameters in both methods, the sensitivity gradients are determined by the symbolic calculus approach and, on the other hand, pure computational strategy based on the Finite Difference Method (FDM). The implementation and results obtained from these two methods demonstrate the basic limitations of the methods, i.e. necessity of closed-form equations for the symbolic approach and numerical instabilities in the FDM simulations. This knowledge is necessary for significant time savings in the extension of this study to the random composite sensitivity analysis where the heterogeneous periodic composites with probabilistically defined material properties are analysed. The probabilistic sensitivity of such structures is defined through the introduction of sensitivity gradients of probabilistic moments of the effective material parameters with respect to the appropriate moments of composite structure parameters − elastic properties of the constituents as well as interface defect data.

Fracture and fatigue − the collection of various fatigue theories with special emphasis placed on the second order perturbation method application are discussed next. The crucial numerical illustration is presented in the case of the Paris Erdogan rule where some of the system input data are treated as random va riables. Therefore, expected values are compared against the deterministic values and standard deviations are added, too. An analogous approach is used to reformulate the well-known fracture criteria applied for composite materials – Tsai–Wu and Tsai–Hill - and to use them in symbolic computations for probabilistic parameters of the composite material fracture parameters. The essential part of this chapter is devoted to the FEM modelling of fracture and fatigue of some composites where analytical solutions are not available. Computational illustrations consist of static fracture of curved composite under shear loading leading to the delamination, fatigue analysis of composite pipe joint as well as thermomechanical fatigue of the curved laminate under thermal and/or static quasistatic load varying in time with constant amplitude. Most of the frequently used theories and equations for fatigue analysis are collected in the appendix to this chapter.

Reliability analysis is included in the Chapter 6 and it consists of a discussion of various order reliability computational approaches together with the Weibull Second Order and Third Moment model (W-SOTM). This methodology is used to compute the reliability index for the composite Hertz contact problem, where elastic spherical inclusion of the reinforcement is loaded by the force to remain in contact with the matrix. Further in this chapter a stochastic process description of the degradation phenomena is also given, which appears to be common for the homogeneous and heterogeneous structures and materials. It can find a broad field of applications together with efficient implementations of stochastic processes (with both spatial and temporal randomness) in the Finite Element Method (or BEM, FDM, meshless as well as hybrid method based) programs.

An application of the wavelet-based multiresolutional approach to composite materials in terms of homogenisation of multiscale media is the extension of previous considerations and concludes the book. The traditional composite materials model consisting of two or three geometrical scales is now rewritten in view of practically infinite number of separate scales (resolutions) that can be linked using interscale wavelet projection (some mathematical transformation). The basic tool necessary for such an analysis development is the basic wavelet basis (a mathematical function varying rapidly in a given geometrical scale), which can be used now to transform between neighboring scales. The homogenised characteristics for the composite can be determined usually in the closed–form equation if and only if the limit of an infinite series of wavelet projections between all geometrical scales exists and is unique. As is illustrated by some wavelet function samples, such an analysis type can be some alternative for the random analysis, because the wavelet functions used in various scale makes, in the coarsest scale, the impression that the relevant material property demonstrates the great level of some kind of uncertainty. It is not underlined clearly that the main limitation of this methodology is that the wavelet projection between the neighbouring scales can be continued through the range of validity of the same physical laws. It is not possible to carry out the passage from the atomistic to the global scale of the composite using the same wavelet projection and, most probably, this is the way that this research area should be extended. The multiresolutional homogenisation is demonstrated for a very general case – a linear ordinary differential equation, which can reflect the linear elastic behaviour of a unidirectional multiscale composite in compression/tension or in bending. On the other hand, this technique can be applied with only small modifications to the unidirectional field problems for heat conduction, seepage flow, electrostatic problems, etc. Further, as is documented by the mathematical derivations, the MRA approach formula reduces to the results obtained in the asymptotic homogenisation technique for the two-scale medium. Since some research is done towards the multiscale analysis and homogenisation for 2D heterogeneous media, the main interest has been directed next to the multiresolutional homogenisation of dynamic and transient problems. Some basic theoretical and computational results are obtained under the assumption that non–stationary and dynamic components of the relevant ODEs can be homogenised independently from the stationary part. In practice it makes possible to calculate effective dynamic structural parameters as the relevant spatial averages for the entire multiscale composite; it is however done for the material properties given *a priori* as some algebraic combination of the elementary wavelets (harmonic, Haar, Gabor, Morlet, Daubechies and Mexican hat functions). Since the homogenisation is the intermediate technique to determine the homogeneous equivalent medium and to replace the real structure with this medium, the results are incorporated next in the classical Finite Element calculus for various boundary value or boundary initial engineering problems. They unambiguously show the limitations of the application of various homogenisation techniques used in engineering computations, i.e. simple spatial averaging, asymptotic approach and multiresolutional method. As can be expected, spatial averaging gives the fastest but least precise approximation for the real structure. The application of the wavelet technique is more recommended to periodic composites having a smaller number of periodicity cells in the Representative Volume Element (RVE), whereas the asymptotic approach gives the best results for increased number of cells in the RVE. Therefore, for most engineering composite structures, where the total number of the periods through their lengths is limited, the proposed multiscale approach seems to be the most efficient. The wavelet functions can be incorporated in the Finite Element Method automatic projection between various scales even for the needs of homogeneous system structural computations – for the fluid flow problems where the profile of the flow is a nonlinear and multiscale complex function (wind pressure profile for high buildings in civil engineering applications). That is why some elementary equations and ideas are collected here and the conjunction of such an analysis with the second order perturbation analysis is presented here to extend the applicability range of traditional wavelet projection on probabilistic analyses, where some input random fields are given using the expected values and the spatial or temporal cross–correlations. The elementary numerical example of cosinusoidal wavelet function implemented in the symbolic package MAPLE demonstrates the computational aspects of this methodology. There is no doubt, however, that the next step will be to make the multiresolutional version of asymptotic homogenisation of the multiscale plane periodic structures where the Daubechies wavelets can find application.

A number of references follow the last chapter. However new valuable conference papers, research and review articles as well as entire books continue to appear on the publishing market. Therefore, it is impossible to appreciate the significant contributions of all the people to this field. The book is completed with the appendix containing the user's manual to the computer code MCCEFF available from the author on the special request. Following the algorithm for data preparation, the reader will be able to solve either deterministic and/or probabilistic homogenisation problems for the fibre−reinforced composite for the rectangular RVE containing a single fibre with the round cross−section. The next part of this appendix is devoted to the batch file for the elastoplastic analysis of the steel−reinforced concrete plate using the commercial FEM system ABAQUS. This file contains the author's comments written in such a manner that the file is ready−to−use by ABAQUS without further processing. Symbolic computation code written in the MAPLE standard concludes the appendix. This script is responsible for a computational mathematic derivation of the homogenised heat conductivity coefficient for the unidirectional multiscale periodic composite structure according to (1) the spatial averaging method, (2) asymptotic homogenisation approach and (3) multiresolutional homogenisation method. It returns for initially specified wavelet functions the values of homogenised parameters, their variability with respect to the contrast parameter and the interface location for two−component RVE. This file can be used without further modifications for sensitivity gradient symbolic computations for the effective parameters returned from these methods with respect to the design parameters mentioned. Probabilistic analysis using Monte Carlo simulation, probabilistic integration technique and perturbation−based analysis is under construction now and will be available also by a special request from the author.

Contents

1 Mathematical Preliminaries

1.1 Probability Theory Elements

1.1.1 Introduction

Probability theory [326,357,365] is a part of theoretical and applied mathematics, which is engaged in establishing the rules governing random events – random games or experimental testing. The definitions, theorems and lemmas given below are necessary to understand the basic equations and computer implementation aspects used in the later numerical analyses presented in the book. They can also be used to calculate many of the closed–form equations applied frequently in applied sciences and engineering practice [19,37,150,201,202,253].

Definition
The variations with n elements for k elements are k elements series where each number $1,2,...,k$ corresponds to the single element from the initial set. The variations can differ in the elements or their order. The total number of all variations with n elements for k is described by the relation

$$V_n^k = \frac{n!}{(n-k)!} = \underbrace{n(n-1)...(n-k+1)}_{k-times} \tag{1.1}$$

Example
Let us consider the three-element set A{X,Y,Z}. Two–element variations of this set are represented as $V_3^2 = 6$: XY, YZ, XZ, YX, ZY, ZX.

Definition
Permutations with n elements are n–element series where each number $1,2,...,n$ corresponds to the single element from the initial n–element set. The difference between permutations is in the element order. The total number of all permutations with n different elements is given by the formula:

$$P_n = V_n^n = 1 \cdot 2 \cdot ... \cdot n = n! \tag{1.2}$$

If among n elements X, Y, Z,... there are identical elements, where X repeats a times, Y appears b times, while Z repeats c times etc., then

$$P_n = \frac{n!}{a!b!c!} \tag{1.3}$$

Example

Let us consider the three element set $A\{X,Y,Z\}$. The following permutations of the set A are available: $P_3 = 6$: XYZ, XZY, YZX, YXZ, ZXY, ZYX.

Definition

The combinations with n elements for k elements are k−elements sets, which can be created by choosing any k elements from the given n−element set, where the order does not play any role. The combinations can differ in the elements only. The total number of all combinations with n for k elements is described by the formula

$$C_n^k = \binom{n}{k} = \frac{n!}{k!(n-k)!} \tag{1.4}$$

In specific cases it is found that

$$C_n^1 = \binom{n}{1} = n \, , \; C_n^n = \binom{n}{n} = 1 \tag{1.5}$$

where

$$\binom{n}{k} = \binom{n}{n-k}, \; \binom{n}{n} = \binom{n}{0} = 1 \tag{1.6}$$

Example

Let us consider a set $A\{X,Y,Z\}$ as before. Two−element variations of this set are the following: XY, XZ and YZ.

The fundamental concepts of probability theory are random experiments and random events resulting from them. A single event, which can result from some random experiment is called elementary event, an and for the single die throw is equivalent to any sum of the dots on a die taken from the set $\{1,...,6\}$. Further, it is concluded that all elementary events corresponding to the random experiment form the elementary events space defined usually as Ω, which various subsets like A and/or B belong to (favouring the specified event or not, for instance).

Definition

A formal notation $\omega \in A$ denotes that the elementary event ω belongs to the event A and is understood in the following way – if ω results from some experiment, then the event A happened too, which ω belongs to. The notation means that the elementary event ω favours the event A.

Definition

The formal notation $A \subset B$, which means that event A is included in the event B is understood such that event A results in the event B since the following implication holds true: if the elementary event ω favours event A, then event ω favours event B, too.

Definition

An alternative of the events $A_1, A_2, ..., A_n$ is the following sum:

$$A_1 \cup A_2 \cup ... \cup A_n = \bigcup_{i=1}^{n} A_i \qquad (1.7)$$

which is a random event consisting of all random elementary events belonging to at least one of the events $A_1, A_2, ..., A_n$.

Definition

A conjunction of the events $A_1, A_2, ..., A_n$ is a product

$$A_1 \cap A_2 \cap ... \cap A_n = \bigcap_{i=1}^{n} A_i \qquad (1.8)$$

which proceeds if and only if any of the events $A_1, A_2, ..., A_n$ proceed.

Definition

Probability is a function P which is defined on the subsets of the elementary events and having real values in closed interval $[0,1]$ such that

(1) $P(\Omega)=1$, $P(\varnothing)=0$;

(2) for any finite and/or infinite series of the excluding events $A_1, A_2, ..., A_n, ...$ $A_i \cap A_j = \varnothing$, there holds for $i \neq j$

$$P\left(\bigcup_i A_i\right) = \sum_i P(A_i) \qquad (1.9)$$

Starting from the above definitions one can demonstrate the following lemmas:

Lemma

The probability of the alternative of the events is equal to the sum of the probabilities of these events.

Lemma

If event B results from event A then

$$P(A) \leq P(B) \tag{1.10}$$

The definition of probability does not reflect however a natural very practical need of its value determination and that is why the simplified Laplace definition is frequently used for various random events.

Definition
If n trials forms the random space of elementary events where each experiment has the same probability equal to $1/n$, then the probability of the m−element event A is equal to

$$P(A) = \frac{m}{n} \tag{1.11}$$

Next, we will explain the definition, meaning and basic properties of the probability spaces. The probability space (Ω, F, P) is uniquely defined by the space of elementary random events Ω, the events field F and probabilistic measure P. The field of events F is the relevant family of subsets of the space of elementary random events Ω. This field F is a non−empty, complementary and countable additive set having σ-algebra structure.

Definition
The probabilistic measure P is a function

$$P : F \to [0,1] \tag{1.12}$$

which is a nonnegative, countable additive and normalized function defined on the fields of random events. The pair (Ω, F) is a countable space, while the events are countable subsets of Ω. The value $P(A)$ assigned by the probabilistic measure P to event A is called a probability of this event.

Definition
Two events A and B are independent if they fulfil the following condition:

$$P(A \cap B) = P(A) \cdot P(B) \tag{1.13}$$

while the events $\{A_1, A_2, ..., A_n\}$ are pair independent, if this condition holds true for any pair from this set.

Definition
Let us consider the probability space (Ω, F, P) and measurable space $\{\Re^n, B_n\}$, where B_n is a class of the Borelian sets. Then, the representation

$$X : \Omega \to \mathfrak{R}^n \tag{1.14}$$

is an n-dimensional random variable or n-dimensional random vector.

Definition
The probability distribution of the random variable X is a function $P_X : B \to [0,1]$
such that

$$\mathop{\forall}_{b \in B} P_X(b) = P(X \in B) \tag{1.15}$$

The probability distribution of the random variable is a probabilistic measure.

Definition
Let us consider the following probability space $(\mathfrak{R}, B, P_X)$. The function
$F_X : \mathfrak{R} \to [0,1]$ defined as

$$F(x) = P_X\big[(-\infty, x)\big] \tag{1.16}$$

is called the cumulative distribution function of the variable X.

Definition
The function $f : \mathfrak{R} \to \mathfrak{R}_+$ has the following properties:
(1) there holds almost everywhere (in each point of the cumulative distribution
function differentiability):

$$\frac{dF(x)}{dx} = f(x) \tag{1.17}$$

(2)
$$f(x) \ge 0 \tag{1.18}$$

(3)
$$\int_{-\infty}^{+\infty} f(x)dx = 1 \tag{1.19}$$

(4) for any Borelian set $b \in B$ the integral $\int_b f(x)dx = P(X \in b)$ is a probability

density function (PDF) of the variable X.

Definition
Let us consider the random variable $X : \Omega \to \mathfrak{R}$ defined on the probabilistic space
(Ω, F, P). The expected value of the random variable X is defined as

$$E[X] = \int_{-\infty}^{+\infty} X(\omega)dP(\omega) \tag{1.20}$$

if only the Lesbegue integral with respect to the probabilistic measure exists and converges.

Lemma

$$\underset{c \in \Re}{\forall}\; E[c] = c \tag{1.21}$$

Lemma

There holds for any random numbers X_i and the real numbers $c_i \in \Re$

$$E\left[\sum_{i=1}^{n} c_i X_i\right] = \sum_{i=1}^{n} c_i E[X_i] \tag{1.22}$$

Lemma

There holds for any independent random variables X_i

$$E\left[\prod_{i=1}^{n} X_i\right] = \prod_{i=1}^{n} E[X_i] \tag{1.23}$$

Definition

Let us consider the following random variable $X : \Omega \to \Re$ defined on the probabilistic space (Ω, F, P). The variance of the variable X is defined as

$$Var(X) = \int_{\Omega} \left(X(\omega) - E[X]\right)^2 dP(\omega) \tag{1.24}$$

and the standard deviation is called the quantity

$$\sigma(X) = \sqrt{Var(X)} \tag{1.25}$$

Lemma

$$\underset{c \in \Re}{\forall}\; Var(c) = 0 \tag{1.26}$$

Lemma

$$\underset{c \in \Re}{\forall}\; Var(cX) = c^2 Var(X) \tag{1.27}$$

Lemma

There holds for any two independent random variables X and Y

$$Var(X \pm Y) = Var(X) + Var(Y) \tag{1.28}$$

$$Var(X \cdot Y) = E^2[X] \cdot Var(Y) + Var(X) \cdot Var(Y) + Var(X) \cdot E^2[Y] \tag{1.29}$$

Definition

Let us consider the random variable $X : \Omega \to \Re$ defined on the probabilistic space (Ω, F, P). A complex function of the real variable $\varphi : \Re \to Z$ such that

$$\varphi(t) = E[\exp(itX)] \tag{1.30}$$

stands for the characteristic function of the variable X.

1.1.2 Gaussian and Quasi-Gaussian Random Variables

Let us consider the random variable X having a Gaussian probability distribution function with m being the expected value and $\sigma > 0$ the standard deviation. The distribution function of this variable is

$$F(x) = \frac{1}{\sqrt{2\pi}} \int_{-\infty}^{x} \exp\left(\frac{-t^2}{2}\right) dt \tag{1.31}$$

where the probability density function is calculated as

$$f(x) = \frac{1}{\sigma\sqrt{2\pi}} \exp\left(-\frac{(x-m)^2}{2\sigma^2}\right) \tag{1.32}$$

The characteristic function for this variable is denoted as

$$\varphi(t) = E[\exp(itX)] = \exp\left(mit - \tfrac{1}{2}\sigma^2 t^2\right). \tag{1.33}$$

If the variable X with the parameters (m, σ) is Gaussian, then its linear transform $Y = AX + B$ with $A, B \in \Re$ is Gaussian, too, and its parameters are equal to $Am+B$ and $|A|\sigma$ for $A \neq 0$, respectively.

Problem

Let us consider the random variable X with the first two moments $E[X]$ and $Var(X)$. Let us determine the corresponding moments of the new variable $Y = X^2$.

Solution

The problem has been solved using three different ways illustrating various methods applicable in this and in analogous cases. The generality of these methods make them available in the determination of probabilistic moments and their parameters for most random variables and their transforms for given or unknown probability density functions of the input frequently takes place in which numerous engineering problems.

I method
Starting from the definition of the variance of a ny random variable one can write

$$Var(Y) = E(Y^2) - E^2(Y) \tag{1.34}$$

Let $Y = X^2$, then

$$Var(X^2) = E((X^2)^2) - E^2(X^2) \tag{1.35}$$

The value of $E[X^4]$ will be determined through integration of the characteristic function for the Gaussian probability density function

$$E[X^4] = \frac{1}{\sigma\sqrt{2\pi}} \int_{-\infty}^{+\infty} x^4 \exp\left(-\frac{(x-m)^2}{2\sigma^2}\right) dx \tag{1.36}$$

where $m = E[X]$ and $\sigma = \sqrt{Var(X)}$ denote the expected value and standard deviation of the considered distribution, respectively. Next, the following standardised variable is introduced

$$t = \frac{x-m}{\sigma}, \text{ where } x = t\sigma + m, dx = \sigma dt \tag{1.37}$$

which gives

$$E[X^4] = \frac{1}{\sqrt{2\pi}} \int_{-\infty}^{+\infty} (t\sigma + m)^4 \exp\left(-\frac{t^2}{2}\right) dt \tag{1.38}$$

After some algebraic transforms of the integrand function it is obtained that

$$E[X^4] = \frac{1}{\sqrt{2\pi}} \int_{-\infty}^{+\infty} (\sigma^4 t^4 + 4\sigma^3 m t^3 + 6\sigma^2 m^2 t^2 + 4\sigma m^3 t + m^4)\, e^{-\frac{t^2}{2}} dt \tag{1.39}$$

and, dividing into particular integrals, there holds

$$E[X^4] = \frac{1}{\sqrt{2\pi}} (\sigma^4 I_1 + 4\sigma^3 m I_2 + 6\sigma^2 m^2 I_3 + 4\sigma m^3 I_4 + m^4 I_5)\, e^{-\frac{t^2}{2}} \tag{1.40}$$

where the components denote

$$I_1 = \int_{-\infty}^{+\infty} t^4 e^{-\frac{t^2}{2}} dt \; ; \; I_2 = \int_{-\infty}^{+\infty} t^3 e^{-\frac{t^2}{2}} dt \; ; \; I_3 = \int_{-\infty}^{+\infty} t^2 e^{-\frac{t^2}{2}} dt \; ;$$

$$I_4 = \int_{-\infty}^{+\infty} t e^{-\frac{t^2}{2}} dt \; ; \; I_5 = \int_{-\infty}^{+\infty} e^{-\frac{t^2}{2}} dt \tag{1.41}$$

It should be mentioned that the values of the odd integrals on the real domain are equal to 0 in the following calculation

$$\int_{-\infty}^{+\infty} f(x)g(x)dx = \int_{-\infty}^{0} f(x)g(x)dx + \int_{0}^{+\infty} f(x)g(x)dx \tag{1.42}$$

If the function $f(x)$ is odd and $g(x)$ is even

$$f(-x)=-f(x), \; g(-x)=g(x), \tag{1.43}$$

then it can be written

$$\int_{-\infty}^{0} f(x)g(x)dx = \int_{0}^{+\infty} f(-x)g(x)dx = -\int_{0}^{+\infty} f(x)g(x)dx . \tag{1.44}$$

Considering that the odd indices integrals are calculated; this results in

$$I_5 = \int_{-\infty}^{+\infty} e^{-\frac{t^2}{2}} dt = \sqrt{2\pi} \tag{1.45}$$

$$I_3 = \int_{-\infty}^{+\infty} t^2 e^{-\frac{t^2}{2}} dt = -\int_{-\infty}^{+\infty} t(te^{-\frac{t^2}{2}})dt = -\int_{-\infty}^{+\infty} td(e^{-\frac{t^2}{2}})$$

$$= -te^{-\frac{t^2}{2}} \Big|_{-\infty}^{+\infty} + \int_{-\infty}^{+\infty} e^{-\frac{t^2}{2}} dt = \sqrt{2\pi} \tag{1.46}$$

$$I_1 = \int_{-\infty}^{+\infty} t^4 e^{-\frac{t^2}{2}} dt = -\int_{-\infty}^{+\infty} t^3 de^{-\frac{t^2}{2}} = -\left[t^3 e^{-\frac{t^2}{2}} \Big|_{-\infty}^{+\infty} - \int_{-\infty}^{+\infty} e^{-\frac{t^2}{2}} dt^3 \right]$$

$$= 3\int_{-\infty}^{+\infty} t^2 e^{-\frac{t^2}{2}} dt = -3\int_{-\infty}^{+\infty} tde^{-\frac{t^2}{2}} = -3\left[te^{-\frac{t^2}{2}} \Big|_{-\infty}^{+\infty} - \int_{-\infty}^{+\infty} e^{-\frac{t^2}{2}} dt \right] = 3\sqrt{2\pi} . \tag{1.47}$$

After simplification the result is

$$E\left[X^4\right] = 3\sigma^4 + 6\sigma^2 m^2 + m^4 = E^4[X] + 6Var(X)E^2[X] + 3Var^2(X) \tag{1.48}$$

$$E\left[X^2\right] = \sigma^2 + m^2 = E^2[X] + Var(X) \tag{1.49}$$

$$Var(X^2) = E\left[X^4\right] - E^2\left[X^2\right] = 2\sigma^2(\sigma^2 + 2m^2)$$
$$= 2Var(X)(Var(X) + 2E^2\left[X\right])$$

(1.50)

II method
Initial algebraic rules can be proved following the method shown below. Using a modified algebraic definition of the variance

$$Var(X^2) = E\left[X^4\right] - E^2\left[X^2\right]$$

(1.51)

and the expected value

$$E\left[X^2\right] = Var(X) + E^2[X]$$

(1.52)

subtracted from the following equation

$$E^2\left[X^2\right] = \left(Var(X) + E^2[X]\right)^2 = Var^2 X + 2Var(X)E^2[X] + E^4[X]$$

(1.53)

we can demonstrate the following desired result:

$$Var(X^2) = E\left[X^4\right] - Var^2(X) - 2Var(X)E^2[X] - E^4[X]$$

(1.54)

III method
The characteristic function for the Gaussian PDF has the following form:

$$\varphi(t) = \exp\left(mit - \tfrac{1}{2}\sigma^2 t^2\right)$$

(1.55)

where

$$\varphi^{(k)}(0) = i^k E\left[X^k\right]; \ k \geq 0$$

(1.56)

and

$$\varphi^{(0)} = \varphi \ ; \ \varphi'(0) = im$$

(1.57)

The mathematical induction rule leads us to the conclusion that

$$\varphi^{(n)}(t) = \left(im - t\sigma^2\right) \cdot \varphi^{(n-1)}(t) - (n-1)\sigma^2 \cdot \varphi^{(n-2)}(t), \ n \geq 2$$

(1.58)

which results in the equations

$$\varphi^{(2)}(0) = -m^2 - \sigma^2 \tag{1.59}$$

$$\varphi^{(3)}(0) = -mi\left(m^2 + 3\sigma^2\right) \tag{1.60}$$

$$\varphi^{(4)}(0) = m^4 + 6m^2\sigma^2 + 3\sigma^4 \tag{1.61}$$

giving the same input equations.

Problem 2

Let us determine the value of the Gaussian integral $\int\limits_{-\infty}^{+\infty}\exp(-x^2)dx$.

Solution
Starting from the obvious fact that

$$\iint\limits_{K_1}\exp\left(-\left(x^2+y^2\right)\right)dxdy$$
$$\leq \iint\limits_{K}\exp\left(-\left(x^2+y^2\right)\right)dxdy \leq \iint\limits_{K_2}\exp\left(-\left(x^2+y^2\right)\right)dxdy \tag{1.62}$$

with K_1 being a circle with radius t and located in the centre of Cartesian coordinates, K denotes the square contour with edge equal to $t\sqrt{2}$, while K_2 stands for a circle with radius $t\sqrt{2}$. The coordinates transform to the polar system given by

$$x = r\cos\varphi \ , \ \ y = r\sin\varphi \tag{1.63}$$

returns

$$\iint\limits_{\substack{0\leq r\leq t\\0\leq\varphi\leq 2\pi}}\exp\left(-r^2\right)rdrd\varphi$$
$$\leq \int\limits_{-t}^{t}\exp\left(-x^2\right)dx\int\limits_{-t}^{t}\exp\left(-y^2\right)dy \leq \iint\limits_{\substack{0\leq r\leq t\sqrt{2}\\0\leq\varphi\leq 2\pi}}\exp\left(-r^2\right)drd\varphi \tag{1.64}$$

Using the observation that

$$\int\limits_{-t}^{t}\exp\left(-x^2\right)dx = \int\limits_{-t}^{t}\exp\left(-y^2\right)dy \tag{1.65}$$

one can determine

$$\int\limits_{0}^{2\pi}d\varphi\int\limits_{0}^{t}\exp\left(-r^2\right)rdr \leq \left[\int\limits_{-t}^{t}\exp\left(-x^2\right)dx\right]^2 \leq \int\limits_{0}^{2\pi}d\varphi\int\limits_{0}^{t\sqrt{t}}\exp\left(-r^2\right)rdr \tag{1.66}$$

Next, considering the rule

$$\int \exp\left(-r^2\right)r\,dr = -\tfrac{1}{2}\exp\left(-r^2\right) \tag{1.67}$$

and the symmetry

$$\int_{-t}^{t} \exp\left(-x^2\right)dx = 2\int_{0}^{t}\exp\left(-x^2\right)dx \tag{1.68}$$

it is obtained finally that

$$\pi\left(1-\exp\left(-t^2\right)\right) \leq 4\left[\int_{0}^{t}\exp\left(-x^2\right)dx\right]^2 \leq \pi\left(1-\exp\left(-2t^2\right)\right) \tag{1.69}$$

Then, a square rooting procedure gives

$$\tfrac{1}{2}\sqrt{\pi}\left(1-\exp\left(-t^2\right)\right) \leq \int_{0}^{t}\exp\left(-x^2\right)dx \leq \tfrac{1}{2}\sqrt{\pi}\left(1-\exp\left(-2t^2\right)\right) \tag{1.70}$$

The three functions theorem and the limiting procedure for $t \to \infty$ allow us to show

$$\lim_{t \to \infty}\int_{0}^{t}\exp\left(-x^2\right)dx = \tfrac{1}{2}\sqrt{\pi} \tag{1.71}$$

with

$$\lim_{t \to \infty}\sqrt{1-\exp\left(-t^2\right)} = \lim_{t \to \infty}\sqrt{1-\exp\left(-2t^2\right)} = 1 \tag{1.72}$$

Lemma (Central Limit Theorem)

For any independent random variables X_i for $i=1,2,...,n$ the following sum $X = \sum_i X_i$ is asymptotically Gaussian where the parameters are equal $m = \sum_i m_i$ and $\sigma^2 = \sum_i \sigma_i^2$, respectively.

Further X be the random variable with P and F being the probability density and distribution functions, respectively and S any given Borelian set such that

$$P = P(S) = P(X \subset S) \tag{1.73}$$

Now, let us consider the following subset $S_0 \subset S$ such that

$$S_0 = \{X \in S : a \leq X \leq b\}; \quad a < b \leq +\infty \tag{1.74}$$

The probability density function defined on a domain S_0 is called a truncated Gaussian and its distribution function is given as

$$F(x | a \leq X \leq b) = \begin{cases} 0, & x < a \\ \frac{F(x) - F(a)}{F(b) - F(a)}; & a \leq x \leq b \\ 1, & x > b \end{cases} \tag{1.75}$$

and its probability density function is equal to

$$f(x | a \leq X \leq b) = \frac{f(x)}{\int\limits_a^b f(t)dt} \tag{1.76}$$

The considered problem of cutting off the probability density function in case of Gaussian or related random variable is very important for engineering applications in probabilistic calculus. Most engineering parameters, being random, must have nonnegative values as Young modulus or heat conductivity coefficients for instance [132]. Other parameters, like the Poisson ratio, are restricted to small intervals only. Then, let us focus on modifications of the presented formula describing the expected values and variances demonstrated for classical Gaussian variates in the case of bounded real domains.

Let us consider the Gaussian variable $N(m, \sigma)$ restricted to the positive values only. According to the above formulae, there holds

$$\lambda = \frac{f\left(-\frac{m}{\sigma}\right)}{1 - F\left(-\frac{m}{\sigma}\right)} \tag{1.77}$$

Then, the first two probabilistic moments for the so modified Gaussian PDF are obtained as

$$E[X] = m + \lambda\sigma \tag{1.78}$$

$$Var(X) = m^2 + \lambda\sigma m + \sigma^2 \tag{1.79}$$

Starting from the derived equations one can calculate the expected values and the variances of the quasi–Gaussian random variables, whose domains are restricted to the specific and bounded intervals resulting from physical interpretation of a specific equilibrium problem.

1.2 Monte Carlo Simulation Method

Monte Carlo simulation is a numerical method based in general on random sampling and statistical estimation [39,44,125] and now there are multiple numerical realizations of the latters as crude simulation, stratified and importance sampling as well as Latin Hypercube Sampling methodology. Nevertheless, the most important part of the method is a reliable random number generator. Monte Carlo simulation (MCS) is, in fact, a numerical method based on random sampling via a random number generator. The most important applications of the MCS technique in engineering of composite materials are: (a) fatigue and/or failure modeling [10,243], (b) modeling of random material properties [73,171,174,175, 191,196,306] and (c) reliability analysis [79]. Random nature of the effective properties calculated in homogenisation problem follows usually randomness of material properties of a composite, which are defined as the input Gaussian variables. To obtain the random sequences of this variable it is necessary to produce first numerically uniform deviates. They are random numbers, which lie within a specified range ([0,1] typically), and each number is as likely to occur as any other in the range. Generation of the uniform distributions is done using a standard FORTRAN library routine, which can be implemented as a linear congruential generator, which generates a sequence of integer numbers $I_1, I_2, I_3, ...,$ each between 0 and m-1, by using the recurrence relation [39]

$$I_{j+1} = a \cdot I_j + c \qquad (\text{mod } m) \qquad (1.80)$$

where m is called the modulus and a, c are positive integers called the multiplier and the increment, respectively. The recurrence (1.80) will possibly repeat itself with a period that is obviously no greater than m. If m, a and c are properly chosen, then the period of recurrence is of maximal length m. The sequence of real numbers between 0 and 1 is returned here by dividing I_{j+1} by m, so that it is strictly less than 1, but occasionally (once in m calls). The linear congruential method is very fast and requires only a few operations per call, but it is not free of sequential correlation on successive calls and the special shuffling routine has to be added to eliminate this disadvantage. Next, the Box–Muller method is implemented to transform these variables to the normalized Gaussian distribution– let us consider for this purpose the transformation between two uniform deviates on (0,1) denoted by x_1, x_2 and two quantities y_1 and y_2 defined as follows

$$y_1 = \sqrt{-2 \cdot \ln x_1} \cdot \cos 2\pi x_2 \qquad (1.81)$$

$$y_2 = \sqrt{-2 \cdot \ln x_1} \cdot \sin 2\pi x_2 \qquad (1.82)$$

Equivalently it can be written that

$$x_1 = \exp\left[-\tfrac{1}{2}\left(y_1^2 + y_2^2\right)\right] \tag{1.83}$$

$$x_2 = \frac{1}{2\pi} \cdot arctg \frac{y_2}{y_1} \tag{1.84}$$

with the Jacobian determinant of the form

$$\frac{\partial(x_1, x_2)}{\partial(y_1, y_2)} = \begin{vmatrix} \frac{\partial x_1}{\partial y_1} & \frac{\partial x_1}{\partial y_2} \\ \frac{\partial x_2}{\partial y_1} & \frac{\partial x_2}{\partial y_2} \end{vmatrix} = -\tfrac{1}{2\pi}\exp\left(-\tfrac{1}{2}(y_1^2 + y_2^2)\right) \tag{1.85}$$

since it is a product of the functions of y_2 and y_1 separately. Finally, we obtain each y is returned as the independent Gaussian variable.

The second part of the simulation procedure is a statistical estimation procedure [29], which enables approximation of probabilistic moments and the relevant coefficients for the given series of output variables and for the specified number of random trials. The equations listed below are implemented in the statistical estimation procedure to compute the probabilistic moments with respect to M, which denotes here the total number of Monte Carlo random trials.

Statistical estimation theory is devoted to determination and verification of statistical estimators computed on a basis of the random trials sets. These estimators are necessary for efficient approximation of the analysed random variable and they are introduced for the random variables, fields and processes to assure their stochastic convergence.

Definition
If there exist a random variable X such that

$$\underset{\varepsilon>0}{\forall}\ \lim_{n\to\infty} P\left(\left|X_n - X\right| < \varepsilon\right) = 1 \tag{1.86}$$

then the series of random variables X_n stochastically converges to X. Let us note that the consistent, unbiased, most effective and asymptotically most effective estimators are available in statistical estimation theory.

Definition
The consistent estimator is each estimator stochastically convergent to the estimated parameter.

Definition
The unbiased estimator fulfils the following condition:

$$E\left[\hat{Q}_n\right] = Q \tag{1.87}$$

Definition
The most effective estimator is the unbiased estimator with the minimal variance.

Definition
The asymptotically most effective estimator of the quantity $\hat{Q}_n$ is the following one:

$$\lim_{n \to \infty} \left(\frac{Var(\hat{Q}_0)}{Var(\hat{Q}_n)} \right) = 1 \tag{1.88}$$

where $Var\left(\hat{Q}_0\right)$ is the most effective variance estimator.

Definition
The expected value estimator of the random variable $X(\omega)$ in an n−element random trial is the average value

$$E[X(\omega)] = \tfrac{1}{n} \sum_{i=1}^{n} X_i = \overline{X} \tag{1.89}$$

It can be proved that this is consistent, unbiased and the most effective estimator for the Gaussian, binomial and Poisson probability distribution.

Definition
The variance estimator for the random variable $X(\omega)$ in an n−element random event is the quantity

$$Var(X(\omega)) = \tfrac{1}{n-1} \sum_{i=1}^{n} \left(X_i - \overline{X}\right)^2 \tag{1.90}$$

It can be demonstrated that this estimator is consistent and unbiased. Using this estimator one can determine standard deviation estimator.

Definition
The standard deviation estimator is equal to

$$S(X(\omega)) = \sqrt{Var(X(\omega))} \tag{1.91}$$

Comment
The variance estimator in the n−element random event can be defined as

$$Var(X(\omega)) = \tfrac{1}{n} \sum_{i=1}^{n} \left(X_i - \overline{X}\right)^2 \tag{1.92}$$

It can be demonstrated that

$$E[Var(X(\omega))] = \tfrac{n-1}{n}\sigma^2 \tag{1.93}$$

which gives the negative bias. The estimator bias is defined as the deviation of this estimator from its value to be determined. There holds

$$E[S_n^2] - \sigma^2 = \tfrac{n-1}{n}\sigma^2 - \sigma^2 = -\tfrac{1}{n}\sigma^2 \tag{1.94}$$

which results in a negative and bias, which is irrelevant since the natural condition for the variance $VarY \geq 0$.

Definition
The estimator of the ordinary kth order probabilistic moment of the random variable $X(\omega)$ in the n−element random trial is given as

$$m_k(X(\omega)) = \tfrac{1}{n}\sum_{i=1}^{n} X_i^k \tag{1.95}$$

Definition
The estimator of the kth order central probabilistic moment is defined as

$$\mu_k(X(\omega)) = m_k\left[(X(\omega)) - m_1(X(\omega))\right] \tag{1.96}$$

Any central moments of odd order are equal to 0 in case of the normalized Gaussian PDF $N(m,\sigma)$, while the first three even moments are given below.

Definition
The estimator of the second order central moment is equal to

$$\mu_2(X(\omega)) = \sigma^2 \tag{1.97}$$

Definition
The estimator of the fourth order central moment is given as

$$\mu_4(X(\omega)) = 3\sigma^4 \tag{1.98}$$

Definition
The estimator of the sixth order central moment is equal to

$$\mu_6(X(\omega)) = \frac{15\sigma^6}{m^3} \tag{1.99}$$

Using the proposed estimators of the central moments of the random variable $X(\omega)$ valid for the $n-$element random event, the following probabilistic coefficients are usually calculated:

Definition
The coefficient of variation for $X(\omega)$ is equal to

$$\alpha(X(\omega)) = \frac{\sigma(X(\omega))}{E[X(\omega)]} \qquad (1.100)$$

Definition
The coefficient of asymmetry for $X(\omega)$ equals to

$$\beta(X(\omega)) = \frac{\mu_3(X(\omega))}{\sigma^3(X(\omega))} \qquad (1.101)$$

Definition
The coefficient of concentration for $X(\omega)$ is equal to

$$\gamma(X(\omega)) = \frac{\mu_4(X(\omega))}{\sigma^4(X(\omega))} \qquad (1.102)$$

which results in $\beta \xrightarrow[n\to\infty]{} 0$ and $\gamma \xrightarrow[n\to\infty]{} 3$ for the Gaussian random variables.

Definition
The estimator of covariance for two random variables $X(\omega)$ and $Y(\omega)$ in a two dimensional $n-$element random trial is defined as

$$Cov(X(\omega),Y(\omega)) = \frac{1}{n-1} \sum_{i=1}^{n} (X_i(\omega) - \overline{X})(Y_i(\omega) - \overline{Y}) \qquad (1.103)$$

Definition
The coefficient of correlation for two variables $X(\omega)$ and $Y(\omega)$ in two dimensional $n-$element random event is equal to

$$\rho_{XY} = \frac{Cov(X(\omega), Y(\omega))}{\sqrt{Var(X(\omega))Var(Y(\omega))}} \qquad (1.104)$$

Remark
Two random variables $X(\omega)$ and $Y(\omega)$ are fully correlated only if $\rho_{XY}=1$ and uncorrelated in case of $\rho_{XY}=0$.

Equations (1.101) and (1.102) are very useful together with the relevant PDF estimator in recognising of the probabilistic distribution function type for the output variables – using the Central Limit Theorem the Gaussian variables can be found. This is very important aspect considering the fact that theoretical considerations in this subject are rather complicated and not always possible.

1.3 Stochastic Second Moment Perturbation Approach

1.3.1 Transient Heat Transfer Problems

The main concept of stochastic second order perturbation technique [263] applied in the next chapters to various transient heat transfer computations can be explained on the example of the following equation [135]:

$$C \cdot \dot{T} + K \cdot T = Q \qquad (1.105)$$

where K, C are some linear stochastic operators equivalent to the heat conductivity and capacity matrices, T is the random thermal response vector for the structure with $\dot{T}$ representing its time derivative, while Q is the admissible heat flux (due to the boundary conditions) applied on the system. To introduce a precise definition of K, for instance, let us consider the Hilbert space $\mathbf{H}$ defined on a real domain D and the probability space (Ω, σ, P), where $x \in D$, $\theta \in \Omega$ and $\Theta : \Omega \to R$. Then, the operator $K(\mathbf{x}; \omega)$ is some stochastic operator defined on $\mathbf{H} \times \Theta$, which means that it is a differential operator with the coefficients varying randomly with respect to one or more independent design random variables of the system; the operator C can be defined analogously. As is known, the analytical solutions to such a class of partial differential equations are available for some specific cases and that is why quite different approximating numerical methods are used (simulation, perturbation or spectral methods as well).

Further, let us denote the vector of random variables of a problem as $\left\{ b^r(x; \theta) \right\}$ and its probability density functions as $g(b^r)$ and $g(b^r, b^s)$, respectively; $r, s = 1, 2, ..., R$ are indexing input random variables. Next, let us introduce integral definition for the expected values of this vector as

$$E[b^r] = \int_{-\infty}^{+\infty} b^r g(b^r) db^r \qquad (1.106)$$

with its covariance in the form

$$Cov(b^r,b^s)= \int\limits_{-\infty}^{+\infty} \int\limits_{-\infty}^{+\infty} (b^r - E[b^r])(b^s - E[b^s]) g(b^r,b^s) db^r db^s \qquad (1.107)$$

Next, all material and physical parameters of Ω as well as their state functions being random fields are extended by the use of stochastic expansion via the Taylor series as follows:

$$K(x;\theta)= K^0(x;\theta)+ \sum_{n=1}^{N}\left\{\frac{\varepsilon^n}{n!} K^{(n)}(x;\theta) \prod_{r=1}^{n} \Delta b^r (\theta)\right\} \qquad (1.108)$$

where ε is some given small perturbation parameter, $\varepsilon\Delta b^r$ denotes the first order variation of Δb^r about its expected value $E[b^r]$ and $K^{(n)}(x;\theta)$ represents the nth order partial derivatives with respect to the random variables determined at the expected values. The variable θ represents here the random event belonging to the corresponding probability space of admissible events (nonnegative, for instance). The second order perturbation approach is now analysed and then the random operator $K(x;\theta)$ is expanded as

$$K(x;\theta) = K^0(x;\theta) + \varepsilon K^{,r}(x;\theta)\Delta b^r + \tfrac{1}{2}\varepsilon^2 K^{,rs}(x;\theta)\Delta b^r \Delta b^s \qquad (1.109)$$

It can be noted that the second order equation is obtained by multiplying the R-variate probability density function $p_R(b_1,b_2,...,b_R)$ by the ε^2-terms and by integrating over the domain of $\mathbf{b}(x;\theta)$. Assuming that the small parameter ε of the expansion is equal to 1 and applying the stochastic second order perturbation methodology to the fundamental deterministic equation (1.105), we find

- zeroth order equations:

$$C^0(x;\theta)\cdot \dot{T}^0(x;\theta)+ K^0(x;\theta)\cdot T^0(x;\theta) = Q^0(x;\theta) \qquad (1.110)$$

- first order equations (for $r=1,...,R$):

$$\begin{aligned} C^{,r}(x;\theta)\cdot \dot{T}^0(x;\theta)+ C^0(x;\theta)\cdot \dot{T}^{,r}(x;\theta) \\ + K^{,r}(x;\theta)\cdot T^0(x;\theta) + K^0(x;\theta)\cdot T^{,r}(x;\theta) = Q^{,r}(x;\theta) \end{aligned} \qquad (1.111)$$

- second order equations (for $r,s=1,...,R$):

$$C^{,rs}(x;\theta)\cdot\dot{T}^0(x;\theta)+2C^{,r}(x;\theta)\cdot\dot{T}^{,s}(x;\theta)+C^0(x;\theta)\cdot\dot{T}^{,rs}(x;\theta)$$
$$+K^{,rs}(x;\theta)\cdot T^0(x;\theta)+2K^{,r}(x;\theta)\cdot T^{,s}(x;\theta)+K^0(x;\theta)\cdot T^{,rs}(x;\theta) \qquad (1.112)$$
$$=Q^{,rs}(x;\theta)$$

It is clear that coefficients for the products of K, C and T are the successive orders of the initial basic deterministic eqn (1.110) and they are taken from the well–known Pascal triangle. As far as the nth order partial differential perturbation-based approach is concerned, then the general statement can be written out using the Leibniz differentiation rule in the following form:

$$\binom{n}{0}C^{(n)}(x;\theta)\cdot\dot{T}^0(x;\theta)+\binom{n}{1}C^{(n-1)}(x;\theta)\cdot\dot{T}^{(1)}(x;\theta)+...$$
$$+\binom{n}{n}C^{(0)}(x;\theta)\cdot\dot{T}^{(n)}(x;\theta)+\binom{n}{0}K^{(n)}(x;\theta)\cdot T^0(x;\theta)$$
$$+\binom{n}{1}K^{(n-1)}(x;\theta)\cdot T^{(1)}(x;\theta)+... \qquad (1.113)$$
$$+\binom{n}{n-1}K^{(1)}(x;\theta)\cdot T^{(n-1)}(x;\theta)+\binom{n}{n}K^{(0)}(x;\theta)\cdot T^{(n)}(x;\theta)$$
$$+\binom{n}{n}K^{(0)}(x;\theta)\cdot T^{(n)}(x;\theta)=Q^{(n)}(x;\theta)$$

The equations from $m=0$ to the specific value of n should be generated to introduce all hierarchical equations system for the nth order perturbation approach. Usually, it is assumed that higher than second order perturbations can be neglected, the system of equations (1.110) – (1.112) constitutes the given equilibrium problem. The detailed convergence studies should be carried out in further extensions of the model with respect to perturbation order, parameter ε and the coefficient of variation of input random variables.

Furthermore, it can be noted that system (1.111) is rewritten for all random parameters of the problem indexed by $r=1,...,R$ (R equations), while system (1.112) gives us generally R^2 equations. The unnecessary equations are eliminated here through multiplying both sides of the highest order equation by the appropriate covariance matrix of input random parameters. There holds

- zeroth order equations:

$$C^0(x;\theta)\cdot\dot{T}^0(x;\theta)+K^0(x;\theta)\cdot T^0(x;\theta)=Q^0(x;\theta) \qquad (1.114)$$

- 1st order equations (for $r=1,...,R$):

$$C^{\cdot r}(x;\theta) \cdot \dot{T}^0(x;\theta) + C^0(x;\theta) \cdot \dot{T}^{\cdot r}(x;\theta)$$
$$+ K^{\cdot r}(x;\theta) \cdot T^0(x;\theta) + K^0(x;\theta) \cdot T^{\cdot r}(x;\theta) = Q^{\cdot r}(x;\theta) \tag{1.115}$$

- second order equations (for $r,s=1,\dots,R$):

$$C^0(x;\theta) \cdot \dot{T}^{(2)}(x;\theta) + K^0(x;\theta) \cdot T^{(2)}(x;\theta)$$
$$= \left\{ Q^{\cdot rs}(x;\theta) - K^{\cdot rs}(x;\theta) \cdot T^0(x;\theta) + 2K^{\cdot r}(x;\theta) \cdot T^{\cdot s}(x;\theta) \right.$$
$$\left. - C^{\cdot rs}(x;\theta) \cdot \dot{T}^0(x;\theta) + 2C^{\cdot r}(x;\theta) \cdot \dot{T}^{\cdot s}(x;\theta) \right\} Cov\!\left(b^r,b^s\right) \tag{1.116}$$

It is observed that solving for the nth order perturbation equations system, the closure of the entire hierarchical system is obtained by nth order correlation of input random vector components b^r and b^s, respectively; for this purpose nth order statistical information about input random variables is however necessary. To obtain the probabilistic solution for the analysed heat flow problem, eqn (1.114) is solved for T^0, eqn (1.115) for first order terms $T^{\cdot r}$ and, finally, eqn (1.116) for $T^{(2)}$. Therefore, using the definition of expected value and applying the second order expansion, it is derived that

$$E\big[T[\mathbf{b}(\mathbf{x};\theta);\mathbf{x}]\big] = \int\limits_{-\infty}^{+\infty} T[\mathbf{b}(\mathbf{x};\theta);\mathbf{x}]\, p_R(\mathbf{b}(\mathbf{x};\theta))\, d\mathbf{b}$$
$$= \int\limits_{-\infty}^{+\infty} \Big\{ T^0[\mathbf{b}(\mathbf{x};\theta);\mathbf{x}] + T^{\cdot r}[\mathbf{b}(\mathbf{x};\theta);\mathbf{x}]\Delta b_r(\mathbf{x}) \tag{1.117}$$
$$+ \tfrac{1}{2} T^{\cdot rs}[\mathbf{b}(\mathbf{x};\theta);x]\Delta b_r(\mathbf{x})\Delta b_s(\mathbf{x}) \Big\} p_R(\mathbf{b}(\mathbf{x};\theta))\, d\mathbf{b}$$

and further

$$T^0(\mathbf{x};\theta)\int\limits_{-\infty}^{+\infty} p_R(\mathbf{b}(\mathbf{x};\theta))\, d\mathbf{b} + T^{\cdot r}(\mathbf{x};\theta)\int\limits_{-\infty}^{+\infty} \Delta b_r(\mathbf{x};\theta)\, p_R(\mathbf{b}(\mathbf{x};\theta))\, d\mathbf{b}$$
$$+ \tfrac{1}{2} T^{\cdot rs}(\mathbf{x};\theta)\int\limits_{-\infty}^{+\infty} \Delta b_r(\mathbf{x};\theta)\Delta b_s(\mathbf{x};\theta)\, p_R(\mathbf{b}(\mathbf{x};\theta))\, d\mathbf{b} \tag{1.118}$$

This result leads us to the following relation for the expected values [135,190]:

$$E\big[T[\mathbf{b}(\mathbf{x};\theta);\mathbf{x}]\big] = T^0[\mathbf{b}(\mathbf{x};\theta);\mathbf{x}] + \tfrac{1}{2} T^{\cdot rs}[\mathbf{b}(\mathbf{x};\theta);\mathbf{x}] S_b^{rs} \tag{1.119}$$

Now, using the perturbation approach, both spatial and temporal cross-covariances can be determined separately. There holds for spatial cross–covariance computed at the specific time moment τ

$$Cov\left(T\left[\mathbf{b}\left(\mathbf{x}^{(1)};\theta\right),\mathbf{x}^{(1)};\tau\right];T\left[\mathbf{b}\left(\mathbf{x}^{(2)};\theta\right),\mathbf{x}^{(2)};\tau\right]\right)=S_T^{ij}\left(\mathbf{x}^{(1)};\mathbf{x}^{(2)};\tau\right)$$

$$=\int_{-\infty}^{+\infty}\left\{T\left[\mathbf{b}\left(\mathbf{x}^{(1)};\theta\right),\mathbf{x}^{(1)};\tau\right]-E\left[T\left[\mathbf{b}\left(\mathbf{x}^{(1)};\theta\right),\mathbf{x}^{(1)};\tau\right]\right]\right\} \tag{1.120}$$

$$\times\left\{T\left[\mathbf{b}\left(\mathbf{x}^{(2)};\theta\right),\mathbf{x}^{(2)};\tau\right]-E\left[T\left[\mathbf{b}\left(\mathbf{x}^{(2)};\theta\right),\mathbf{x}^{(2)};\tau\right]\right]\right\}p_R\left(\mathbf{b}(\mathbf{x};\theta)\right)d\mathbf{b}$$

which gives as a result

$$S_T^{ij}\left(\mathbf{x}^{(1)};\mathbf{x}^{(2)};\theta;\tau\right)=T^{,r}\left(\mathbf{x}^{(1)};\theta;\tau\right)T^{,s}\left(\mathbf{x}^{(2)};\theta;\tau\right)S_b^{rs}\left(\mathbf{x}^{(1)};\mathbf{x}^{(2)};\theta;\tau\right) \tag{1.121}$$

Alternatively, one can compute the time cross–covariances in the case where the input random process varies in time (and does not depend on spatial variables). It is obtained for time moments τ_1 and τ_2 by the use of analogous definitions that

$$Cov\left(T\left[\mathbf{b}(\mathbf{x};\theta),\mathbf{x};\tau_1\right];T\left[\mathbf{b}(\mathbf{x};\theta),\mathbf{x};\tau_2\right]\right)=S_T^{ij}(\mathbf{x};\tau_1;\tau_2)$$

$$=\int_{-\infty}^{+\infty}\left\{T\left[\mathbf{b}(\mathbf{x};\theta),\mathbf{x};\tau_1\right]-E\left[T\left[\mathbf{b}(\mathbf{x};\theta),\mathbf{x};\tau_1\right]\right]\right\} \tag{1.122}$$

$$\times\left\{T\left[\mathbf{b}(\mathbf{x};\theta),\mathbf{x};\tau_2\right]-E\left[T\left[\mathbf{b}(\mathbf{x};\theta),\mathbf{x};\tau_2\right]\right]\right\}p_R\left(\mathbf{b}(\mathbf{x};\theta)\right)d\mathbf{b}$$

which yields

$$S_T^{ij}(\mathbf{x};\theta;\tau_1;\tau_2)=T^{,r}(\mathbf{x};\theta;\tau_1)T^{,s}(\mathbf{x};\theta;\tau_2)S_b^{rs}(\mathbf{x};\theta;\tau_1;\tau_2) \tag{1.123}$$

It is important to underline that the perturbation methodology at the present stage does not allow for computational modeling of the boundary–initial problems where the input parameters are full stochastic processes varying in space and time.

1.3.2 Elastodynamics with Random Parameters

Generally, the following problem is solved now [56,181,198]:

$$M\ddot{u}+C\dot{u}+Ku=f \tag{1.124}$$

where M, C and K are linear stochastic operators, u is the random structural response, while f is the admissible excitation of this system. The definitions of the matrices as random operators are introduced analogously to the considerations included in Sec. 1.3.1. Usually, such operators are identified as mass, damping and stiffness matrices in structural dynamics applications. As is known, the analytical solutions for such a class of partial differential equations are available for some specific cases, since quite different approximating numerical methods are used;

various mathematical approaches to the solution of that problem are reported and presented in [233,249,324,326]. However the second order perturbation second central probabilistic moment approach is documented below.

The stochastic second order Taylor series based extension [208] of the basic deterministic equation (1.124) of the problem leads by equating of the same order terms for $\tau \in [0,\infty)$ to

- zeroth order equations:

$$M^0\left(b^0\right)\ddot{u}^0\left(b^0;\tau\right)+ C^0\left(b^0\right)\dot{u}^0\left(b^0;\tau\right)+ K^0\left(b^0\right)u^0\left(b^0;\tau\right)= f^0\left(b^0;\tau\right) \tag{1.125}$$

- first order equations (for $r=1,\ldots,R$):

$$\begin{aligned}
&M^{,r}\left(b^0\right)\ddot{u}^0\left(b^0;\tau\right)+ C^{,r}\left(b^0\right)\dot{u}^0\left(b^0;\tau\right)+ K^{,r}\left(b^0\right)u^0\left(b^0;\tau\right)\\
&+ M^0\left(b^0\right)\ddot{u}^{,r}\left(b^0;\tau\right)+ C^0\left(b^0\right)\dot{u}^{,r}\left(b^0;\tau\right)+ K^0\left(b^0\right)u^{,r}\left(b^0;\tau\right)\\
&= f^{,r}\left(b^0;\tau\right)
\end{aligned} \tag{1.126}$$

- second order equations (for $r,s=1,\ldots,R$):

$$\begin{aligned}
&M^{,rs}\left(b^0\right)\ddot{u}^0\left(b^0;\tau\right)+ C^{,rs}\left(b^0\right)\dot{u}^0\left(b^0;\tau\right)+ K^{,rs}\left(b^0\right)u^0\left(b^0;\tau\right)\\
&+ 2M^{,r}\left(b^0\right)\ddot{u}^{,s}\left(b^0;\tau\right)+ 2C^{,r}\left(b^0\right)\dot{u}^{,s}\left(b^0;\tau\right)+ 2K^{,r}\left(b^0\right)u^{,s}\left(b^0;\tau\right)\\
&+ M^0\left(b^0\right)\ddot{u}^{,rs}\left(b^0;\tau\right)+ C^0\left(b^0\right)\dot{u}^{,rs}\left(b^0;\tau\right)+ K^0\left(b^0\right)u^{,rs}\left(b^0;\tau\right)\\
&= f^{,rs}\left(b^0;\tau\right)
\end{aligned} \tag{1.127}$$

Therefore, the generalized nth order partial differential perturbation–based equation of motion can be proposed as

$$\begin{aligned}
&\sum_{k=0}^{n}\binom{n}{k}M^{,n-k}\left(b^0(x;\theta)\right)\ddot{u}^{,k}\left(b^0(x;\theta);\tau\right)\\
&+ \sum_{k=0}^{n}\binom{n}{k}C^{,n-k}\left(b^0(x;\theta)\right)\dot{u}^{,k}\left(b^0(x;\theta);\tau\right)\\
&+ \sum_{k=0}^{n}\binom{n}{k}K^{,n-k}\left(b^0(x;\theta)\right)u^{,k}\left(b^0(x;\theta);\tau\right)= f^{,n}\left(b^0(x;\theta);\tau\right)
\end{aligned} \tag{1.128}$$

where the operators $M^{,n}, C^{,n}, K^{,n}$ denote nth order partial derivatives of mass, damping and stiffness matrices with respect to the input random variables determined at the expected values of these variables, respectively. The vectors $f^{,n}\left(b^0;\tau\right)$, $\ddot{u}^{,n}\left(b^0;\tau\right)$, $\dot{u}^{,n}\left(b^0;\tau\right)$, $u^{,n}\left(b^0;\tau\right)$ represent analogous nth order partial derivatives of external excitation, accelerations, velocities as well as displacements of the system.

Let us note that the stochastic hierarchical equations of motion for desired perturbation order m can be obtained from eqn (1.128) by successive expansion and substitution of n by the natural numbers $0,1,\ldots,m$, which returns the system of $(m+1)$ equations. Then zeroth order solution is obtained from the first equation; then, inserting the zeroth order solution into the second equation (of the first order), the first order solution can be determined. An analogous procedure is repeated to determine all orders of the structural response, which are finally used in the calculation of the response probabilistic moments.

Assuming that higher than second order perturbations can be neglected, this equation system constitutes the equilibrium problem. The detailed convergence studies should be carried out in further extensions of the model with respect to perturbation order, parameter θ and coefficient of variation of input random variables. If higher than the second probabilistic moment approach is considered, then the coefficients of assymetry, concentration, etc., also influence final effectiveness of the perturbation–based solution.

Analogously to the stochastic expansion of (1.105), the first and second order equations are modified and it is found that

• zeroth order equations:

$$M^0\!\left(b^0\right)\ddot{u}^0\!\left(b^0;\tau\right)+C^0\!\left(b^0\right)\dot{u}^0\!\left(b^0;\tau\right)+K^0\!\left(b^0\right)u^0\!\left(b^0;\tau\right)=f^0\!\left(b^0;\tau\right) \qquad (1.129)$$

• first order equations (for $r=1,\ldots,R$):

$$\begin{aligned}
&M^0\!\left(b^0\right)\ddot{u}^{,r}\!\left(b^0;\tau\right)+C^0\!\left(b^0\right)\dot{u}^{,r}\!\left(b^0;\tau\right)+K^0\!\left(b^0\right)u^{,r}\!\left(b^0;\tau\right)\\
&=f^{,r}\!\left(b^0;\tau\right)-\left\{M^{,r}\!\left(b^0\right)\ddot{u}^0\!\left(b^0;\tau\right)+C^{,r}\!\left(b^0\right)\dot{u}^0\!\left(b^0;\tau\right)+K^{,r}\!\left(b^0\right)u^0\!\left(b^0;\tau\right)\right\}
\end{aligned} \qquad (1.130)$$

• second order equations (for $r,s=1,\ldots,R$):

$$\begin{aligned}
&M^0\!\left(b^0\right)\ddot{u}^{(2)}\!\left(b^0;\tau\right)+C^0\!\left(b^0\right)\dot{u}^{(2)}\!\left(b^0;\tau\right)+K^0\!\left(b^0\right)u^{(2)}\!\left(b^0;\tau\right)\\
&=\left\{f^{,rs}\!\left(b^0;\tau\right)-M^{,rs}\!\left(b^0\right)\ddot{u}^0\!\left(b^0;\tau\right)-C^{,rs}\!\left(b^0\right)\dot{u}^0\!\left(b^0;\tau\right)-K^{,rs}\!\left(b^0\right)u^0\!\left(b^0;\tau\right)\right.\\
&\left.-2M^{,r}\!\left(b^0\right)\ddot{u}^{,s}\!\left(b^0;\tau\right)-2C^{,r}\!\left(b^0\right)\dot{u}^{,s}\!\left(b^0;\tau\right)-2K^{,r}\!\left(b^0\right)u^{,s}\!\left(b^0;\tau\right)\right\}Cov\!\left(b^r,b^s\right)
\end{aligned} \qquad (1.131)$$

Let us observe that looking for the nth order perturbation approach, the closure of hierarchical equations is obtained by the nth order correlation of input random process components b^r and b^s, respectively; nth order statistical information about input random variables is however necessary for this purpose.

To obtain the probabilistic solution for the considered equilibrium problem, (1.129) is solved for u^0 (and its time derivatives $\dot{u}^0$ and $\ddot{u}^0$, respectively), next (1.130) for first order terms of $u^{,r}$ and, finally, (1.131) for $u^{(2)}$. Two probabilistic moment characterisations of all the state functions for the boundary value problem starts from the expected value of the structural displacement vector components. Using its definition

$$E[u(t)] = \int\limits_{-\infty}^{+\infty} u(t) p_R(\mathbf{b}(\mathbf{x};\theta))\, d\mathbf{b} \tag{1.132}$$

the second order accurate expectations are equal to

$$E[u(t)] = u^0(t) + \tfrac{1}{2} u^{,rs} S_b^{rs} = u^0(t) + \tfrac{1}{2} u^{(2)} \tag{1.133}$$

In quite a similar manner the second moment probabilistic characteristics are obtained. Defining the time cross−correlation function as

$$Cov(u(t_1), u(t_2)) = \int\limits_{-\infty}^{+\infty} \{ u(t_1) - E[u(t_1)] \}\{ u(t_2) - E[u(t_2)] \} p_R(\mathbf{b}(\mathbf{x};\theta))\, d\mathbf{b} \tag{1.134}$$

it is found that

$$Cov(u(t_1); u(t_2)) = u^{,r}(t_1) u^{,s}(t_2) Cov(b^r, b^s) \tag{1.135}$$

which completes the two−moment characterization of the perturbation−based solution for the dynamic equilibrium problem (1.124). The entire solution simplifies in the case of free vibrations when the following equations are to be solved:

$$[K - \Omega_{(\alpha)} M]\, \Phi = 0 \tag{1.136}$$

$\Omega_{(\alpha)}$ and Φ are the eigenvalues and eigenvectors, respectively and $\alpha = 1,...,N$ denotes the total number of degrees of freedom of a structure. The second order expansion leads to the following equation system:

$$[K^0 - \Omega_{(\alpha)}^0 M^0]\, \Phi^0 = 0 \tag{1.137}$$

$$[K^0 - \Omega_{(\alpha)}^0 M^0]\, \Phi^{,r} = -[K^{,r} - \Omega_{(\alpha)}^{,r} M^0 - \Omega_{(\alpha)}^0 M^{,r}]\, \Phi^0 \tag{1.138}$$

$$\begin{aligned}
[K^0 - \Omega_{(\alpha)}^0 M^0]\, \Phi^{(2)} = \\
-\big\{ [K^{,rs} - \Omega_{(\alpha)}^{,rs} M^0 - 2\Omega_{(\alpha)}^{,r} M^{,s} - \Omega_{(\alpha)}^0 M^{,rs}]\, \Phi^0 \\
- 2[K^{,r} - \Omega_{(\alpha)}^{,r} M^0 - \Omega_{(\alpha)}^0 M^{,r}]\, \Phi^{,s} \big\} Cov(b^r, b^s)
\end{aligned} \tag{1.139}$$

To determine the probabilistic moments of the eigenvectors, up to the second order derivatives with respect to input random variables are to be determined first. While zeroth order quantities are obtained directly from the relation (1.137), the methodology of first order terms calculation is definitely more complicated.

Equation (1.138) is transformed for this purpose by multiplying by the transposed zeroth order eigenvector, which gives

$$\Phi^{0^T}\left[K^0 - \Omega^0_{(\alpha)}M^0\right]\Phi^{,r} - \Phi^{0^T}\Omega^r_{(\alpha)}M^0\Phi^0 = -\Phi^{0^T}\left[K^{,r} - \Omega^0_{(\alpha)}M^{,r}\right]\Phi^{,r} \qquad (1.140)$$

Since Φ^0 is diagonal and K^0 and M^0 are symmetric, (1.140) is modified as

$$\left[\Phi^{0^T}\left[K^0 - \Omega^0_{(\alpha)}M^0\right]\Phi^{,r}\right]^T = \Phi^{,r^T}\left[K^0 - \Omega^0_{(\alpha)}M^0\right]\Phi^{,r} = 0 \qquad (1.141)$$

Let us observe that $\Omega^{,r}$ is diagonal and therefore

$$\Phi^{0^T}\Omega^r_{(\alpha)}M^0\Phi^0 = \Omega^r_{(\alpha)}\Phi^{0^T}M^0\Phi^0 \qquad (1.142)$$

which finally results in

$$\Omega^r_{(\alpha)} = \Phi^{0^T}\left[K^{,r} - \Omega^0_{(\alpha)}M^{,r}\right]\Phi^0 \qquad (1.143)$$

Next, using an analogous technique in the case of the second order equation, it is derived that

$$\begin{aligned}
&\Phi^{0^T}\left[K^0 - \Omega^0_{(\alpha)}M^0\right]\Phi^{(2)} - \Phi^{0^T}\Omega^{(2)}_{(\alpha)}M^0\Phi^0 \\
&= -\Phi^{0^T}\left[K^{,rs} - 2\Omega^r_{(\alpha)}M^{,s} - \Omega^0_{(\alpha)}M^{,rs}\right]\Phi^0 Cov\left(b^r,b^s\right) \\
&\quad - \Phi^{0^T}\left[K^{,r} - \Omega^r_{(\alpha)}M^0 - \Omega^0_{(\alpha)}M^{,r}\right]\Phi^{,s} Cov\left(b^r,b^s\right)
\end{aligned} \qquad (1.144)$$

which finally implies

$$\begin{aligned}
\Phi^{(2)} &= \Phi^{0^T}\left[K^{,rs} - 2\Omega^r_{(\alpha)}M^{,s} - \Omega^0_{(\alpha)}M^{,rs}\right]\Phi^0 Cov\left(b^r,b^s\right) \\
&\quad + 2\Phi^{0^T}\left[K^{,r} - \Omega^r_{(\alpha)}M^0 - \Omega^0_{(\alpha)}M^{,r}\right]\Phi^{,s} Cov\left(b^r,b^s\right)
\end{aligned} \qquad (1.145)$$

The next problem is to determine the first and second order derivatives of the eigenvectors. Basically, the eigenvector derivative is expressed as a linear combination of all the eigenvectors in the original system. Equations describing the coefficients of the linear combination are formed using the $M-$orthonormality and $K-$orthogonality conditions. Starting from (1.138), the αth eigenpair is determined as

$$\left[K^0 - \Omega^0_{(\alpha)}M^0\right]\Phi^{,r}_{(\alpha)} = -\left[K^{,r} - \Omega^r_{(\alpha)}M^0 - \Omega^0_{(\alpha)}M^{,r}\right]\Phi^0_{(\alpha)} \qquad (1.146)$$

and (1.143) in the following form:

$$\Omega^{,r}_{(\alpha)} = \Phi^0_{(\alpha)} \left[K^{,r} - \omega^0_{(\alpha)} M^{,r} \right] \Phi^0_{(\alpha)} \tag{1.147}$$

It yields by substitution

$$\left[K^0 - \Omega^0_{(\alpha)} M^0 \right] \Phi^{,r}_{(\alpha)} = R^r_{(\alpha)} \tag{1.148}$$

with $R^r_{(\alpha)}$ being equal to

$$R^r_{(\alpha)} = -\left[K^{,r} - \Phi^0_{(\alpha)} \left(K^{,r} - \Omega^0_{(\alpha)} M^{,r} \right) \Phi^0_{(\alpha)} M^0 - \Omega^0_{(\alpha)} M^{,r} \right] \Phi^0_{(\alpha)} \tag{1.149}$$

Further, it is assumed that the αth first order eigenvector $\Phi^r_{(\alpha)}$ can be expressed as a linear combination of all the zeroth order eigenvectors as

$$\Phi^{,r}_{(\alpha)} = \Phi^0 a^r_{(\alpha)} \tag{1.150}$$

The complete description of the coefficients $a^r_{(\alpha)}$ is given by the following formula:

$$a^r_{(\alpha)} = \begin{cases} \dfrac{\Phi^0_{(\alpha)} R^r_{(\hat{\alpha})}}{\omega^0_{(\alpha)} - \omega^0_{(\hat{\alpha})}}, & \textit{for } \alpha \neq \hat{\alpha} \\[2ex] -\dfrac{1}{2} \Phi^0_{(\hat{\alpha})} M^{,r} \Phi^0_{(\hat{\alpha})}, & \textit{for } \alpha = \hat{\alpha} \end{cases} \tag{1.151}$$

Similarly as above, the second order eigenvector $\Phi^{(2)}_{(\alpha)}$ is approximated by a linear combination of all the zeroth order eigenvectors

$$\Phi^{(2)}_{(\alpha)} = \Phi^0 a^{(2)}_{(\alpha)} \tag{1.152}$$

Then, one can show the following result [208]:

$$a^{(2)}_{(\hat{\alpha})} = \begin{cases} \dfrac{\Phi^0_{(\alpha)} R^{(2)}_{(\hat{\alpha})}}{\omega^0_{(\alpha)} - \omega^0_{(\hat{\alpha})}}, & \textit{for } \alpha \neq \hat{\alpha} \\[2ex] -\left(\dfrac{1}{2} \Phi^0_{(\hat{\alpha})} M^{,rs} \Phi^0_{(\hat{\alpha})} + 2\Phi^0_{(\hat{\alpha})} M^{,r} \Phi^{,s}_{(\hat{\alpha})} + a^r_{(\hat{\alpha})} a^s_{(\hat{\alpha})} \right) Cov\left(b^r, b^s \right) \\ \textit{for } \alpha = \hat{\alpha} \end{cases} \tag{1.153}$$

Finally, the first two probabilistic moments of the eigenvalues and eigenvectors are found in a typical way, which completes the solution of the second order second central probabilistic moment eigenvalue and eigenvector problem.

Summing up the applications of the stochastic perturbation methodology it should be pointed out that the main disadvantage is dependence between the assumed order of the expansion, interrelations between input probabilistic characteristics and overall precision of such a computational methodology. The method found its numerous applications in structural engineering [88,208,237], in homogenisation [162,164,192] as well as in fluid dynamics computations [184]. Computational implementation in conjunction with Finite Element Method both in displacement [208] and stress versions [186], Boundary Element Method [51,185] as well as with Finite Difference Method [187,198] are available now, whereas the scaled the Boundary–Finite Element Method has no such extension [369].

Nevertheless, the perturbation method can be very useful after successful implementation in symbolic computations programs, which will enable automatic perturbation–based extension of up to nth order [178] for any variational equation [25,297] as well as ordinary or partial differential equations solutions [68,90]. The application of the perturbation method in stochastic processes [319,326] modelling needs its essential improvements, because now the randomness of an input cannot be introduced both in space and time.

2 Elasticity Problems

Numerical experiments devoted to multi-component and multiscale media modelling are still one of the most important part of modern computational mechanics and engineering [98,161,272,312]. The main idea of this chapter in this context is to present a general approach to numerical analysis of elastostatic problems in 1D and 2D heterogeneous media [105,274,300,317] and the homogenisation method of periodic linear elastic engineering composite structures exhibiting randomness in material parameters [32,83,356,372,375]. As is shown below, the effective elasticity tensor components of such structures are obtained as the closed–form equations in the deterministic approach, which enables a relatively easy extension to the stochastic analysis by the application of the second order perturbation second central probabilistic moment analysis. On the other hand, the Monte Carlo simulation approach is employed to solve the cell problem. As is known from numerous books and articles in this area, the main difficulty in homogenisation is the lack of one general model valid for various composite structures; different nature homogenised constitutive relations are derived for beams, plates, shells etc. and even for the same type of engineering structure different effective relations are fulfilled for composites with constituents of different types (with ceramic, metal or polymer matrices and so forth). That is why numerous theoretical and numerical homogenisation models of composites are developed and applied in engineering practice.

All the theories in this field can be arbitrarily divided, considering especially the method and form of the final results, into two essentially different groups. The first one contains all the methods resulting in closed form equations characterizing upper and lower bounds [108,138,156,285,339] or giving direct approximations of the effective material tensors [122,123,248]. In alternative, so–called cell problems are solved to calculate, on the basis of averaged stresses or strains, the final global characteristics of the composite in elastic range [11,214,304,383], for thermoelastic analysis [117], for composites with elasto–plastic [50,57,58,146,332] or visco–elasto–plastic components [366], in the case of fractured or porous structures [38,361] or damaged interfaces [224,252,358]. The very recently even multiscale methods [236,340] and models have been worked out to include the atomistic scale effects in global composite characteristics [67,145]. The results obtained for the first models are relatively easy and fast in computation. However, usually these approximations are not so precise as the methods based on the cell problem solutions. In this context, the decisive role of symbolic computations and the relevant computational tools (MAPLE, MATHEMATICA, MATLAB, for instance) should be underlined [64,111,268]. By using the MAPLE program and any closed form equations for effective characteristics of composites as well as thanks to the stochastic second order perturbation technique (in practice of any finite order), the probabilistic moments of these characteristics can be derived and computed. The great value of such a computational technique lies in its usefulness in stochastic sensitivity studies. The closed form probabilistic moments of the homogenised tensor make it possible to derive explicitly the sensitivity gradients with respect to the expected values and standard deviations of the original material properties of a composite.

Probabilistic methods in homogenisation [116,120,141,146,259,287,378] obey (a) algebraic derivation of the effective properties, (b) Monte−Carlo simulation of the effective tensor, (c) Voronoi−tesselations of the RVE together with the relevant FEM studies, (d) the moving−window technique. The alternative stochastic second order approach to the cell problem solution, where the SFEM analysis should be applied to calculate the effective characteristics, is displayed below. Various effective elastic characteristics models proposed in the literature are extended below using the stochastic perturbation technique and verified numerically with respect to probabilistic material parameters of the composite components. The entire homogenisation methodology is illustrated with computational examples of the two−component heterogeneous bar, fibre−reinforced and layered unidirectional composites as well as the heterogeneous plate. Thanks to these experiments, the general computational algorithm for stochastic homogenisation is proposed by some necessary modifications with comparison to the relevant theoretical approach.

Finally, it is observed that having analytical expressions for the effective Young modulus and their probabilistic moments, the model presented can be extended to the deterministic and stochastic structural sensitivity analysis for elastostatics or elastodynamics of the periodic composite bar structures. It can be done assuming ideal bonds between different homogeneous parts of the composites or even considering stochastic interface defects between them and introducing the interphase model due to the derivations carried out or another related microstructural phenomena both in linear an nonlinear range. In the same time, starting from the deterministic description of the homogenised structure, the effective behaviour related to any external excitations described by the stochastic processes can be obtained.

2.1 Composite Model. Interface Defects Concept

The main object of the considerations is the random periodic composite structure Y with the Representative Volume Element (RVE) denoted by Ω. Let us assume that Ω contain perfectly bonded, coherent and disjoint subsets being homogeneous (for classical fibre−reinforced composites there are two components, for instance) and let us assume that the scale between corresponding geometrical diameters of Ω and the whole Y is described by some small parameter $\varepsilon > 0$; this parameter indexes all the tensors rewritten for the geometrical scale connected with Ω. Further, it should be mentioned that random periodic composites are considered; it is assumed that for an additional ω belonging to a suitable probability space there exists such a homothety that transforms Ω into the entire composite Y. In the random case this homothety is defined for all probabilistic moments of input random variables or fields considered. Next, let us introduce two different coordinate systems: $\mathbf{y} = (y_1, y_2, y_3)$ at the microscale of the composite and $\mathbf{x} = (x_1, x_2, x_3)$ at the macroscale. Then, any periodic state function F defined on Y can be expressed as

$$F^{\varepsilon}(\mathbf{x}) = F\left(\frac{\mathbf{x}}{\varepsilon}\right) = F(\mathbf{y}) \tag{2.1}$$

This definition allows a description of the macro functions (connected with the macroscale of a composite) in terms of micro functions and vice versa. Therefore, the elasticity coefficients (being homogenised) can be defined as

$$C_{ijkl}^{\varepsilon}(\mathbf{x}) = C_{ijkl}(\mathbf{y}) \tag{2.2}$$

Random fields under consideration are defined in terms of geometrical as well as material properties of the composite. However the periodic microstructure as well as its macrogeometry is deterministic. Randomising different composite properties, the set of all possible realisations of a particular introduced random field have to be admissible from the physical and geometrical point of view, which is partially explained by the below relations. Let each subset Ω_a contain linear–elastic and transversely isotropic materials where Young moduli and Poisson coefficients are truncated Gaussian random variables with the first two probabilistic moments specified. There holds

$$0 < e(x; \omega) < \infty \tag{2.3}$$

$$E[e(x; \omega)] = \begin{cases} e_1; \ x \in \Omega_1 \\ e_2; \ x \in \Omega_2 \end{cases} \tag{2.4}$$

$$Cov(e_i(x; \omega); e_j(x; \omega)) = \begin{bmatrix} Vare_1 & 0 \\ 0 & Vare_2 \end{bmatrix}; \ i, j = 1, 2 \tag{2.5}$$

$$-1 < v(x; \omega) < \tfrac{1}{2} \tag{2.6}$$

$$E[v(x; \omega)] = \begin{cases} v_1; \ x \in \Omega_1 \\ v_2; \ x \in \Omega_2 \end{cases} \tag{2.7}$$

$$Cov(v_i(x; \omega); v_j(x; \omega)) = \begin{bmatrix} Var \ v_1 & 0 \\ 0 & Var \ v_2 \end{bmatrix}; \ i, j = 1, 2 \tag{2.8}$$

Moreover, it is assumed that there are no spatial correlations between Young moduli and Poisson coefficients and the effect of Gaussian variables cutting–off in the context of (2.3) and (2.6) does not influence the relevant probabilistic moments. This assumption will be verified computationally in the numerical experiments; a discussion on the cross–property correlations has been done in [315]. Further, the random elasticity tensor for each component material can be defined as

$$C_{ijkl}(x;\omega) = \delta_{ij}\delta_{kl}\frac{v(x;\omega)}{(1+v(x;\omega))(1-2v(x;\omega))}e(x;\omega)$$
$$+\left(\delta_{ik}\delta_{jl}+\delta_{il}\delta_{jk}\right)\frac{1}{2(1+v(x;\omega))}e(x;\omega) \qquad ; i,j,k,l = 1,2 \qquad (2.9)$$

Considering all the assumptions posed above, the random periodicity of Y can be understood as the existence of such a translation which, applied to Ω, enables to cover the entire Y (as a consequence, the probabilistic moments of $e(x;\omega)$ and $v(x;\omega)$ are periodic too). Thus, let us adopt Y as a random composite if relevant properties of the RVE are Gaussian random variables with specified first two probabilistic moments; these variables are bounded to probability spaces admissible from mechanical and physical point of view.

Let us note that the probabilistic description of the elasticity simplifies significantly if the Poisson coefficient is assumed to be a deterministic function so that

$$v(x) = v_a, \text{ for } a=1,2,...,n; \ x\in \Omega_a \qquad (2.10)$$

Finally, the random elasticity tensor field $C_{ijkl}(x;\omega)$ is represented as follows:

$$C_{ijkl}(x;\omega)$$
$$= e(x;\omega)\left\{\delta_{ij}\delta_{kl}\frac{v(x)}{(1+v(x))(1-2v(x))}+\left(\delta_{ik}\delta_{jl}+\delta_{il}\delta_{jk}\right)\frac{1}{2(1+v(x))}\right\} \qquad (2.11)$$

Because of the linear relation between the elasticity tensor components and the Young modulus these components have the truncated Gaussian distribution and can thus be derived uniquely from their first two moments as

$$E\left[C_{ijkl}(x;\omega)\right] = A_{ijkl(a)}(x)\cdot E\left[e_a(x;\omega)\right]$$
$$\text{for } i,j,k,l=1,2, \ a=1,2,...,n; \ x\in \Omega_a \qquad (2.12)$$

and

$$Var\left(C_{ijkl}(x;\omega)\right) = A_{ijkl(a)}(x)A_{ijkl(a)}(x)Var\left(e_a(x;\omega)\right)$$
$$\text{for } i,j,k,l=1,2, \ a=1,2,...,n; \ x\in \Omega_a, \qquad (2.13)$$

with no sum over repeating indices at the right hand side.

There holds

$$A_{ijkl}(x) = \delta_{ij}\delta_{kl}\frac{v(x)}{(1+v(x))(1-2v(x))} + (\delta_{ik}\delta_{jl} + \delta_{il}\delta_{jk})\frac{1}{2(1+v(x))}$$
$$i,j,k,l=1,2$$

(2.14)

General methodology leading to the final results of the effective elasticity tensor is to rewrite either strain energy (or complementary energy, for instance) or equilibrium equations for a homogeneous medium and the heterogeneous one. Next, the effective parameters are derived by equating corresponding expressions for the homogeneous and for the real structure. This common methodology is applied below, particular mathematical considerations are included in the next sections and only the final result useful in further general stochastic analysis is shown. The expected values for the effective elasticity tensor in the most general case can be obtained by the second order perturbation based extension as [162,208]

$$E\left[C_{ijkl}^{(eff)}(\mathbf{y})\right] = \int_{-\infty}^{+\infty}\left(C_{ijkl}^{(eff)0}(\mathbf{y}) + \Delta b^r C_{ijkl}^{(eff),r}(\mathbf{y}) + \tfrac{1}{2}\Delta b^r \Delta b^s C_{ijkl}^{(eff),rs}\right)p_R(\mathbf{b})\,d\mathbf{b}$$

(2.15)

Using classical probability theory definitions and theorems it is obtained that

$$\int_{-\infty}^{+\infty}p_R(b(\mathbf{y}))\,db = 1, \quad \int_{-\infty}^{+\infty}\Delta b p_R(b(\mathbf{y}))\,db = 0$$

(2.16)

$$\int_{-\infty}^{+\infty}\Delta b^r \Delta b^s p_R(b(\mathbf{y}))\,db = Cov(b^r,b^s); \quad 1 \le r,s \le R$$

(2.17)

Therefore

$$E\left[C_{ijkl}^{(eff)}(\mathbf{y})\right] = C_{ijkl}^{(eff)0}(\mathbf{y}) + \tfrac{1}{2}C_{ijkl}^{(eff),rs}Cov(b^r,b^s)$$

(2.18)

Further, the covariance matrix $Cov\left(C_{ijkl}^{(eff)};C_{pqmn}^{(eff)}\right)$ of the effective elasticity tensor is calculated using its integral definition

$$Cov\left(C_{ijkl}^{(eff)};C_{pqmn}^{(eff)}\right)$$
$$= \int_{-\infty}^{+\infty}\left(C_{ijkl}^{(eff)} - C_{ijkl}^{(eff)0}\right)\left(C_{pqmn}^{(eff)} - C_{pqmn}^{(eff)0}\right)g(b_i,b_j)\,db_i db_j$$

(2.19)

whereas inserting the second order perturbation expansion it is found that

$$= \int_{-\infty}^{+\infty} \left\{ \left(C_{ijkl}^{(eff)0} + \Delta b_r C_{ijkl}^{(eff),r} + \tfrac{1}{2} \Delta b_r \Delta b_s C_{ijkl}^{(eff),rs} - C_{ijkl}^{(eff)0} \right) \right.$$

$$\left. \left(C_{mnpq}^{(eff)0} + \Delta b_r C_{mnpq}^{(eff),r} + \tfrac{1}{2} \Delta b_r \Delta b_s C_{mnpq}^{(eff),rs} - C_{mnpq}^{(eff)0} \right) \right\} g\left(b_i, b_j\right) db_i db_j \qquad (2.20)$$

After all algebraic transformations and neglecting terms of order higher than second, there holds

$$Cov\left(C_{ijkl}^{(eff)} ; C_{pqmn}^{(eff)} \right)$$

$$= \int_{-\infty}^{+\infty} \Delta b_r C_{ijkl}^{(eff),r} \Delta b_s C_{mnpq}^{(eff),s} \, g\left(b_i, b_j\right) db_i db_j = C_{ijkl}^{(eff),r} C_{mnpq}^{(eff),s} Cov\left(b_r, b_s\right). \qquad (2.21)$$

Then, starting from two–moment characterization of the effective elasticity tensor and the corresponding homogenisation models presented in (2.15) − (2.21), the stochastic second order probabilistic moment analysis of a particular engineering composites can be carried out. In the general case, these equations lead to a rather complicated description of probabilistic moments for the effective elasticity tensor particular components.

In the theory of elasticity the continuum is usually uniquely represented by its geometry and elastic properties; most often a character of these features is considered as deterministic. It has been numerically proved for the fibre composites that the influence of the elastic properties randomness on the deterministically represented geometry can be significant. The most general model of the linear elastic medium, and intuitively the nearest to the real material, is based on the assumption that both its geometry and elasticity are random fields or stochastic processes. The phenomenon of random, both interface [5,27,131,200, 225,242] and volumetric [74,316,342,353,388], non-homogeneities occur mainly in the composite materials. While the interface defects (technological inaccuracies, matrix cracks, reinforcement breaks or debonding) are important considering the fracturing of such composites, the volume heterogeneities generally follow the discrete nature of many media. The existing models of stochastic media (based on various kinds of geometrical tesselations) do not make it possible to analyse such problems and that is why a new formulation is proposed.

The main idea of the proposed model is a transformation of the stochastic medium into some deterministic media with random material parameters, more useful in the numerical analysis. Such a transformation is possible provided the probabilistic characteristics of geometric dimensions and total number of defects occuring at the interfaces are given, assuming that these random fields are Gaussian with non-negative or restricted values only. All non–homogeneities introduced are divided into two groups: the stochastic interface defects (SID), which have non–zero intersections with the interface boundaries, and the volumetric stochastic defects (VSD) having no common part with any interface or external composite boundary. Further, the interphases are deterministically constructed around all interface boundaries using probabilistic bounds of geometric dimensions of the SID considered. Finally, the stochastic geometry is replaced by random elastic characteristics of composite constituents thanks to a probabilistic modification of the spatial averaging method (PAM). Let us note that the formulation proposed for including the SID in the interphase region has its origin in micro–mechanical approach to the contact problems rather than in the existing interface defects models.

Having so defined the composite with deterministic geometry and stochastic material properties, the stochastic boundary–value problem can be numerically solved using either the Monte Carlo simulation method, which is based on computational iterations over input random fields, or the SFEM based on second–order perturbation theory or based on spectral decomposition. The perturbation–based method has found its application to modeling of fibre–reinforced composites and, in view of its computational time savings, should be preferred.

Finally, let us consider the material discontinuities located randomly on the boundaries between composite constituents (interfaces) as it is shown in Figs. 2.1 and 2.2.

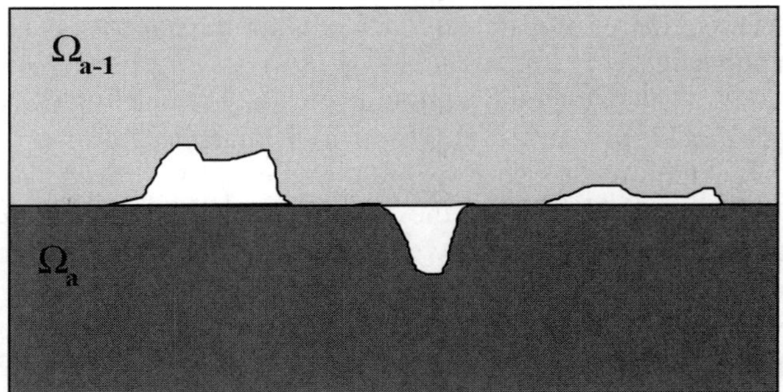

Figure 2.1. Interface defects geometrical sample

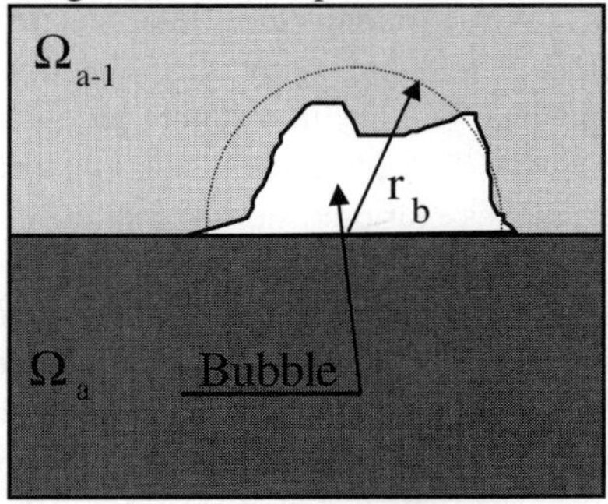

Figure 2.2. A single interface defect geometric idealization

Numerical model for such nonhomogeneities is based on the assumption that [193,194]:

(1) there is a finite number of material defects on all composite interfaces; the total number of defects considered is assumed as a random parameter (with nonnegative values only) defined by its first two probabilistic moments;
(2) interface defects are approximated by the semi-circles (bubbles) lying with their diameters on the interfaces; the radii of the bubbles are assumed to be the next random parameter of the problem defined by the expected value and the variance;
(3) geometric dimensions of every defect belonging to any Ω_a are small in comparison with the minimal distance between the $\Gamma_{(a-2,a-1)}$ and $\Gamma_{(a-1,a)}$ boundaries for $a=3,...,n$ or with Ω_1 geometric dimensions;
(4) all elastic characteristics specified above are assumed equal to 0 if $x \in D_a$, for $a=1,2,...,n$.

It should be underlined that the model introduced approximates the real defects rather precisely. In further investigations the semi−circle shape of the defects should be replaced with semi−elliptical [353] and their physical model should obey nucleation and growth phenomena [345,346] preserving a random character. However to build up the numerical procedure, the bubbles should be appropriately averaged over the interphases, which they belong to. Probabilistic averaging method is proposed in the next section to carry out this smearing.

Let us consider the stochastic material non−homogeneities contained in some $\Omega_a \subset \Omega$. The set of the defects considered D_a can be divided into three subsets D'_a, D''_a and D'''_a, where D'_a contains all the defects having a non-zero intersection with the boundary $\Gamma_{(a-1,a)}$, D''_a having zero intersection with $\Gamma_{(a-1,a)}$ and $\Gamma_{(a,a+1)}$, and D'''_a having a non-zero intersection with $\Gamma_{(a,a+1)}$. Further, all the defects belonging to subsets D'_a and D'''_a are called the stochastic interface defects (SID) and those belonging to D''_a the volumetric stochastic defects (VSD). Let us consider such Ω'_a, Ω''_a and Ω'''_a, where $\Omega_a = \Omega'_a \cup \Omega''_a \cup \Omega'''_a$, that with probability equal to 1, there holds $D'_a \subset \Omega'_a$, $D''_a \subset \Omega''_a$ and $D'''_a \subset \Omega'''_a$ (cf. Figure 2.3).

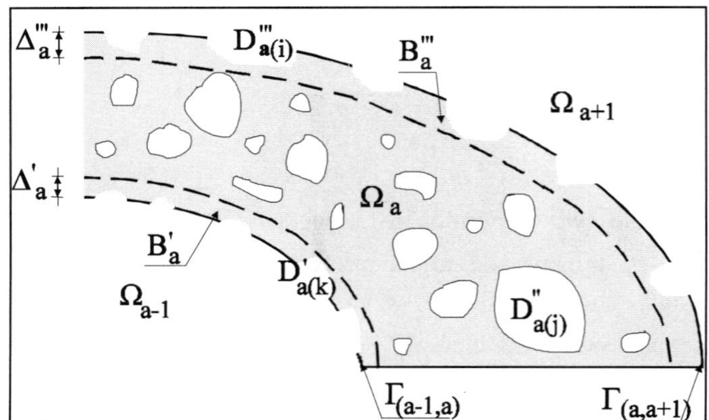

Figure 2.3. Interphase schematic representation

The subsets $\Omega'_a, \Omega''_a, \Omega'''_a$ can be geometrically constructed using probabilistic moments of the defect parameters (their geometric dimensions and total number). To provide such a construction let us introduce random fields $\Delta'_a(x;\omega)$ and $\Delta'''_a(x;\omega)$ as upper bounds on the norms of normal vectors defined on the boundaries $\Gamma_{(a-1,a)}$ and $\Gamma_{(a,a+1)}$ and the boundaries of the SID belonging to D'_a, and D'''_a, respectively. Next, let us consider the upper bounds of probabilistic distributions of $\Delta'_a(x;\omega)$ and $\Delta'''_a(x;\omega)$ given as follows:

$$\Delta'_a = E\left[\Delta'_a(x;\omega)\right] + 3\sqrt{Var\left(\Delta'_a(x;\omega)\right)} \tag{2.22}$$

$$\Delta'''_a = E\left[\Delta'''_a(x;\omega)\right] + 3\sqrt{Var\left(\Delta'''_a(x;\omega)\right)} \tag{2.23}$$

Thus, Ω'_a, Ω'''_a can be expressed in the following form:

$$\Omega'_a = \left\{ P(x_i) \in \Omega_a : d(P, \Gamma_{(a-1,a)}) \le \Delta'_a \right\} \tag{2.24}$$

$$\Omega'''_a = \left\{ P(x_i) \in \Omega_a : d(P, \Gamma_{(a,a+1)}) \le \Delta'''_a \right\} \tag{2.25}$$

where $i=1,2$ and $d(P,\Gamma)$ denotes the distance from a point P to the contour Γ. Let us note that Ω''_a can be obtained as

$$\Omega''_a = \Omega_a - \Omega'_a \cup \Omega'''_a \tag{2.26}$$

Deterministic spatial averaging of properties Y_a on continuous and disjoint subsets Ω_a of Ω is employed to formulate the probabilistic averaging method. The averaged property $Y^{(av)}$ characterizing the region Ω is given by the following equation [65,129]:

$$Y^{(av)} = \frac{\sum_{a=1}^{n} Y_a |\Omega_a|}{|\Omega|} \; ; \; x \in \Omega \tag{2.27}$$

where $|\Omega|$ is the two–dimensional Lesbegue measure of Ω. Deterministic averaging can be transformed to the probabilistic case only if Ω is defined deterministically, and Y_a and Ω_a are uncorrelated random fields. The expected value of probabilistically averaged $Y^{(pav)}(\omega)$ on Ω can be derived as

$$E\left[Y^{(pav)}(\omega)\right] = \frac{1}{|\Omega|} \sum_{a=1}^{n} E[Y_a(\omega)]\, E\big[|\Omega_a(\omega)|\big] \tag{2.28}$$

and, similarly, the variance as

$$Var\left(Y^{(pav)}(\omega)\right) = \frac{1}{|\Omega|^2} \sum_{a=1}^{n} Var(Y_a(\omega))\, Var\big(|\Omega_a(\omega)|\big) \tag{2.29}$$

Using the above equations Young moduli are probabilistically averaged over all Ω_a regions and their $\Omega'_a, \Omega''_a, \Omega'''_a$ subsets. Finally, a primary stochastic geometry of the considered composite is replaced by the new deterministic one. In this way, the $n-$component composite having m interfaces with stochastic interface defects on both sides of each interface and with volume non-homogeneities can be transformed to a $n+m-$component structure with deterministic geometry and probabilistically defined material parameters. More detailed equations of the PAM can be derived for given stochastic parameters of interface defects (if these defects can be approximated by specific shapes – circles, hexagons or their halves for instance).

Let us suppose that there is a finite element number of discontinuities in the matrix region located on the fibre–matrix interface. These discontinuities are approximated by bubbles – semicircles placed with their diameters on the interface, see Figure 2.4. The random distribution of the assumed defects is uniquely defined by the expected values and variances of the total number and radius of the bubbles; it is shown below, there is a sufficient number of parameters to obtain a complete characterization of semicircles averaged elastic constants.

Using (2.28) and (2.29) one can determine the expected value and the variance of the effective Young modulus e_k, the terms included in the covariance matrix of this modulus and also the Poisson ratio. It yields for the expected value

$$E[e_{2c}] = E\left[\frac{S_{\Omega_{2c}} - S_b}{S_{\Omega_{2c}}} \cdot e_2\right] = E[e_2] \cdot \left(1 - \frac{1}{S_{\Omega_{2c}}} \cdot E[S_b]\right) \tag{2.30}$$

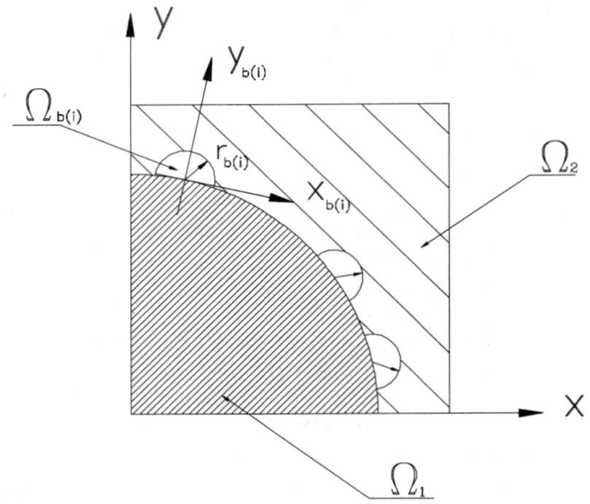

Figure 2.4. Bubble interface defects in the fibre−reinforced composite

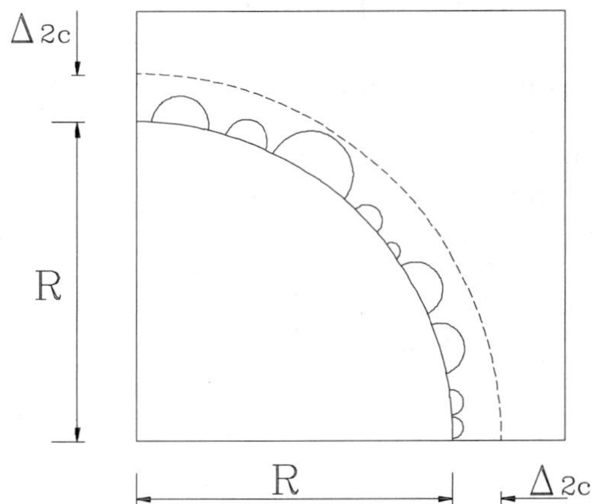

Figure 2.5. Interphase for bubble interface defects

As can be easily seen in the above relation, there holds

$$S_{\Omega_{2c}} = \pi\left\{\left(R + E\left[r_{\overline{\Omega}}\right] + 3\sqrt{Var\left[r_{\overline{\Omega}}\right]}\right)^2 - R^2\right\}$$

(2.31)

In a similar way the variance is derived as

$$Var\left[e_{2c}\right] = Var\left[\left(1 - \frac{S_b}{S_{\Omega_{2c}}}\right)\cdot e_2\right]$$

(2.32)

It can be shown that this equation could have the following form:

$$Var[e_{2c}] = \left\{ 1 - \frac{1}{S_{\Omega_{2c}}} E[S_b] \right\}^2 Var[e_2]$$
$$+ \frac{1}{S_{\Omega_{2c}}^2} Var[S_b] Var[e_2] + \frac{1}{S_{\Omega_{2c}}^2} Var[S_b] E^2[e_2] \tag{2.33}$$

which, neglecting moments of higher than second order, can be reduced to

$$Var[e_{2c}] = \left\{ 1 - \frac{1}{S_{\Omega_{2c}}} E[S_b] \right\}^2 Var[e_2] + \frac{1}{S_{\Omega_{2c}}^2} Var[S_b] E^2[e_2] \tag{2.34}$$

Now the distribution parameters S_b have to be found. As can be seen

$$S_b = \tfrac{1}{2}\pi (r_b)^2 M_b \tag{2.35}$$

where M_b is the number of $\Omega_{b(i)}$ regions found in Ω_{2c} (according to Figures 2.4 and 2.5) and is equal to

$$M_b = 2\pi R m_b \tag{2.36}$$

Therefore, using fundamental properties of random variables it is obtained that

$$E[M_b] = 2\pi R \cdot E[m_b] \tag{2.37}$$

and

$$Var[M_b] = 4\pi^2 R^2 \cdot Var[m_b] \tag{2.38}$$

From the definition of the expected value one derives

$$E[S_b] = \tfrac{\pi}{2} E\left[(r_b)^2 M_b \right] = \tfrac{\pi}{2} \left\{ E^2[r_b] + Var[r_b] \right\} E[M_b] \tag{2.39}$$

Finally, the variance of S_b is found as

$$Var[S_b] = Var\left[\tfrac{\pi}{2} (r_b)^2 M_b \right] = \tfrac{\pi^2}{4} Var\left[(r_b)^2 M_b \right] \tag{2.40}$$

It can be shown that this expression may be transformed into the form:

$$Var[S_b] = \frac{\pi^2}{4}\left(E^2[r_b] + Var[r_b]\right)^2 Var[M_b]$$
$$+ \frac{\pi^2}{2} Var[r_b]\left(E^2[M_b] + Var[M_b]\right)\left(2E^2[r_b] + Var[r_b]\right)$$

(2.41)

Substituting the equations describing S_b distribution parameters into the relations describing the expected value and variance of the e_k modulus, we can similarly derive the data necessary for numerical analysis.

Using analogous equations, the stochastic interface defects in the fibre region can be approximated. So, let us assume a finite number of these discontinuities inserted into the contact zone. As already established, the fibre material has good resistance to degradation (much better than the matrix) and because of this, the defects in the Ω_1 region can be approximated as teeth with their sharp sides directed towards the fibre centre. A single discontinuity is, from the geometrical point of view, the superposition of two circular quadrants with the same radius (Figure 2.6). There holds

$$E[e_{1c}] = E[e_1]\left(1 - \frac{1}{S_{\Omega_{1c}}} E[S_t]\right)$$

(2.42)

and

$$Var[e_{1c}] = \left\{1 - \frac{1}{S_{\Omega_{1c}}} E[S_t]\right\}^2 Var[e_1]$$
$$+ \frac{1}{S_{\Omega_{1c}}^2} Var[S_t] Var[e_1] + \frac{1}{S_{\Omega_{1c}}^2} Var[S_t] E^2[e_1]$$

(2.43)

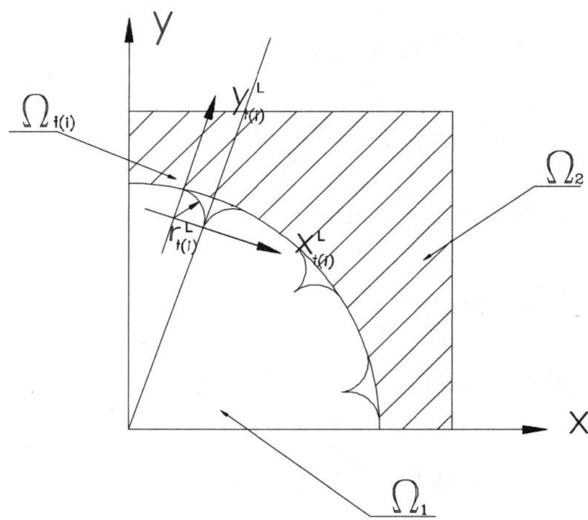

Figure 2.6. Teeth interface defects in fibre-reinforced composite

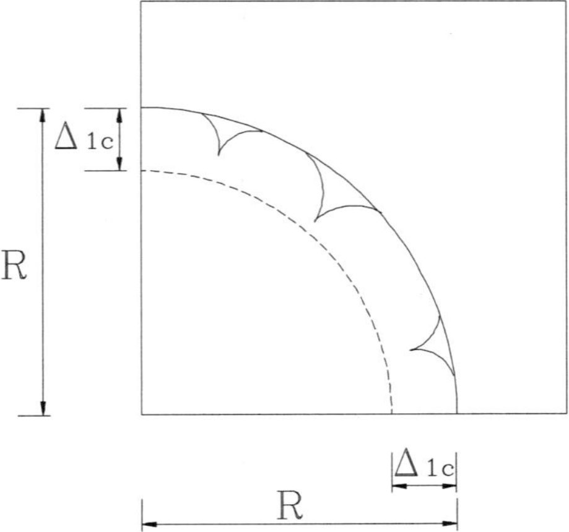

Figure 2.7. Interphase for teeth interface defects

The relations describing the discontinuity parameters will have a different form

$$S_t = \left(2 - \tfrac{\Pi}{2}\right)(r_t)^2 M_t \tag{2.44}$$

so that

$$E[S_t] = \left(2 - \tfrac{\pi}{2}\right) E\left[(r_t)^2 M_t\right] = \left(2 - \tfrac{\pi}{2}\right)\left\{E^2[r_t] + Var[r_t]\right\} E[M_t] \tag{2.45}$$

and, finally

$$\begin{aligned}
Var[S_t] &= \left(2 - \tfrac{\pi}{2}\right)^2 \left(E^2[r_t] + Var[r_t]\right)^2 Var[M_t] \\
&\quad + 2\cdot\left(2 - \tfrac{\pi^2}{2}\right)^2 Var[r_t]\left(E^2[M_t] + Var[M_t]\right)\left(2E^2[r_t] + Var[r_t]\right)
\end{aligned} \tag{2.46}$$

The Poisson ratio for the fibre interphase region can be obtained in analogous way. Finally, the covariance matrix of the Young modulus for this composite takes the following form:

$$Cov\left(e^{(i)}, e^{(j)}\right) =$$

$$\begin{bmatrix}
Var[e_1] & Cov[e_1, e_{1c}] & 0 & 0 \\
Cov[e_1, e_{1c}] & Var[e_{1c}] & 0 & 0 \\
0 & 0 & Var[e_{2c}] & Cov[e_{2c}, e_2] \\
0 & 0 & Cov[e_{2c}, e_2] & Var[e_2]
\end{bmatrix} \tag{2.47}$$

Zeroing of the corresponding covariance matrix components can be achieved from the assumed mutual independence of the Young modulus in the fibre, its contact zone and associated regions for the matrix.

Special purpose numerical procedure has been implemented to check the influence of the interface defects parameters on the effective elastic parameters of the interphase. The expected values of the interface discontinuities in the matrix and fibre contact zone were assumed as 4, 10, 20 and 40 with the width of the observed interface varying between 4.0E-3 and 2.0E-2. The results of these computations are presented in Figures 2.8 to 2.13: the expected values of the homogenised Young modulus functions are given in Figures 2.8 and 2.9, the averaged Poisson ratio functions in Figures 2.10 and 2.11 and the variances of the Young modulus functions in Figures 2.12 and 2.13. All of these variables are marked on the vertical axis and the expected values of the interface defects radius are shown in the horizontal ones.

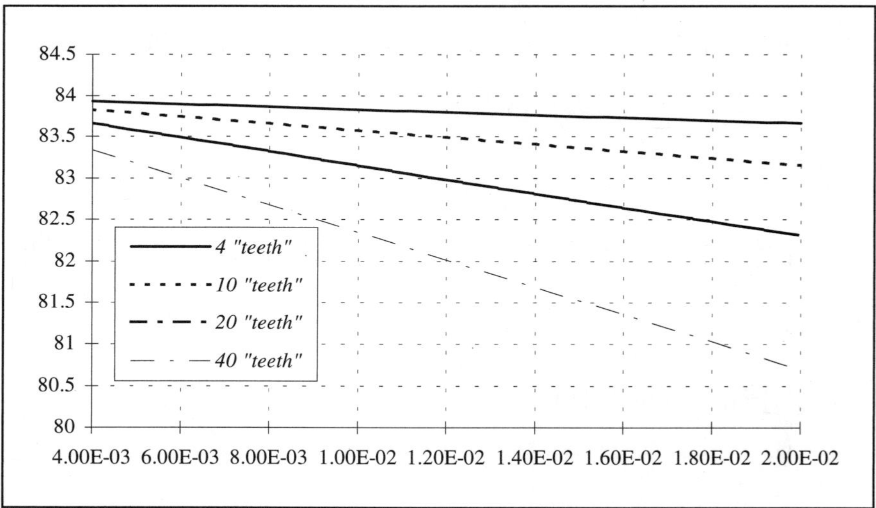

Figure 2.8. Expected values of probabilistically averaged Young modulus in fibre

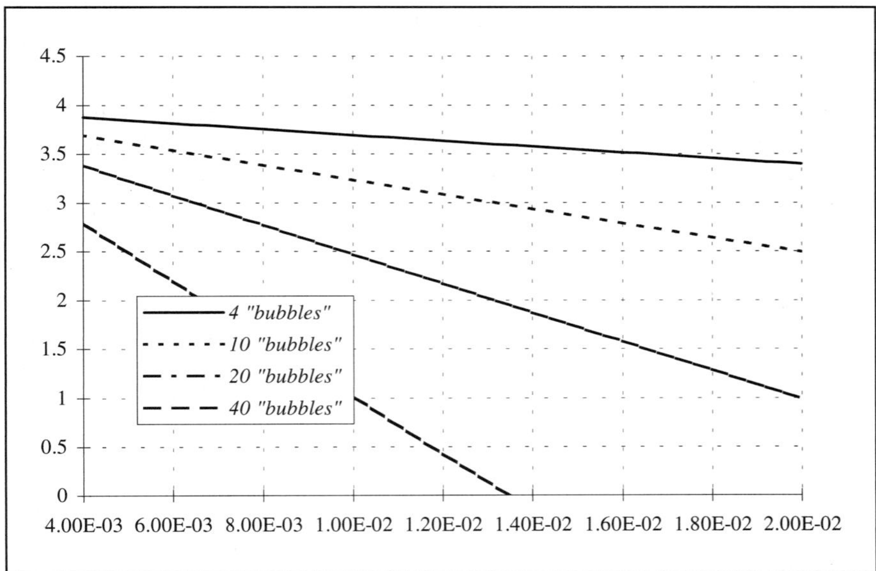

Figure 2.9. Expected values of probabilistically averaged Young modulus in matrix

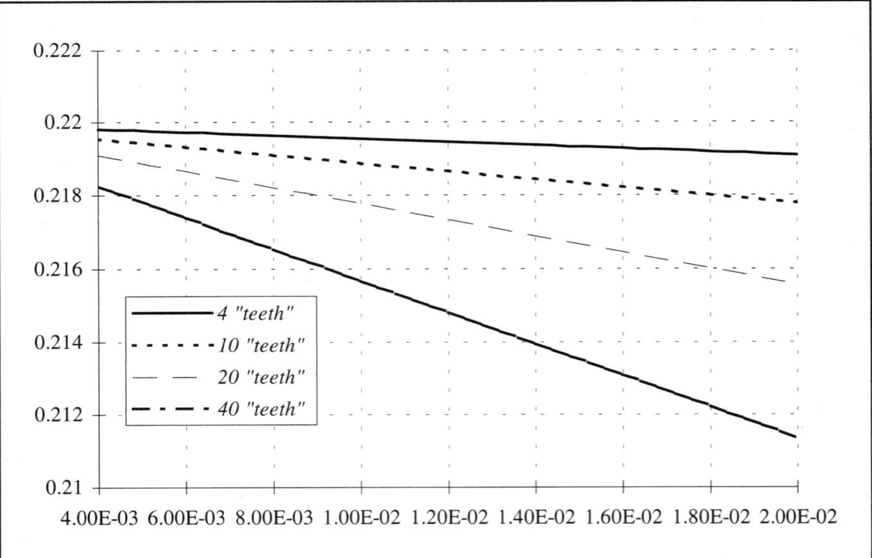

Figure 2.10. Probabilistically averaged Poisson ratio in fibre

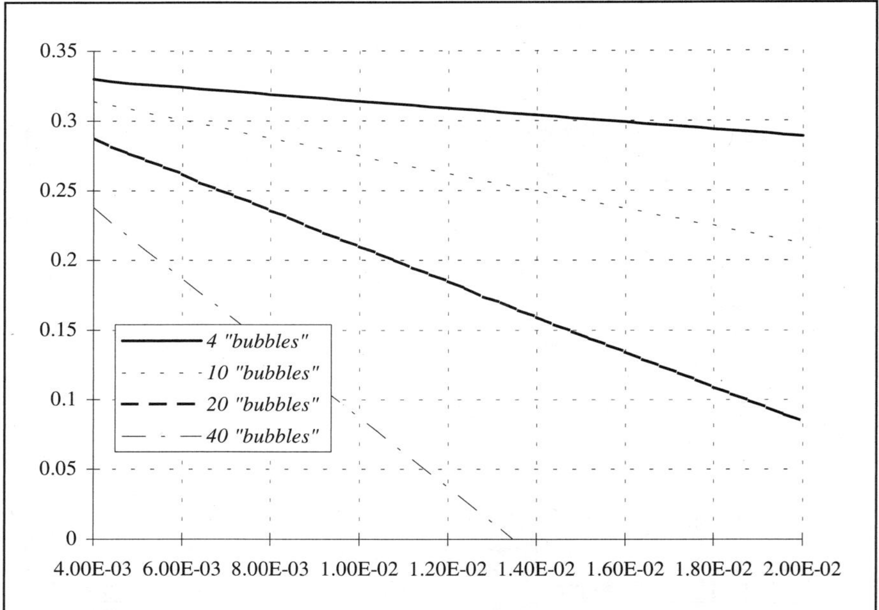

Figure 2.11. Probabilistically averaged Poisson ratio in matrix

As is expected, the resulting expected values of the homogenised Young modulus both in the matrix and the fibre regions, and similarly the Poisson ratio, are linear functions of the contact zone widths. The variances of the averaged Young modulus are second or higher order functions of this variable and this order is directly dependent on the number of interface defects.

Comparing Figures 2.8 with 2.9 and 2.12 with 2.13 it can be seen that the Young modulus in the matrix contact zone is, for the present problem, much more sensitive to variation of its parameters than the same modulus in the fibre interphase. Larger coefficient of variation of this modulus is obtained in the matrix interface region rather than in the fibre contact zone. On the other hand, the homogenised elastic properties are derived by averaging their values in both regions. Thus, greater changes in these properties can be expected in the matrix because of the larger volume of bubbles related to the fibre teeth.

Another interesting effect (cf. Figures 2.12 and 2.13) is the increase of variances of the homogenised Young modulus in the matrix contact zone for increasing width of this zone and the number of bubbles. The reverse effect occurs for the fibre side of the interface and its teeth. This is due to the fact that bubbles occupy more than half of a volume of the corresponding contact zone, and teeth less than a half.

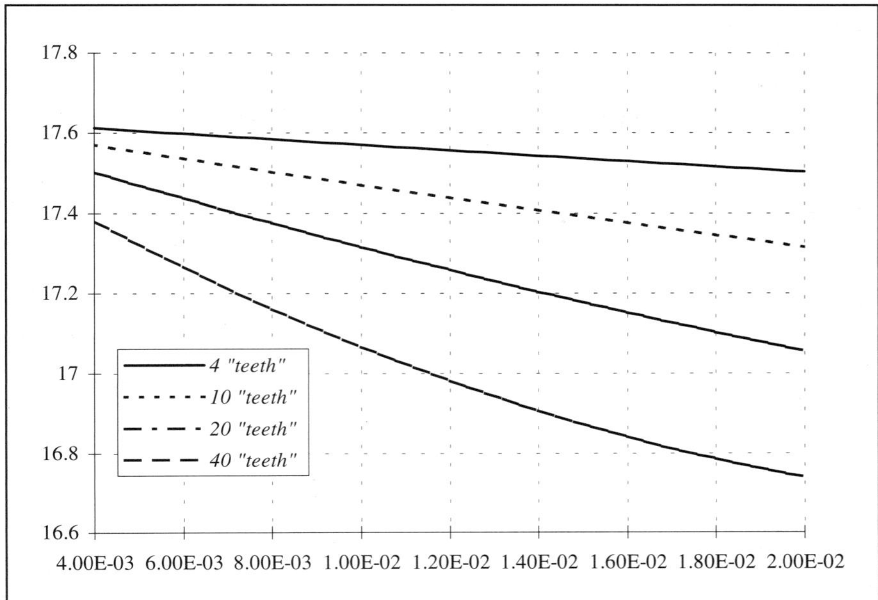

Figure 2.12. Variances of probabilistically averaged Young modulus in fibre

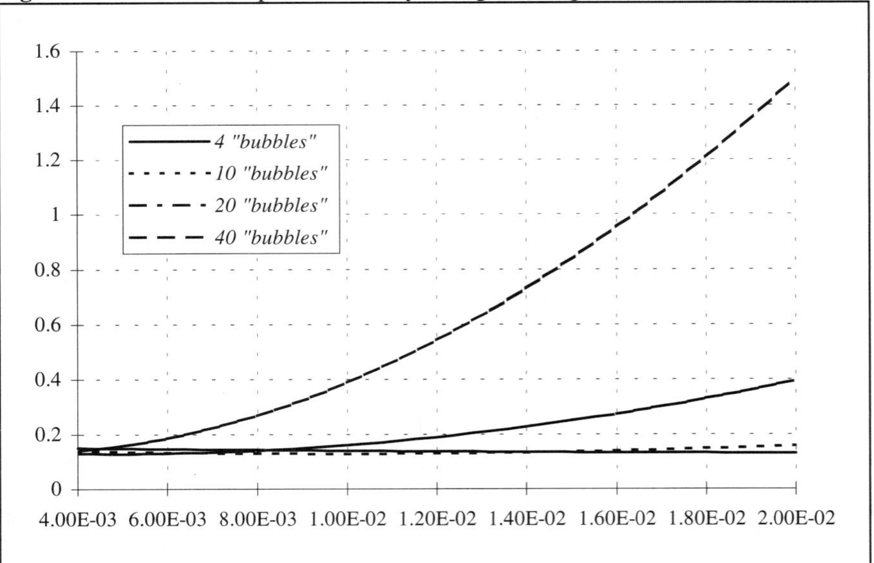

Figure 2.13. Variances of probabilistically averaged Young modulus in matrix

2.2 Elastostatics of Some Composites

Elastic properties and geometry of Ω so defined result in the random displacement field $u_i(x;\omega)$ and random stress tensor $\sigma_{ij}(x;\omega)$ satisfying the classical boundary–value problem typical for the linear elastostatics problem. Let us assume that there are the stress and displacement boundary conditions, $\partial\Omega_t$ and $\partial\Omega_u$ respectively, defined on Ω. Let C_{ijkl} be a random function of C^1 class defined on the entire Ω region. Let ρ denote the mass density of a material contained in Ω and ρf_i denote the vector of body forces per a unit volume. The boundary–differential equation system describing this equilibrium problem can be written as follows

$$\sigma_{ij}(x;\omega) = C_{ijkl}(x;\omega)\,\varepsilon_{kl}(x;\omega) \tag{2.48}$$

$$\varepsilon_{ij}(x;\omega) = \tfrac{1}{2}\left(\frac{\partial u_i(x;\omega)}{\partial x_j} + \frac{\partial u_j(x;\omega)}{\partial x_i}\right) \tag{2.49}$$

$$\sigma_{ij,j}(x;\omega) + \rho(\omega)\,f_i = 0 \tag{2.50}$$

$$E[u_i(x;\omega)] = E[\hat{u}_i(x;\omega)]; \quad x \in \partial\Omega_u \tag{2.51}$$

$$Var(u_i(x;\omega)) = 0; \quad x \in \partial\Omega_u \tag{2.52}$$

$$E[\sigma_{ij}(x;\omega)]n_j = E[t_i(x;\omega)]; \quad x \in \partial\Omega_t \tag{2.53}$$

$$Var(\sigma_{ij}(x;\omega))n_j = 0; \quad x \in \partial\Omega_t \tag{2.54}$$

for $a=1,2,...,n$ and $i,j,k,l=1,2$.

Generally, the equation system posed above is solved using the well–established numerical methods. However it should first be transformed to the variational formulation. Such a formulation, based on the Hamilton principle, is presented in the next section. To have the formulation better illustrated, an example of the periodic superconducting coil structure is employed. The stochastic non–homogeneities simulate the technological innacuracies of placing the superconducting cable in the RVE. Its periodicity cell in that case is subjected to horizontal uniform tension on its vertical boundaries to analyse the influence of the stochastic defects on the probabilistic moments of horizontal displacements. The stochastic variations of these displacements with respect to the thickness of the interphase constructed are verified numerically. Stochastic computational experiments are performed using the ABAQUS system and the program POLSAP specially adapted for this purpose.

2.2.1 Deterministic Computational Analysis

The main idea of the numerical experiments provided in this section is to illustrate the horizontal displacements fields and the shear stresses obtained for the deterministic problem of uniform extension of the periodicity cell quarter. Both Young modulus and Poisson ratio are assumed here as deterministic functions; for the purpose of the tests, the program ABAQUS [1] is used together with its graphical postprocessor. The periodicity cell quarter has been discretised by 224 rectangular 4−node plane strain isoparametric finite elements according to Figure 2.14.

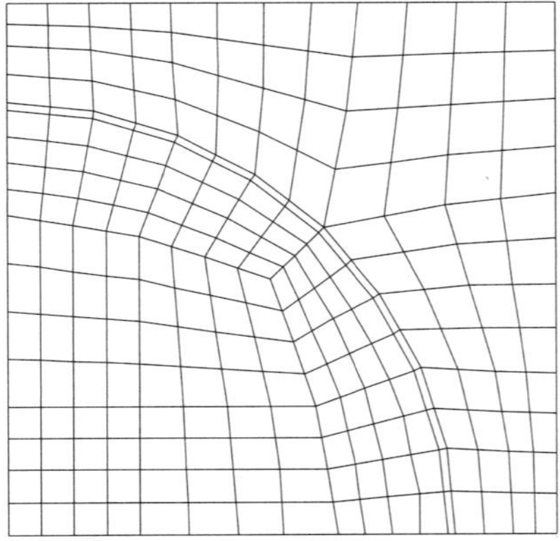

Figure 2.14. Discretisation of the fibre-reinforced composite cell quarter

The symmetry displacement boundary conditions are applied on the horizontal edges of the quarter as well as on the left horizontal edge, while the uniform extension is applied on the right vertical edge of the RVE. The standard deviations of the composite component Young moduli are taken as $\sigma(e_1) = 4.2$ GPa, $\sigma(e_2) = 0.4$ GPa and the stochastic interface defect data are approximated by the following values: $E[n]=3$, $\sigma(n)=0.05$ $E[n]=0.15$, $E[r]=0.02$ R, $\sigma(r)=0.1$ $R=8.0E-4$. Probabilistically averaged values of the interphase elastic characteristics are obtained from these parameters as follows $E[e_k]=3.82$ GPa, $Var(e_k)=1.48$ GPa, $v_k=0.324$ with the interphase thickness equal to $\Delta_k=0.0104$. Four numerical experiments have been carried out for these parameters taking the values collected in Table 2.1.

Table 2.1. Young modulus values of the interphase for particular tests

Test number	1	2	3	4
e_k	e_2	$E[e_k]$	$E[e_k]-3\cdot\sigma(e_k)$	$E[e_k]+3\cdot\sigma(e_k)$

Horizontal displacement fields and the shear stress fields for particular experiments are presented in Figure 2.15 and 2.19 (test no 1), Figure 2.16 and 2.20 for test no 2, Figure 2.17, 2.21 for test no 3 and Figure 2.18 for test no 4.

Comparing these results, it is seen that a decrease of the Young modulus value lower than its expected value results in a jump of the horizontal displacements field within and around the interphase. This effect can be interpreted as the possibility of debonding of the composite components caused by the worsening of the interphase elastic parameters, which confirms the usefulness of the presented mathematical–numerical model in the interphase phenomena analysis. It should be underlined that in other models of interphase defects and contact effects at the interface, the horizontal displacements have discontinuous character too. On the other hand, increasing the Young modulus above its expected value does not introduce any sensible differences in comparison with the traditional deterministic model for the perfect interface.

Analysing the shear stresses fields $\sigma_{12}(x_i)$ collected in Figures 2.19 and 2.21 a jump of the respective values of stresses at the boundary between the fibre and the interphase region is observed in all cases. In the case of tests no. 1, 2 and 4 the shear stress fields have quite similar characters differing one from another in the interface regions placed near the horizontal and vertical edges of the periodicity cell quarter. The $\sigma_{12}(x_i)$ field obtained for test no. 3 has decisively different character: for almost the entire interface the jump of stresses between the matrix, interphase and fibre regions is visible. It may confirm the previous thesis based on the displacement results dealing with the usefulness of the model proposed for the analysis of the interface phenomena.

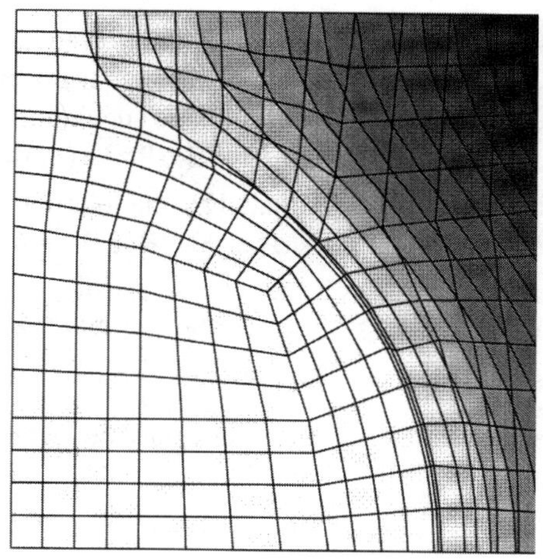
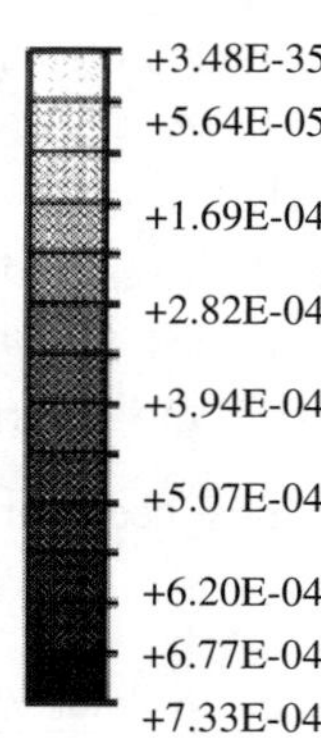

Figure 2.15. Horizontal displacements for test no. 1

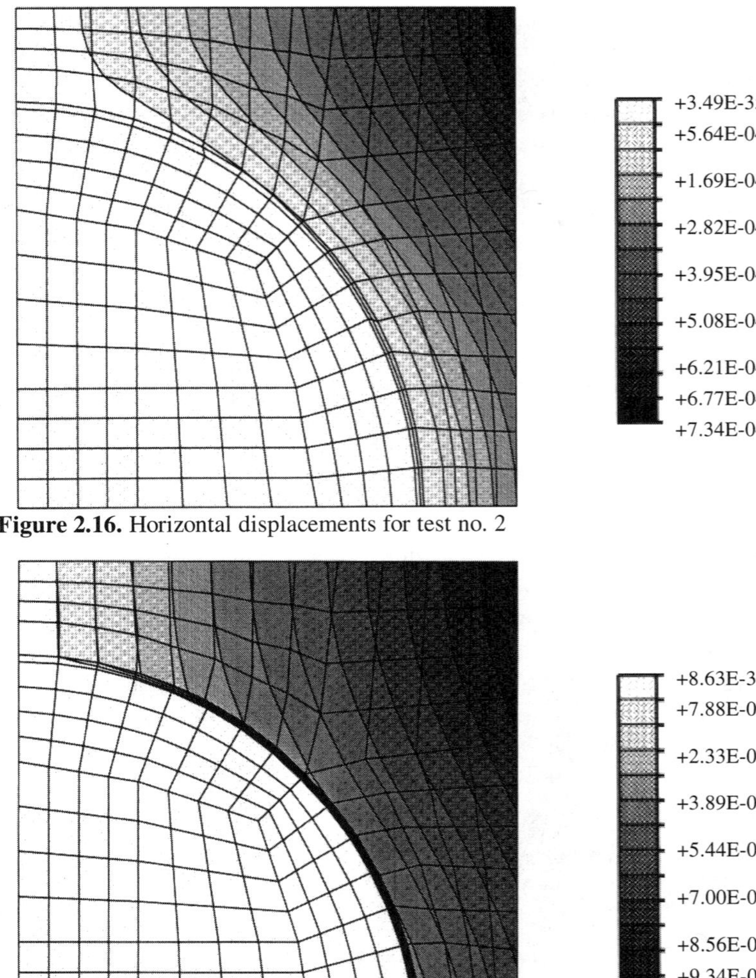

Figure 2.16. Horizontal displacements for test no. 2

Figure 2.17. Horizontal displacements for test no. 3

Figure 2.18. Horizontal displacements for test no. 4

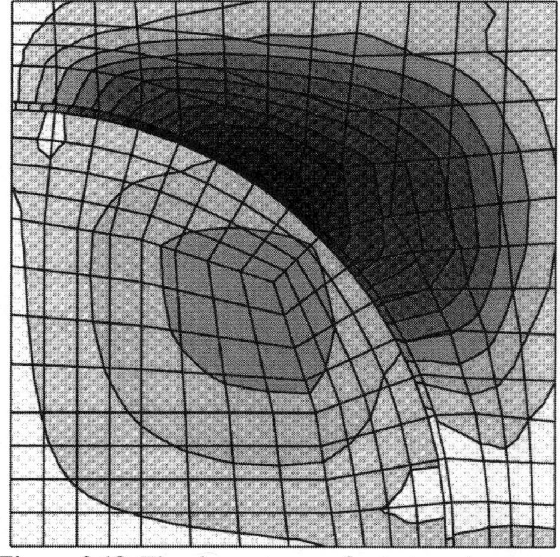
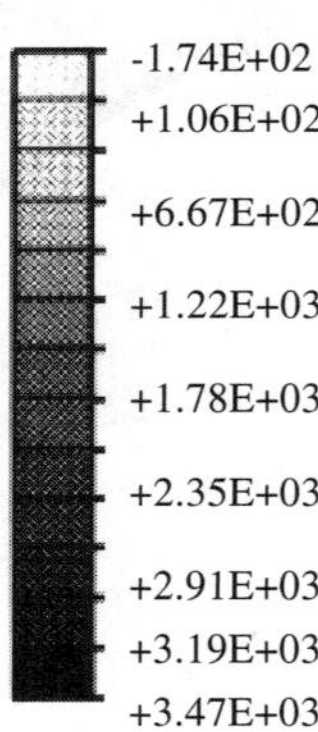

Figure 2.19. The shear stresses for test no. 1

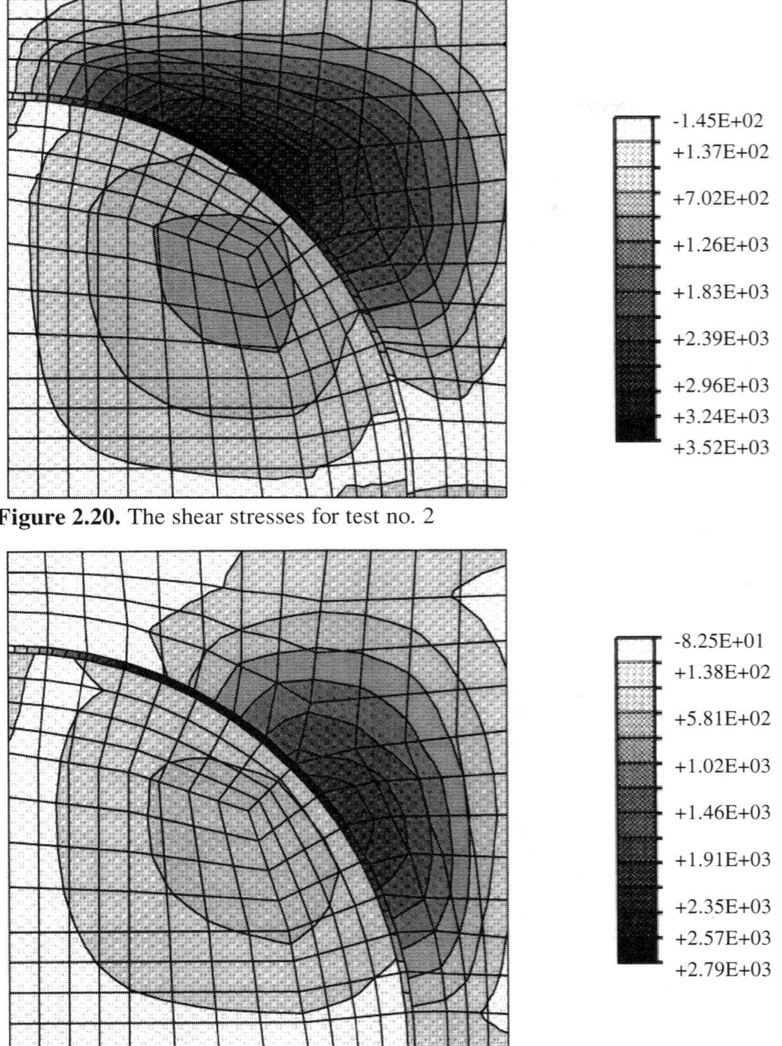

Figure 2.20. The shear stresses for test no. 2

Figure 2.21. The shear stresses for test no. 3

The general purpose of the computational experiments performed is to verify the stochastic elastic behaviour of the composite materials with respect to probabilistic moments of the input random variables: both the Young moduli of the constituents as well as the stochastic interface defects parameters. The starting point for such analyses is a verification of the probabilistically averaged Young modulus in the interphase (example 1). This has been done by the use of the special FORTRAN subroutine, while the next tests have been carried out using the 4−node isoparametric rectangular plane strain element of the system POLSAP. Material parameters of the composite constituents are taken in examples 1 to 3 as

$E(e_1) = 84.0$ GPa, $v_1 = 0.22$, $\sigma(e_1) = 4.2$ GPa, $E(e_2) = 4.0$ GPa, $\sigma(e_2) = 0.4$ GPa, $v_2 = 0.34$ (expected values and standard deviations of the Young modulus and Poisson ratio, respectively).

2.2.2 Random Composite without Interface Defects

The main aim of the numerical analysis is to verify numerically the elastic behaviour of a fibre composite when the Young modulus of composite components is Gaussian random variable. Moreover, numerical tests are carried out to state in what way, for various contents of fibre (with round section) in a periodicity cell, the random material properties of reinforcement and matrix influence the displacement and stress distribution in the cell. A quarter of a fibre composite periodicity cell is tested in numerical analysis and its discretisation is shown in Figure 2.22.

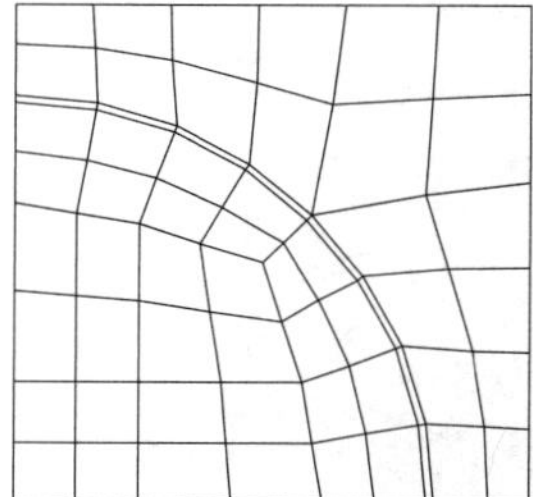

Figure 2.22. Discretisation of the periodicity cell quarter

Numerical implementation enabling the computations is made using a 4−node rectangular plane element of the program POLSAP (Plane Strain/Stress and Membrane Element). The composite structure is subjected to the uniform tension (100 kN/m) on a vertical cell boundary (60 finite elements with 176 degrees of freedom). Vertical displacements are fixed on the remaining cell external boundaries and the plane strain analysis is performed. Twelve numerical tests are carried out assuming the fibre contents of 30, 40 and 50 % and the resulting coefficients of variation are taken from Table 2.2.

Table 2.2. Coefficients of variation for different numerical tests

Test number	$\alpha(e_1)$	$\alpha(e_2)$
1	0.10	0.10
2	0.10	0.05
3	0.05	0.10
4	0.05	0.05

Each time the first two probabilistic moments of the displacements are observed at the interface and on the tensioned vertical edge. In the case of stress expectations, location and maximum value of reduced stress are examined. Figures 2.23 and 2.24 demonstrate radial displacement coefficients of variation of points belonging to the fibre–matrix boundary, which depend on the angle β describing their locations on this boundary.

The results of test no. 1 (Table 2.2) are presented in Figure 2.23, and the next figure shows the results of test no. 3; results of the remaining tests (no. 2 and 4) agree with them respectively. In both cases coefficients of variation for $\theta = 90°$ are omitted on the graphs because of their large values. For fibre contents equal to 50%, they are approximately 1.5 times greater than for $\theta = 0°$ (disproportion of the data would give an illegible picture). Therefore, it can be concluded that the randomness of displacements on the considered boundary depends mainly on the random character of fibre elastic properties, which means

$$\alpha[u(x)] \cong \alpha[e_1]; x \in \partial\Omega_{1,2} \tag{2.55}$$

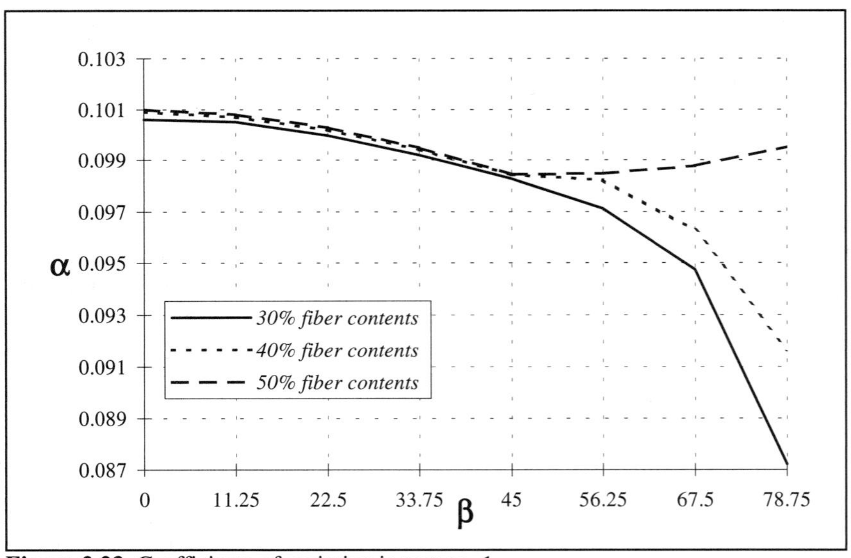

Figure 2.23. Coefficients of variation in test no. 1

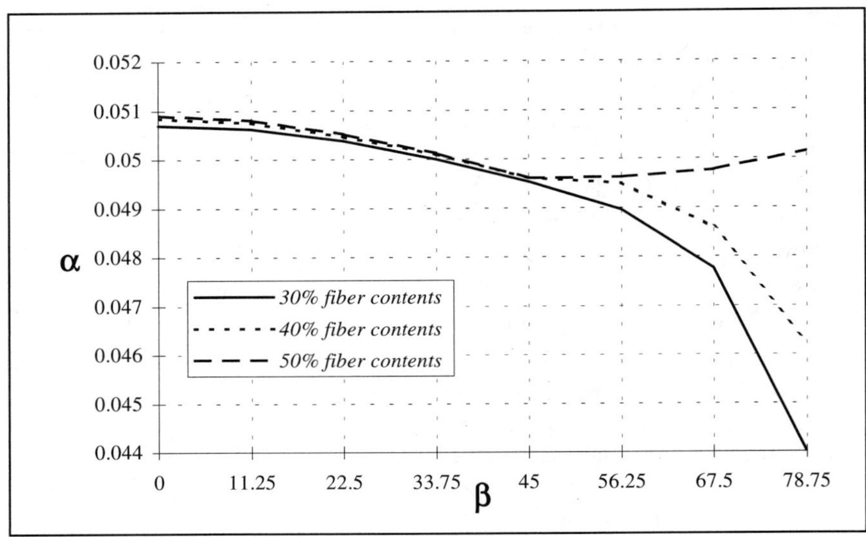

Figure 2.24. Coefficients of variation in test no. 3

Fibre contents in periodicity cell influence also displacement coefficients of variation on $\partial\Omega_{12}$. This influence is visible especially at $0° \leq \theta \leq 45°$. For 40% contents this decrease is not so sharp, and for 50% plane fraction the tendency is the opposite: the coefficient increases up to about 1.5 times of the value obtained at $\theta = 0°$. In a physical way it may be interpreted as increasing the random measure of uncertainty about displacements perpendicular to the fibre boundary of the points belonging to its upper part with increasing fibre radius.

Figures 2.25−2.26 show displacement coefficients of the variation of horizontal points belonging to a vertical, uniformly tensioned edge of periodicity cell obtained in tests no. 1, 2, 3 and 4 respectively. Real numbers in decreasing order denote height on the vertical tensioned edge on the horizontal axes of these figures.

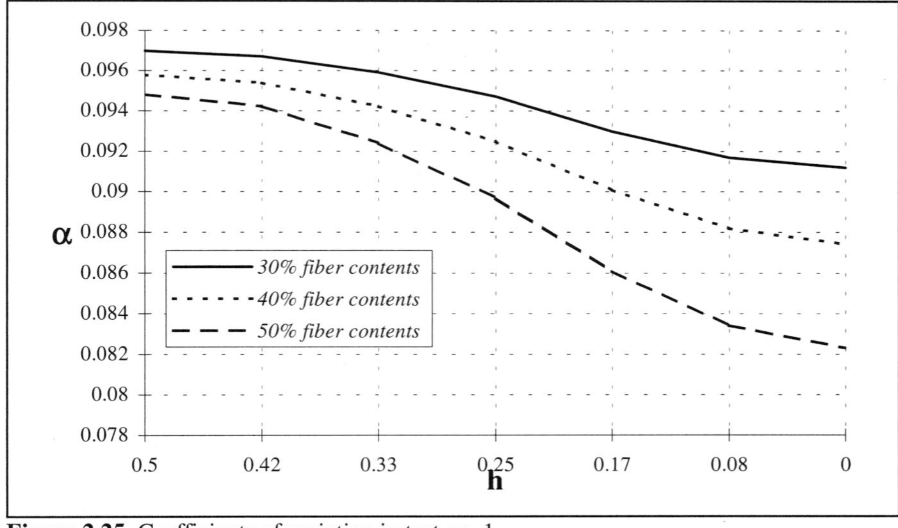

Figure 2.25. Coefficients of variation in test no. 1

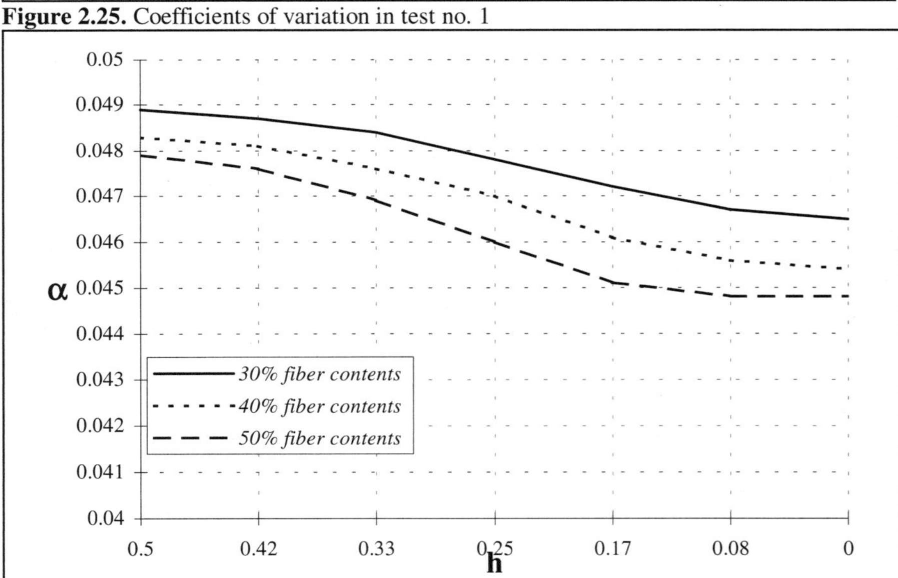

Figure 2.26. Coefficients of variation in test no. 2

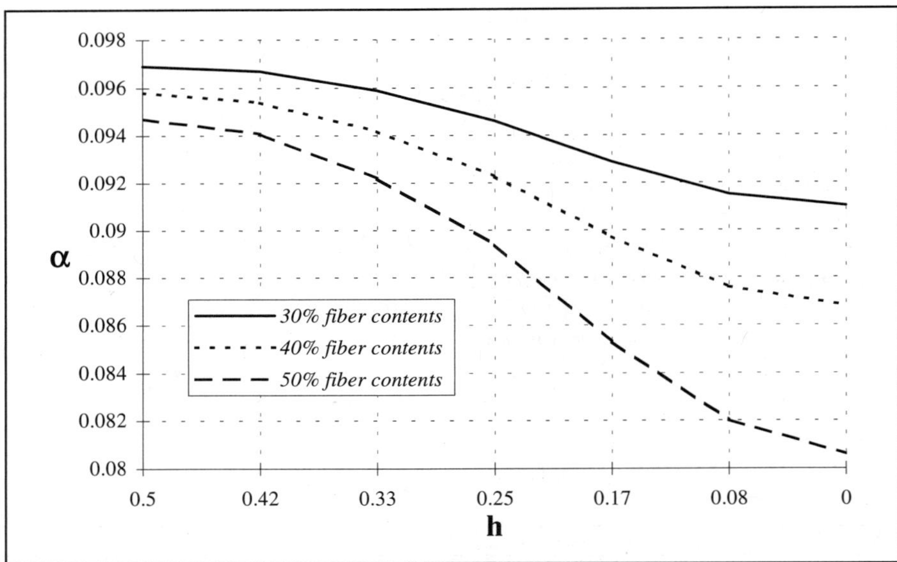

Figure 2.27. Coefficients of variation in test no. 3

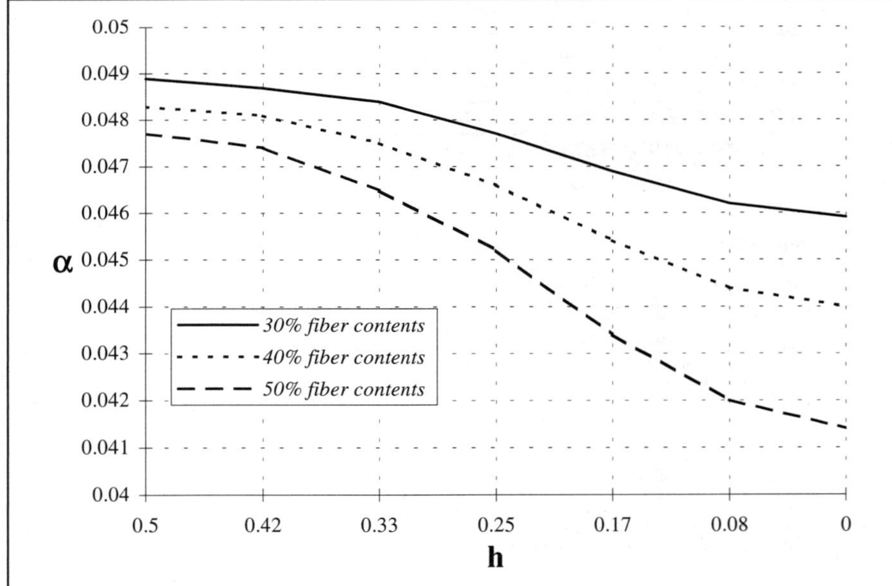

Figure 2.28. Coefficients of variation in test no. 4

The main conclusion from these results is that the random character of the matrix elastic properties influences the randomness of displacements at the tensioned edge of the composite specimen tested. Analogously to the previous observations it can be written that

$$\alpha[u(x)] \cong \alpha[e_2] \; ; \; x \in \partial\Omega_{\hat{\sigma}} \tag{2.56}$$

Let us note that the random character of fibre stiffness has rather secondary influence here. The curves describing displacement variation coefficients on the edge are becoming less and less sharp together with an increase of the coefficients of variation of the fibre Young modulus. Increase of fibre contents in the periodicity cell, as expected, in all cases decreases variation coefficients of tensioned edge displacements, which physically can be interpreted as increasing stiffening of periodicity cell by the fibre.

Now, let us analyse expected values of maximum stresses (in MPa) in fibre and matrix specified in Figure 2.29. Darker bars show the maximal stresses in the matrix region, while lighter bars denote the fibre region, respectively.

Generally, it can be observed that the difference between the obtained expected values and the results of deterministic tests is approximately equal to the computational error. This difference would undoubtedly be much bigger if the formula describing these values included a component connected with elasticity tensor derivatives. The present version of computer program includes only the first two components, which correspond with expected values of displacement functions.

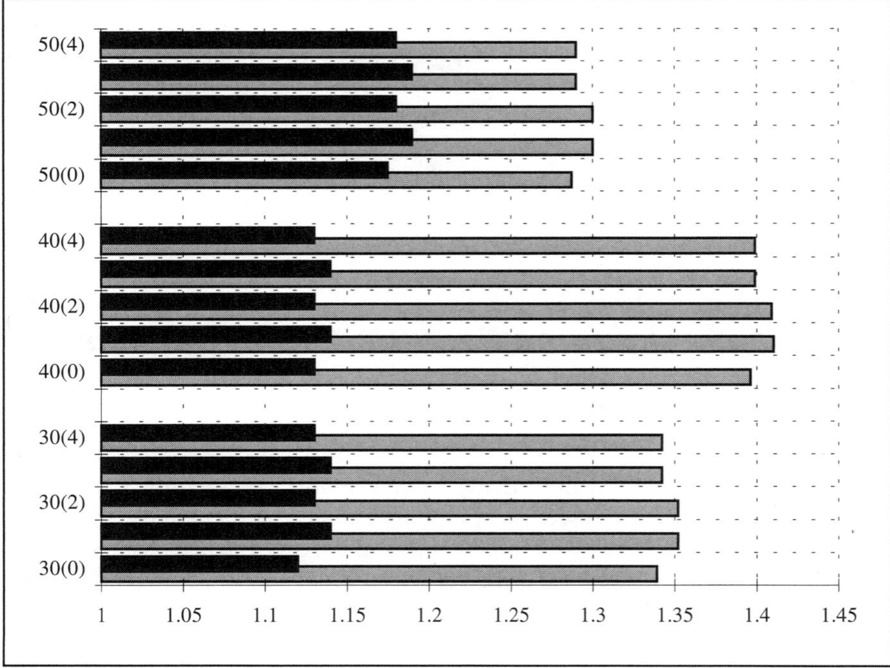

Figure 2.29. Expected values of maximal stresses

The results obtained lead to the conclusion that the most important factor influencing the value of maximum stresses is unquestionably the fibre radius, cf.

Figure 2.29. In the case of the matrix region, maximum stresses increase approximately in direct proportion to fibre radius increment

$$E[\sigma_{max}(x)] \approx R \, ; \; x \in \Omega_2 \qquad (2.57)$$

To get an analogous relation for maximum stress appearing in the fibre, it is necessary to make a more precise numerical analysis. In tested examples with plane fractions of 30, 40 and 50% extremum appeared at 40% contents of fibre in the periodicity cell. Another factor, which influences the expected values of maximum stresses within a given material, is its coefficient of variation for the Young modulus. The following relation can be formulated:

$$E[\sigma_{max}(x)] \approx \alpha[e_i] \, ; \; x \in \Omega_i \qquad (2.58)$$

Finally, it can be observed that there is a third–order influence of stronger material random changes of elastic features on maximum stresses in the matrix, especially with decreasing fibre contents in the RVE.

In the context of the present numerical analysis of maximum stresses it should be added that, apart from changes in the expected values of these stresses, a change of their locations was observed. In order to state the relation between the location of changes in the direction of the stress functions extremum and fibre radius increment it would be necessary to consider a wider range of this radius variation (equivalent to, for example, a surface fraction of 10 to 60%) with simultaneously increasing the number of tests (each 1 to 5% for example). The most essential thing would be, however, creating a mesh much more precise than the one used in the above tests, especially near the composite interface, where we have, of interest to us, maximum stresses.

2.2.3 Fibre–reinforced Composite with Stochastic Interface Defects

The subject of the third numerical example is the fibre–reinforced periodic composite, which has been discretised in Figure 2.30 as a cell quarter with smaller contact zones on the left and with larger ones on the right. The composite structure is subjected to uniform tension on the vertical cell boundary. Six numerical tests have been performed assuming interphases with different values of the total number of defects (in turn: 0%, 25% and 50% of the interface length). In each test, the first two probabilistic moments of the displacements are observed on the phase boundary and on the vertical edge subjected to tension and the coefficient of variation for displacements.

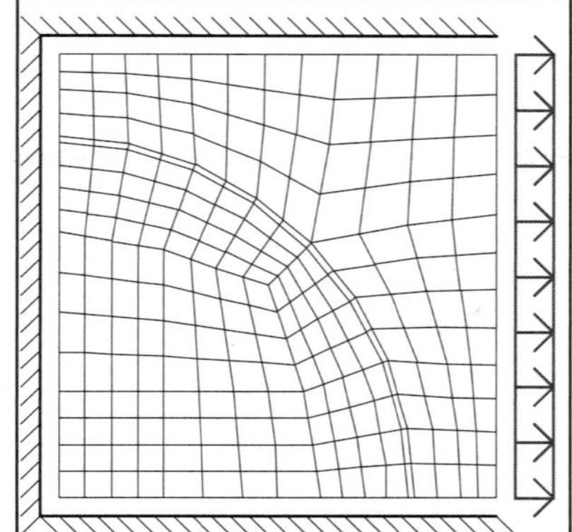

Figure 2.30. Quarter periodicity cell mesh for the SFEM analysis

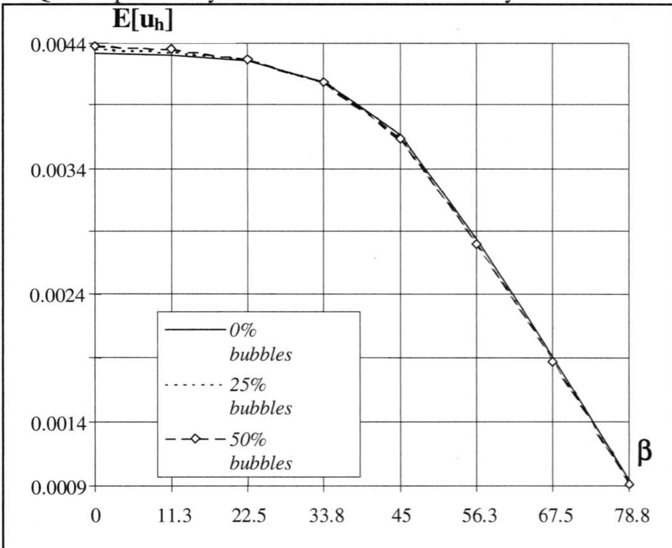

Figure 2.31. Expected values of horizontal displacement at the interface

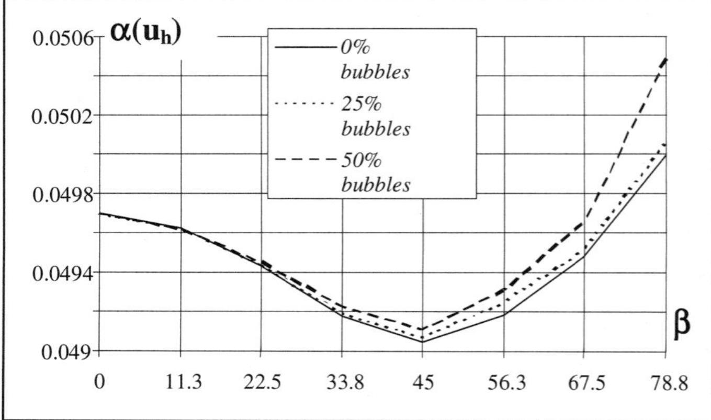

Figure 2.32. Coefficients of variation of horizontal displacements at the interface

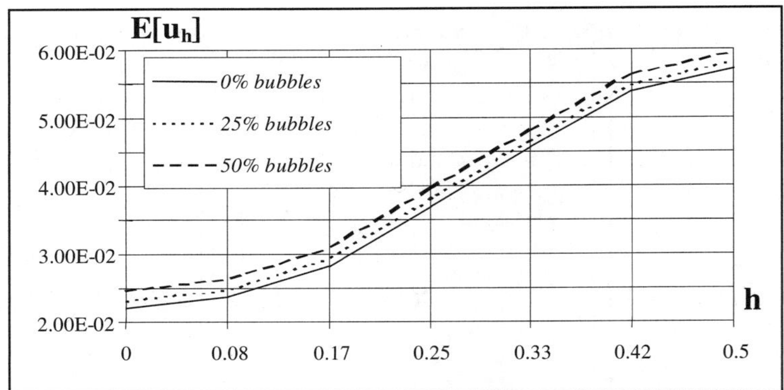

Figure 2.33. Expected values of horizontal displacements at the tensioned edge

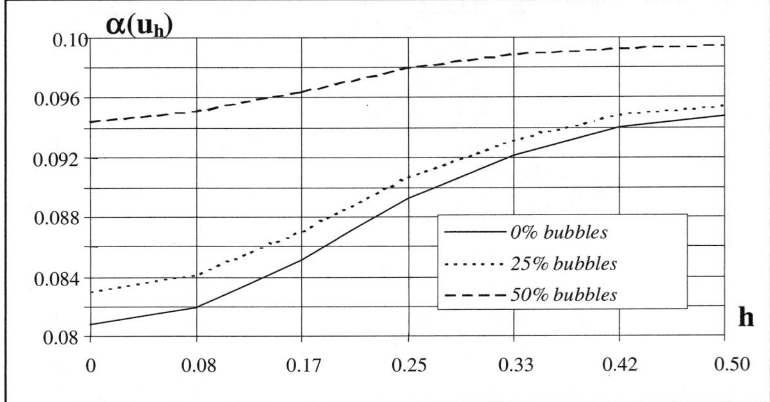

Figure 2.34. Coefficients of variation of horizontal displacements at the tensioned edge

The expected values of the displacements and their coefficients of variation are placed on the vertical axes of all figures. The angle β, which determines the location of a point on the fibre−matrix interface with respect to the x or y− coordinate on the tensioned edge, and which is marked on the vertical axes.

A further general observation is a direct proportionality between the number of interface defects and the volume of the contact zone as well as the expected values or coefficients of variation of these displacements. Small differences occur in the interface expected values of displacements for larger values of the angle β.

By comparing the coefficients of variation of the interface displacements (Figure 2.32 and 2.34) quite different forms for the relation between these coefficients and the angle β are observed. The model with a large contact zone shows a high sensitivity to the number of defects and the changes for the small contact zone are proportional. In the case of the coefficients of variation of the tensioned edge horizontal displacement both the models give approximately reversed effects. For example a small contact zone causes larger coefficients for smaller β values than for the larger ones (Figure 2.32). For both sizes of the contact zones the changes in the coefficient α are inversely proportional to the increase in the number of discontinuities and show some similarity.

Finally, in both models the expected values of the displacement are quite similar with respect to their locations. In the large contact zone (Figure 2.31 and 2.33) the differences between the obtained expected values of displacements for 0%, 25% and 50% of discontinuities are more significant.

2.2.4 Stochastic Interface Defects in Laminated Composite

The two–component layered composite has been tested in this example. The discretisation into 72 finite elements and 233 degrees of freedom as well as the mixed boundary conditions is shown in Figure 2.35. Both layers have been uniformly extended in the opposite directions to verify the influence of interphase between them on the overall behaviour of the structure.

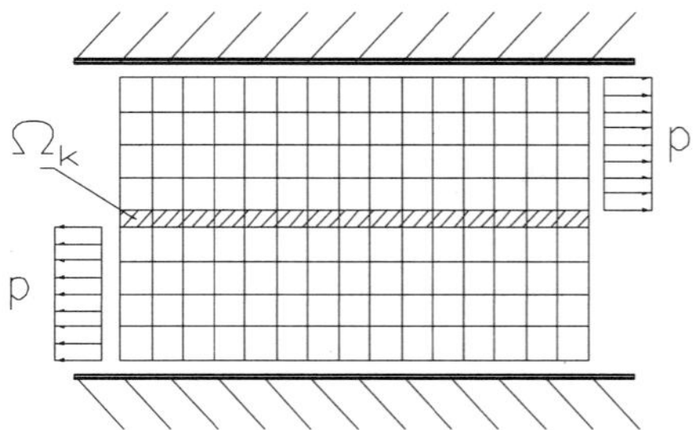

Figure 2.35. Two–layer laminate in the computational shear test

Ten numerical experiments have been carried out in the example: the deterministic problem (test–d) and the stochastic one without interface defects (test–s). In the next experiments the standard deviations of the defects are taken as $\sigma[r] = 0.1 \cdot E[r]$, $\sigma[n] = 0.1 \cdot E[n]$, and the expected values are shown in Table 2.3 (contribution of the boundary occupied by bubbles to the total boundary is given in brackets).

Table 2.3. The expected values of the interface defects tested

	Test 1	Test 2	Test 3	Test 4	Test 5	Test 6	Test 7	Test 8
$E[r]$	5.0E-2	5.0E-2	5.0E-2	5.0E-2	1.0E-1	1.0E-1	1.0E-1	1.0E-1
$E[n]$	5 (10%)	10 (20%)	15 (30%)	20 (40%)	5 (20%)	10 (40%)	15 (60%)	20 (80%)

The results of the analyses have been collected in Table 2.3, which shows the expected values and the coefficients of variation of the displacements and are generally consistent with those obtained experimentally (in the range of expected values). The increases of the expected values in comparison to the results obtained in test−d and test−s are included also in this table. The coefficients of variation of the horizontal displacements for smaller and greater interphase are presented in Figure 2.36 and 2.37 as a function of the location of a point on the Ω_2 boundary.

On the horizontal axis the height of the point (h) in decreasing order is presented: the coordinate 2.5 denotes the point belonging to the interface and Ω_1 region on the extended Ω_2 boundary, while the coordinate 5.0 denotes the point belonging to the upper Ω_2 boundary.

Table 2.4. The expected values and coefficients of variation of the displacements tested

	Test-d	test-s	test 1	test 2	test 3	test 4	test 5	test 6	test 7	Test 8
$E[q]$ (E-2)	1.924 2.610	1.939 2.633	2.049	2.089	2.134	2.188	2.686	2.844	3.065	3.408
$\Delta E[q]$ (%)	-0.8 -0.9	0.0 0.0	5.7	7.7	10.1	12.8	2.0	8.0	16.4	29.4
$\alpha[q]$	-	0.082	0.078	0.080	0.083	0.089	0.088	0.098	0.120	0.158

Generally, all the results computed show that the most sensitive region to the input random parameters is the point located on the weaker material (matrix) and the interphase on the extended Ω_2 boundary. Moreover, analysing the increases of the expected values of horizontal displacements on the tensioned boundary the significant influence of the stochastic interface defects introduced can be observed. In all tests performed the displacements obtained are greater than for the composites without defects between the composite constituents.

Moreover, the increases of the displacements analysed increase faster than the increases of the total length of the boundary occupied by the bubbles, which follows the stochastic nonlinearity of the model presented. The diagrams of the displacements have analogous characteristics to those obtained for the coefficients of variation presented later. However, considering the large disproportions between the values computed near the interphase and outside it, these graphs have been omitted.

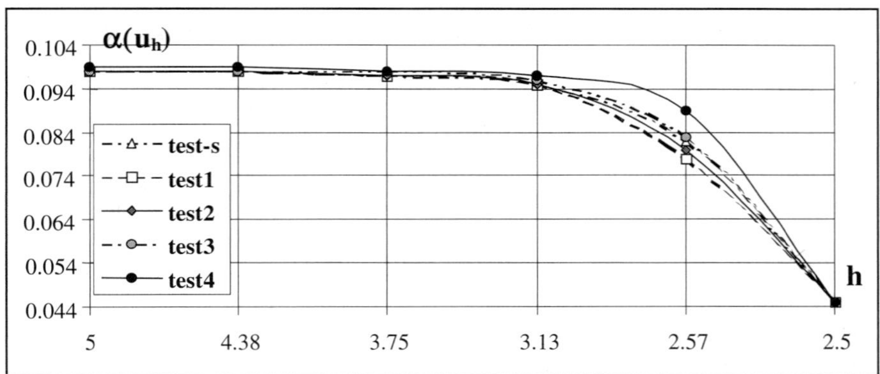

Figure 2.36. Coefficients of variation of horizontal displacements for shear test (I)

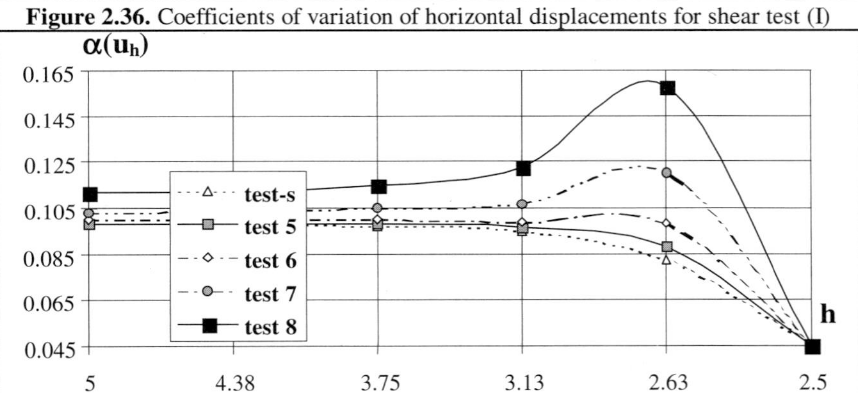

Figure 2.37. Coefficients of variation of horizontal displacements for shear test (II)

Comparing the coefficients of variation of the horizontal displacements it is seen that, especially in case of tests no. 5 to 8 (the interphase twice as large as for tests 1 to 4) the significant increase of these displacements is about 95% in case of test no. 8. These increases are analogous to the increases of expected values greater for displacements rather than the corresponding increases of total length of interface boundary occupied by the bubbles.

As can be expected, the statistical response of the laminate should depend on the contrast between stronger and weaker layer material properties, interphase elastic parameters, the total number of layers in the composite etc. Essentially different situation can be observed when both material properties and external load are introduced as random variables [273].

2.2.5 Superconducting Coil Cable Probabilistic Analysis

The main ideas of the experiment [193] are as follows: (i) comparison of the stochastic behaviour of the superconducting coil cable in the original geometry with the model in which the technological nonhomogeneities have been probabilistically averaged; (ii) verification of the model sensitivity to the assumed thickness of the interphase introduced.

The example of the RVE analysed is presented in Figures 2.38 and 2.39 (all geometric dimensions are given in mm). A single discontinuity is modelled by complementing two circle quarters (teeths with their sharp sides directed to the interior of the superconducting cable). Their radii are equal to 1.5 mm for defects on the interface superconducting cable–tube and 2.0 mm for defects on the interface cable–jacket. The periodicity cell is subjected to a horizontal uniform tension on its vertical boundaries; due to symmetry only one quarter of this cell is employed. Displacement boundary conditions on all the remaining external boundaries are assumed to satisfy the symmetry conditions.

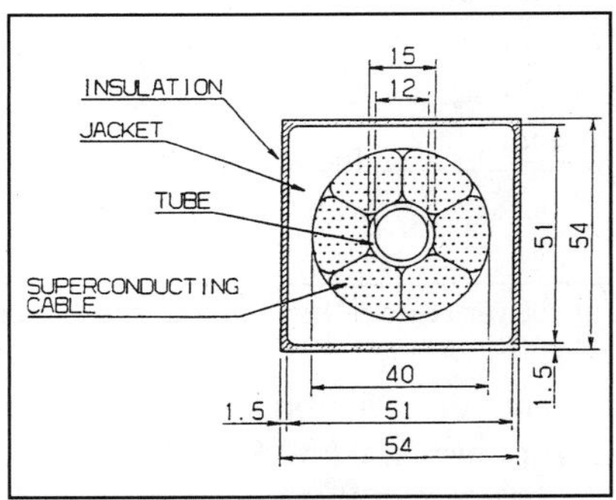

Figure 2.38. Superconducting coil cable RVE geometry

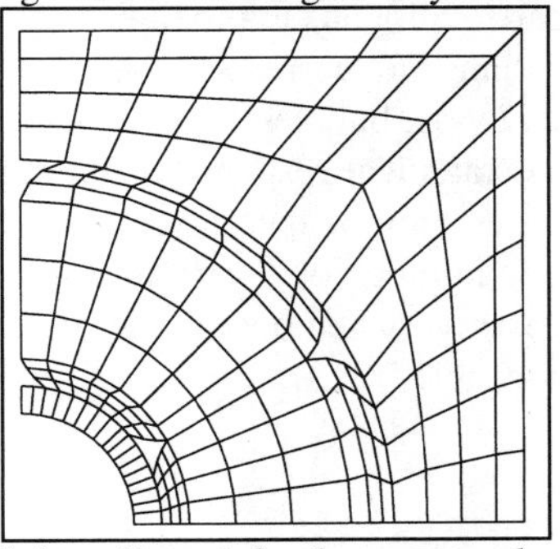

Figure 2.39. Quarter periodicity cell mesh for the superconductor

The elastic properties and their probabilistic characteristics of the RVE components, the expected values and the standard deviations of Young moduli, Poisson ratios and Kirchhoff moduli are collected in Table 2.5.

Table 2.5.   Elastic characteristics of composite components

Material	$E[e]$ [GPa]	$\sigma(e)$ [GPa]	ν	G [GPa]
Tube	205.0	8.0	0.265	81.0
Superconductor	182.0	0.0	0.300	70.0
Jacket	126.0	12.0	0.311	48.0
Insulation	36.0	0.0	0.210	11.0

The following tests are performed: deterministic test including defects non–averaged (test 1), an experiment without defects (test 2), an experiment with defects averaged in the interphase (test 3) or over the finite elements which they belong to (test 4). The first two probabilistic moments of the displacements are observed in each test on the interface determined by a radius equal to 9.0 mm (between the lower superconductor interphase and the superconductor region). Four additional tests are performed in test 3 to verify the results variations with respect to the interphase thicknesses: test 3A, where the thickness is equal to the expected value of the relevant geometric dimensions of interface defects, test 3D with thickness given by eqn (2.22) and tests 3B and 3C with the intermediate thicknesses.

The results of these computations due to tests numbered 1 to 4 are presented in Figures 2.40 and 2.41: the expected values of the horizontal displacements and their coefficients of variation. The first two moments are marked on the vertical axes of these figures, while the angle β, which determines the location of a point on the interface considered with respect to the x–coordinate on the horizontal axes. The results of tests 3A to 3D are collected in Table 2.6 presented below the figures. The expected values of displacements observed (in mm) are given in the upper row of each table cell and the coefficients of variation in the lower one.

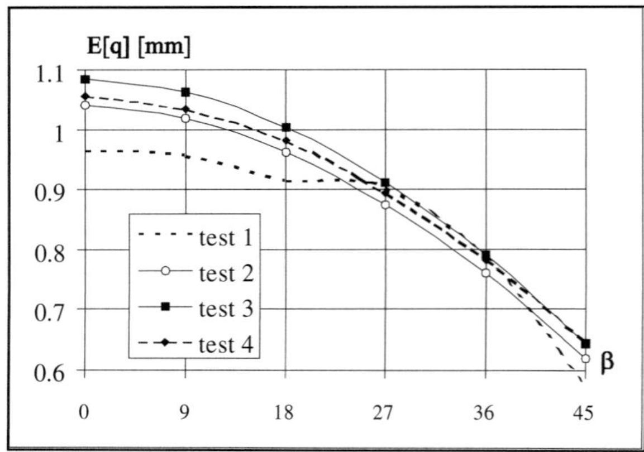

Figure 2.40. Expected values of horizontal displacements at the tensioned edge

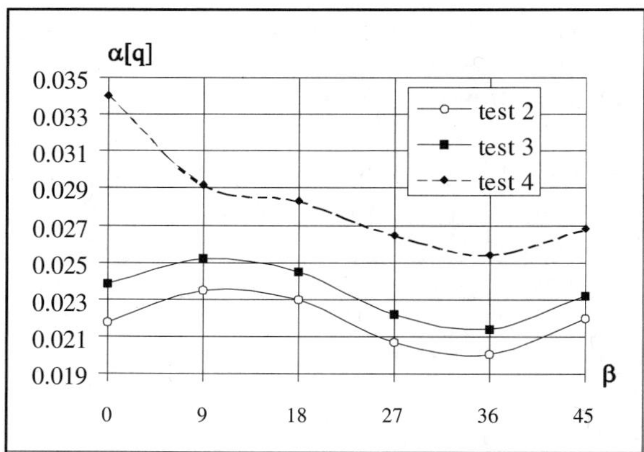

Figure 2.41. Coefficients of variation of horizontal displacements at the tensioned edge

Table 2.6. The expected values and the coefficients of variation of horizontal displacements

β [°]	Test 3A	Test 3B	Test 3C	Test 3D
0	1.066	1.069	1.078	1.085
	0.0241	0.0237	0.0235	0.0233
9	1.047	1.053	1.057	1.062
	0.0239	0.0238	0.0234	0.0232
18	0.985	0.993	0.994	1.003
	0.0236	0.0234	0.0231	0.0230
27	0.895	0.897	0.908	0.910
	0.0239	0.0235	0.0234	0.0231
36	0.783	0.784	0.784	0.790
	0.0241	0.0238	0.0235	0.0232
45	0.631	0.634	0.639	0.645
	0.0212	0.0212	0.0213	0.0214

Generally, it can be observed that in all stochastic tests the expected values of horizontal displacements are greater than the corresponding values obtained from deterministic tests, which follow equation (1.134). The greatest expected values of displacements observed are obtained for test 4: from 50% (for $\beta \approx 0°$) to 100% (for $\beta \approx 80°$) greater than in the remaining tests. Analogous observation can be done for the coefficients of variation. Generally, these facts follow the great variances of the Young moduli in finite elements containing defects averaged in comparison to the remaining elements.

On the basis of these results it can be stated that the observed probabilistic moments of displacements are strongly sensitive to the scale of the composite structure, which probabilistic averaging is applied in. A rapid decrease of the total area of the region averaged results in a significant increase of the effective Young modulus and much smaller increases of the expected values for the displacements. Further, the expected values obtained in test 2 (without including interface defects in any form) give the most exact results of the horizontal displacements computed in the deterministic model. However, for $\beta \approx 30°$, which corresponds to the defects location, the best approximation is obtained in test 3 (with interphase zones introduced).

Finally, let us consider the stochastic variations of the interface horizontal displacements to the interphase zone thicknesses illustrated by the results collected in Table 2.6. It can be observed that increasing thickness causes a small increase of the horizontal displacements and a decrease of the coefficients of variation. The decrease (or increase) has a linear character and the maximum increment is no greater than 2% of the values considered. It confirms the possibility of using the model presented in the numerical analysis of stochastic non−homogeneities (especially interface defects) in composite materials. To verify the applicability of the model presented this sensitivity should be discussed as a function of interface defects and elastic properties of composite component stochastic parameters.

Let us note that the SFEM methodology can be applied in further analyses for numerical modelling of random both uncoupled and coupled thermal, electric or magnetic fields in various superconducting structures [221,385]. A common application of the stochastic perturbation technique with computational plasticity algorithms will enable us to perform modelling of interface debonding in the case of laminates and fibre-reinforced composites, which will essentially extend our knowledge of composites behaviour in relation to the existing models [251,384,386].

2.3 Homogenisation Approach

Homogenisation methods present some specific approach to such computational analysis of composite materials, where the homogeneous medium equivalent to the real composite is proposed. The assumptions decisive for these methods are introduced in the context of numerous equivalence criteria; usually it is assumed that internal energies per unit mass stored in both systems are to be equal. A concept of the Representative Volume Element (RVE) for the composite is most frequently used together with the corresponding assumption on a scale parameter relating dimensions of the RVE to the entire composite – it has to tend to 0, which is usually unrealistic for most of engineering composites. It is evident now that the spatial distribution of the reinforcement (uniaxially periodic, with rectangular, hexagonal, triangular periodicity or completely random according to Gaussian or Poisson distributions) is of decisive importance for the effective material tensors [52]. There exist some mathematical approaches, where the scale parameter is assumed to be some small and positively defined [370]. It gives a less restrictive model, but such an approach has no general corresponding FEM computational implementations in the existing software. The essential differences between these two methodologies are especially apparent in homogenisation of dynamic and transient heat transfer problems, where dispersive effects are observed under the last assumption only.

Most of the homogenisation methods have one common point – the necessity of use of the so–called homogenisation functions. These functions are the solutions of the cell problem on the RVE under periodic boundary conditions, where some additional conditions can be imposed on external boundaries or/and interfaces between the composite constituents. Some exceptions can be obtained for the 1D homogenisation problems, where effective thermal (and/or elastic) parameters may be derived directly. Let us note that if some further assumptions on composite microgeometry are introduced (a composite has a specified number of components in the periodicity cell and the shapes and/or location of the components are given), then the closed form equations for the effective material properties for either 2 or 3D structures can be derived [6,65,253].

2.3.1 Unidirectional Periodic Structures

Let us consider a unidirectional heterogeneous bar in unstressed and unstrained state, with periodic structure and with elastic properties piecewise constant. An example of the structure under considerations is presented below (Figures 2.42 and 2.43).

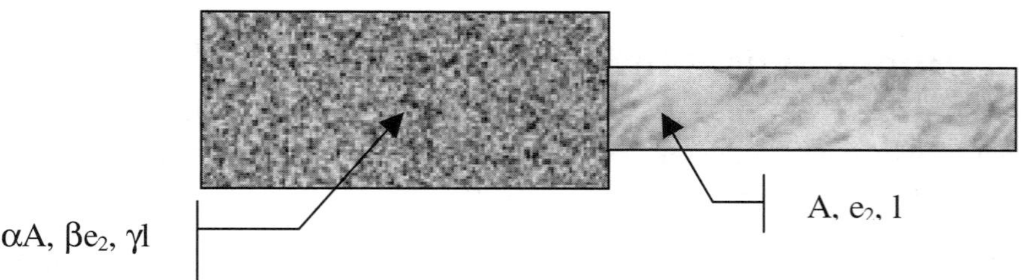

Figure 2.42. RVE of two−component composite bar

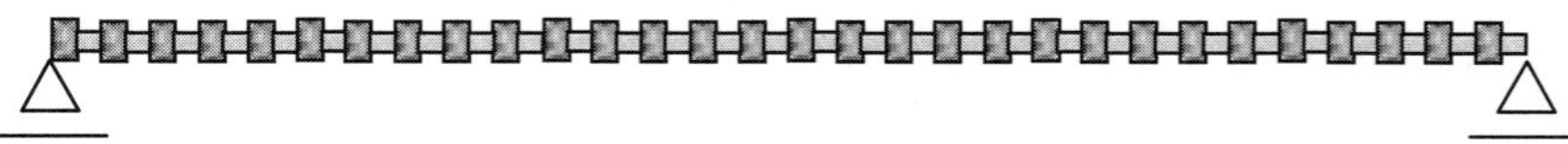

Figure 2.43. Unidirectional periodic two−component composite beam

Using the parameter ε the displacements and stresses are asymptotically expanded in the bar as follows [30,43,133,308]:

$$u^{\varepsilon}(\mathbf{x}) = u^{0}(\mathbf{x},\mathbf{y}) + \varepsilon u^{1}(\mathbf{x},\mathbf{y}) + \varepsilon^{2} u^{2}(\mathbf{x},\mathbf{y}) + \ldots \tag{2.59}$$

and

$$\sigma^{\varepsilon}(\mathbf{x}) = \sigma^{0}(\mathbf{x},\mathbf{y}) + \varepsilon \sigma^{1}(\mathbf{x},\mathbf{y}) + \varepsilon^{2} \sigma^{2}(\mathbf{x},\mathbf{y}) + \ldots \tag{2.60}$$

where $u^{(i)}(\mathbf{x},\mathbf{y})$, $\sigma^{(i)}(\mathbf{x},\mathbf{y})$ are periodic, too; the coordinate $\mathbf{x}$ is introduced in the macro scale (Figure 2.43), with $\mathbf{y}$ in the micro scale (cf. Figure 2.42). Introducing these expansions into classical Hooke law it is found that

$$\sigma^{\varepsilon}(\mathbf{x}) = \sigma^{0}(\mathbf{x},\mathbf{y}) + \varepsilon \sigma^{1}(\mathbf{x},\mathbf{y}) + \varepsilon^{2} \sigma^{2}(\mathbf{x},\mathbf{y}) + \ldots$$
$$= e(y)\left\{ \frac{\partial u^{0}(x,y)}{\partial x} + \frac{1}{\varepsilon}\frac{\partial u^{0}(x,y)}{\partial y} + \varepsilon\frac{\partial u^{1}(x,y)}{\partial x} + \frac{\partial u^{1}(x,y)}{\partial y} + \ldots \right\} \tag{2.61}$$

whereas the equilibrium equation

$$\frac{\partial \sigma^{\varepsilon}}{\partial x} + \gamma^{\varepsilon} = 0 \tag{2.62}$$

results in

$$\frac{\partial \sigma^0}{\partial \mathbf{x}} + \frac{1}{\varepsilon}\frac{\partial \sigma^0}{\partial \mathbf{y}} + \varepsilon\frac{\partial \sigma^1}{\partial \mathbf{x}} + \frac{\partial \sigma^1}{\partial \mathbf{y}} + \varepsilon^2\frac{\partial \sigma^2}{\partial \mathbf{x}} + \varepsilon\frac{\partial \sigma^2}{\partial \mathbf{y}} + ... + \gamma(\mathbf{y}) = 0 \qquad (2.63)$$

Hence, the following zeroth, first and second order constitutive equations are derived:

$$0 = e(\mathbf{y})\frac{\partial u^0}{\partial \mathbf{y}} \qquad (2.64)$$

$$\sigma^0 = e(\mathbf{y})\left\{\frac{\partial u^0}{\partial \mathbf{x}} + \frac{\partial u^1}{\partial \mathbf{y}}\right\} \qquad (2.65)$$

$$\sigma^1 = e(\mathbf{y})\left\{\frac{\partial u^1}{\partial \mathbf{x}} + \frac{\partial u^2}{\partial \mathbf{y}}\right\} \qquad (2.66)$$

Applying an analogous methodology, the equilibrium equation is expanded as

$$\frac{\partial \sigma^0}{\partial \mathbf{y}} = 0 \qquad (2.67)$$

$$\frac{\partial \sigma^0}{\partial x} + \frac{\partial \sigma^1}{\partial y} + \gamma(y) = 0 \qquad (2.68)$$

It is observed that zeroth order displacements and stresses depend on the macroscale coordiante only $u^0 = u^0(\mathbf{x})$ and $\sigma^0 = \sigma^0(\mathbf{x})$, so that it can be written that

$$\sigma^0(\mathbf{x}) = e(\mathbf{y})\left(\frac{\partial u^0(\mathbf{x})}{\partial \mathbf{x}} + \frac{\partial u^1(\mathbf{x},\mathbf{y})}{\partial \mathbf{y}}\right) \qquad (2.69)$$

Integrating both sides of (2.69) over the periodicity cell of a bar, there holds

$$\sigma^0(x) = \left(|\Omega| / \int_\Omega \frac{dy}{e(y)}\right)\frac{\partial u^0(x)}{\partial x} \qquad (2.70)$$

which leads to the following description of the homogenised (effective) Young modulus

$$e^{(eff)} = \frac{|\Omega|}{\int_\Omega \frac{dy}{e(y)}} \qquad (2.71)$$

Such a formulation makes it possible to derive the closed form equations for the expected values and covariances of the homogenised Young moduli using classical definitions of probabilistic moments or by an application of perturbation theory. It is possible to derive such equations for particular engineering examples only if the bar has a geometrical characteristics piecewise constant within its length. Let us consider further the RVE built up with n piecewise constant components defined on Ω by the use of design parameters (e_i, A_i, l_i) where e_i=const. for $y \in l_i$ and such that $e_i \neq e_j$ for $i,j=1,...,n$. Hence, the integration in formula (2.71) can be replaced by

$$e^{(eff)} = \frac{1}{\displaystyle\sum_{i=1}^{n} \frac{A_i l_i}{e_i}} \qquad (2.72)$$

where the variables A_i, l_i denote the cross-sectional area and the length of the ith structural element. After some algebraic transformations relation (2.72) can be transformed to

$$e^{(eff)} = \frac{\displaystyle\prod_{i=1}^{n} e_i}{\displaystyle\sum_{i=1}^{n} A_i l_i e_1 e_2 ... e_{i-1} e_{i+1} ... e_n} \qquad (2.73)$$

which can be efficiently implemented in any FEM computer program. Let us note that an analogous procedure can be applied successfully for the transient heat transfer problem Young moduli are to be replaced here by the relevant coefficients of heat conduction.

If the general beam structure is to be homogenised, the equilibrium and constitutive equations should be enriched with transversal effect components but, for the composite beams having constant Poisson ratio within its length and various Young moduli, the formulation posed above is quite sufficient for the needs of computational analyses. Moreover, it should be underlined that the homogenisation model for 2D and 3D problems is carried out similarly but the effective elasticity tensor is to be introduced instead of the Young modulus only. As a result, it is not possible to derive any closed form algebraic equations describing the effective properties of a composite, which significantly complicates numerical analysis. On the other hand, the randomness in multidimensional composite structures appears usually in their geometry, too, which must be implemented in the FEM analysis using some special finite element types.

Finally, considering further applications of the homogenisation approach in the elastodynamics of engineering structures, the effective mass density of a composite can be derived according to the spatial averaging method as [28,265]

$$\rho^{(eff)} = \frac{1}{|\Omega|} \int_\Omega \rho(y) dy .$$

(2.74)

Let us mention that this relation is used for any space configuration and periodicity conditions of a composite. Since that, having a homogenised elastostatic problem, especially in random case, further extension to the elastodynamic analysis in the context of a stochastic second order perturbation technique does not seem to be very complicated. The expected values for the effective Young modulus can be obtained by the second order perturbation second probabilistic moment analysis as [162]

$$E\left[e^{(eff)}(\mathbf{y})\right] = \int_{-\infty}^{+\infty} \left(e^{(eff)0}(\mathbf{y}) + \Delta b^r e^{(eff),r}(\mathbf{y}) + \tfrac{1}{2}\Delta b^r \Delta b^s e^{(eff),rs}(\mathbf{y})\right) p_R(b) db$$

(2.75)

Using classical probability theory definitions and theorems

$$\int_{-\infty}^{+\infty} p_R(b(\mathbf{y})) db = 1, \quad \int_{-\infty}^{+\infty} \Delta b p_R(b(\mathbf{y})) db = 0$$

(2.76)

$$\int_{-\infty}^{+\infty} \Delta b^r \Delta b^s p_R(b(\mathbf{y})) db = Cov(b^r, b^s); \quad 1 \le r, s \le R$$

(2.77)

one can determine that

$$E\left[e^{(eff)}(\mathbf{y})\right] = e^{(eff)0}(\mathbf{y}) + \tfrac{1}{2} e^{(eff),rs}(\mathbf{y}) Cov(b^r, b^s)$$

(2.78)

Further, using the analogous methodology the covariance matrix for the effective Young modulus $Cov\left(e^{(eff)}\right)$ is derived

$$Cov\left(e_i^{(eff)}; e_j^{(eff)}\right) = \int_{-\infty}^{+\infty} \left(e_i^{(eff)}(\mathbf{y}) - e_i^{(eff)0}(\mathbf{y})\right)\left(e_j^{(eff)}(\mathbf{y}) - e_j^{(eff)0}(\mathbf{y})\right) g(b_i, b_j) db_i db_j$$

and, using the classical perturbation approach, there holds

$$= \int_{-\infty}^{+\infty} \left\{ \left(e_i^{(eff)0}(\mathbf{y}) + \Delta b_r e_i^{(eff),r}(\mathbf{y}) + \tfrac{1}{2}\Delta b_r \Delta b_s e_i^{(eff),rs}(\mathbf{y}) - e_i^{(eff)0}(\mathbf{y})\right) \right.$$
$$\left. \times \left(e_j^{(eff)0}(\mathbf{y}) + \Delta b_u e_j^{(eff),u}(\mathbf{y}) + \tfrac{1}{2}\Delta b_u \Delta b_v e_j^{(eff),uv}(\mathbf{y}) - e_j^{(eff)0}(\mathbf{y})\right) \right\} g(b_i, b_j) db_i db_j$$

After all possible algebraic transformations and by neglecting the terms of order greater than the second, it is obtained that

$$Cov\left(e_i^{(eff)};e_j^{(eff)}\right)=\int\limits_{-\infty}^{+\infty}\Delta b_r e_i^{(eff),r}(\mathbf{y})\Delta b_s e_i^{(eff),s}(\mathbf{y})\,g\left(b_i,b_j\right)db_i db_j \tag{2.79}$$

$$=e_i^{(eff),r}(\mathbf{y})e_i^{(eff),s}(\mathbf{y})\,Cov\left(b_r,b_s\right)$$

For the particular case of the two–component composite structure there holds

$$e^{(eff)}=\frac{\Omega}{\dfrac{A_1 l_1}{e_1}+\dfrac{A_2 l_2}{e_2}}=\frac{\left(A_1 l_1+A_2 l_2\right)e_1 e_2}{e_2 A_1 l_1+e_1 A_2 l_2} \tag{2.80}$$

Let us consider the case of a 1D bar structure with two homogeneous components having deterministically defined geometry (cross–sections and lengths) and with Young moduli assumed to be the input random variables. The zeroth, first and second order derivatives of the effective elasticity with respect to the Young moduli of the composite constituents are obtained by analytical derivation:

- zeroth order components

$$e^{(eff)0}=\frac{\left(A_1 l_1+A_2 l_2\right)E[e_1]E[e_2]}{E[e_2]A_1 l_1+E[e_1]A_2 l_2} \tag{2.81}$$

- first order components

$$\frac{\partial e^{(eff)}}{\partial e_1}=\frac{A_1 l_1\left(A_1 l_1+A_2 l_2\right)E^2[e_2]}{\left(E[e_2]A_1 l_1+E[e_1]A_2 l_2\right)^2}\qquad \frac{\partial e^{(eff)}}{\partial e_2}=\frac{A_2 l_2\left(A_1 l_1+A_2 l_2\right)E^2[e_1]}{\left(E[e_2]A_1 l_1+E[e_1]A_2 l_2\right)^2} \tag{2.82}$$

- second order components

$$\frac{\partial^2 e^{(eff)}}{\partial e_1^2}=\frac{-2A_1 l_1 A_2 l_2\left(A_1 l_1+A_2 l_2\right)E^2[e_2]}{\left(E[e_2]A_1 l_1+E[e_1]A_2 l_2\right)^3} \tag{2.83}$$

$$\frac{\partial^2 e^{(eff)}}{\partial e_2^2}=\frac{-2A_1 l_1 A_2 l_2\left(A_1 l_1+A_2 l_2\right)E^2[e_1]}{\left(E[e_2]A_1 l_1+E[e_1]A_2 l_2\right)^3} \tag{2.84}$$

$$\frac{\partial^2 e^{(eff)}}{\partial e_1 \partial e_2}=\frac{2A_1 l_1 A_2 l_2\left(A_1 l_1+A_2 l_2\right)E[e_1]E[e_2]}{\left(E[e_2]A_1 l_1+E[e_1]A_2 l_2\right)^3} \tag{2.85}$$

Then, the resulting covariance matrix of the effective elastic behaviour for the two component composite structure is described as follows:

$$\begin{cases} Cov\left(e_1^{(eff)}, e_1^{(eff)}\right) = Cov(e_1, e_1)\dfrac{\partial e^{(eff)}}{\partial e_1}\dfrac{\partial e^{(eff)}}{\partial e_1} \\[2mm] Cov\left(e_1^{(eff)}, e_2^{(eff)}\right) = Cov(e_1, e_2)\dfrac{\partial e^{(eff)}}{\partial e_1}\dfrac{\partial e^{(eff)}}{\partial e_2} \\[2mm] Cov\left(e_2^{(eff)}, e_2^{(eff)}\right) = Cov(e_2, e_2)\dfrac{\partial e^{(eff)}}{\partial e_2}\dfrac{\partial e^{(eff)}}{\partial e_2} \end{cases} \tag{2.86}$$

To obtain the stochastic finite element model let us introduce the displacement field approximation. The zeroth, first and second order stiffness matrices for the homogenised bar structures may be written out by analogy to the previous considerations:

- zeroth order stiffnesses

$$\mathbf{K}^{(eff)0} = \left(\frac{e^{(eff)}A}{l}\right)^0 \begin{bmatrix} 1 & -1 \\ -1 & 1 \end{bmatrix} = \sum_{j=1}^{m} e^{(eff)0}\frac{A^{(m)}}{l^{(m)}}\begin{bmatrix} 1 & -1 \\ -1 & 1 \end{bmatrix} \tag{2.87}$$

with m denoting the total number of bar intervals with constant cross−sectional area $A^{(m)}$;

- first order stiffnesses

$$\begin{aligned} \mathbf{K}^{(eff),e} &= \frac{\partial \mathbf{K}}{\partial e^{(eff)}} = \frac{A}{l}\begin{bmatrix} 1 & -1 \\ -1 & 1 \end{bmatrix} \\[2mm] \mathbf{K}^{(eff),A} &= \frac{\partial \mathbf{K}^{(eff)}}{\partial A} = \frac{e^{(eff)}}{l}\begin{bmatrix} 1 & -1 \\ -1 & 1 \end{bmatrix} \\[2mm] \mathbf{K}^{(eff),l} &= \frac{\partial \mathbf{K}^{(eff)}}{\partial l} = -\frac{e^{(eff)}A}{l^2}\begin{bmatrix} 1 & -1 \\ -1 & 1 \end{bmatrix} \end{aligned} \tag{2.88}$$

- second order stiffness

$$\mathbf{K}^{(eff),e^{(eff)}e^{(eff)}} = \frac{\partial^2 \mathbf{K}^{(eff)}}{\partial \left(e^{(eff)}\right)^2} = \mathbf{0}, \quad \mathbf{K}^{(eff),AA} = \frac{\partial^2 \mathbf{K}^{(eff)}}{\partial A^2} = \mathbf{0}$$

$$\mathbf{K}^{(eff),ll} = \frac{\partial^2 \mathbf{K}^{(eff)}}{\partial l^2} = -\frac{2e^{(eff)}A}{l^3}\begin{bmatrix} 1 & -1 \\ -1 & 1 \end{bmatrix}$$

$$\mathbf{K}^{(eff),e^{(eff)}A} = \frac{\partial^2 \mathbf{K}^{(eff)}}{\partial e^{(eff)}\partial A} = \frac{1}{l}\begin{bmatrix} 1 & -1 \\ -1 & 1 \end{bmatrix}$$

$$\mathbf{K}^{(eff),e^{(eff)}l} = \frac{\partial^2 \mathbf{K}^{(eff)}}{\partial e^{(eff)}\partial l} = -\frac{A}{l^2}\begin{bmatrix} 1 & -1 \\ -1 & 1 \end{bmatrix}$$

$$\mathbf{K}^{(eff),Al} = \frac{\partial^2 \mathbf{K}^{(eff)}}{\partial A\partial l} = -\frac{e^{(eff)}}{l^2}\begin{bmatrix} 1 & -1 \\ -1 & 1 \end{bmatrix}$$

$$(2.89)$$

Hence, the canonical set of the second order SFEM equations can be rewritten as follows:

$$\mathbf{K}^{(eff)0}\mathbf{q}^{(eff)0} = \mathbf{Q}^{(eff)0} \tag{2.90}$$

$$\mathbf{K}^{(eff)0}\mathbf{q}^{(eff),r} = -\mathbf{K}^{(eff),r}\mathbf{q}^{(eff)0} \tag{2.91}$$

$$\mathbf{K}^{(eff)0}\mathbf{q}^{(eff)(2)} = -2\mathbf{K}^{(eff),r}\mathbf{q}^{(eff),s}\mathbf{Cov}\left(e_r^{(eff)},e_r^{(eff)}\right) \tag{2.92}$$

which makes it possible to compute $\mathbf{q}^{(eff)0}$, $\mathbf{q}^{(eff),r}$ and $\mathbf{q}^{(eff),rs}$ and to calculate the first probabilistic moments of displacements as

$$E\left[\mathbf{q}^{(eff)}\right]= \mathbf{q}^{(eff)0} + \tfrac{1}{2}\mathbf{q}^{(eff),rs}\mathbf{Cov}\left(e_r^{(eff)},e_r^{(eff)}\right) \tag{2.93}$$

$$\mathbf{Cov}\left(\mathbf{q}^{(eff)r},\mathbf{q}^{(eff)s}\right)= \mathbf{q}^{(eff),r}\mathbf{q}^{(eff),s}\mathbf{Cov}\left(e_r^{(eff)},e_r^{(eff)}\right) \tag{2.94}$$

The expected values and cross−covariances of the stresses are obtained in comparison to the heterogeneous model as

$$E\left[\sigma_{ij}^{(eff)e}\right]$$
$$= \left\{ C_{ijkl}^{(eff)(e)0}(q^{(eff)0} + \tfrac{1}{2}q^{(eff),rs}) + C_{ijkl}^{(eff)(e),r}q^{(eff),s} \right\}B_{kl}^{(e)}Cov(e_r^{(eff)},e_s^{(eff)}) \tag{2.95}$$

and

$$Cov\left(\sigma_{ij}^{(eff)e},\sigma_{ij}^{(eff)f}\right)= B_{kl}^{(e)}B_{mn}^{(f)}Cov(e_r^{(eff)},e_s^{(eff)})$$
$$\times\left\{ C_{ijkl}^{(eff)(e)0}C_{ijmn}^{(eff)(f)0}q^{(eff),r}q^{(eff),s} + C_{ijkl}^{(eff)(e),r}C_{ijmn}^{(eff)(f),s}q^{(eff)0}q^{(eff)0} \right. \tag{2.96}$$
$$\left. C_{ijkl}^{(eff)(e),r}C_{ijmn}^{(eff)(f)0}q^{(eff),s}q^{(eff)0} + C_{ijkl}^{(eff)(e)0}C_{ijmn}^{(eff)(f),r}q^{(eff),s}q^{(eff)0} \right\}$$

The first computational example deals with Young moduli defined as deterministic function and cross−sectional area being a random field, while in the second Young moduli of the constituents are only randomised. Due to the homogenisation method presented, the effective Young modulus is obtained in the form of a random field in both cases. Since the fact that homogenisation is only the intermediate tool to analyse composite structures, the expected values and standard deviations of displacements for homogenised structures are compared against those obtained for real, multi-component structure models.

The results of these analyses make it possible to modify the theoretically established probabilistic homogenisation algorithm to approximate expected values as well as covariances in the most efficient way. Neglecting the fact that effective material characterisation presented above is derived assuming periodicity of a composite, we try to use this approach in composites having small number of the RVEs on their lengths.

The first numerical experiment deals with the homogenisation of a beam clamped at both sides and subjected to uniformly distributed vertical static load (see Figure 2.44), analogously to the computational illustration demonstrated in [208].

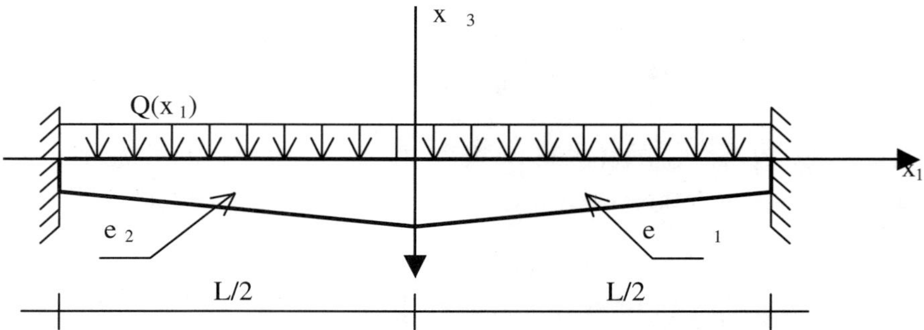

Figure 2.44. Clamped beam homogenised

Young moduli of the composite beam constituents discretised here by the use of 100 finite elements, are assumed to be deterministic variables, so that $e_1/e_2=\{1.00;\ 1.25;\ ...;\ 3.00\}$, while $e_2=2.0$ GPa and $v_1=v_2=0.30$. The mesh nodes are numbered sequentially from the left to right edge. The cross–sectional area of the beam A_r is an input random field defined as

$$E[A_r] = A^0\left(1.0 + \frac{0.3}{L} x_r\right) \quad r = 1,...,50 \quad \text{and} \quad A^0 = 5.0 \times 10^{-3}$$

$$\mu(A_r, A_s) = \exp\left(-\frac{|x_r - x_s|}{\lambda}\right) \quad \lambda = 0.10\,; \alpha = 0.07;\ r, s = 1,...,100$$

Other data are taken as follows:

$$Q(x_1) = f + \gamma A^0 \text{ for } f = 49.61 \text{ and } \gamma = 7.7126$$

while

$$I_{x2} = I_{x3} = \beta(A_r)^2; \ I_{x1} = I_{x2} + I_{x3}; \ \beta = \tfrac{1}{6}; \ L = 1.0$$

It is observed that starting from deterministically defined Young moduli the effective Young modulus random field is obtained as a result of the cross–sectional area randomness.

The main purpose of the SFEM–based tests is to verify the variability of the two–moment statistical response of the structure with respect to probabilistic input random fields. The results of the analysis are collected in Figures 2.45–2.48. The first two figures report expected values (vertical axes) as functions of location around the midpoint of a beam (horizontal axes); variable NN denotes here the node number where node 51 is the central point. The models outlined in the legend correspond to different composite configurations related to e_1/e_2 value – model 2R is equivalent to computational analysis of the beam in its real heterogeneous configuration with the Young moduli relation taken as 1.25. Thee data labelled as model 2H denote SFEM analysis results for the same homogenised model. The data obtained for model 1 denote the homogeneous beam with $e_1 = e_2$, while 'j' from 'model jR' or 'model jH' is equivalent to the relation taken from the set $\{1.00; 1.25; \ldots; 3.00\}$, accordingly.

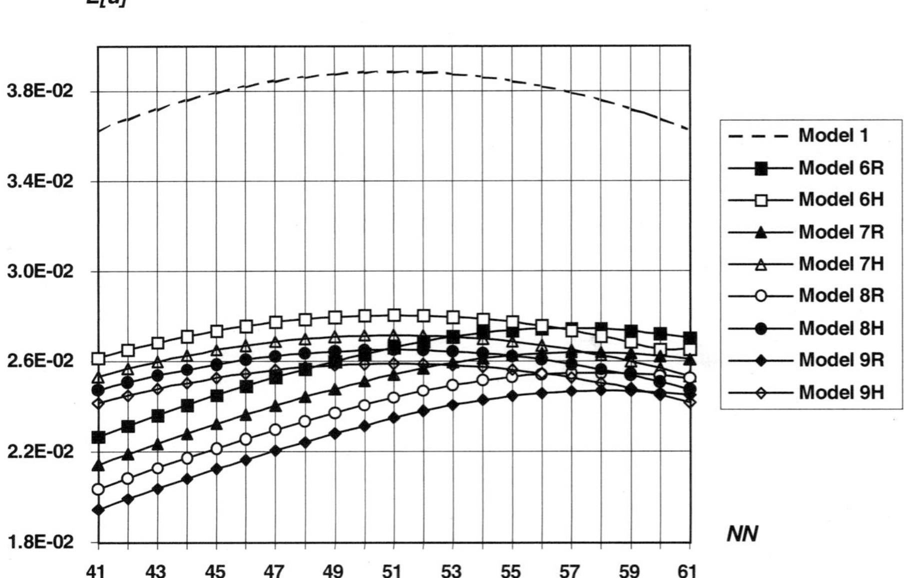

Figure 2.45. Expected values of the beam displacements

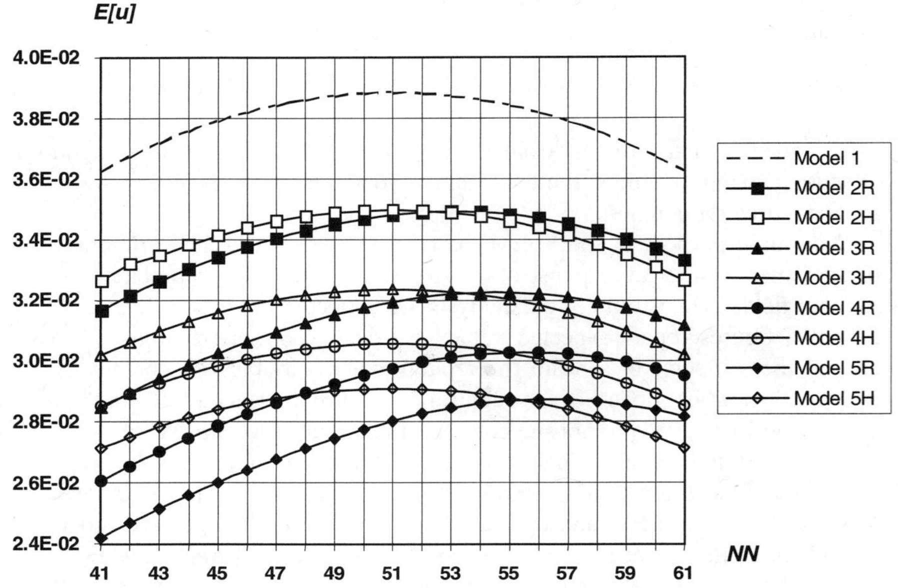

Figure 2.46. Expected values of the beam displacements

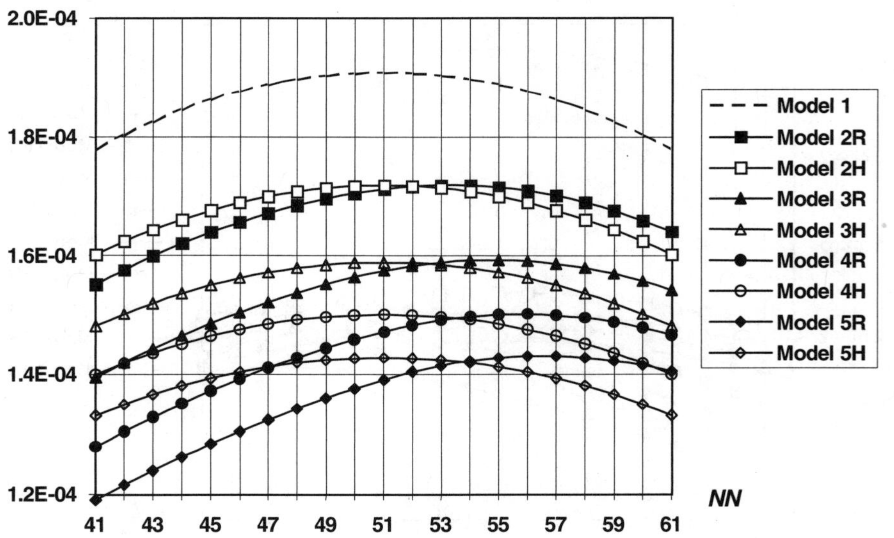

Figure 2.47. Standard deviations of the beam displacements

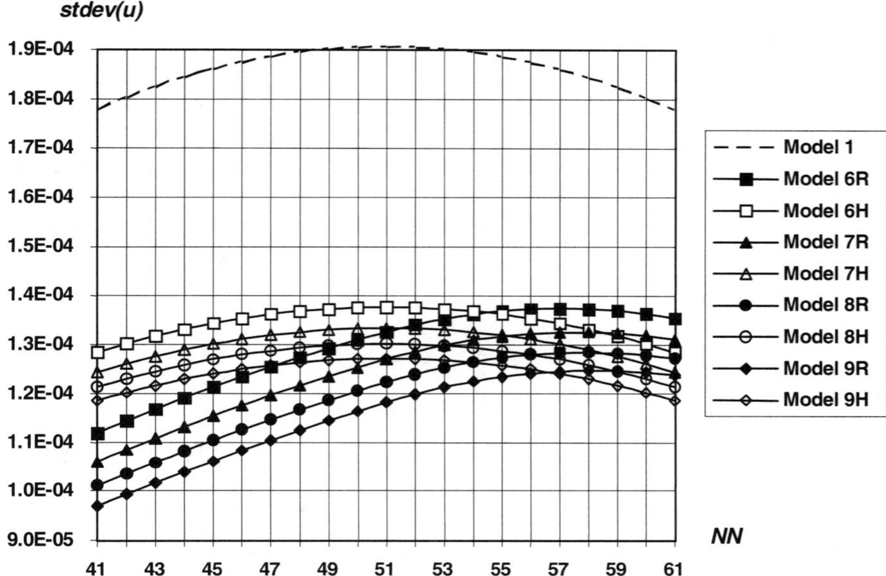

Figure 2.48. Standard deviations of the beam displacements

Analysing the results presented in Figures 2.45 and 2.46 as well as 2.47 and 2.48 it is seen that the homogenised structure approximates the real one with satisfactory precision, which is observed especially for smaller values of the relation e_1/e_2. It can be seen that this approximation effectiveness has the same character for the expected values and standard deviations of displacements analysed. It is characteristic that while probabilistic moments of structural displacements are symmetric for symmetric boundary conditions imposed on the homogenised beam then for a real composite beam this field has not the symmetric character at all with greater values under the weaker part of a beam. Further, relating standard deviations to the corresponding expected values, it is observed that output coefficients of variance for displacements are equal to 0.05 (in real and homogenised beam) which, taking into account limitations of the perturbation technique, enable one to confirm the usefulness of this methodology for such an analysis. It should be underlined that neglecting the bending effects in homogenisation procedure has no effect on the differences observed because the Poisson ratio of both composite components is the same while the 3D beam finite element used is quite appropriate for that analysis.

Two-component linear elastic composite bar is built up with two homogeneous components with the following material and geometrical data: $E[e_1]=3000$, $A_1=4$, $l_1=15$, $E[e_2]=2500$, $A_2=2$, $l_2=10$ are considered (see Figure 2.49). The covariance matrix of Young moduli variables is assumed to be equal:

$$Cov(e_r, e_s) = \begin{bmatrix} 90,000 & 75,000 \\ symm. & 62,500 \end{bmatrix} \times 10^3$$

while the external loads Q_1=200 and Q_2=250 are applied to the structure:

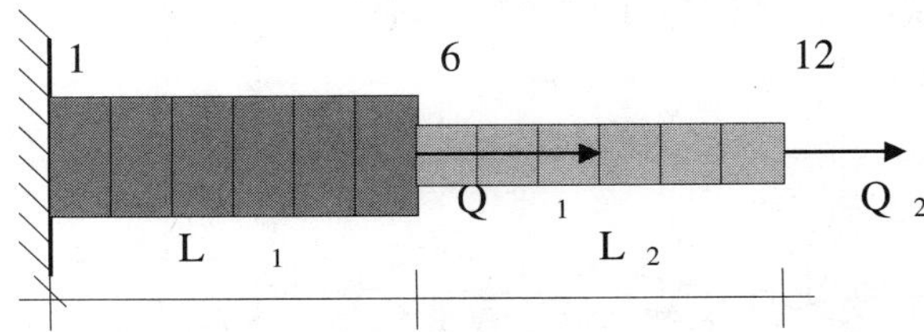

Figure 2.49. Two−component bar structure

The expected value and the covariance matrix of the effective Young modulus are calculated first and next, probabilistic moments of displacements and stresses for the original composite are computed. We compare these results against those determined for the homogenised structure. The input data and the results of computations are collected in Table 2.7 given below – the components of covariance matrix are equivalent to 10% standard deviation of the input Young moduli according to the following relation:

$$Cov(e_i, e_j) = \begin{bmatrix} \sigma_1^2 & \sigma_1\sigma_2 \\ symm. & \sigma_2^2 \end{bmatrix}$$

Table 2.7.
Probabilistic data and intermediate results for computational experiments

Model no	Input data (1st probabilistic moments)	Input data (2nd probabilistic moments)
1	$E[e_1,e_2]$={3000,2500}	$Cov(e_r, e_s) = \begin{bmatrix} 90,000 & 75,000 \\ symm. & 62,500 \end{bmatrix} \times 10^3$
2	$E[e^{(eff)}]$=2857,1437	$Cov(e_r^{(eff)}, e_s^{(eff)}) = \begin{bmatrix} 41,649 & 16,659 \\ symm. & 6,663 \end{bmatrix} \times 10^3$
3	$E[e^{(eff)}]$=2857,1437	$Cov(e_r^{(eff)}, e_s^{(eff)}) = \begin{bmatrix} 90,000 & 75,000 \\ symm. & 62,500 \end{bmatrix} \times 10^3$

Next, the first two probabilistic moments of horizontal displacements are analysed along the bar. The results obtained for the stiffer part show better approximation by model 2 (with covariance matrix homogenised), while for a weaker part by model 3 (with original covariance matrix). Quite a different situation is observed for the standard deviations – those resulting from model 3 approximate those obtained in model 1 very well, while the results of model 2 are definitely smaller.

Table 2.8. Expected values and standard deviations of beam displacements

Node Number (NN)	Expected values			Standard deviations		
	Model 1	Model 2	Model 3	Model 1	Model 2	Model 3
1	0	0	0	0	0	0
2	0.0789	0.0825	0.0829	7.81E-03	5.89E-03	8.61E-03
3	0.1578	0.1649	0.1659	1.56E-02	1.18E-02	1.72E-02
4	0.2367	0.2474	0.2488	2.34E-02	1.77E-02	2.58E-02
5	0.3156	0.3298	0.3318	3.13E-02	2.35E-02	3.45E-02
6	0.3945	0.4123	0.4147	3.91E-02	2.94E-02	4.31E-02
7	0.4734	0.4947	0.4976	4.69E-02	3.53E-02	5.17E-02
8	0.5786	0.586	0.5895	5.73E-02	3.80E-02	5.97E-02
9	0.6838	0.6772	0.6813	6.77E-02	4.06E-02	6.76E-02
10	0.7891	0.7684	0.7732	7.81E-02	4.32E-02	7.56E-02
11	0.8943	0.8596	0.865	8.85E-02	4.59E-02	8.36E-02
12	0.9995	0.9509	0.9569	9.90E-02	4.85E-02	9.16E-02
13	1.1045	1.0421	1.0487	0.1094	0.0511	9.95E-02

Taking into account the results of computational experiments presented in Table 2.8, the following algorithm is proposed to model strictly periodic composite beams using homogenisation–based SFEA.

<u>Input random variables definition</u>

$$E[b_r], \; Cov\left(b^r, b^s\right)$$

<u>Initial boundary value problem</u>

$$\sigma^{\varepsilon}_{ij,j} + \gamma^{\varepsilon} = 0$$

$\rightarrow$ solve:

$$\mathbf{K}^{(\varepsilon)0}\mathbf{q}^{(\varepsilon)0} = \mathbf{Q}^0$$

$$\mathbf{K}^{(\varepsilon)0}\mathbf{q}^{(\varepsilon),r} = -\mathbf{K}^{(\varepsilon),r}\mathbf{q}^{(\varepsilon)0}$$

$$\mathbf{K}^{(\varepsilon)0}\mathbf{q}^{(\varepsilon)(2)} = \left(-2\mathbf{K}^{(\varepsilon),r}\mathbf{q}^{(\varepsilon),s} - \mathbf{K}^{(\varepsilon),rs}\mathbf{q}^{(\varepsilon)0}\right)Cov\left(b^r, b^s\right)$$

<u>Evaluation of effective Young moduli parameters</u>

$$E\left[e^{(eff)}\right]Cov\left(e^{(eff)r}, e^{(eff)s}\right)$$

<u>Homogenised boundary value problem</u>:

$$\sigma^{(eff)}_{ij,j} + \gamma^{(eff)} = 0$$

<u>1st SFEM solution</u> (*zeroth order homogenised displacements*):
$\rightarrow$ solve:

$$\mathbf{K}^{(eff)0}\mathbf{q}^{(eff)0}_{[1]} = \mathbf{Q}^0$$

$$\mathbf{K}^{(eff)0}\mathbf{q}^{(eff),r}_{[1]} = -\mathbf{K}^{(eff),r}\mathbf{q}^{(eff)0}_{[1]}$$

$$\mathbf{K}^{(eff)0}\mathbf{q}_{[1]}^{(eff)(2)} = \left(-2\mathbf{K}^{(eff),r}\mathbf{q}_{[1]}^{(eff),s} - \mathbf{K}^{(eff),rs}\mathbf{q}_{[1]}^{(eff)0}\right)Cov\left(e^{(eff)r},e^{(eff)s}\right)$$

$\underline{2^{nd}\ \text{SFEM solution}}$ (*first and second order homogenised displacements*)
$\rightarrow$ solve:

$$\mathbf{K}^{(eff)0}\mathbf{q}_{[2]}^{(eff)0} = \mathbf{Q}^0$$

$$\mathbf{K}^{(eff)0}\mathbf{q}_{[2]}^{(eff),r} = -\mathbf{K}^{(eff),r}\mathbf{q}_{[2]}^{(eff)0}$$

$$\mathbf{K}^{(eff)0}\mathbf{q}_{[2]}^{(eff)(2)} = \left(-2\mathbf{K}^{(eff),r}\mathbf{q}_{[2]}^{(eff),s} - \mathbf{K}^{(eff),rs}\mathbf{q}_{[2]}^{(eff)0}\right)Cov\left(b^r,b^s\right)$$

$\underline{\text{Final evaluation of displacements probabilistic moments}}$

$$E\left[q_\beta^{(eff)}\right] = q_{\beta[1]}^{(eff)0} + q_{\beta[2]}^{(eff)(2)}$$

$$Cov\left(q_\alpha^{(eff)},q_\beta^{(eff)}\right) = q_{\alpha[2]}^{(eff),r}\,q_{\beta[2]}^{(eff),s}\,Cov\left(b^r,b^s\right)$$

Figure 2.50. Algorithm of homogenisation-based SFEM analysis

It should be underlined that such a stochastic second order homogenisation scheme has its basis in the computational observations only. However its results are in good agreement with those observed for the real composite model subjected to the same boundary conditions.

2.3.2 2D and 3D Composites with Uniaxially Distributed Inclusions

This class of composites is equivalent to all 2D and 3D periodic heterogeneous structures where isotropic homogeneous constituents are distributed periodically along the x_3 axis, which in practice is observed in case of the periodic laminates. Further, it should be mentioned that the effective elasticity tensor components valid for these structures can be reduced to the periodic bar structure shown above only if the 1D case is considered. The following system of partial differential equations is considered here to calculate probabilistic moments of the effective elasticity tensor [159]:

$$\left(C_{ijkl}\left(\frac{x_3}{\delta}\right)u_{k,l}^\delta\right)_{,j} = f_i(\mathbf{x}),\quad \mathbf{u}^\delta(\mathbf{x}) = \mathbf{u}^o(\mathbf{x}),\quad \mathbf{x}\in\partial\Omega. \tag{2.97}$$

According to the general theory, the homogenised formulation of the problem has the following form:

$$\left(C_{ijkl}^{(eff)}u_{k,l}\right)_{,j} = f_i(\mathbf{x}),\quad \mathbf{u}(\mathbf{x}) = \mathbf{u}^o(\mathbf{x}),\quad \mathbf{x}\in\partial\Omega. \tag{2.98}$$

where the effective coefficients $C_{ijkl}^{(eff)}$ are given by the formula. The homogenisation functions $\chi^{kl}(\mathbf{y})$ are determined as the solution of the local problem on the RVE

$$\frac{\partial}{\partial y_j}\left(C_{ijkl}(y_3)\frac{\partial}{\partial y_l}\left(\chi_k^{mn}\right)+C_{ijmn}(y_3)\right)=0\,;\; x\in\Omega \tag{2.99}$$

for any $\chi^{mn}(\mathbf{y})$ periodic on the RVE. Since the heterogeneity distribution is observed along y_3 axis only, a solution should be of the form $\chi^{mn}(\mathbf{y})=\chi^{mn}(y_3)$. It yields the following problem for determination of $\chi^{mn}(y_3)$

$$\frac{\partial}{\partial y_3}\left(C_{i3k3}(y_3)\frac{\partial}{\partial y_3}\left(\chi_k^{mn}\right)+C_{i3mn}(y_3)\right)=0,\quad x\in\Omega \tag{2.100}$$

for any $\chi^{mn}(y_3)$ being periodic on the RVE. Therefore, (2.100) is ordinary differential equations system, which can be solved explicitly as

$$C_{i3k3}(y_3)\,\chi_{k,3}^{mn}+C_{i3mn}(y_3)=A_i\,. \tag{2.101}$$

If the elasticity tensor components C_{i3k3} are invertible, then

$$\chi_{k,3}^{mn}=-\left\{C_{k3j3}\right\}^{-1}C_{j3mn}+\left\{C_{k3j3}\right\}^{-1}A_j \tag{2.102}$$

The periodicity condition results in $\left\langle \chi_{,3}^{mn}\right\rangle_\Omega=0$ which introduced in (2.102) yields

$$0=-\left\langle\left\{C_{k3j3}\right\}^{-1}C_{j3mn}\right\rangle_\Omega+\left\langle\left\{C_{k3j3}\right\}^{-1}\right\rangle_\Omega A_j \tag{2.103}$$

Therefore

$$A_i=\left\langle\left\{C_{i3k3}\right\}^{-1}\right\rangle_\Omega^{-1}\left\langle\left\{C_{k3j3}\right\}^{-1}C_{j3mn}\right\rangle_\Omega \tag{2.104}$$

and there holds

$$\chi_{k,3}^{mn}=-\left\{C_{k3j3}\right\}^{-1}C_{j3mn}+\left\{C_{k3j3}\right\}^{-1}\left\langle\left\{C_{j3q3}\right\}^{-1}\right\rangle_\Omega^{-1}\left\langle\left\{C_{q3p3}\right\}^{-1}C_{p3mn}\right\rangle_\Omega \tag{2.105}$$

Taking into account that the state functions depend on y_3 axis only, the effective parameters are expressed as

$$C_{ijkl}^{(eff)} = \left\langle C_{ijkl} + C_{ijm3}\chi_{m,3}^{kl} \right\rangle_\Omega \tag{2.106}$$

Finally, the homogenised elasticity tensor components are given by

$$C_{ijkl}^{(eff)} = \left\langle C_{ijkl} \right\rangle_\Omega - \left\langle C_{ijm3}\left\{C_{m3p3}\right\}^{-1}C_{p3kl} \right\rangle_\Omega$$
$$+ \left\langle C_{ijm3}\left\{C_{m3p3}\right\}^{-1} \right\rangle_\Omega \left\langle \left\{C_{p3n3}\right\}^{-1} \right\rangle_\Omega^{-1} \left\langle \left\{C_{n3q3}\right\}^{-1}C_{q3kl} \right\rangle_\Omega \tag{2.107}$$

In case of isotropic and linear elastic constituent materials of this composite, it is obtained after some algebraic manipulation [159,177]

$$C_{1111}^{(eff)} = C_{2222}^{(eff)} = \left\langle \frac{(1-v)e}{(1+v)(1-2v)} \right\rangle_\Omega - \left\langle \frac{(1-2v)e}{1-v^2} \right\rangle_\Omega + \frac{\left\langle \frac{1-2v}{1-v} \right\rangle_\Omega^2}{\left\langle \frac{(1+v)(1-2v)}{(1-v)e} \right\rangle_\Omega} \tag{2.108}$$

$$C_{3333}^{(eff)} = \frac{1}{\left\langle \frac{(1+v)(1-2v)}{(1-v)e} \right\rangle_\Omega} \tag{2.109}$$

$$C_{1133}^{(eff)} = C_{3311}^{(eff)} = C_{2233}^{(eff)} = C_{3322}^{(eff)} = \frac{\left\langle \frac{1-2v}{1-v} \right\rangle_\Omega}{\left\langle \frac{(1+v)(1-2v)}{(1-v)e} \right\rangle_\Omega} \tag{2.110}$$

$$C_{1122}^{(eff)} = C_{2211}^{(eff)} = \left\langle \frac{e}{(1-v)} \right\rangle_\Omega - \left\langle \frac{(1-2v)e}{1-v^2} \right\rangle_\Omega + \frac{\left\langle \frac{1-2v}{1-v} \right\rangle_\Omega}{\left\langle \frac{(1+v)(1-2v)}{(1-v)e} \right\rangle_\Omega} \tag{2.111}$$

$$C_{1212}^{(eff)} = C_{2121}^{(eff)} = \left\langle \frac{e}{(1+v)} \right\rangle_\Omega, \quad C_{1212}^{(eff)} = C_{2121}^{(eff)} = \left\langle \frac{1}{\frac{1+v}{e}} \right\rangle_\Omega \tag{2.112}$$

while the remaining components are equal to 0. The layered structure analysed in this experiment has material characteristics corresponding to a glass–epoxy composite: $E[e_1]= 84.0$ GPa, $\sigma(e_1)= 8.4$ GPa, $v_1=0.22$ and $E[e_2]= 4.0$ GPa, $\sigma(e_2)= 0.4$ GPa, $v_2=0.34$; the volume ratios are taken as equal. The results of computational analysis are collected as deterministic quantities, expected values and coefficients of variation computed for the particular components in Table 2.9 below.

Table 2.9. Effective materials characteristics

Effective elasticity tensor components	Deterministic	Expected value	Variation
$C_{1111}=C_{2222}$	29.2316 GPa	29.2260 GPa	0.0767
C_{3333}	10.4662 GPa	10.4566 GPa	0.0954
$C_{1133}=C_{3311}=C_{2233}=C_{3322}$	6.1479 GPa	6.1424 GPa	0.0954
$C_{1122}=C_{2211}$	34.3657 GPa	34.3601 GPa	0.0794
$C_{1212}=C_{2121}$	50.7785 GPa	50.7785 GPa	0.0936
$C_{2323}=C_{3232}$	51.5489 GPa	51.5608 GPa	0.0968

Comparing the results presented in Table 2.9 it is seen that there is no difference between the deterministic result and the corresponding expected values for effective tensor components, while the coefficient of variation has values generally smaller or almost equal to the corresponding input variables value 0.1. To verify the variability of the tensor with respect to input Young moduli expected values, the MAPLE plot3d option for $E\!\left[C_{2323}^{(eff)}\right]$ and $\alpha\!\left(C_{2323}^{(eff)}\right)$ has been applied; the remaining components show almost the same tendencies. The range of variability for both the composite components Young moduli is taken as $\pm10\%$ of the original values and, as can be observed in Figures 2.51 and 2.52, Young modulus of the weaker material appears to be the decisive parameter for the overall elastic characteristics of this composite in terms of a homogenisation method applied. Further, it can be noticed that an increase of the coefficient of variation $\alpha\!\left(C_{2323}^{(eff)}\right)$ results from decreasing matrix Young modulus, while the inverse relation is observed in case of $E\!\left[C_{2323}^{(eff)}\right]$.

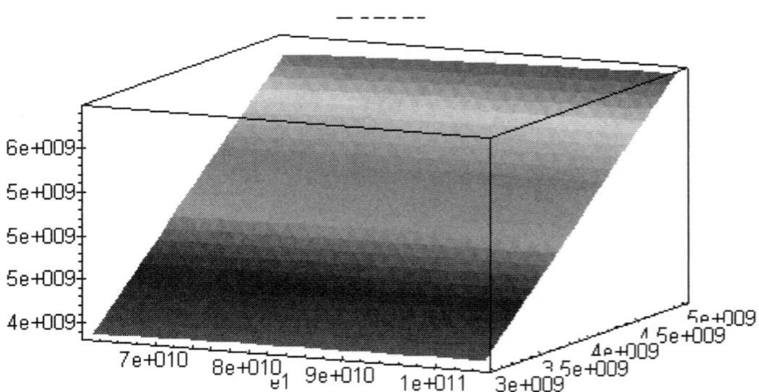

Figure 2.51. Expected values for C_{2323} component

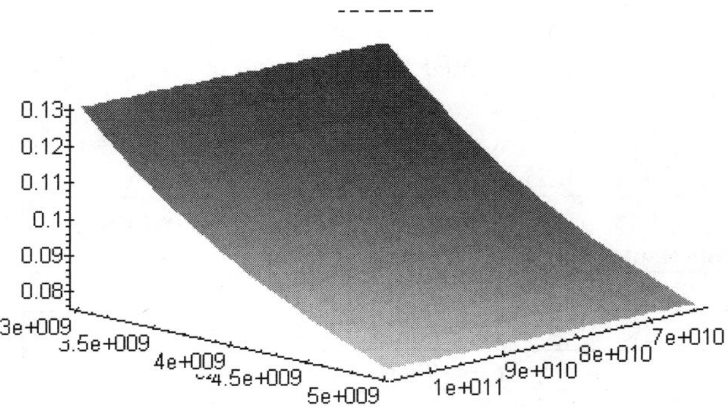

Figure 2.52. Coefficients of variation for C_{2323} component

It should be underlined that the model for one dimensionally distributed inhomogeneity is valid after some minor simplifications for the heat conduction homogenisation of the same composites, since probabilistic numerical algorithm has a quite general character.

2.3.3 Fibre-Reinforced Composites

2.3.3.1 Algebraic Equations for Homogenised Characteristics

It should be emphasised that the homogenisation procedure can be applied to the fibre−reinforced composite with anisotropic consituents, too. The effective elasticity tensor in terms of different transverse and longitudinal Young moduli and Poisson ratios can be calculated explicitly using the Mori−Tanaka or the self−consistent analytical homogenisation technique as follows [18,31]:

$$
\begin{Bmatrix} \sigma_{11} \\ \sigma_{22} \\ \sigma_{33} \\ \sigma_{23} \\ \sigma_{31} \\ \sigma_{12} \end{Bmatrix} = \begin{Bmatrix} n & l & l & 0 & 0 & 0 \\ & k+m & k-m & 0 & 0 & 0 \\ & & k+m & 0 & 0 & 0 \\ & & & m & 0 & 0 \\ & sym. & & & p & 0 \\ & & & & & p \end{Bmatrix} \begin{Bmatrix} \varepsilon_{11} \\ \varepsilon_{22} \\ \varepsilon_{33} \\ \varepsilon_{23} \\ \varepsilon_{31} \\ \varepsilon_{12} \end{Bmatrix}
\tag{2.113}
$$

where the following description for the constants k, l, m, n and p is applied:

$$p = \frac{2c_f p_m p_f + c_m \left(p_m p_f + p_m^2 \right)}{2c_f p_m + c_m \left(p_f + p_m \right)} \tag{2.114}$$

$$m = \frac{m_m m_f \left(k_m + 2m_m \right) + k_m m_m \left(c_f m_f + c_m m_m \right)}{k_m m_m + \left(k_m + 2m_m \right)\left(c_f m_m + c_m m_f \right)} \tag{2.115}$$

$$k = \frac{c_f k_f \left(k_m + m_m \right) + c_m m_m \left(k_f + m_m \right)}{c_f \left(k_m + m_m \right) + c_m \left(k_f + m_m \right)} \tag{2.116}$$

$$l = \frac{c_f l_f \left(k_m + m_m \right) + c_m l_m \left(k_f + m_m \right)}{c_f \left(k_m + m_m \right) + c_m \left(k_f + m_m \right)} \tag{2.117}$$

$$n = c_m n_m + c_f n_f + \left(l - c_f l_f - c_m l_m \right)\frac{l_f - l_m}{k_f - k_m} \tag{2.118}$$

There holds for matrix and fibre

$$\left\{ \begin{aligned} k &= -\left[\frac{1}{G_T} - \frac{4}{E_T} + \frac{4v_L^2}{E_L} \right]^{-1} \\ l &= 2kv_L \\ n &= E_L + 4kv_L^2 = E_L + \frac{l^2}{k} \\ m &= G_T, p = G_L \end{aligned} \right. \tag{2.119}$$

where c_f and c_m denote fibre and matrix volume fractions of a composite measured in the direction transverse to the fibres. The indices L and T denote longitudinal and transversal elastic characteristics for the components. It can be observed that closed form relations for effective elasticity tensor components are obtained in this case without the necessity of a cell problem solution.

Two alternative ways of fibre–reinforced composite homogenisation have been proposed below. Since the fact that the computational illustration for the SFEM solution of the cell problem is shown in [192], then only the second order perturbation based model is discussed here. The composite taken to illustrate probabilistic moments of relevant material properties is exactly the same as in the previous example. The final equations for the effective characteristics for a layered and fibre–reinforced composite do not contain any shape parameters – different forms of the reinforcement lead, according to some mathematical considerations, to different equations rewritten however for the same parameters: material properties and volume ratios of the constituents only. That is why such a comparative studies, especially in terms of the random spaces of the material properties analysed, are important.

The deterministic and the corresponding expected values as well as coefficients of variation are collected in Table 2.10 for the components of the effective tensor k, l, m, n and p, separately. Generally, it can be observed that, as previously noted, expected values are almost equal to relevant deterministic quantities and the resulting coefficients of variation are almost equal to the corresponding input probabilistic coefficients. Further, comparing the data collected in Tables 2.9 and 2.10 it can be noted that the layered structure has greater effective elastic characteristics than the fibre-reinforced composite with the same constituents – this observation is very important considering practical applications and optimisation of composites.

Table 2.10. Effective materials characteristics

Effective elasticity tensor components	Deterministic	Expected value	Variation
k	6.8350 GPa	6.8216 GPa	0.0902
l	5.2983 GPa	5.2898 GPa	0.0909
m	3.5892 GPa	3.5840 GPa	0.0927
n	46.9052 GPa	46.9000 GPa	0.0938
p	4.0195 GPa	4.0121 GPa	0.0907

Further, see Figures 2.53–2.62, the parameter variability of the expected values of the effective parameters k, l, m, n and p (Figures 2.53, 2.55, 2.57, 2.59 and 2.61) as well as their variances (Figures 2.54, 2.56, 2.58, 2.60 and 2.62) is computed with respect to expected values of the Young moduli of the components. It is seen that the expected values of all these parameters show greater sensitivity with respect to stronger material Young moduli; all the changes are significant especially for decreasing values of both moduli. As can be predicted from these figures, the sensitivity gradients of all the parameters have positive signs – an increase of any effective constant k, l, m, n and p results from the increase of Young moduli of fibre or/and matrix. In further computational studies, the probabilistic moments so computed may be applied in the FEM–based probabilistic computational simulation for an engineering composite by using the Monte Carlo simulation technique or, as is done in the first example, the SFEM approach.

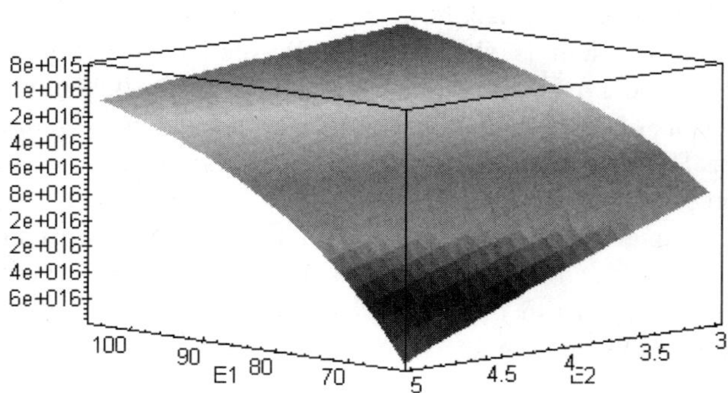

Figure 2.53. Expected values of the component k

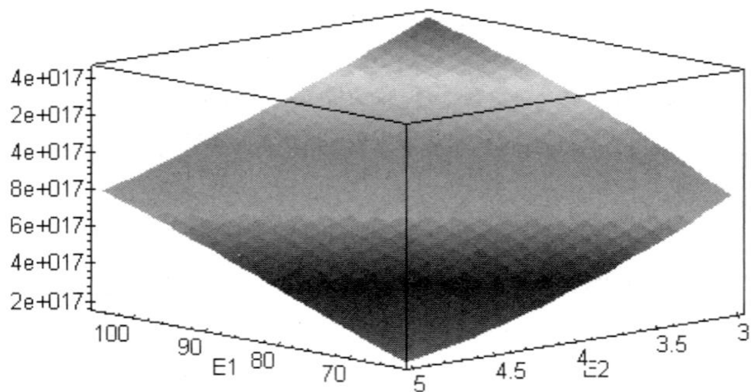

Figure 2.54. Variance of the component k

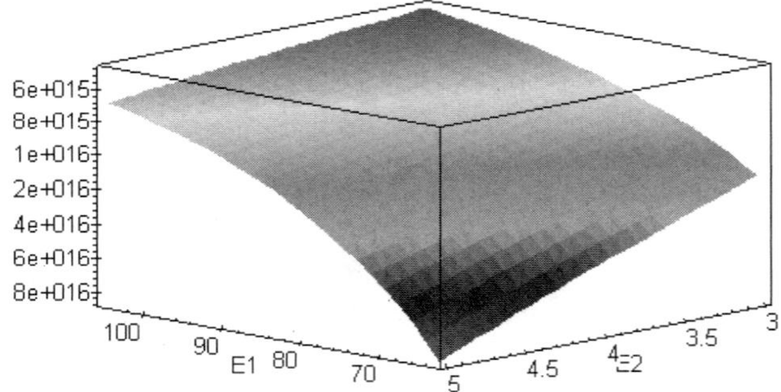

Figure 2.55. Expected values of the component l

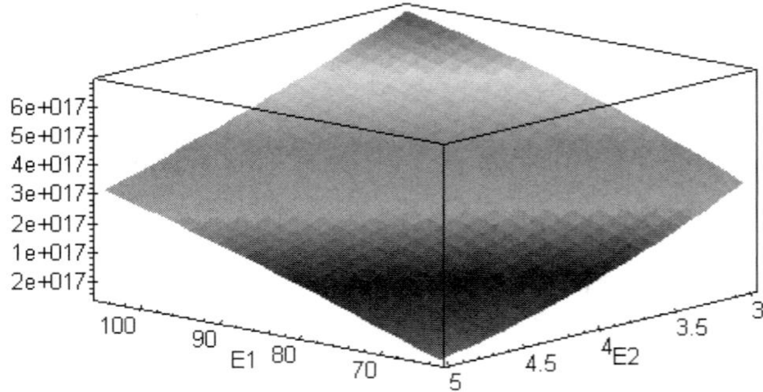

Figure 2.56. Variance of the component l

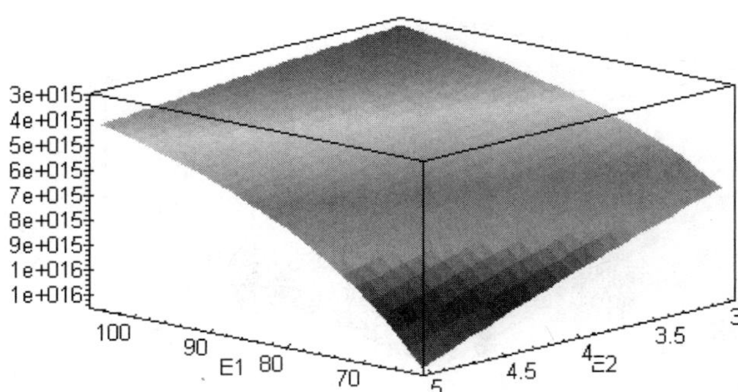

Figure 2.57. Expected values of the component m

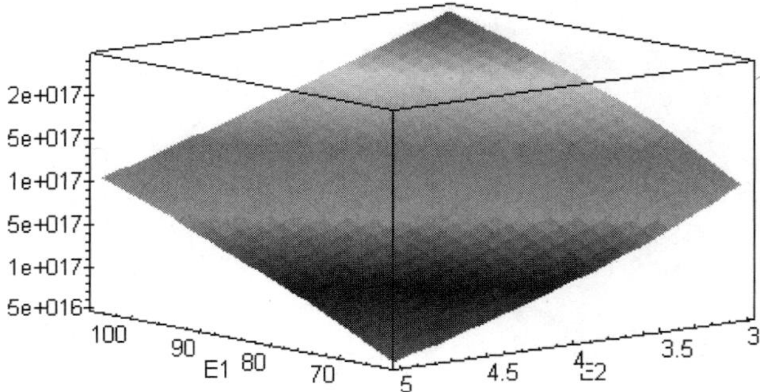

Figure 2.58. Variance of the component m

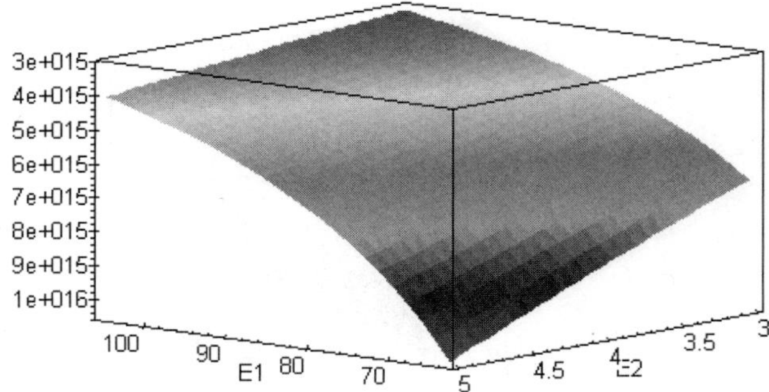

Figure 2.59. Expected values of the component n

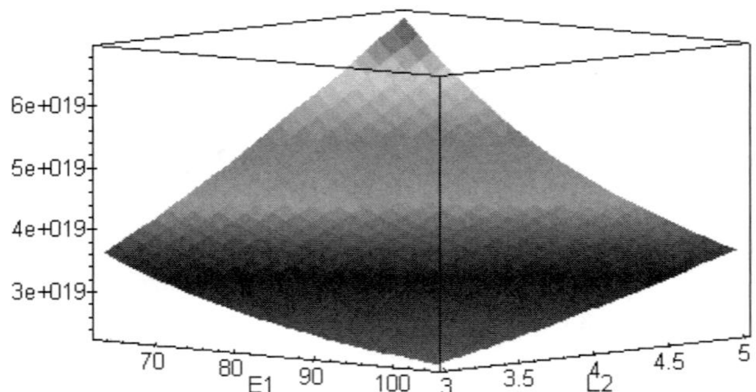

Figure 2.60. Variance of the component n

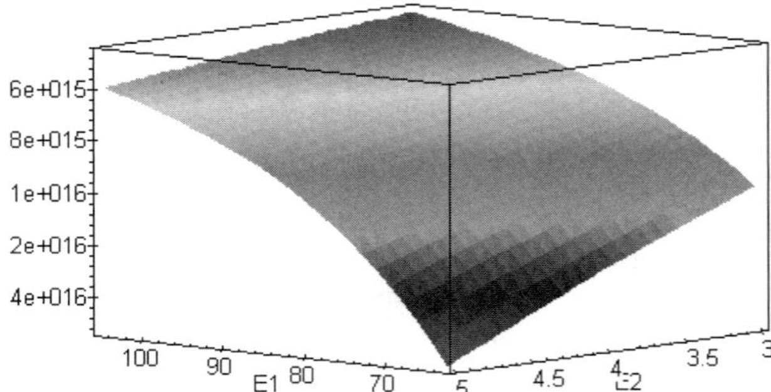

Figure 2.61. Expected values of the component p

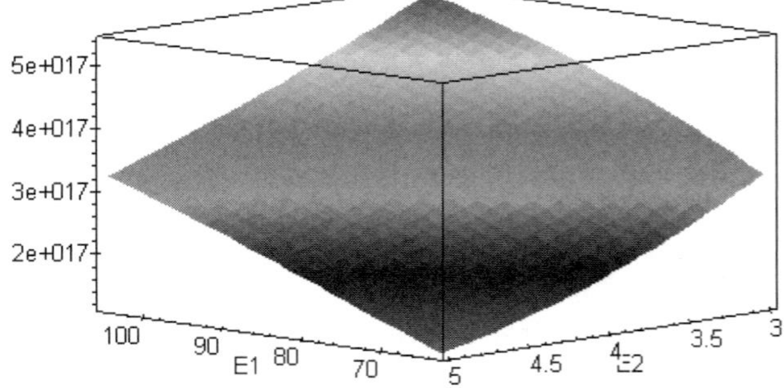

Figure 2.62. Variance of the component p

2.3.3.2 Asymptotic Homogenisation Method

2.3.3.2.1 Deterministic Approach to the Problem

The homogenisation of the n−component periodic composites in the context of linear deterministic elastostatic problem is studied here; the effective modules method worked out previously for two−component heterogeneous media is now extended on the n-component composites to homogenise multi-component materials in general form. The approach proposed enables particularly, as is demonstrated, to calculate effective elastic characteristics for composites with some interphases between the constituents. As it is known, the interphase in engineering practice may be considered as the next homogeneous component of the composite with small volume in comparison to the rest of a structure that increases contact between reinforcement and matrix and can be crucial for the composite macro−behaviour [59,255,270,314]. One of the interphase computational modelling method is based on the special (both elastic and elastoplastic) interface finite elements [238,260,318].

On the other hand, there are some approaches in the mechanics of composite materials, where the interphase is the hypothetical region containing all interface defects that appear between the original components of a composite. Usually, the interphase is introduced with thickness and material parameters constant within its region; ultrasonic emission seems to be the most efficient experimental method in this field. Numerical studies based on this formulation and collected in this chapter show the sensitivity of the periodic composite effective parameters to strengthening and weakening, in the context of elastic parameters, of the interphase. Due to the fact that the observations correspond with engineering practice, it may confirm the usefulness of the method to homogenise n−component heterogeneous media.

Very important aspect of the method proposed is that the effective modules method in present extended version enables to homogenize the composite materials with the microdefects appearing in the constituents − they have the dimensions relatively small with comparison to the components. Next, we observe that the method presented can be relatively easily transformed to the probabilistic case where material properties as well as the periodicity cell geometry may be treated as random; the Monte Carlo simulation method is the most recommended technique. This formulation may be used to formulate and to compute the deterministic or stochastic sensitivity, in a phenomenological or structural sense, to both material and geometrical parameters of the composite that enable one to find out the most decisive parameters for the entire computational homogenisation procedure.

The linear problem of elasticity is formulated for the n−component composite shown in Figure 2.64 with the Representative Volume Element given in Figure 2.63 as follows:

$$\begin{cases} \dfrac{\partial \sigma_{ij}^{\varepsilon}}{\partial x_j} = 0 \\[2mm] \sigma_{ij}^{\varepsilon} n_j = p_i;\ \mathbf{x} \in \partial\, \Omega_{\sigma} \\[1mm] u_i^{\varepsilon} = 0;\ \mathbf{x} \in \partial\, \Omega_u \\[1mm] \sigma_{ij}^{\varepsilon} = C_{ijkl}^{\varepsilon}(\mathbf{x})\varepsilon_{kl}^{\varepsilon} \\[1mm] \varepsilon_{kl}^{\varepsilon} = \tfrac{1}{2}\left(u_{k,l}^{\varepsilon} + u_{l,k}^{\varepsilon} \right) \end{cases} \qquad i,j,k,l=1,2 \qquad\qquad (2.120)$$

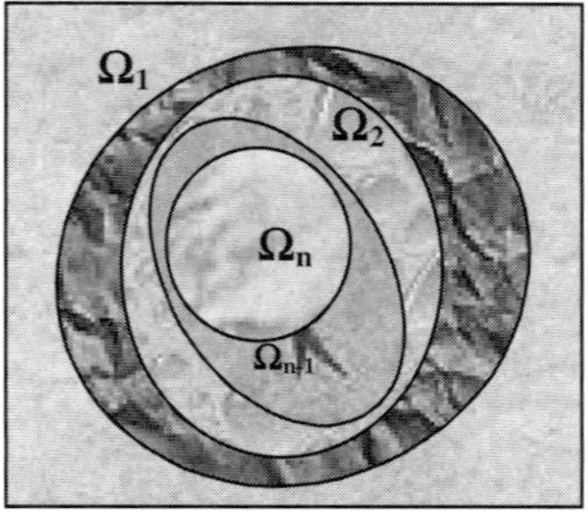

Figure 2.63. Cross−section of periodic composite structure

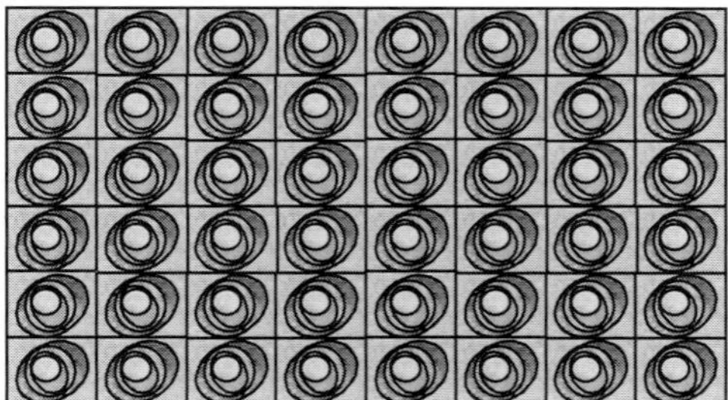

Figure 2.64. The RVE of plane composite

Let us assume that all interfaces of the composite are perfect in the sense that

$$\left[u_i^{\varepsilon} \right] = 0, \qquad \left[\sigma_{ij} n_j^{\Gamma} \right] = 0 \qquad\qquad (2.121)$$

where the symbol [.] denotes a jump of the respective function values at the interface. The homogenisation problem is to find the limit of solution $\mathbf{u}^{\varepsilon}$ with ε tending to 0. For this purpose let us consider a bilinear form $a^{\varepsilon}(\mathbf{u}, \mathbf{v})$ defined as follows:

$$a^{\varepsilon}(\mathbf{u}, \mathbf{v}) = \int\limits_{\Omega} C_{ijkl}\left(\tfrac{\mathbf{x}}{\varepsilon}\right) \varepsilon_{ij}(\mathbf{u})\, \varepsilon_{kl}(\mathbf{v}) d\Omega \qquad\qquad (2.122)$$

and the linear form:

$$L(\mathbf{v}) = \int_\Omega F_i v_i \, d\Omega + \int_{\partial\Omega_\sigma} p_i v_i \, d(\partial\Omega) \qquad (2.123)$$

both in a Hilbert space

$$V = \left\{ \mathbf{v} \,\middle|\, \mathbf{v} \in \left(H^1(\Omega)\right)^3, \mathbf{v}\big|_{\partial\Omega_u} = 0 \right\} \qquad (2.124)$$

characterised by the norm

$$\|\mathbf{v}\|^2 = \int_\Omega \varepsilon_{ij}(\mathbf{v})\varepsilon_{ij}(\mathbf{v}) \, d\Omega \qquad (2.125)$$

A variational statement equivalent to the equilibrium problem (2.120) is to find $\mathbf{u}^\varepsilon \in V$ fulfilling the equation

$$a^\varepsilon\left(\mathbf{u}^\varepsilon, \mathbf{v}\right) = L(\mathbf{v}) \qquad (2.126)$$

for any $\mathbf{v} \in V$. Let us introduce for this purpose a space of periodic functions $P(\Omega) = \left\{ \mathbf{v}, \mathbf{v} \in \left(H^1(\Omega)\right)^3 \right\}$ so that the trace of $\mathbf{v}$ is equal on opposite sides of Ω. Let us denote for any $\mathbf{u}, \mathbf{v} \in P(\Omega)$

$$a_y(\mathbf{u}, \mathbf{v}) = \int_\Omega C_{ijkl}(\mathbf{y})\varepsilon_{ij}(\mathbf{u})\varepsilon_{kl}(\mathbf{v}) \, d\Omega \qquad (2.127)$$

and introduce a homogenisation function $\chi_{(ij)k} \in P(\Omega)$ as a solution for the local problem on a periodicity cell:

$$a_y\left(\left(\chi_{(ij)k} + y_j \delta_{ki}\right)\mathbf{n}_k, \mathbf{w}\right) = 0 \qquad (2.128)$$

for any $\mathbf{w} \in P(\Omega)$; δ_{ki} denotes the Kronecker delta while $\mathbf{n}_k$ is the unit coordinate vector. Assuming finally that:

$$C_{ijkl} \in L^\infty(\Re^3) \qquad (2.129)$$

$$C_{ijkl} = C_{klij} = C_{jikl} \qquad (2.130)$$

$$\exists C_0 > 0; \; C_{ijkl}\xi_{ij}\xi_{kl} \ge C_0 \xi_{ij}\xi_{ij}, \quad \forall_{i,j}\xi_{ij} = \xi_{ji} \qquad (2.131)$$

we may introduce a homogenisation theorem as follows:

Homogenisation theorem
The solution $\mathbf{u}^\varepsilon$ of problem (2.126) converges weakly in space V

$$\mathbf{u}^{\varepsilon} \to \mathbf{u} \tag{2.132}$$

if the tensor $C_{ijkl}^{\varepsilon}(\mathbf{y})$ is Ω–periodic and its components fulfil conditions (2.129–2.131). Solution $\mathbf{u}$ is the unique one for the problem

$$\mathbf{u} \in V: \quad D(\mathbf{u}, \mathbf{v}) = L(\mathbf{v}) \tag{2.133}$$

for any $\mathbf{v} \in V$ and

$$D(\mathbf{u}, \mathbf{v}) = \int_{Y} D_{ijkl} \varepsilon_{ij}(\mathbf{u}) \varepsilon_{kl}(\mathbf{v}) dY \tag{2.134}$$

where

$$D_{ijkl} = \frac{1}{|\Omega|} a_{y} \left(\left(\chi_{(ij)p} + y_{i}\delta_{pj} \right) \mathbf{n}_{p}, \left(\chi_{(kl)q} + y_{l}\delta_{qk} \right) \mathbf{n}_{q} \right) \tag{2.135}$$

As a result of this theorem, a limit for $\varepsilon \to 0$ gives a homogeneous elastic material described by the tensor [163]:

$$C_{ijkl}^{(eff)} = \frac{1}{|\Omega|} \int_{\Omega} \left(C_{ijkl}(\mathbf{y}) + C_{ijmn}(\mathbf{y}) \varepsilon_{mn}^{y}\left(\chi_{(kl)}(\mathbf{y}) \right) \right) d\Omega \tag{2.136}$$

The most important result is that neither the local problem nor the statement (2.136) really depend on the stress boundary conditions since that solution obtained has a general character. To show formally these results, the local problem is rewritten in its differential form

$$\frac{\partial}{\partial x_{j}} \left(C_{ijkl}\left(\tfrac{\mathbf{x}}{\varepsilon} \right) \varepsilon_{kl}\left(u_{i}^{\varepsilon} \right) \right) + F_{i} = 0 \; ; \; \tfrac{x}{\varepsilon} = \mathbf{y} \in \Omega \; ; \; u_{i}^{\varepsilon} = 0 \text{ for } \mathbf{y} \in \partial \Omega \tag{2.137}$$

Next, similarly to the stochastic perturbation approach, an asymptotic expansion is employed in terms of the parameter ε as follows:

$$u_{i}^{\varepsilon}(\mathbf{x}) = u_{i}^{(0)}(\mathbf{x}, \mathbf{y}) + \varepsilon u_{i}^{(1)}(\mathbf{x}, \mathbf{y}) + \varepsilon^{2} u_{i}^{(2)}(\mathbf{x}, \mathbf{y}) + ... \tag{2.138}$$

where $u_{i}^{(m)}(\mathbf{x}, \mathbf{y})$ are periodic in $\mathbf{y}$ with a periodicity cell Ω. The main idea of this expansion is presented schematically in Figure 2.65: to better illustrate the meaning of (2.133) only a quarter of the composite is shown.

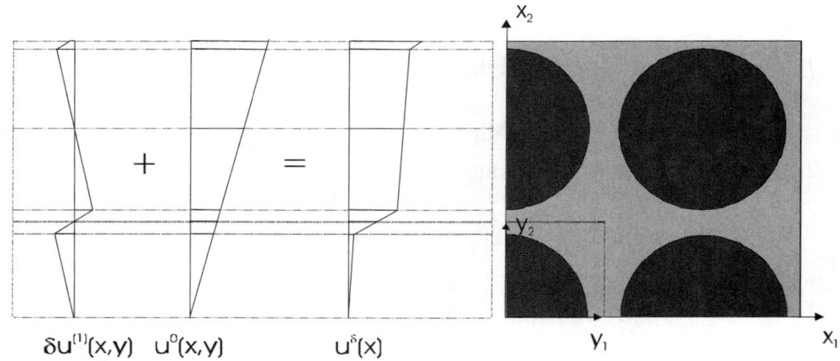

Figure 2.65. First order asymptotic expansion of displacements in a composite

Let us note that differentiation separates the coordinates $\mathbf{x}$ and $\mathbf{y}$, so that

$$\varepsilon_{ij}(\mathbf{v}) = \varepsilon_{ij}^{x}(\mathbf{v}) + \tfrac{1}{\varepsilon}\varepsilon_{ij}^{y}(\mathbf{v}) \qquad (2.139)$$

where the strain tensors $\varepsilon_{ij}^{x}(\mathbf{v})$, $\varepsilon_{ij}^{y}(\mathbf{v})$ correspond to small deformations

$$\varepsilon_{ij}^{x}(\mathbf{v}) = \frac{1}{2}\left(\frac{\partial v_i}{\partial x_j} + \frac{\partial v_j}{\partial x_i}\right), \quad \varepsilon_{ij}^{y}(\mathbf{v}) = \frac{1}{2}\left(\frac{\partial v_i}{\partial y_j} + \frac{\partial v_j}{\partial y_i}\right). \qquad (2.140)$$

Thus, (2.132) can be rewritten as follows:

$$\left(\varepsilon^{-2}L_1 + \varepsilon^{-1}L_2 + L_3\right)\left(u_i^{(0)} + \varepsilon u_i^{(1)} + \varepsilon^2 u_i^{(2)} + ...\right) + F_i = 0 \qquad (2.141)$$

where

$$L_1 u_i = \frac{\partial}{\partial y_j}\left(C_{ijkl}(\mathbf{y})\frac{\partial u_k}{\partial y_l}\right) \qquad (2.142)$$

$$L_2 u_i = C_{ijkl}(\mathbf{y})\frac{\partial}{\partial x_j}\left(\frac{\partial u_k}{\partial y_l}\right) + \frac{\partial}{\partial y_j}\left(C_{ijkl}(\mathbf{y})\right)\frac{\partial u_k}{\partial x_l} \qquad (2.143)$$

$$L_3 u_i = C_{ijkl}(\mathbf{y})\frac{\partial}{\partial x_j}\left(\frac{\partial u_k}{\partial x_l}\right) \qquad (2.144)$$

Next, we equate to 0 the terms with the same order of ε, obtaining an infinite sequence of equations. The relations adequate to its zeroth, first and second orders can be written as

$$L_1 u_i^{(0)} = 0 \tag{2.145}$$

$$L_1 u_i^{(1)} + L_2 u_i^{(0)} = 0 \tag{2.146}$$

$$L_1 u_i^{(2)} + L_2 u_i^{(1)} + L_3 u_i^{(0)} + F_i = 0 \tag{2.147}$$

The displacements fields $u_i^{(0)}$, $u_i^{(1)}$ and $u_i^{(2)}$ can be found from these equations recurrently only if $\mathbf{x}$ and $\mathbf{y}$ are independent variables. Let us note also that the equation

$$L_1 u_i + P_i = 0 \tag{2.148}$$

with u_i being Ω–periodic function has a unique solution for

$$\langle P \rangle = \frac{1}{|\Omega|} \int_\Omega P_i 1_i \, d\mathbf{y} = 0 \tag{2.149}$$

Further, if the unique solution $\mathbf{u}(\mathbf{x},\mathbf{y})$; $\mathbf{x} \in \Omega$ of (2.148) is constant then for all $\mathbf{x}$ (where $\mathbf{x}$ plays here the role of parameter) we have $u_0 = const$. Considering this fact it can be obtained that

$$u_i^{(0)}(\mathbf{x},\mathbf{y}) = u_i(\mathbf{x}) \tag{2.150}$$

which can be observed in Figure 2.65 as well. It can be observed that the first term of the expansion of $\mathbf{u}$ does not depend on the micro variable $\mathbf{y}$ and can be considered as a mean displacement altered by the higher order terms only. Thus, (2.146) takes the following form:

$$L_1 u_i^{(1)}(\mathbf{x},\mathbf{y}) + \frac{\partial}{\partial y_j}\left(C_{ijkl}(\mathbf{y})\right)\frac{\partial}{\partial x_l}\left(u_k^{(0)}(\mathbf{x})\right) = 0 \tag{2.151}$$

The solution is obtained by separation of $\mathbf{x}$ and $\mathbf{y}$

$$u_i^{(1)}(\mathbf{x},\mathbf{y}) = \chi_{(kl)i}(\mathbf{y})\frac{\partial}{\partial x_l}\left(u_k^{(0)}(\mathbf{x})\right) + u_i(\mathbf{x}) \tag{2.152}$$

The last two equations give the formulation for the Ω–periodic functions $\chi_{(kl)i}(\mathbf{y})$

$$\frac{\partial}{\partial y_j}\left(C_{ijmn}(\mathbf{y})\frac{\partial \chi_{(kl)m}(\mathbf{y})}{\partial y_n}\right) + \frac{\partial}{\partial y_j}\left(C_{ijkl}(\mathbf{y})\right) = 0 \tag{2.153}$$

which completes our consideration of general homogenisation method for linear elastostatic problems.

It is relatively easy to see that the local problems for homogenisation functions $\chi_{(kl)i}(\mathbf{y})$ reduce to the equations given above for any region Ω_a where $1 \leq a \leq n$ for the so−called fibre−like composite materials where one component is placed into the next one, etc. Let us denote by $\Gamma_{(k-1,k)}$ the interface between components Ω_{a-1} and Ω_a. Then the following conditions are true for $a=2,...,n$ and $x \in \Gamma_{(a-1,a)}$:

$$\left[\chi_i^{kl} \right] = 0 \tag{2.154}$$

and

$$\sigma_{ij}\left(\chi_{(pq)}\right)n_j = \left[C_{pqij} \right]_{\Gamma_{(a-1,a)}} n_j = F_{(pq)i}\Big|_{\Gamma_{(a-1,a)}} \tag{2.155}$$

$$\left[C_{pqij} \right]_{\Gamma_{(a-1,a)}} = C_{pqij}^{(a)} - C_{pqij}^{(a-1)}; \quad \mathbf{x} \in \Gamma_{(a-1,a)} \tag{2.156}$$

Summing up all the considerations on the homogenisation problem (2.126), we compute the effective elasticity tensor components given by (2.136) using the homogenisation functions $\chi_{(kl)i}$ being a solution of a classical well−posed boundary value problem with periodicity conditions on the external boundaries of Ω. The stress boundary conditions are equal to the difference of constitutive tensor components at the particular composite interface. The variational formulation necessary for a finite element formulation of the local problem can be introduced as follows:

$$\sum_{a=1}^{n} \int_{\Omega_a} C_{ijkl}\varepsilon_{kl}\left(\chi_{(pq)}\right)\varepsilon_{ij}(\mathbf{v})\,d\Omega = -\sum_{a=2}^{n} \int_{\Gamma_{(a-1,a)}} \sigma_{ij}\left(\chi_{(pq)}\right)n_j v_i\,d\Gamma + \int_{\Omega} f_i v_i\,d\Omega \tag{2.157}$$

which by neglecting body forces leads to

$$\sum_{a=1}^{n} \int_{\Omega_a} C_{ijkl}\varepsilon_{kl}\left(\chi_{(pq)}\right)\varepsilon_{ij}(\mathbf{v})\,d\Omega = -\sum_{a=2}^{n} \int_{\Gamma_{(a-1,a)}} F_{(pq)i}v_i\,d\Gamma \tag{2.158}$$

Having determined the homogenisation functions for the n−component composite, the effective elasticity tensor components from (2.136) are calculated as the result.

The general configuration of the n−component composite denotes that there are m interfaces in the periodicity cell where $m \in N$ and $m \geq n-1$. It can be observed that for coherent components, as was assumed at first, the case of $m=n$-1 (minimum value of m) is equivalent to the fibre−like composite characterised in the previous section or the composite where n-1 components are embedded into one matrix. In that case the variational formulation of the homogenisation problem has the following form:

$$\sum_{a=1}^{n} \int_{\Omega_a} C_{ijkl} \varepsilon_{kl}\left(\chi_{(pq)}\right) \varepsilon_{ij}(\mathbf{v}) \, d\Omega = -\sum_{a=2}^{n-1} \int_{\Gamma_{(1,a)}} F_{(pq)i} v_i \, d\Gamma \qquad (2.159)$$

Moreover, it can be seen that the n-component composite in a general configuration generates, due to the component permutation scheme, the bounded set of $(n\text{-}1)!$ various effective elasticity tensors. If some components are disjoint, the total number of these subsets must be included in the permutation procedure. It would be interesting to calculate, due to the homogenisation method presented, the upper and lower bounds of the effective elasticity tensor components for such a set of permutations.

Next, it is observed that in the general case the effective elasticity tensor components can be calculated by the following modification of (2.159):

$$\sum_{a=1}^{n} \int_{\Omega_a} C_{ijkl} \varepsilon_{kl}\left(\chi_{(pq)}\right) \varepsilon_{ij}(\mathbf{v}) \, d\Omega = -\sum_{r=1}^{m} \int_{\Gamma_r} F_{(pq)} v \, d\Gamma \qquad (2.160)$$

where the RHS summation is carried out along all interfaces detected in the composite periodicity cell. Further, if any interface shows some finite number of nonsmoothness, the integration over such contour to be replaced with the sum of integrals defined on partially smooth curves composing the interface.

Finally, it is observed that the effective modules method of homogenisation formulated by (2.158) − (2.159) enables one to calculate effective properties for the composites including microdefects or interface defects; it can be done by equating the appropriate material characteristics to 0 for these regions. For this purpose, the computational procedure applied in numerical experiments can be linked with the program for digital processing of composite cross−section images.

Now let us consider the Finite Element Method discretisation of the homogenisation problem. Let us introduce the following approximation of homogenisation functions $\chi_{(rs)i}$ $(i,r,s=1,2)$ at any point of the considered continuum Ω in terms of a finite number of generalised coordinates $q_{(rs)\alpha}$ and the shape functions $\varphi_{i\alpha}$

$$\chi_{(rs)i} = \varphi_{i\alpha} q_{(rs)\alpha}, \quad i,r,s = 1,2, \quad \alpha = 1,...,N \qquad (2.161)$$

In the same way the strain $\varepsilon_{ij}(\chi_{(rs)})$ and stress $\sigma_{ij}(\chi_{(rs)})$ tensors are rewritten as

$$\varepsilon_{ij}(\chi_{(rs)}) = B_{ij\alpha} q_{(rs)\alpha} \qquad (2.162)$$

$$\sigma_{ij(rs)} = \sigma_{ij}(\chi_{(rs)}) = C_{ijkl}\varepsilon_{kl}(\chi_{(rs)}) = C_{ijkl} B_{kl\alpha} q_{(rs)\alpha} \qquad (2.163)$$

where $B_{kl\alpha}$ represents the shape functions derivatives. Introducing (2.162) – (2.163) into the virtual work equation in its variational form it is found that

$$\int_\Omega \delta\chi_{(rs)i,j} C_{ijkl}\chi_{(rs)k,l}\, d\Omega = \sum_{p=2}^{m} \int_{\Gamma_p} \delta\chi_{(rs)i} \left[F_{(rs)i}\right]_{\Gamma_p} d\Gamma \qquad \text{(no sum on } r,s\text{)} \qquad (2.164)$$

Furthermore, let us define the composite global stiffness matrix as

$$K_{\alpha\beta} = \sum_{e=1}^{E} K_{\alpha\beta}^{(e)} = \sum_{e=1}^{E} \int_{\Omega_e} C_{ijkl} B_{ij\alpha} B_{kl\beta}\, d\Omega \qquad (2.165)$$

Using this notation in (2.164) and minimising the variational statement with respect to the generalised coordinates we arrive at

$$K_{\alpha\beta} q_{(rs)\alpha} = Q_{(rs)\alpha} \qquad (2.166)$$

with $Q_{(rs)\alpha}$ being the external load vector containing the boundary forces given by (2.155) – (2.156), which is employed to determine the homogenisation function $\chi_{(rs)i}$ in three numerical tests for $r,s=1,2$. To ensure the symmetry conditions on a periodicity cell, the orthogonal displacements and rotations for every nodal point belonging to the external boundaries of Ω are fixed. For the functions $\chi_{(rs)i}$ so defined we compute the stresses $\sigma_{ij}(\chi_{(rs)})$ and average this tensor numerically over the region Ω according to the formula (2.136).

The fibre–reinforced glass–epoxy composite example with an interphase between the fibre and the matrix is analysed in computational experiments [163]. The microgeometry of the periodicity cell is shown in Figure 2.66, while material characteristics of the constituents are collected in Table 2.11.

The weaker interphase in our tests may simulate any boundary defects appearing in fibre–reinforced composites that are caused by the difference in thermal stresses during the fabrication process in metal matrix composites (MMC) for instance. On the other hand, a stronger interphase model homogenised numerically is equivalent to the case when some layer between the fibre and matrix is introduced to enforce component interface bonding strength.

Generally, 11 groups of computational experiments are performed to compute the effective elastic and thermal characteristics for the composite considered. Material properties are increased in the interphase starting from 50% of additional matrix characteristics with increments equal to 10% for each of the next test group. Thus for the 6th group the interphase equivalent to the matrix is obtained and for the 11th the material properties of the interphase are equal to 150% of the matrix parameters; the results of this analysis are presented in Table 2.12.

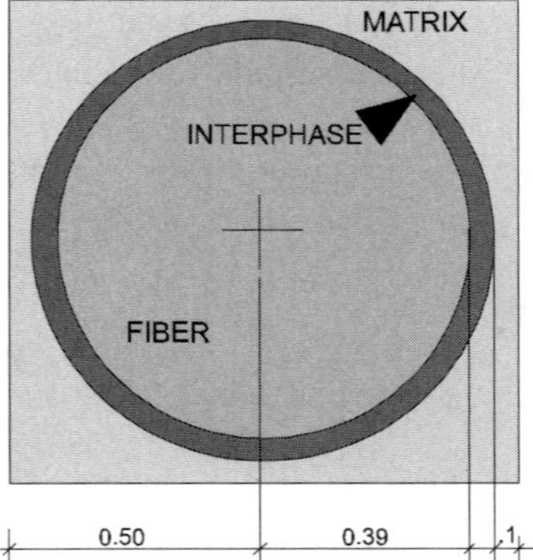

Figure 2.66. Microgeometry of the periodicity cell

Table 2.11. Material data for composite components

Material	e	ν
Glass fibres	72.38	0.200
Epoxy matrix	2.75	0.350

Table 2.12. Effective elastic and thermal parameters

Test no	$C_{1111}^{(\mathit{eff})}$	$C_{1122}^{(\mathit{eff})}$	$C_{1212}^{(\mathit{eff})}$
1	8.566	3.122	14.577
2	8.815	3.209	14.580
3	9.020	3.278	14.582
4	9.197	3.337	14.584
5	9.338	3.391	14.586
6	9.474	3.445	14.588
7	9.610	3.503	14.589
8	9.761	3.572	14.591
9	9.949	3.681	14.593
10	10.619	4.218	14.594
11	11.399	4.940	14.596

Analysing these results it can be concluded that all effective parameters show some sensitivity to the improved interphase and its material parameters. The greatest sensitivity is obtained for $C_{1122}^{(\mathit{eff})}$ and $C_{1111}^{(\mathit{eff})}$ components, while the smallest for $C_{1212}^{(\mathit{eff})}$. To obtain more realistic results it will be valuable to introduce anisotropy in the equivalent parameters of the interphase; in that case the sensitivity of the $C_{1212}^{(\mathit{eff})}$ component increases significantly. However, neglecting these disproportions the results computed lead us to the conclusion that the improved homogenisation method confirms the crucial role of the interphase on the overall characteristics of the composite structure, which is observed in engineering practice. Moreover, the variability resulting from computational experiments confirms generally the usefulness of the homogenisation method proposed. Other series of computational tests are done to the visualisation of the homogenisation functions as well as the resulting stresses and various numerical error estimators.

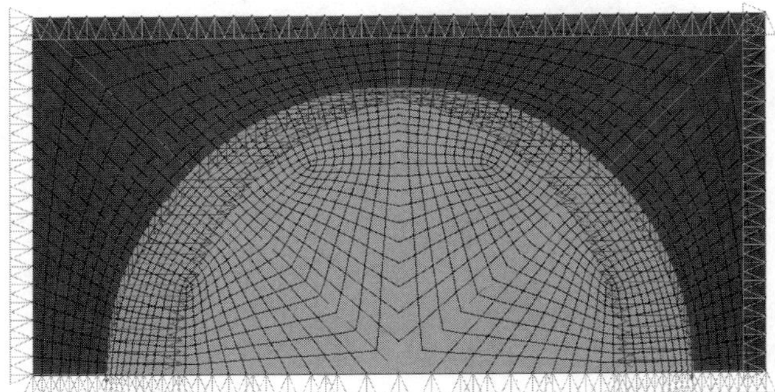

Figure 2.67. Boundary conditions for homogenisation problems χ^{11}

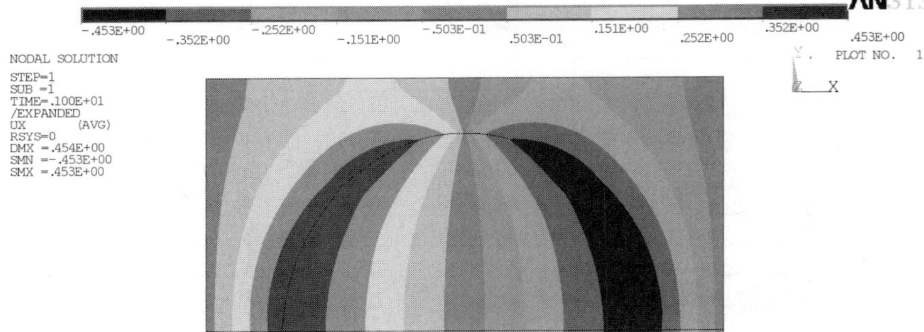

Figure 2.68. Horizontal components of the homogenisation function χ^{11}

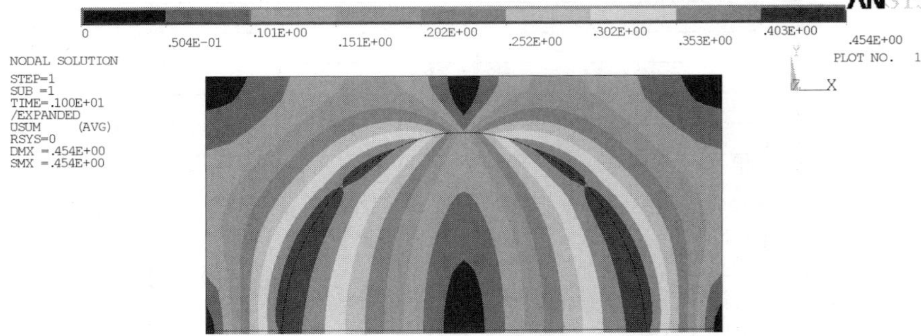

Figure 2.69. Vertical components of the homogenisation function χ^{11}

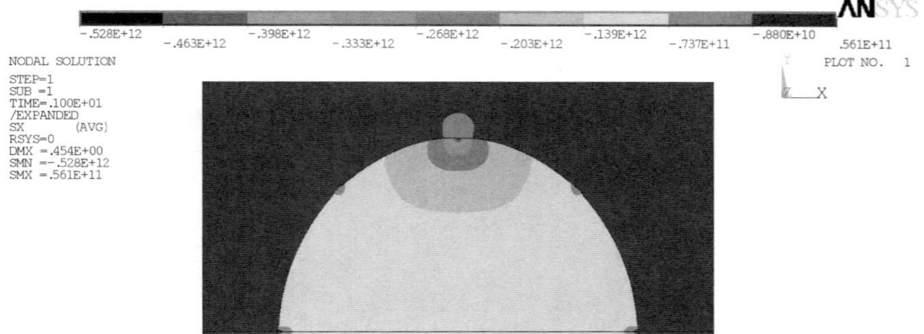

Figure 2.70. Horizontal stresses in the homogenisation problem χ^{11}

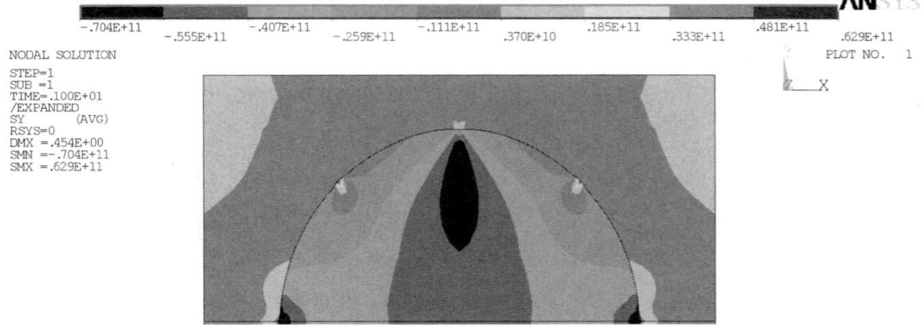

Figure 2.71. Vertical stresses in the homogenisation problem χ^{11}

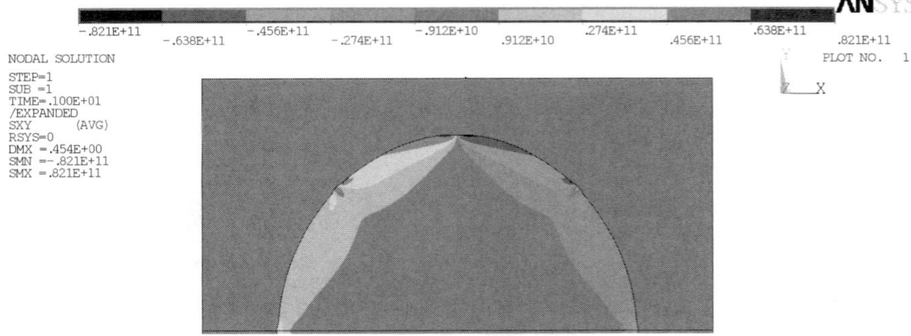

Figure 2.72. Shear stresses in the homogenisation problem χ^{11}

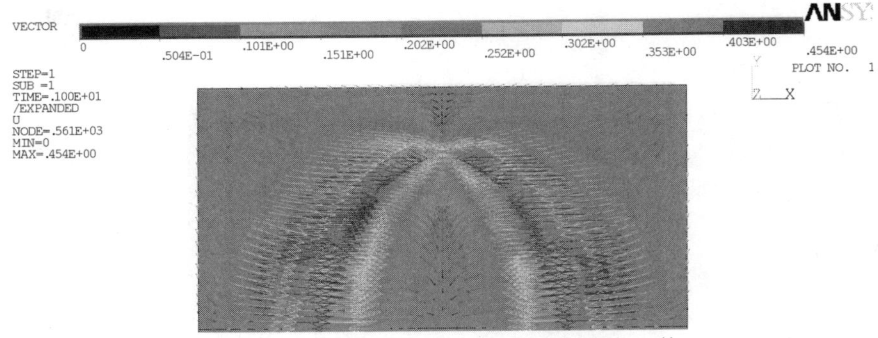

Figure 2.73. Vortex visualization of the homogenisation function χ^{11}

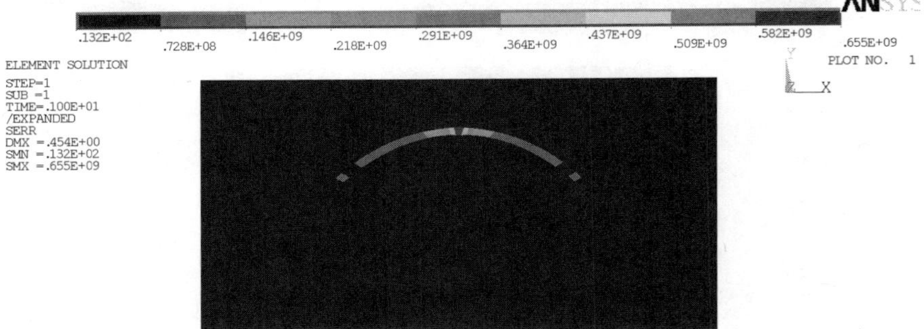

Figure 2.74. Relative error of the stresses determination in the problem χ^{11}

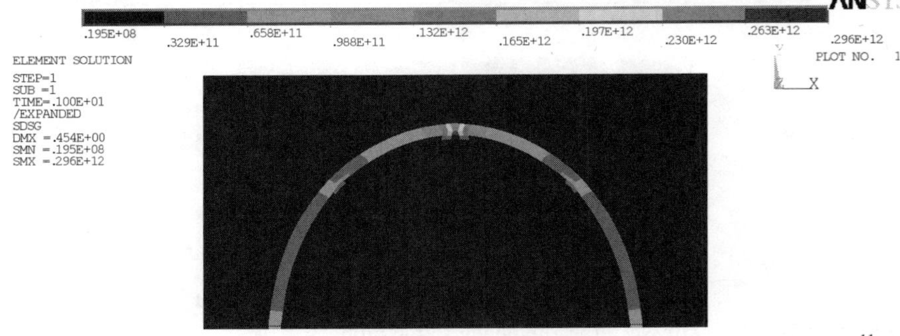

Figure 2.75. Relative error for strain determination in the homogenisation problem χ^{11}

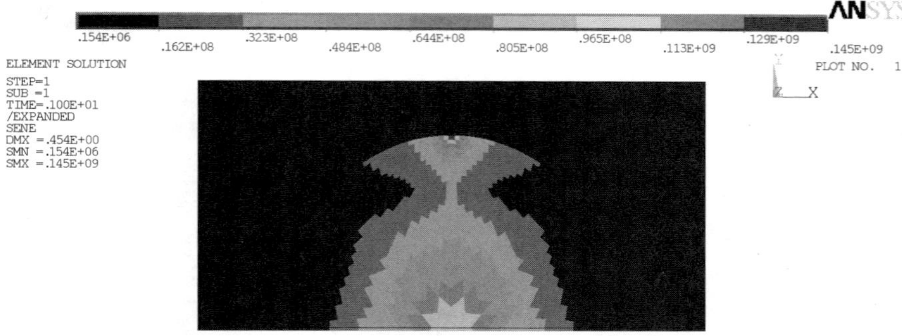

Figure 2.76. Relative error of the strain energy determination χ^{11}

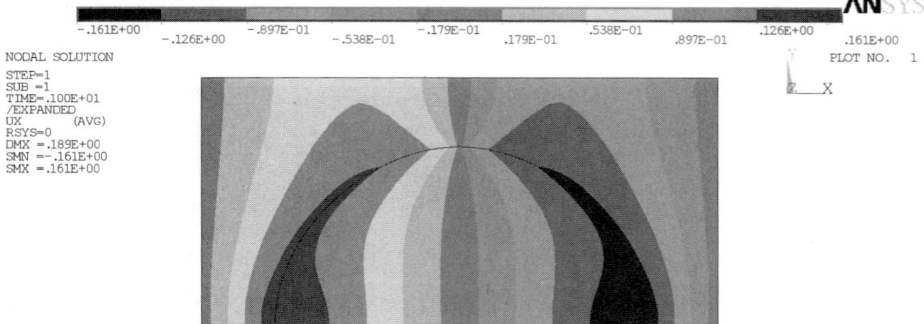

Figure 2.77. Horizontal components of the homogenisation function χ^{12}

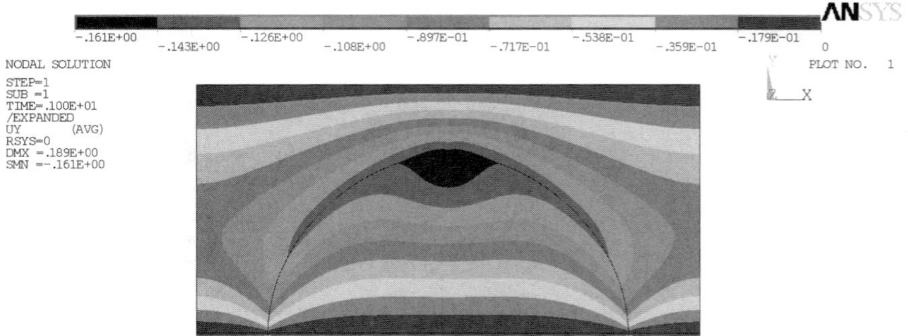

Figure 2.78. Vertical components of the homogenisation function χ^{12}

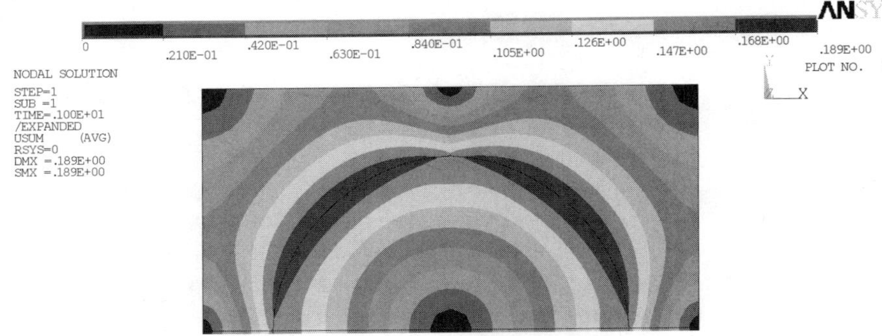

Figure 2.79. Total values of the homogenisation function χ^{12}

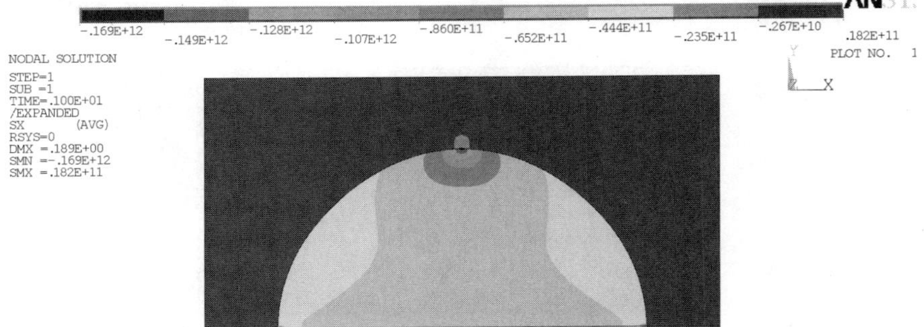

Figure 2.80. Horizontal stresses in the homogenisation problem χ^{12}

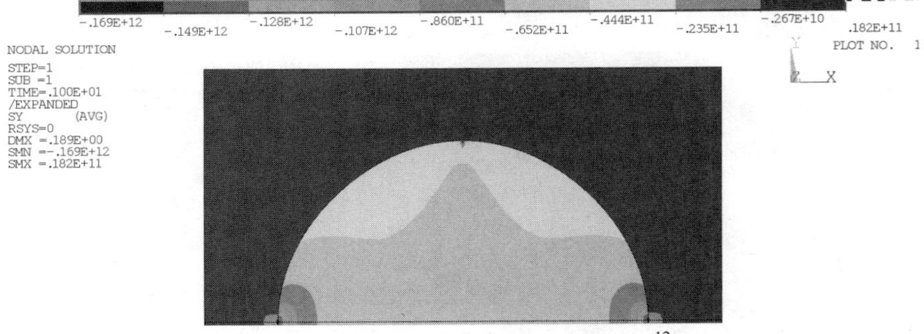

Figure 2.81. Vertical stresses in the homogenisation problem χ^{12}

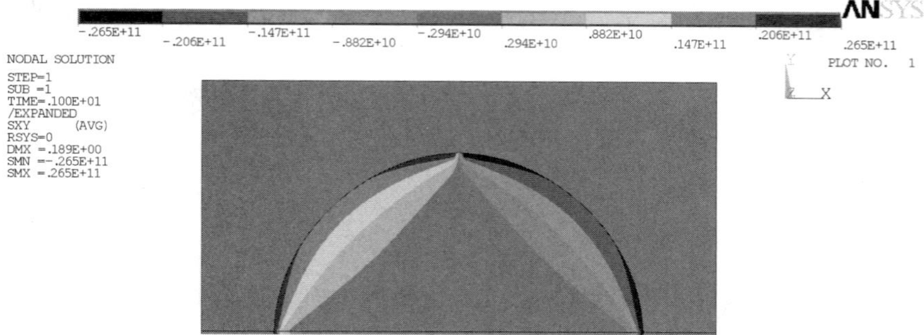

Figure 2.82. Shear stresses in the homogenisation problem χ^{12}

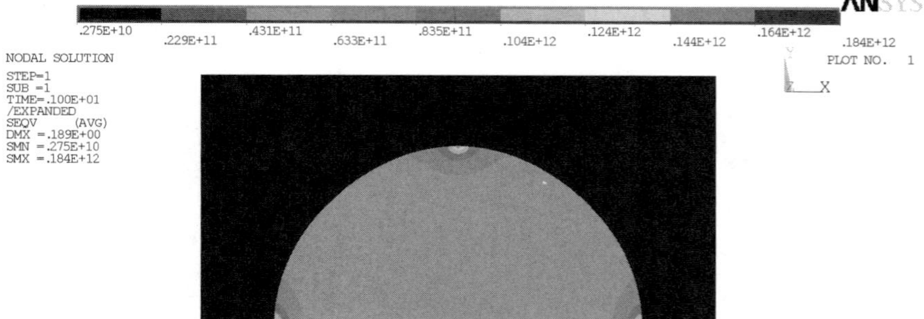

Figure 2.83. Equivalent von Mises stresses in the homogenisation problem χ^{12}

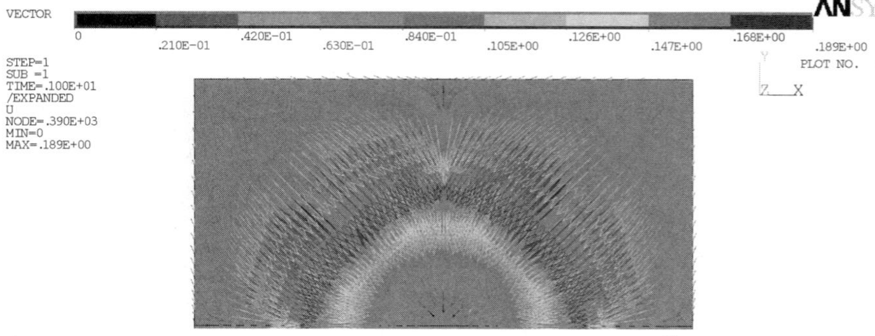

Figure 2.84. Vortex visualization of the homogenisation function χ^{12}

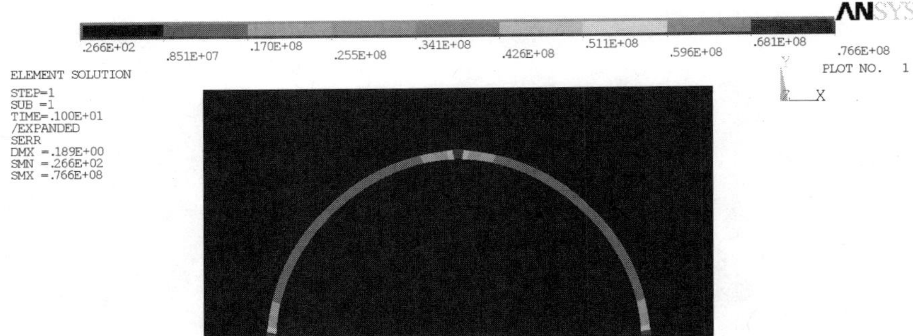

Figure 2.85. Relative error of the stresses determination in the problem χ^{12}

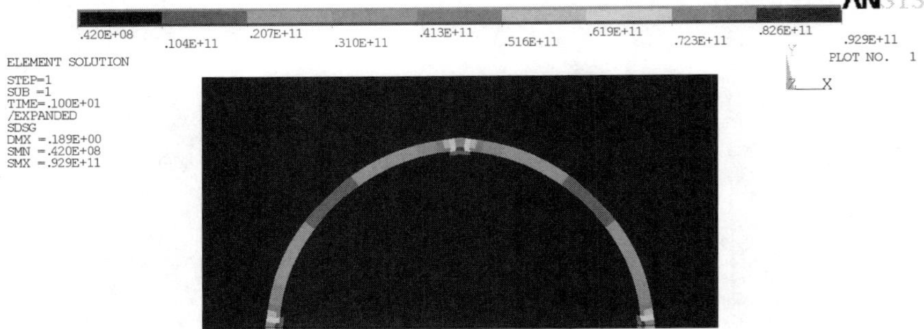

Figure 2.86. Relative error of the strain determination in the problem χ^{12}

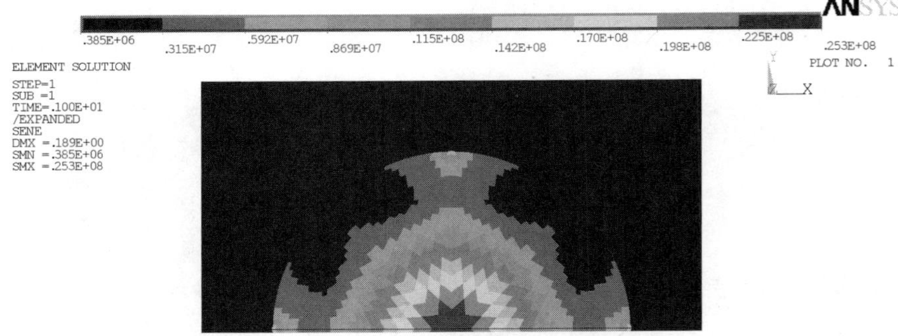

Figure 2.87. Relative error of the strain energy determination χ^{12}

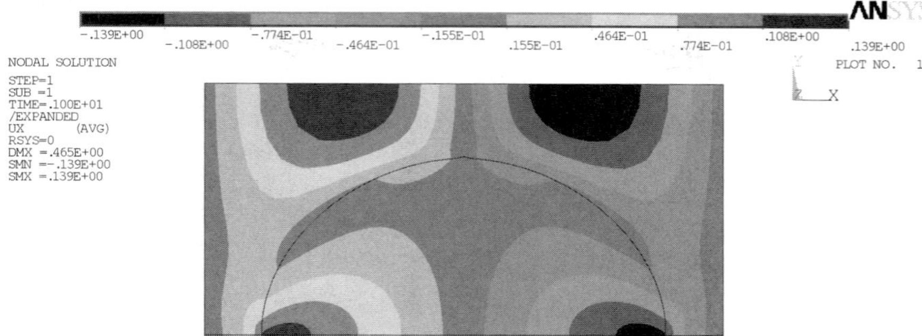

Figure 2.88. Horizontal components of the homogenisation function χ^{22}

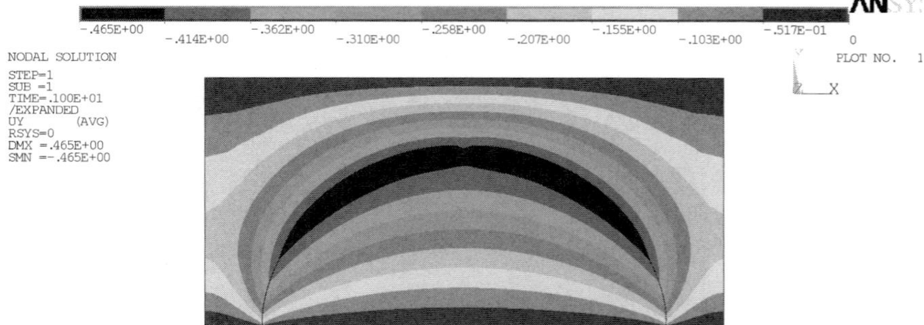

Figure 2.89. Vertical components of the homogenisation function χ^{22}

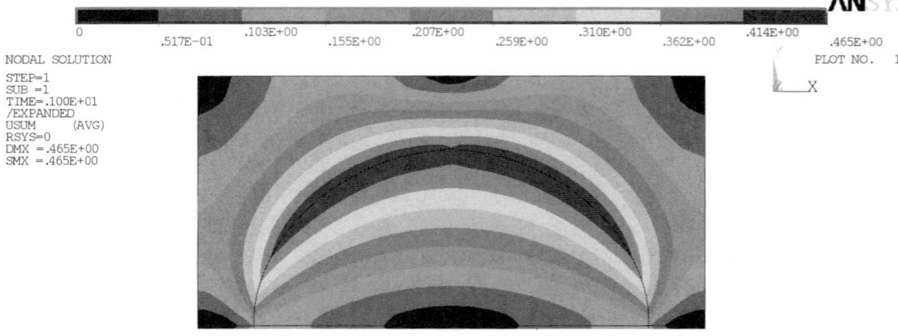

Figure 2.90. Total values of the homogenisation function χ^{22}

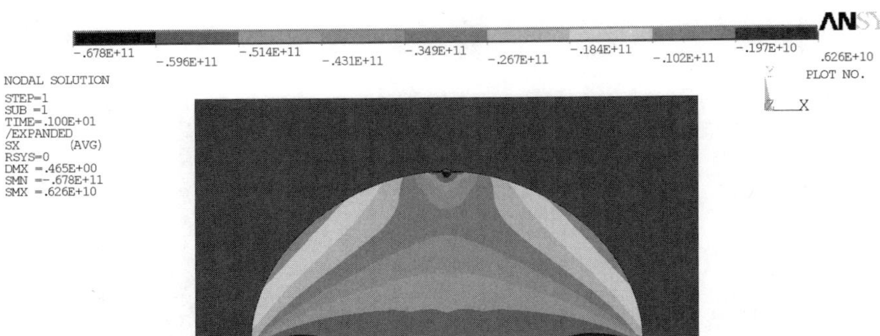

Figure 2.91. Horizontal stresses in the homogenisation problem χ^{22}

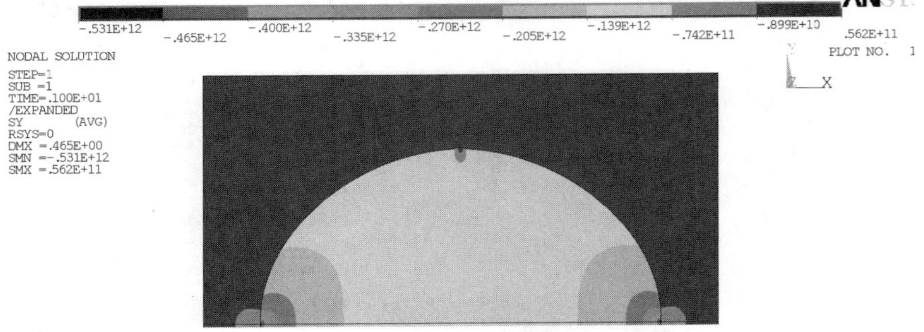

Figure 2.92. Vertical stresses in the homogenisation problem χ^{22}

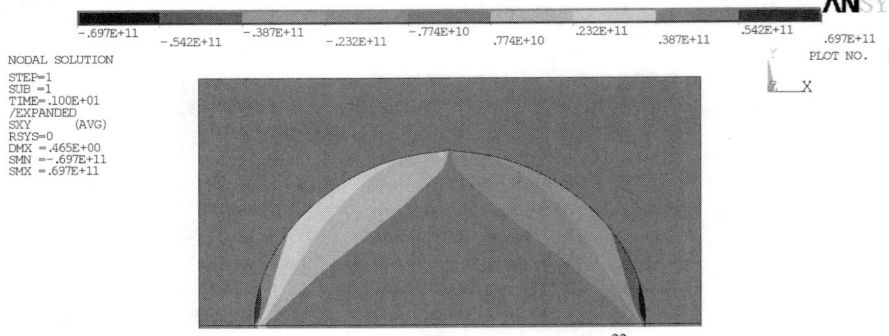

Figure 2.93. Shear stresses in the homogenisation problem χ^{22}

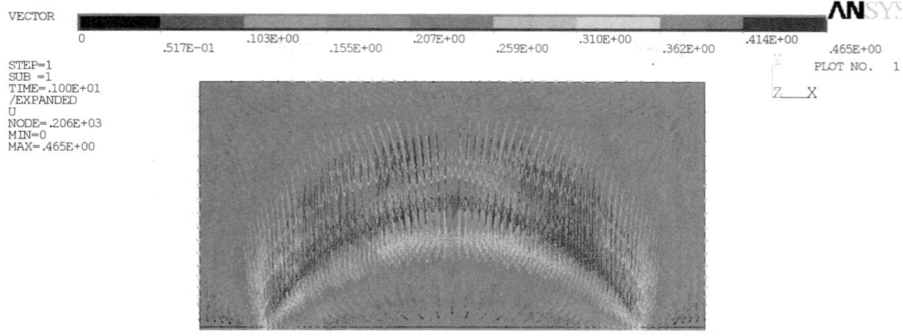

Figure 2.94. Vortex visualization of the homogenisation function χ^{22}

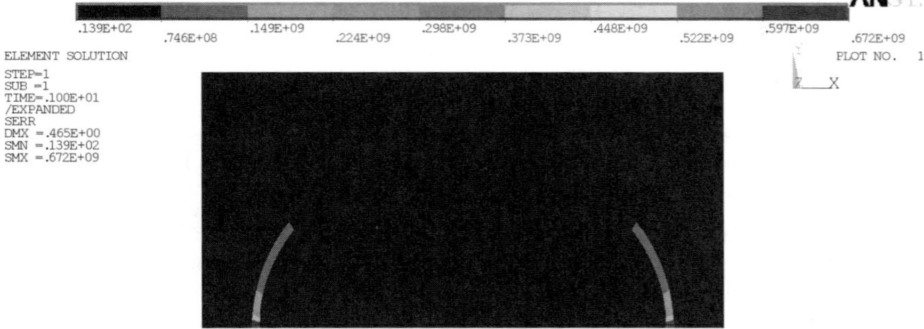

Figure 2.95. Relative error of the stresses determination in the problem χ^{22}

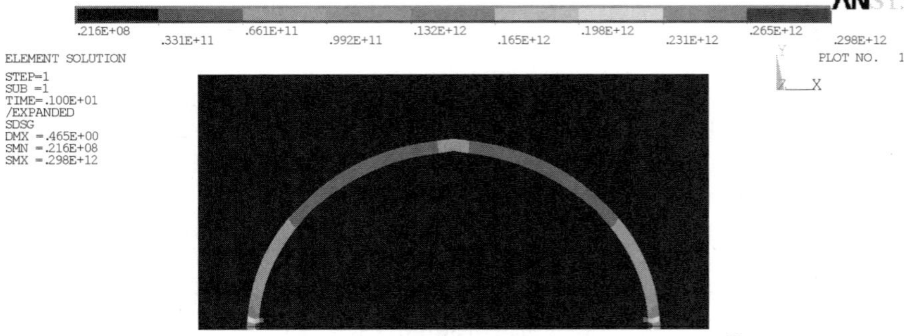

Figure 2.96. Relative error of the strain determination in the problem χ^{22}

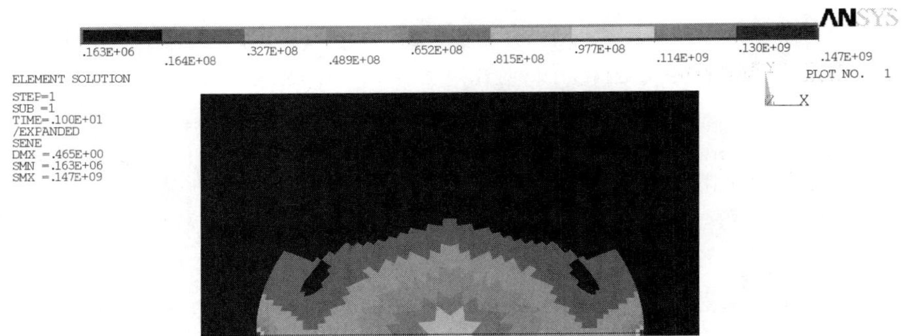

Figure 2.97. Relative error of the strain energy determination χ^{22}

The results of the computational analysis carried out in this section show that the effective properties of the composite and, at the same time, the overall behaviour of the composite, in the context of the homogenisation method, are sensitive to the interphase between the constituents and its material parameters. It should be underlined that the interphase, improved in the example presented above, has small total area in the comparison to the fibre and matrix. It can be expected that the previous, simplified approach (upper and lower bounds or direct approximations of effective properties cited above) do not enable us to arrive at such effects.

Considering the assumption that the scale factor between the RVE and the whole composite structure tends to 0 in our analysis and, on the other hand, that this quantity in real composites is small but differs from 0, the sensitivity of the effective characteristics to this parameter are to be calculated in the next analyses based on this approach. To carry out such studies, the scale parameter has to be introduced in the equations describing effective properties and next, due to the well-known sensitivity analysis methods, the influence of the scale parameter ε relating composite micro– and macrostructure may be shown. In the analogous way we can study the sensitivity of the effective characteristics of the composite to the component material parameters but there is no need in this case to introduce any extra components into the equations cited above.

Further mathematical and computational extensions of the model presented should be provided to include in the constitutive tensor the components responsible for the thermal expansion [228,311]. Having computed the effective characteristics on the basis of Young moduli, Poisson ratios, coefficient of thermal expansion and heat conduction coefficient [106,163,347] it will be possible to provide the coupled temperature–displacement FE analyses of periodic composite materials. At the same time it will be valuable to work out the problem presented in the context of viscoelastic or elastoviscoplastic material models of the composite constituents [74,368]. It will enable us to approximate computationally the fracture and failure phenomena in composites resulting from the interface defects or partial debonding using the homogenisation approach.

2.3.3.2.2 Monte Carlo Simulation Analysis

Starting from the formula describing the effective elasticity tensor components, their expected values are derived using the basic theorems on the random variables as follows [191]:

$$E\left[C_{ijkl}^{(eff)}(x;\omega)\right] = E\left[\left\langle \sigma_{kl}(\chi^{ij}(x;\omega)\right\rangle_{\Omega}\right] + E\left[\left\langle C_{ijkl}(x;\omega)\right\rangle_{\Omega}\right] \tag{2.167}$$

The expressions for the variances (and generally covariances) have a more complicated form than the expectations because the averaged stresses and elasticity tensor are correlated variables. Therefore

$$Var\left(C_{ijkl}^{(eff)}(x;\omega)\right) = Var\left(\left\langle \sigma_{kl}(\chi^{ij}(x;\omega)\right\rangle_{\Omega}\right)$$
$$+ 2Cov\left(\left\langle \sigma_{kl}(\chi^{ij}(x;\omega)\right\rangle_{\Omega}, \left\langle C_{ijkl}(x;\omega)\right\rangle_{\Omega}\right) + Var\left(\left\langle C_{ijkl}(x;\omega)\right\rangle_{\Omega}\right) \tag{2.168}$$

The random homogenisation fields $\chi^{ij}(x,\omega)$ for general composites, similar to the deterministic ones, are calculated only numerically. The following probabilistic stress boundary conditions are imposed on the boundary $\Gamma_{(a-1,a)}$ to find the homogenisation functions:

$$E\left[F_{(pq)i}(\omega)\big|_{\Gamma_{(a-1,a)}}\right]$$
$$= E\left[\lambda(\omega)\big|_{\Gamma_{(a-1,a)}}\delta_{pq}n_i\right] + E\left[\mu(\omega)\big|_{\Gamma_{(a-1,a)}}\left(n_p\delta_{qi}+n_q\delta_{pi}\right)\right] \tag{2.169}$$

$$Var\left(F_{(pq)i}(\omega)\big|_{\Gamma_{(a-1,a)}}\right)$$
$$= Var\left(\lambda(\omega)\big|_{\Gamma_{(a-1,a)}}\delta_{pq}n_i\right) + Var\left(\mu(\omega)\big|_{\Gamma_{(a-1,a)}}\left(n_p\delta_{qi}+n_q\delta_{pi}\right)\right) \tag{2.170}$$

where $\lambda(\omega)$ and $\mu(\omega)$ are the Lame constants. If Young moduli of composite components are considered as input random variables then the expected values and variances of boundary forces are obtained by separating the RHS into those components corresponding to Ω_{a-1} and Ω_a, respectively. After some algebraic transformations there holds

$$E\left[F_{(pq)i}(x;\omega)\right] = B_{pqi}(v_a)\cdot E\left[e_a\right] - B_{pqi}(v_{a-1})\cdot E\left[e_{a-1}\right] \tag{2.171}$$

where the operator $B_{pqi}(v(\mathbf{x}))$ similar to the tensor A_{ijkl} introduced by eqn (2.14) is defined as

$$B_{pqi}(v(\mathbf{x})) = \delta_{pq} n_i \frac{v(\mathbf{x})}{(1+v(\mathbf{x}))(1-2v(\mathbf{x}))} + (n_p \delta_{qi} + n_q \delta_{pi}) \frac{1}{2(1+v(\mathbf{x}))} \qquad (2.172)$$

and their variances are equal to

$$Var\left(F_{(pq)i}(\mathbf{x};\omega)\right) = \left\{B_{pqi}(v_a)\right\}^2 \cdot Var(e_a) + \left\{B_{pqi}(v_{a-1})\right\}^2 \cdot Var(e_{a-1})$$
$$\text{(no sum on } p,q,i) \qquad (2.173)$$

Finally, probabilistic moments of the effective characteristics are derived using statistical estimation methods, according to which the expected values and the relevant covariances (computed using the unbiased estimator) of the effective elasticity tensor components are obtained as

$$E\left[C_{ijpq}^{(eff)}\right] = \tfrac{1}{M} \sum_{j=1}^{M} C_{ijpq}^{(eff)\bar{j}} \qquad (2.174)$$

$$Cov\left(C_{ijpq}^{(eff)}(\omega), C_{rsuv}^{(eff)}(\omega)\right) = \tfrac{1}{M} \sum_{j=1}^{M} \left(C_{ijpq}^{(eff)\bar{j}} - E\left[C_{ijpq}^{(eff)}\right]\right)\left(C_{rsuv}^{(eff)\bar{j}} - E\left[C_{rsuv}^{(eff)}\right]\right) \qquad (2.175)$$

where $C_{ijpq}^{(eff)\bar{j}}(\omega)$, $\bar{j}=1,...,M$ are random series of the tensor components obtained as a result of the generation of numerical random values.

The homogenisation problem presented is implemented into the program MCCEFF, which is based on the Monte Carlo simulation technique. The implementation of the MCS has been selected from among many other probabilistic methods, because this method consists of computer generation of random variables in the mechanical problem (cf. Figure 2.98) and computing the sequence of deterministic solutions associated with each variable generated; similar engineering software is also available [47]. Considering the fact that a composite structure has a relatively small number of degrees of freedom, a crude random sampling method is used in the computations (contrary to the Random Importance or Stratified Sampling methods) [73,125,139].

<table>
<tr><td style="text-align:center">Define N, m, a, c, E[e], σ(e), E[v], σ(v)</td></tr>
</table>

$$\downarrow$$

<table>
<tr><td style="text-align:center">Generate uniform distribution $\{I_1,...,I_N\} \in (0, m-1)$</td></tr>
<tr><td style="text-align:center">Do for k=1,N
$I_k = aI_{k-1} + c(\bmod m)$
Enndo</td></tr>
</table>

$$\downarrow$$

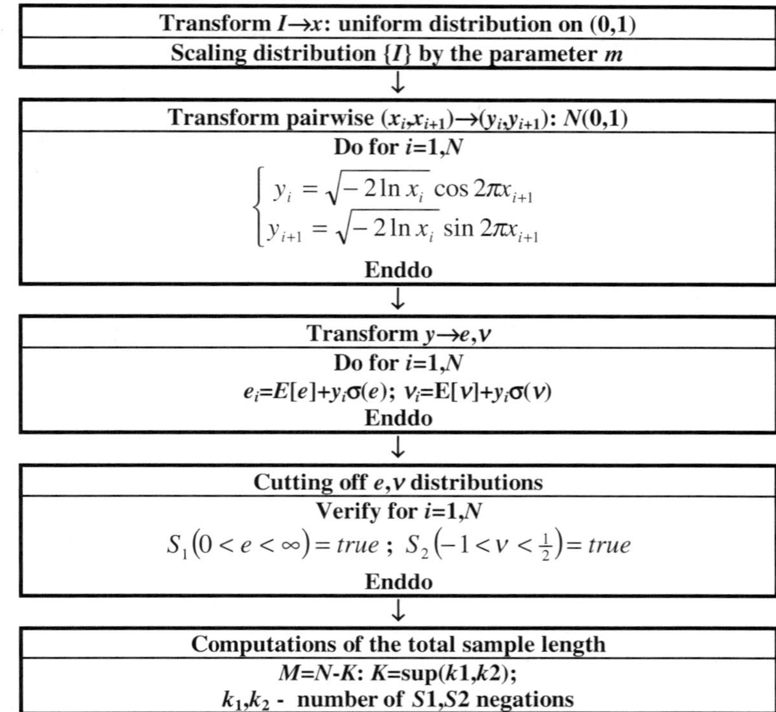

Figure 2.98. Algorithm for random numbers generation

However, the most important reason for the MCS application is that the accuracy of the output variable probabilistic moments estimation does not depend on the input variable coefficient of variation (as for the SFEM), but on the total number of iterations performed. Taking into account the estimator convergence studies and some theoretical considerations, the total number of random trials M has been taken as equal to 1,000. The flowchart of the program used for probabilistic homogenisation is shown in Figure 2.99. As presented, the program makes it possible to discretise automatically the RVE on the basis of the main cell geometrical parameters, visualisation of the mesh introduced, random generation of the input random variables and iterative computations of the homogenisation functions as well as statistical estimators of either upper and lower bounds or direct effective characteristics of the elasticity tensor components.

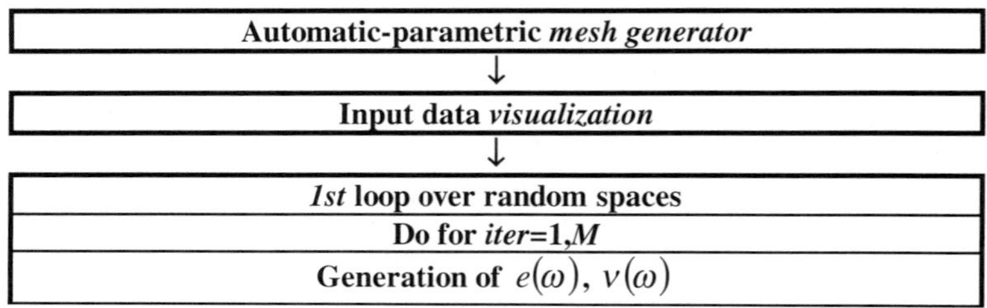

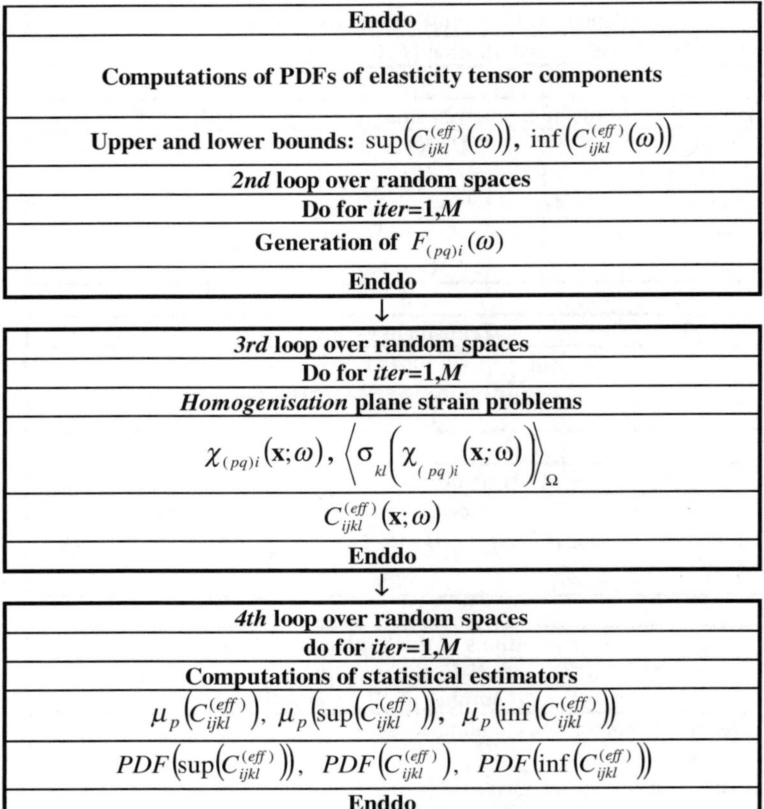

Figure 2.99. Algorithm for the MCS simulation of homogenisation procedure

Numerical analysis of probabilistic homogenisation of the fibre composite with stochastic interface defects has been performed using the MCCEFF system described above. Internal automatic generator for the square RVE with a centrally located round fibre occupying about 50% of the RVE with interface defects has been used (the influence of fibre radius variation on the stochastic displacements and stress fields has been discussed previously). Considering greater composite sensitivity to the matrix defects (bubbles), only composites having such discontinuities have been homogenised. The elastic constants for the fibre material have been taken as follows: $E[e_1]$=84 GPa, v_1=0.22 and the coefficient of Young modulus variation $\alpha(e_1)$=0.1, and for matrix: $E[e_2]$=4 GPa, v_2=0.34. Interface defect parameters have been taken in such a way that the coefficients of variation of these parameters were equal to 0.1 in all tests: $\sigma(r) = 0.1 \cdot E[r]$ and $\sigma(n) = 0.1 \cdot E[n]$.

The main aim of the numerical experiments performed was a numerical verification of the presented mathematical approach to homogenisation of composites with stochastic interface defects. Considering large number of parameters in this approach it was necessary to analyse the probabilistic sensitivity of the effective elasticity tensor components. It was done with respect to the expected values of the interface defect number and volume and the coefficient of matrix Young moduli variation as design parameters. Finally, 132 simulations have been performed (with 1000 iterations each) with the following remaining input values: $E[r]=R\{0.03,0.04,0.05\}$ and $E[n]$ has been assumed as equivalent to the percentage ratio of the boundary where the defects are located to the total interface length from 10% to 60% every 5%. The coefficient of matrix Young modulus variation for tests No 1–4 has been taken as 0.100, 0.075, 0.050, 0.025, respectively.

Probabilistic moments of the effective elasticity tensor obtained as a result of the simulations are compared in Figures 2.100–2.119. The expected values of $C_{1111}^{(eff)}(\omega)$ are shown in such a way that the test results are presented in increasing order in the relevant figures. The coefficients of variation of $C_{1212}^{(eff)}(\omega)$ are neglected in the sensitivity analysis because this random variable is a function of random fluctuations of the fibre Young modulus. In all the collected figures the ratio of interface discontinuities (DB) to the entire boundary is marked on the horizontal axes, while the expected values $E\left[C_{ijkl}^{(eff)}(\omega)\right]$ or the coefficients of variation $\alpha\left(C_{ijkl}^{(eff)}(\omega)\right)$ are displayed on the vertical axes, respectively.

A decrease of the expected values of $C_{ijkl}^{(eff)}(\omega)$ with an increase of the interface defects number is observed with generally small differences in comparison with the composite with perfect interface. For an increase of the parameter DB from 10% to 60%, the decrease considered is about 10% for $E\left[C_{1111}^{(eff)}(\omega)\right]$ and $E\left[C_{1122}^{(eff)}(\omega)\right]$ components, while for $E\left[C_{1212}^{(eff)}(\omega)\right]$ it is only 1%. The low sensitivity of the values for $E\left[C_{ijkl}^{(eff)}(\omega)\right]$ obtained with respect to the coefficient of the matrix Young modulus variation seems to be very important, as well. Moreover, it can be noted that for an increase of the expected values of the interface defects, the values of $E\left[C_{1111}^{(eff)}(\omega)\right]$ and $E\left[C_{1122}^{(eff)}(\omega)\right]$ increase too, and $E\left[C_{1212}^{(eff)}(\omega)\right]$ – decreases. Finally, the increasing DB implies a decrease in the differences of $E\left[C_{1111}^{(eff)}(\omega)\right]$ and $E\left[C_{1122}^{(eff)}(\omega)\right]$ obtained for different defects values, while for $E\left[C_{1212}^{(eff)}(\omega)\right]$ these differences increase with the increasing total number of the defects.

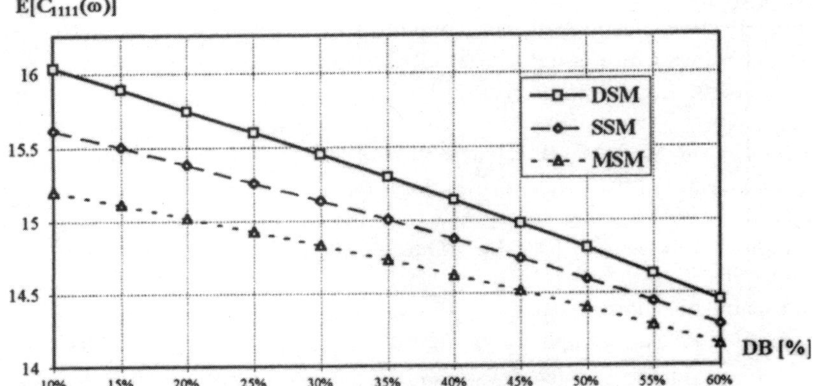

Figure 2.100. Expected values $E\left[C_{1111}^{(eff)}(\omega)\right]$ in test 1

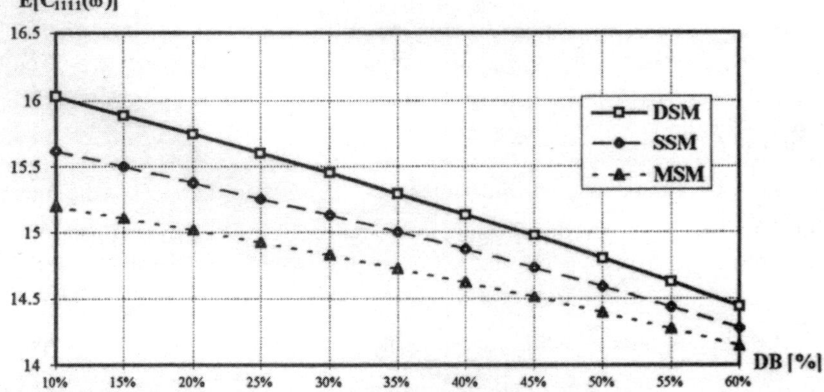

Figure 2.101. Expected values $E\left[C_{1111}^{(eff)}(\omega)\right]$ in test 2

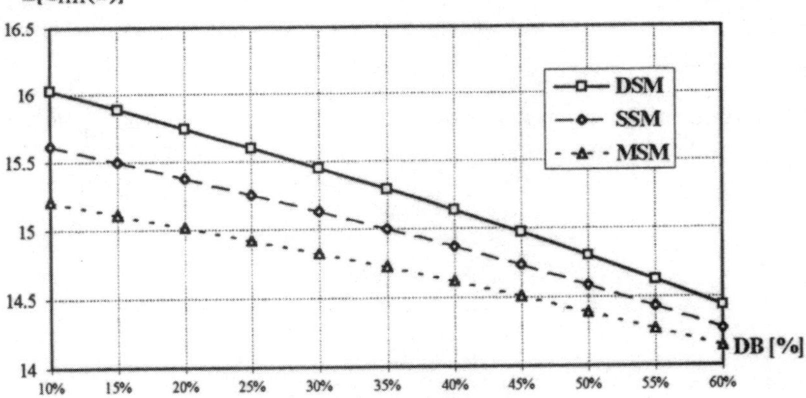

Figure 2.102. Expected values $E\left[C_{1111}^{(eff)}(\omega)\right]$ in test 3

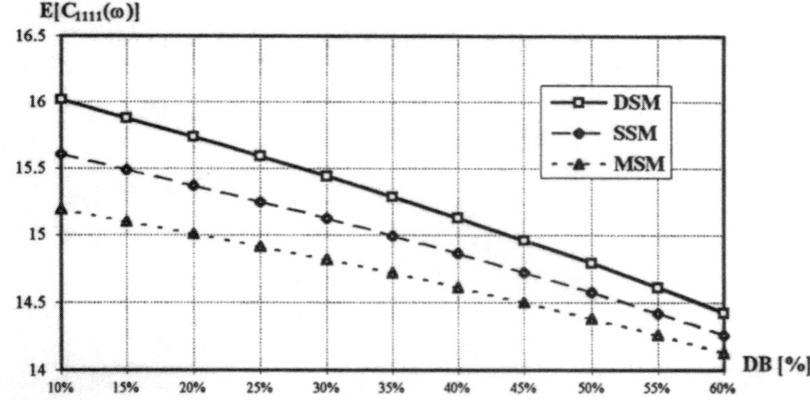

Figure 2.103. Expected values $E\left[C_{1111}^{(eff)}(\omega)\right]$ in test 4

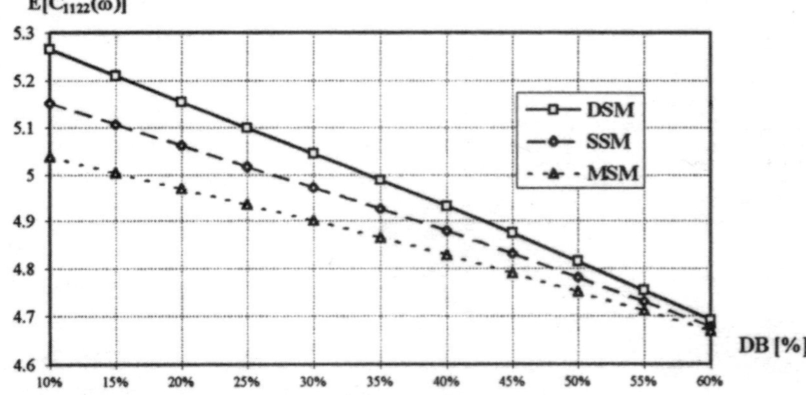

Figure 2.104. Expected values $E\left[C_{1122}^{(eff)}(\omega)\right]$ in test 1

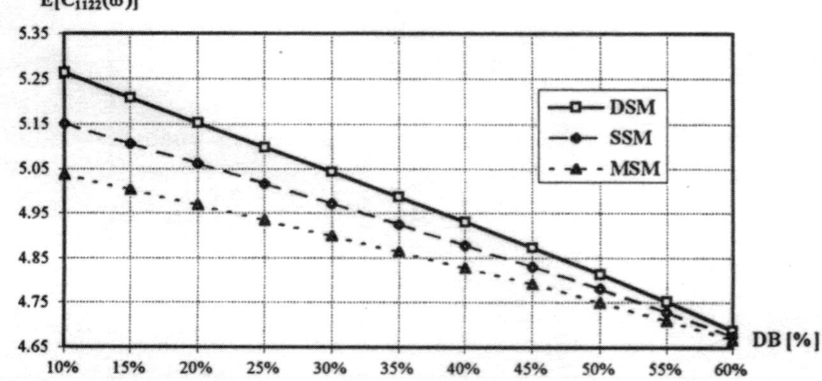

Figure 2.105. Expected values $E\left[C_{1122}^{(eff)}(\omega)\right]$ in test 2

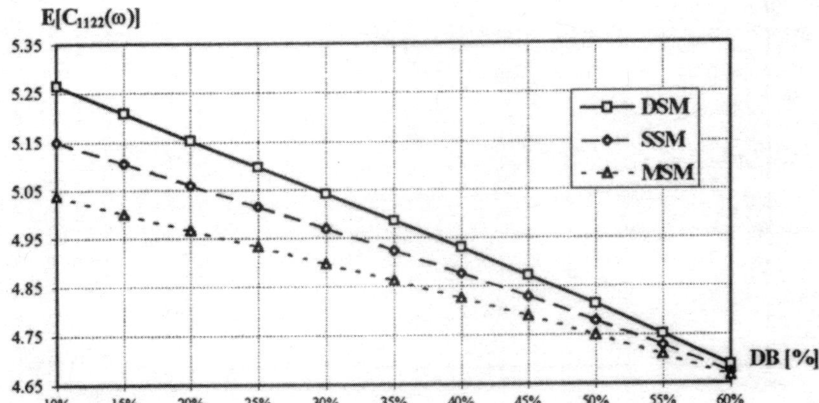

Figure 2.106. Expected values $E\left[C_{1122}^{(eff)}(\omega)\right]$ in test 3

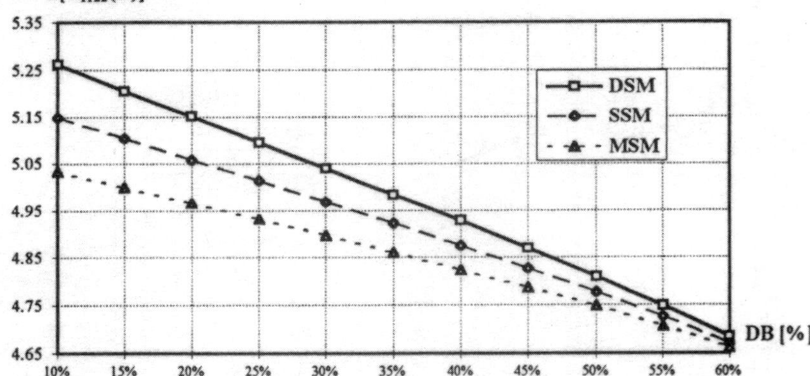

Figure 2.107. Expected values $E\left[C_{1122}^{(eff)}(\omega)\right]$ in test 4

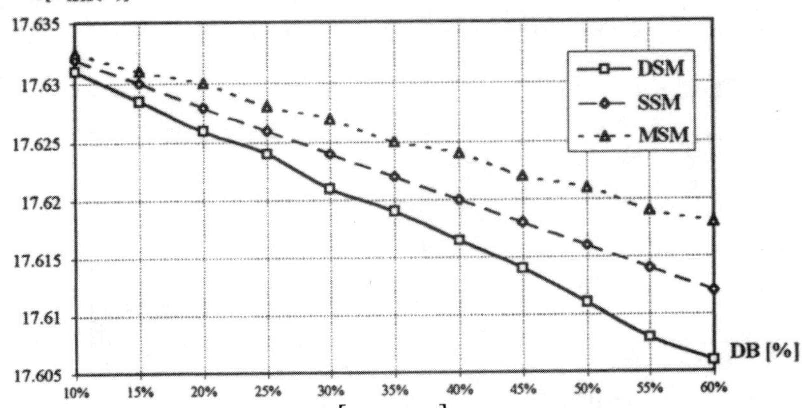

Figure 2.108. Expected values $E\left[C_{1212}^{(eff)}(\omega)\right]$ in test 1

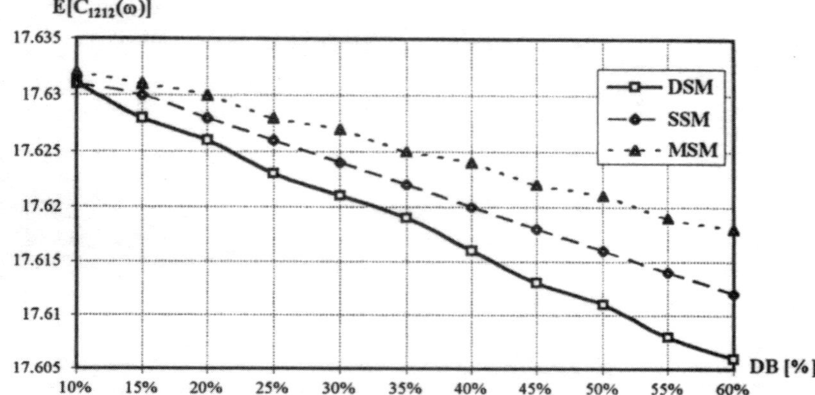

Figure 2.109. Expected values $E\left[C_{1212}^{(eff)}(\omega)\right]$ in test 2

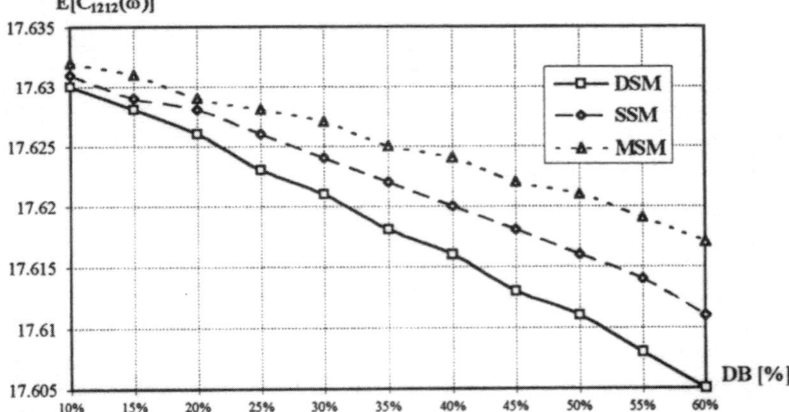

Figure 2.110. Expected values $E\left[C_{1212}^{(eff)}(\omega)\right]$ in test 3

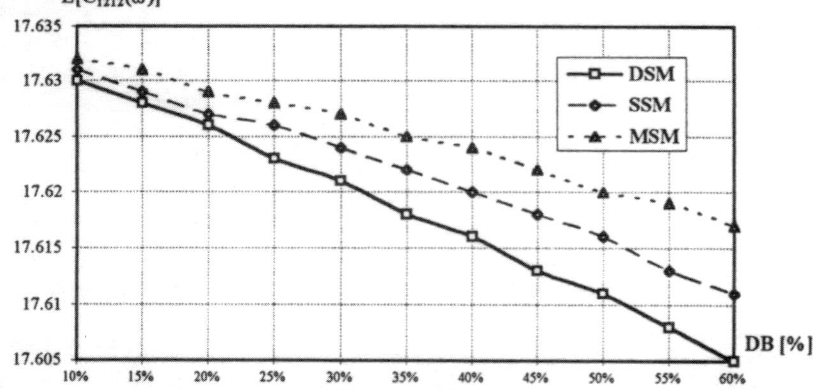

Figure 2.111. Expected values $E\left[C_{1212}^{(eff)}(\omega)\right]$ in test 4

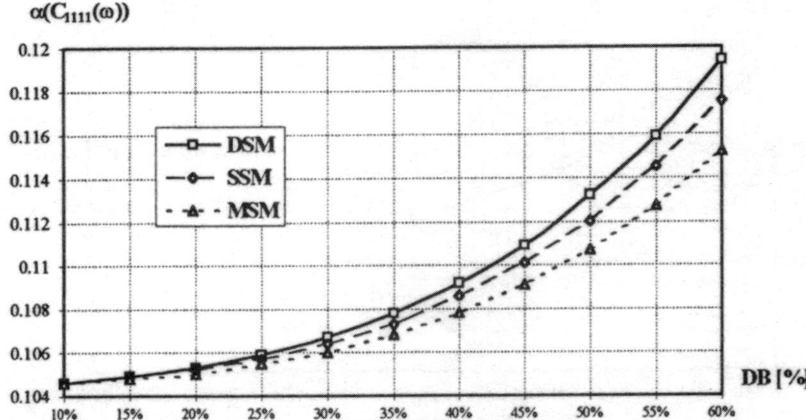

Figure 2.112. Coefficients of variation $\alpha\left(C_{1111}^{(eff)}(\omega)\right)$ in test 1

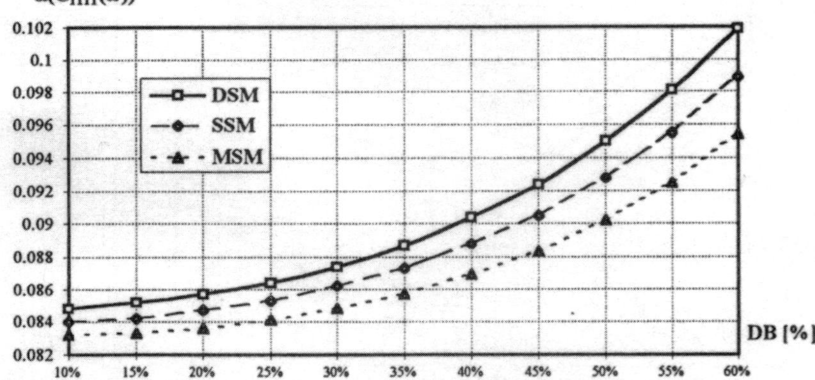

Figure 2.113. Coefficients of variation $\alpha\left(C_{1111}^{(eff)}(\omega)\right)$ in test 2

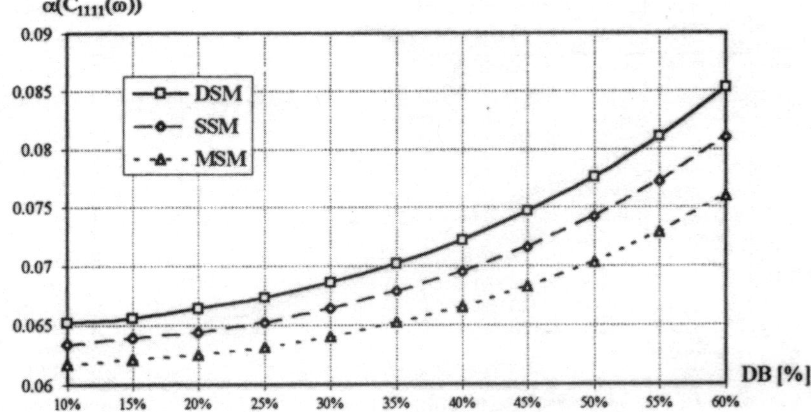

Figure 2.114. Coefficients of variation $\alpha\left(C_{1111}^{(eff)}(\omega)\right)$ in test 3

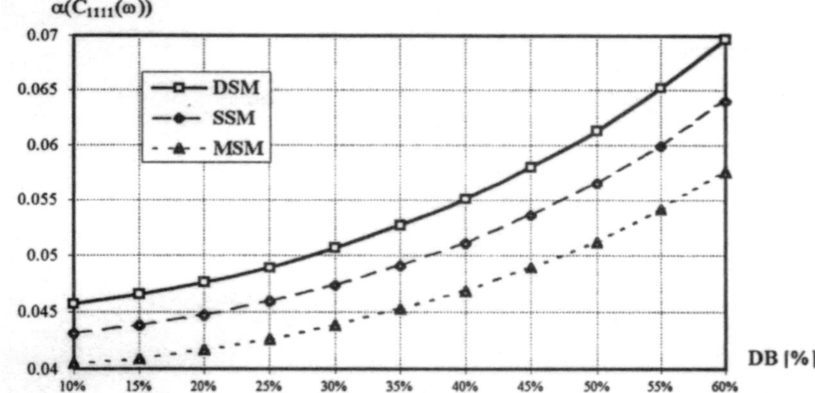

Figure 2.115. Coefficients of variation $\alpha\!\left(C_{1111}^{(eff)}(\omega)\right)$ in test 4

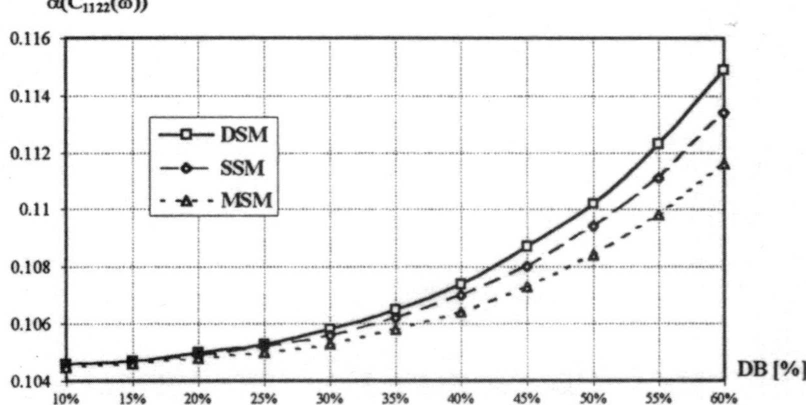

Figure 2.116. Coefficients of variation $\alpha\!\left(C_{1122}^{(eff)}(\omega)\right)$ in test 1

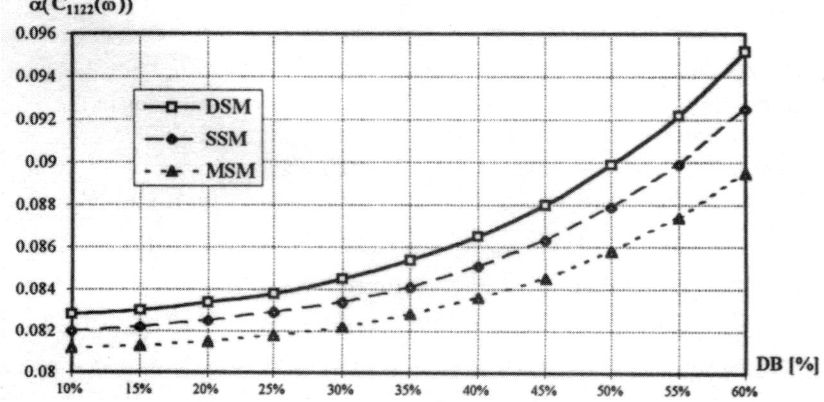

Figure 2.117. Coefficients of variation $\alpha\!\left(C_{1122}^{(eff)}(\omega)\right)$ in test 2

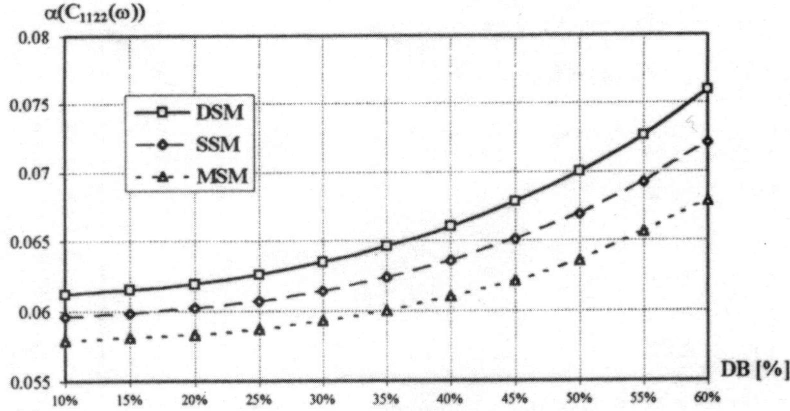

Figure 2.118. Coefficients of variation $\alpha\left(C_{1122}^{(eff)}(\omega)\right)$ in test 3

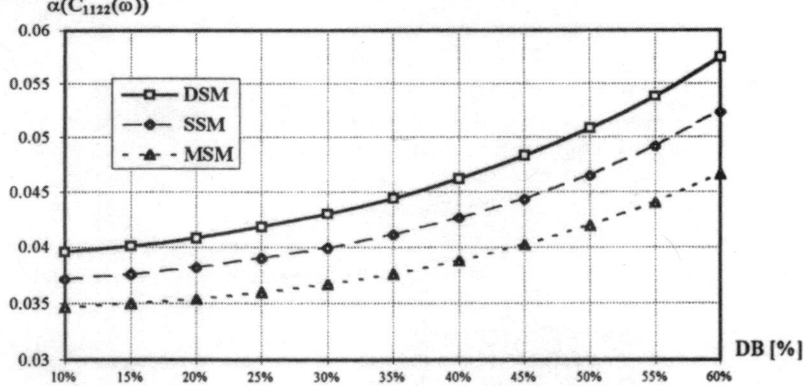

Figure 2.119. Coefficients of variation $\alpha\left(C_{1122}^{(eff)}(\omega)\right)$ in test 4

Analysing the coefficients of variation $\alpha\left(C_{ijkl}^{(eff)}(\omega)\right)$, a nonlinear increase of these coefficients with a DB increase can be observed in all tests. This dependence has a character similar to the behaviour of the coefficient of variation of the Young modulus obtained during the interphase probabilistic averaging. Moreover, all results are in the range of [0.00,0.12] for all the numerical tests, being negligibly greater than the maximum value of the input parameter $\alpha(e_2)$. Furthermore, the correlation of interface defect value increases and an $\alpha\left(C_{ijkl}^{(eff)}(\omega)\right)$ increase is observed, and in opposition to the expected values, the coefficients of the $C_{ijkl}^{(eff)}(\omega)$ tensor variation are sensitive to $\alpha(e_2)$ changes. Together with the decreasing coefficients of the matrix Young modulus variation the following changes are observed:

– decrease of $\alpha\left(C_{1111}^{(eff)}(\omega)\right)$ and $\alpha\left(C_{1122}^{(eff)}(\omega)\right)$;

– increase of differences between these coefficients obtained for particular values of interface defects;

– significantly faster increase of $\alpha\left(C_{ijkl}^{(eff)}(\omega)\right)$ (from 10% in test no 1 to about 30% in test no 4).

The coefficients $\alpha\left(C_{1212}^{(eff)}(\omega)\right)$ (not considered in the analysis) show total non-sensitivity to analysed parameters.

Further, taking into account that all the results obtained from the Monte Carlo simulations, e.g. the first two probabilistic moments of the effective elasticity tensor, are only statistical estimators of the real values of these parameters, the numerical sensitivity of these estimators to the number of iterations should be analysed. Such an analysis is performed on the periodicity cell taking the total number of random trials as $N=5$, 10, 25, 50, 100, 250, 500, 1000, 2500, 5000 and 10000, respectively.

Only the probabilistic parameters of $C_{1111}^{(eff)}(\omega)$ are shown, because variations of the other component moments of $C_{ijkl}^{(eff)}(\omega)$ are quite similar to those presented. The total numbers of random number sampling are marked on the horizontal axes, while the analysed values of $C_{ijkl}^{(eff)}(\omega)$ are on the vertical axes. The functions describing convergence of particular estimators obtained in the numerical experiments enable us to verify the correctness of the simulations performed and come up with an optimum number of the samples for estimation of any probabilistic coefficient and/or moment for the tensor $C_{ijkl}^{(eff)}(\omega)$.

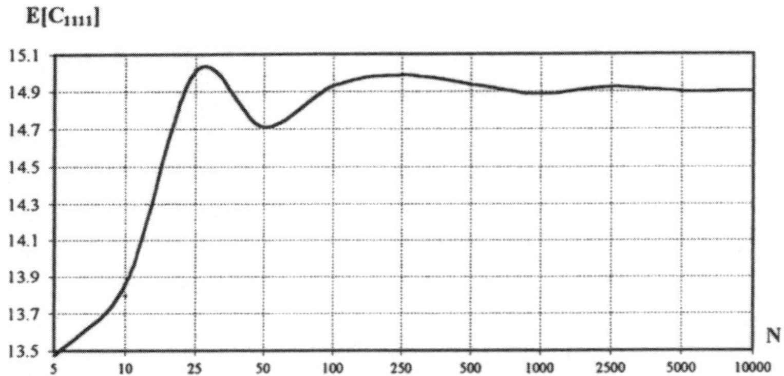

Figure 2.120. Statistical convergence of the expected value $E\left[C_{1111}^{(eff)}(\omega)\right]$

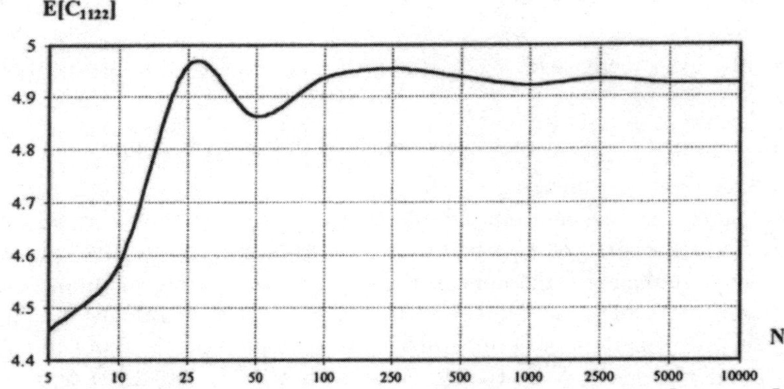

Figure 2.121. Statistical convergence of the expected value $E\left[C_{1122}^{(eff)}(\omega)\right]$

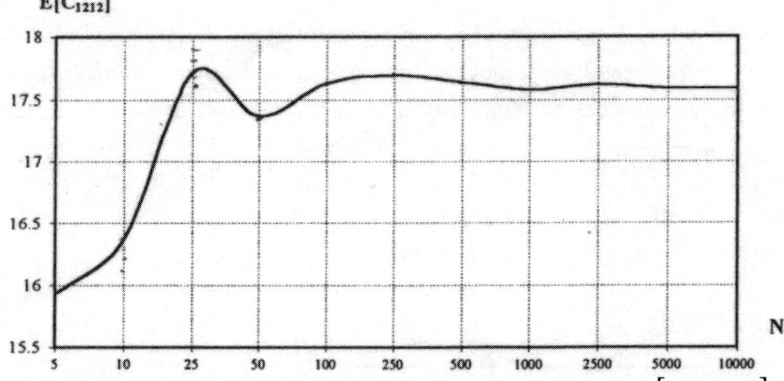

Figure 2.122. Statistical convergence of the expected value $E\left[C_{1212}^{(eff)}(\omega)\right]$

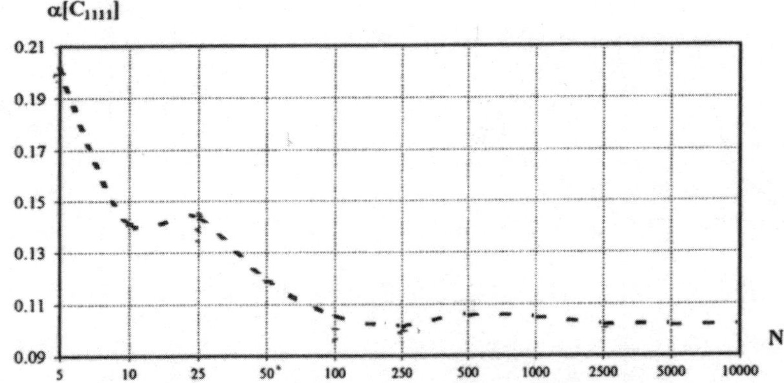

Figure 2.123. Statistical convergence of coefficient of variation $\alpha\left(C_{1111}^{(eff)}(\omega)\right)$

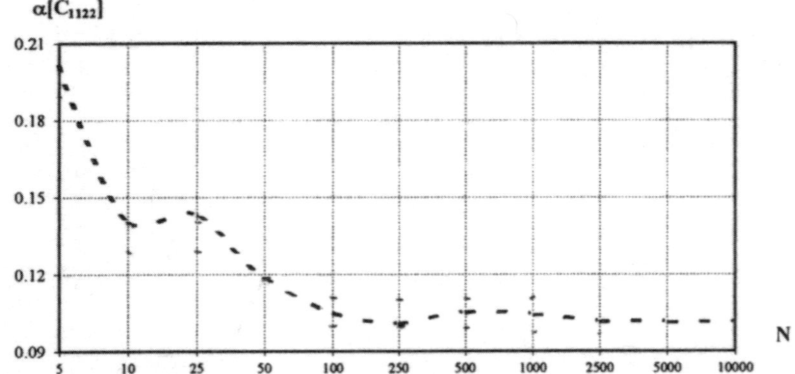

Figure 2.124. Statistical convergence of coefficient of variation $\alpha\left(C_{1122}^{(eff)}(\omega)\right)$

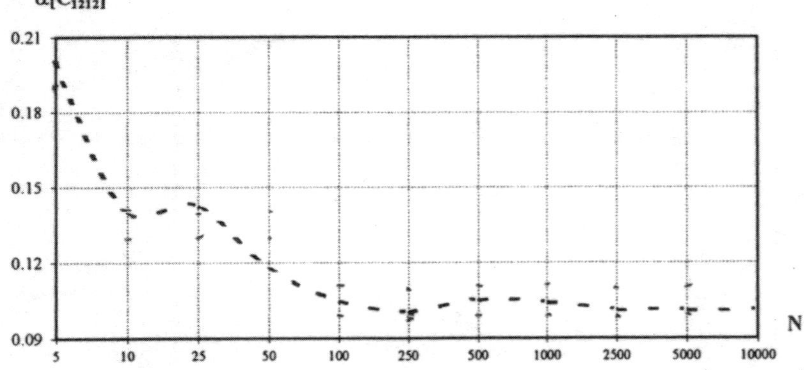

Figure 2.125. Statistical convergence of coefficient of variation $\alpha\left(C_{1212}^{(eff)}(\omega)\right)$

It is seen from the analysis of the expected values of $C_{ijkl}^{(eff)}(\omega)$ that the estimator convergence character is described by a curve of similar shape in all the tests. This curve gradually increases from a minimum at $N=5$ to a maximum at about $N=30$ to oscillate with asymptotic convergence to the value approximated. It is important that in practice for $N=100$ estimator gives quite a good estimation with satisfactory accuracy. Taking for example $N=1000$, computational error resulting from statistical estimation is negligibly small in comparison with the estimated value.

Convergence of $\alpha\left(C_{ijkl}^{(eff)}(\omega)\right)$ estimators has quite a different character than for $E\left[C_{ijkl}^{(eff)}(\omega)\right]$ estimators described above. From the maximum obtained for $N=5$ the curve describing the estimator as a function of the total number of iterations decreases between two inflection points for about $N=10$ and $N=30$, then for about $N=100$ it starts to converge asymptotically to the approximated quantity. Analogous to the expected values the shape of the analysed curves is quite similar each time for different tests and different effective elasticity tensor components. Finally, a good approximation is obtained for $N=100$, while for $N=1000$ the computational error is negligibly small.

As can be seen in Figures 2.126 and 2.127, the total number of random trials necessary in the simulation for precise enough determination of the PDF for $C_{1111}^{(eff)}(\omega)$ is even greater than, for example 5,000–10,000.

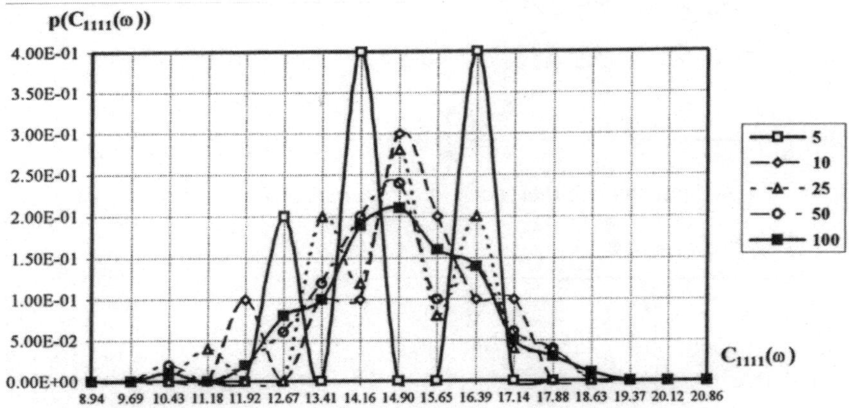

Figure 2.126. Statistical convergence of PDF of $C_{1111}^{(eff)}(\omega)$

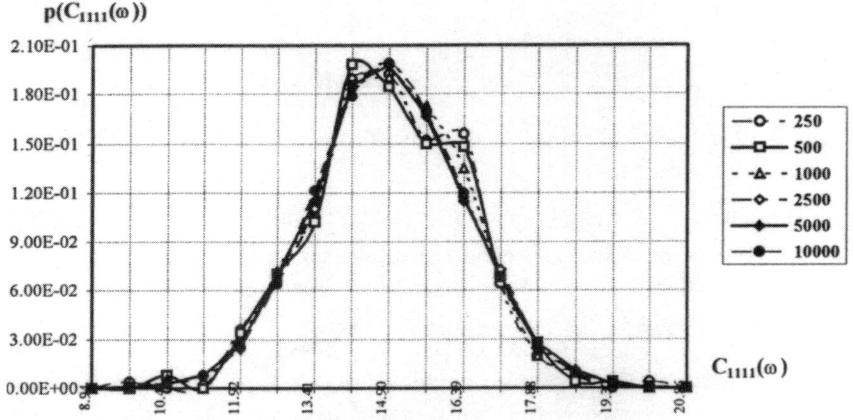

Figure 2.127. Statistical convergence of PDF of $C_{1111}^{(eff)}(\omega)$

The main idea behind performing further numerical experiments is to compute the expected values and variances (or the coefficients of variation) of the effective elasticity tensor components for the RVE of the superconducting coil cable [199,221]. Next aim is to check the variability of the effective characteristic probabilistic moments with respect to the moments of the input random variables.

Probabilistic effective characteristics are compared with the appropriate upper and lower bounds probabilistic moments for the same composite specimen.

Due to the internal horizontal and vertical symmetry of the RVE, only a simple quarter of the periodicity cell has been analysed in the homogenisation procedure for the discretisation of this cell shown before.

Elastic characteristics and their probabilistic moments of the RVE components in the form of the expected values and the standard deviations of Young moduli and Poisson ratios as well as of the Kirchhoff moduli are collected in Table 2.14.

Table 2.14. Probabilistic moments of the elastic characteristics of the superconductor

Material	$E[e]$ [GPa]	$\sigma(e)$ [GPa]	$E[v]$	$\sigma(v)$	$E[G]$ [GPa]	$\sigma(G)$ [GPa]
Tube	205.0	8.0	0.265	0.010	81.0	2.0
Superconductor (test 1)	130.0	0.0	0.340	0.000	70.0	0.0
Superconductor (test 2)	46.8	0.0	0.122	0.000	25.2	0.0
Jacket	126.0	12.0	0.311	0.012	48.0	6.0
Insulation	36.0	0.0	0.210	0.000	11.0	0.0

Three groups of computational experiments have been performed. It is assumed that all elastic characteristics are equal to those specified in Table 2.14 in the first and second groups of computations (tests 1 and 2), while the elastic parameters of the superconducting strands are omitted in the last test. The strand volume fraction in the plane considered is assumed in test 1 as equal to 100%, while in the test 2 it is assumed equal to 36% (approximately the real value). The elastic characteristics of the strands for the second case are calculated using of spatial averaging only. These characteristics can be derived by some homogenisation approach (Mori–Tanaka or self–consistent, for instance) if only the longitudinal elastic modulae are measured statistically.

The results of numerical analyses are presented in Tables 2.15–2.20. Upper and lower bounds as well as the effective elastic properties for test 1 are collected in Tables 2.15 and 2.16, respectively, for test 2 they are outlined in Tables 2.17 and 2.18, while for test 3 they are outlined in Tables 2.19–2.20. Deterministic values of the effective elasticity tensor and their up to fourth order probabilistic characteristics (expected values, coefficients of variation, asymmetry and concentration) are shown for all these tests.

Table 2.15. Effective elasticity tensor components [GPa] in test 1

Effective characteristics	$C^{(eff)}_{1111}$	$C^{(eff)}_{1212}$	$C^{(eff)}_{1122}$
Deterministic values	154.94	68.85	43.67
$E[C]$	154.27	68.52	43.94
$\alpha(C)$	5.56e-2	5.44e-2	5.76e-2
$\beta(C)$	-2.06e-1	-2.41e-1	9.98e-2
$\gamma(C)$	3.27	3.29	3.15

Table 2.16. Upper and lower bounds for effective elasticity tensor [GPa] in test 1

Effective characteristics	$C_{1111}^{(eff)}$		$C_{1212}^{(eff)}$		$C_{1122}^{(eff)}$	
Upper and lower bounds presented						
	sup(C)	inf(C)	sup(C)	inf(C)	Sup(C)	inf(C)
Deterministic values	163.49	146.47	75.56	63.27	43.97	41.60
E[C]	163.60	146.18	75.81	63.16	43.89	41.51
$\alpha(C)$	6.89e-2	5.76e-2	9.78e-2	8.14e-2	4.42e-2	3.95e-2
$\beta(C)$	1.79e-7	-1.04e-7	3.32e-7	-1.12e-8	-1.15e-7	-2.51e-7
$\gamma(C)$	3.09	3.06	3.20	3.02	3.07	3.17
Voigt-Reuss bounds						
Deterministic values	171.49	130.33	80.95	52.63	45.27	38.85
E[C]	171.88	129.97	81.43	52.46	45.23	38.76
$\alpha(C)$	6.78e-2	4.72e-2	9.29e-2	6.60e-2	4.54e-2	3.45e-2
$\beta(C)$	3.23e-7	-2.51e-7	5.15e-7	-1.75e-7	-2.30e-8	-2.50e-7
$\gamma(C)$	3.26	3.17	3.54	3.09	3.03	3.30

Table 2.17. Effective elasticity tensor components [GPa] in test 2

Effective characteristics	$C_{1111}^{(eff)}$	$C_{1212}^{(eff)}$	$C_{1122}^{(eff)}$
Deterministic values	102.33	36.47	33.69
E[C]	102.50	36.69	33.49
$\alpha(C)$	5.83e-2	5.90e-2	6.38e-2
$\beta(C)$	-1.86e-1	-1.92e-1	-9.96e-2
$\gamma(C)$	3.23	3.25	3.15

Table 2.18. Upper and lower bounds for effective elasticity tensor [GPa] in test 2

Effective characteristics	$C_{1111}^{(eff)}$		$C_{1212}^{(eff)}$		$C_{1122}^{(eff)}$	
Upper and lower bounds presented						
	sup(C)	inf(C)	sup(C)	inf(C)	sup(C)	inf(C)
Deterministic values	100.24	82.24	35.21	22.74	32.52	29.75
E[C]	100.37	82.05	35.45	22.68	32.46	29.69
$\alpha(C)$	8.18e-2	4.11e-2	1.40e-1	5.84e-2	4.99e-2	3.46e-2
$\beta(C)$	2.12e-7	-2.38e-7	4.16e-7	-1.58e-7	-9.73e-8	-2.89e-7
$\gamma(C)$	3.16	3.15	3.38	3.08	3.06	3.21
Voigt-Reuss bounds						
Deterministic values	113.11	71.80	43.86	16.64	34.63	27.58
E[C]	113.50	71.65	44.34	16.61	34.58	27.52
$\alpha(C)$	1.03e-2	2.48e-2	1.71e-1	2.57e-2	5.94e-2	2.46e-2
$\beta(C)$	3.23e-7	-4.13e-7	5.15e-7	-4.02e-7	-2.30e-8	-4.17e-7
$\gamma(C)$	3.26	3.40	3.54	3.38	3.03	3.41

Table 2.19. Effective elasticity tensor components [GPa] in test 3

Effective Characteristics	$C^{(eff)}_{1111}$	$C^{(eff)}_{1212}$	$C^{(eff)}_{1122}$
Deterministic values	75.07	30.15	25.89
$E[C]$	75.09	30.29	25.38
$\alpha(C)$	9.29e-2	1.06e-1	6.94e-2
$\beta(C)$	-1.14e-1	-6.40e-2	-9.97e-2
$\gamma(C)$	3.16	3.15	3.17

Tab. 2.20. Upper and lower bounds for effective elasticity tensor [GPa] in test 3

Effective characteristics	$C^{(eff)}_{1111}$		$C^{(eff)}_{1212}$		$C^{(eff)}_{1122}$	
Upper and lower bounds presented						
	sup(C)	inf(C)	sup(C)	inf(C)	sup(C)	inf(C)
Deterministic values	73.50	4.02	30.47	4.64e-2	21.51	1.984
$E[C]$	73.34	4.02	30.37	4.64e-2	21.49	1.98
$\alpha(C)$	1.03e-1	2.34e-3	1.57e-1	3.34e-2	6.56e-3	2.75e-3
$\beta(C)$	2.42e-7	-5.64e-7	4.30e-7	5.57e-7	-7.35e-8	-5.67e-7
$\gamma(C)$	3.182	3.730	3.398	3.704	3.049	3.729
Voigt-Reuss bounds						
Deterministic values	94.84	2.55	41.27	1.23e-2	26.79	1.27
$E[C]$	95.23	2.55	41.74	1.23e-2	26.74	1.27
$\alpha(C)$	1.22e-1	9.72e-4	1.81e-1	3.55e-2	7.67e-2	1.15e-3
$\beta(C)$	3.23e-7	-5.80e-7	5.15e-7	5.79e-7	-2.30e-8	-5.89e-7
$\gamma(C)$	3.26	3.77	3.54	3.76	3.03	3.77

First a general observation, which agrees with engineering intuition, is that the deterministic quantities and expected values for upper and lower bounds and effective elasticity tensor components are greater for test 1 (composite including superconductor) than for test 2 (the cell without superconducting strands). Further, it is seen that the results of deterministic analyses approximate very well the expected values obtained in probabilistic simulations and that deterministic results are generally lower than the approximated expectations.

Analysing the coefficients of variation of all variables computed it is characteristic that the results of test 1 are significantly smaller than the input coefficients and the coefficients resulting from test 2. It is caused mainly by the fact that some of the input elastic characteristics including superconductor have the coefficients of variation equal to 0. Considering that the superconductor occupies a significant part of the periodicity cell, the coefficients α resulting from test 2 are in the range of those characterising the elastic properties of composite components. It should be outlined at the moment that probabilistic moments of effective characteristics of order higher than the second are in general in the range of the corresponding characteristics of the input elastic parameters in the probabilistic homogenisation of elastostatic problems.

Observing characteristics of the third and fourth order it may be concluded that the upper and lower bounds of the effective tensor in both tests have symmetric probability density functions, while the effective characteristics PDFs show some asymmetry. Finally, it can be observed that the coefficients of concentration are approximately equal to the value corresponding to the Gaussian variable probability distribution function.

Considering these observations we can treat the probability density functions of the effective elastic characteristics as Gaussian, which enables us to characterise uniquely these distributions using only their first two probabilistic moments. This conclusion is very important in the context of the SFEM implementation of the problem where only the first two moments of the state functions can be computed and, furthermore, all odd moments are equal to 0.

2.3.2.2.3 Stochastic Perturbation Approach to the Homogenisation

The homogenisation technique presented in the preceding sections is combined now with the stochastic second order perturbation second central probabilistic moment method. To rewrite the stochastic version of the variational formulation of the homogenisation problem, the interface forces equivalent to the stress interface conditions should be stochastically perturbed first. It is known from the classical theory of homogenisation that in case of ideal bonds between the fibre and matrix, the interface load components are obtained in the form of the following difference, cf. (2.155)

$$F_{(pq)i} = F_{(pq)i}^{(2)} - F_{(pq)i}^{(1)}$$
(2.176)

Taking into account the general Taylor series expansion it is found that

$$F_{(pq)i} = \left(F_{(pq)i}\right)^0 + \theta \left(F_{(pq)i}\right)^{,r} \Delta b^r + \tfrac{1}{2}\theta^2 \left(F_{(pq)i}\right)^{,rs} \Delta b^r \Delta b^s$$
(2.177)

Rewriting the forces $F_{(pq)i}^{(t)}$ for $t=0,1,2$, comparing the respective terms of zeroth, first and second order, it is obtained after some additional algebra that

$$\left(F_{(pq)i}\right)^0 = \left(F_{(pq)i}^{(2)}\right)^0 - \left(F_{(pq)i}^{(1)}\right)^0$$
(2.178)

$$\left(F_{(pq)i}\right)^{,r} = \left(F_{(pq)i}^{(2)}\right)^{,r} - \left(F_{(pq)i}^{(1)}\right)^{,r}$$
(2.179)

$$\left(F_{(pq)i}\right)^{,rs} = \left(F_{(pq)i}^{(2)}\right)^{,rs} - \left(F_{(pq)i}^{(1)}\right)^{,rs}$$
(2.180)

Thus, the stochastic version of minimum potential energy principle for the homogenisation problem has the following form:
- a single zeroth order equation:

$$\sum_{a=1,2} \int_{\Omega_a} \delta v_{i,j} C^0_{ijkl} \left(\chi_{(pq)k,l} \right)^p d\Omega = - \int_{\partial\Omega_{12}} \delta v_i \left(F_{(pq)i} \right)^p d(\partial\,\Omega) \qquad (2.181)$$

- R first order equations:

$$\sum_{a=1,2} \int_{\Omega_a} \delta v_{i,j} C^0_{ijkl} \left(\chi_{(pq)k,l} \right)^{,r} d\Omega \qquad (2.182)$$

$$= - \int_{\partial\Omega_{12}} \delta v_i \left(F_{(pq)i} \right)^{,r} d(\partial\Omega) - \sum_{a=1,2} \int_{\Omega_a} \delta v_{i,j} C^{,r}_{ijkl} \left(\chi_{(pq)k,l} \right)^0 d\Omega$$

- a single second order equation:

$$\left(\sum_{a=1,2} \int_{\Omega_a} \delta v_{i,j} C^0_{ijkl} \left(\chi_{(pq)k,l} \right)^{,rs} d\Omega \right) Cov(b^r, b^s)$$

$$= - \left(\int_{\partial\Omega_{12}} \delta v_i \left(F_{(pq)i} \right)^{,rs} d(\partial\Omega) \right) Cov(b^r, b^s) \qquad (2.183)$$

$$- \left(2 \sum_{a=1,2} \int_{\Omega_a} \delta v_{i,j} C^{,r}_{ijkl} \left(\chi_{(pq)k,l} \right)^{,s} d\Omega + \sum_{a=1,2} \int_{\Omega_a} \delta v_{i,j} C^{,rs}_{ijkl} \left(\chi_{(pq)k,l} \right)^0 d\Omega \right)$$

$$\times Cov(b^r, b^s)$$

If the Young moduli of fibre and matrix are the components of the input random variable vector then there holds

$$\frac{\partial \left(C_{ijkl} \left(e(\mathbf{x};\omega); \mathbf{x} \right) \right)}{\partial e_a} = \psi^{(a)} A^{(a)}_{ijkl}(x), \quad \text{for } a=1,2 \qquad (2.184)$$

where $A^{(a)}_{ijkl}$ is the tensor given by (2.14) and calculated for the elastic characteristics of the respective material indexed by a, whereas $\psi^{(a)}$ is the characteristic function. Thus, the first order derivatives of the elasticity tensor with respect to the input random variable vector are obtained as

$$\frac{\partial \left(C_{ijkl} \left(e(\mathbf{x};\omega); \mathbf{x} \right) \right)}{\partial e_a} = \left\{ \Psi^{(1)} A^{(1)}_{ijkl}, \Psi^{(2)} A^{(2)}_{ijkl} \right\} \qquad (2.185)$$

Hence, the second order derivatives have the form

$$\frac{\partial^2 \left(C_{ijkl}\left(e(\mathbf{x};\omega),\mathbf{x}\right)\right)}{\partial e_a^2} = \psi^{(a)} \frac{\partial A_{ijkl}^{(a)}(x)}{\partial e_a} = 0, \quad \text{for } a=1,2 \tag{2.186}$$

while mixed second order derivatives can be written as

$$\frac{\partial^2 \left(C_{ijkl}\left(e(\mathbf{x};\omega),\mathbf{x}\right)\right)}{\partial e_1 \partial e_2} = \psi^{(1)} \frac{\partial A_{ijkl}^{(1)}(\mathbf{x})}{\partial e_2} = \psi^{(2)} \frac{\partial A_{ijkl}^{(2)}(\mathbf{x})}{\partial e_1} = 0 \tag{2.187}$$

Considering the above, all components of the second order derivatives of the stiffness matrixes $K_{\alpha\beta}^{(pq)}$ in this problem are equal to 0. Moreover, since the assumption of the uncorrelation of input random variables

$$Cov\left(e_1;e_2\right) = \begin{bmatrix} Vare_1 & 0 \\ 0 & Vare_2 \end{bmatrix} \tag{2.188}$$

thus, the first and second partial derivatives of the vectors $F_{(pq)i}^{(a)}$ with respect to the random variables vector are calculated as

$$\frac{\partial F_{(pq)i}^{(a)}}{\partial e_a} = \frac{\partial C_{ijpq}^{(a)}}{\partial e_a} n_j = A_{ijpq}^{(a)} n_j, \quad \mathbf{x} \in \partial \Omega_a, \quad a=1,2 \tag{2.189}$$

and

$$\frac{\partial^2 F_{(pq)i}^{(a)}}{\partial e_a^2} = \frac{\partial^2 C_{ijpq}^{(a)}}{\partial e_a^2} n_j = \frac{\partial A_{ijpq}^{(a)}}{\partial e_a} n_j = 0, \quad \mathbf{x} \in \partial \Omega_a, \quad a=1,2 \tag{2.190}$$

After all these simplifications, the set of equations (2.181) – (2.183) can be written in the following form:
- a single zeroth order equation:

$$\sum_{a=1,2} \int_{\Omega_a} \delta v_{i,j} C_{ijkl}^0 \left(\chi_{(pq)k,l}\right)^0 d\Omega = - \int_{\partial \Omega_{12}} \delta v_i \left(F_{(pq)i}\right)^0 d(\partial\Omega) \tag{2.191}$$

- R first order equations:

$$\sum_{a=1,2} \int_{\Omega_a} \delta v_{i,j} C_{ijkl}^0 \left(\chi_{(pq)k,l}\right)^{,r} d\Omega = - \int_{\partial \Omega_{12}} \delta v_i \left[A_{pqij}\right] n_j d(\partial\Omega)$$
$$- \sum_{a=1,2} \int_{\Omega_a} \delta v_{i,j} A_{ijkl}^{(a)} \left(\chi_{(pq)k,l}\right)^0 d\Omega \tag{2.192}$$

- a single second order equation:

$$\sum_{a=1,2} \int_{\Omega_a} \delta v_{i,j} C_{ijkl}^0 \left(\chi_{(pq)k,l} \right)^{(2)} d\Omega$$

$$= - \sum_{a=1,2} \int_{\Omega_a} \delta v_{i,j} C_{ijkl}^{,r} \left(\chi_{(pq)k,l} \right)^{,s} d\Omega \, Cov(b^r,b^s)$$

$$(2.193)$$

where

$$\left(\chi_{(pq)k,l} \right)^{(2)} = -\tfrac{1}{2} \left(\chi_{(pq)k,l} \right)^{,rs} Cov(b^r,b^s)$$

$$(2.194)$$

It should be noted that $(2.191) - (2.194)$ give the set of fundamental variational equations of the homogenisation problem due to the second order stochastic perturbation method. Next, these equations will be discretised by the use of classical finite element technique and, as a result, the zeroth, first and second order algebraic equations are derived. Further, let us introduce the following discretisation of the homogenisation function and its derivatives with respect to the random variables using the classical shape functions $\varphi_{i\alpha}(\mathbf{x})$:

$$\left(\chi_{(pv)i}(\mathbf{x}) \right)^0 = \varphi_{i\alpha}(\mathbf{x}) \cdot \left(q_{(pv)\alpha} \right)^0, \quad \mathbf{x} \in \Omega, \quad p,v=1,2 \qquad (2.195)$$

$$\left(\chi_{(pv)i}(\mathbf{x}) \right)^{,r} = \varphi_{i\alpha}(\mathbf{x}) \cdot \left(q_{(pv)\alpha} \right)^{,r}, \quad \mathbf{x} \in \Omega, \quad p,v=1,2 \qquad (2.196)$$

$$\left(\chi_{(pv)i}(\mathbf{x}) \right)^{,rs} = \varphi_{i\alpha}(\mathbf{x}) \cdot \left(q_{(pv)\alpha} \right)^{,rs}, \quad \mathbf{x} \in \Omega, \quad p,v=1,2 \qquad (2.197)$$

where $i=1,2$; $r,s=1,...,R$; $\alpha=1,...,N$ (N is the total number of degrees of freedom employed in the region Ω). In an analogous way, the approximation of the strain tensor components is introduced as

$$\varepsilon_{ij}^0 \left(\chi_{(pv)}(\mathbf{x}) \right) = B_{ij\alpha}(\mathbf{x}) \left(q_{(pv)\alpha} \right)^0, \quad \mathbf{x} \in \Omega \qquad (2.198)$$

$$\varepsilon_{ij}^{,r} \left(\chi_{(pv)}(\mathbf{x}) \right) = B_{ij\alpha}(\mathbf{x}) \left(q_{(pv)\alpha} \right)^{,r}, \quad \mathbf{x} \in \Omega \qquad (2.199)$$

$$\varepsilon_{ij}^{,rs} \left(\chi_{(pv)}(\mathbf{x}) \right) = B_{ij\alpha}(\mathbf{x}) \left(q_{(pv)\alpha} \right)^{,rs}, \quad \mathbf{x} \in \Omega \qquad (2.200)$$

where $B_{ij\alpha}(\mathbf{x})$ is the typical FEM shape functions derivatives

$$B_{ij\alpha}(\mathbf{x}) = \tfrac{1}{2} [\varphi_{i\alpha,j}(\mathbf{x}) + \varphi_{j\alpha,i}(\mathbf{x})], \quad \mathbf{x} \in \Omega \qquad (2.201)$$

Introducing equations stated above to the zeroth, first and second order statements of the homogenisation problem represented by $(2.191) - (2.194)$, the stochastic formulation of the problem can be discretised through the following set of algebraic linear (in fact deterministic) equations:

$$K^0 q^0_{(pv)} = Q^0_{(pv)} \tag{2.202}$$

$$K^0 q^{,r}_{(pv)} = Q^0_{(pv)} - K^{,r} q^0_{(pv)} \tag{2.203}$$

$$K^0 q^{(2)}_{(pv)} = -K^{,r} q^{,s}_{(pv)} Cov(b^r, b^s) \tag{2.204}$$

where

$$q^{(2)}_{(pv)} = \tfrac{1}{2} q^{,rs}_{(pv)} Cov(b^r, b^s) \tag{2.205}$$

and $\mathbf{K}$, $\mathbf{q}_{(pv)}$, $\mathbf{Q}_{(pv)}$ denote the global stiffness matrix, generalised coordinates vectors of the homogenisation functions and external load vectors, correspondingly. Considering the plane strain nature of the homogenisation problem, the global stiffness matrix and its partial derivatives with respect to the random variables of the problem can be rewritten as follows:

$$K^0_{\alpha\beta} = \sum_{e=1}^{E} \int_{\Omega_e} C^0_{ijkl} B_{ij\alpha} B_{kl\beta} \, d\Omega$$

$$= \sum_{e=1}^{E} \frac{e(1-v)}{(1+v)(1-2v)} \int_{\Omega_e} \begin{bmatrix} 1 & \frac{v}{1-v} & 0 \\ & 1 & 0 \\ symm & & \frac{1-2v}{2(1-v)} \end{bmatrix} B_{ij\alpha} B_{kl\beta} \, d\Omega \tag{2.206}$$

$$K^{,r}_{\alpha\beta} = \sum_{e=1}^{E} \int_{\Omega_e} C^{,r}_{ijkl} B_{ij\alpha} B_{kl\beta} \, d\Omega$$

$$= \sum_{e=1}^{E} \frac{(1-v)}{(1+v)(1-2v)} \int_{\Omega_e} \begin{bmatrix} 1 & \frac{v}{1-v} & 0 \\ & 1 & 0 \\ symm & & \frac{1-2v}{2(1-v)} \end{bmatrix} B_{ij\alpha} B_{kl\beta} \, d\Omega \tag{2.207}$$

$$K^{,rs}_{\alpha\beta} = \sum_{e=1}^{E} \int_{\Omega_e} C^{,rs}_{ijkl} B_{ij\alpha} B_{kl\beta} \, d\Omega \tag{2.208}$$

as far as Young moduli are randomised only. Computing from the above equations successively the zeroth order displacement vector $q^{(0)}_{(pv)}$ from (2.202), first order displacement vector $q^{,r}_{(pv)}$ from (2.203) and the second order displacement vector $q^{(2)}_{(pv)}$ from (2.204) − (2.205), the expected values of the homogenisation function can be derived as

$$E\big[q_{(pv)}\big] = q^0_{(pv)} + \tfrac{1}{2} q^{,rs}_{(pv)} Cov(b^r, b^s) \tag{2.209}$$

Their covariance matrix can be determined in the form

$$Cov\big(q_{(pv)r}, q_{(pv)s}\big) = q^{;r}_{(pv)} q^{;s}_{(pv)} Cov(b^r, b^s) \tag{2.210}$$

where α, β are indexing all the degrees of freedom of the RVE. Then, the expected values of the stress tensor components can be expressed as

$$E\big[\sigma_{ij}^{(e)}\big] = \Big\{ C_{ijkl}^{(e)0}\,(q_{(pv)}^0 + \tfrac{1}{2} q_{(pv)}^{rs}) + C_{ijkl}^{(e),r}\, q_{(pv)}^{;s} \Big\} B_{kl}^{(e)}\, Cov(b^r, b^s) \tag{2.211}$$

while its covariances − from the following equation:

$$\begin{aligned}
Cov\big(\sigma_{ij}^{(e)}, \sigma_{ij}^{(f)}\big) &= B_{kl}^{(e)} B_{mn}^{(f)} Cov(b^r, b^s) \\
&\Big\{ C_{ijkl}^{(e)0} C_{ijmn}^{(f)0} q_{(pv)}^{;r} q_{(pv)}^{;s} + C_{ijkl}^{(e),r} C_{ijmn}^{(f),s} q_{(pv)}^0 q_{(pv)}^0 \\
&+ C_{ijkl}^{(e),r} C_{ijmn}^{(f)0} q_{(pv)}^{;s} q_{(pv)}^0 + C_{ijkl}^{(e)0} C_{ijmn}^{(f),r} q_{(pv)}^0 q_{(pv)}^{;s} \Big\}
\end{aligned} \tag{2.212}$$

where $i,j,k,l,g,h,p,v=1,2;\ 1 \le d, f \le E$ standing for the finite elements numbers in the cell mesh. In accordance with the probabilistic homogenisation methodology, the expected values of the elasticity tensor components can be found starting from (2.136) as

$$E\big[C_{ijpq}^{(eff)}\big] = \frac{1}{|\Omega|} \int_{\Omega} \Big(E\big[C_{ijpq}\big] + E\big[C_{ijkl}\varepsilon_{kl}\big(\chi_{(pq)}\big)\big] \Big)\, d\Omega \tag{2.213}$$

The second term in this integral can be extended using second order perturbation method as follows:

$$\begin{aligned}
&E\big[C_{ijkl}\varepsilon_{kl}\big(\chi_{(pq)}\big)\big] \\
&= \int_{-\infty}^{+\infty} \Big(C_{ijkl}^0 + \Delta b^r C_{ijkl}^{;r} + \tfrac{1}{2}\Delta b^r \Delta b^s C_{ijkl}^{;rs} \Big) p_R\big(\mathbf{b}(x)\big)\, d\mathbf{b} \\
&\times \int_{-\infty}^{+\infty} \Big(\big(\chi_{(pq)k,l}\big)^0 + \Delta b^u \big(\chi_{(pq)k,l}\big)^{,u} + \tfrac{1}{2}\Delta b^u \Delta b^v \big(\chi_{(pq)k,l}\big)^{,uv} \Big) p_R\big(\mathbf{b}(x)\big)\, d\mathbf{b}
\end{aligned} \tag{2.214}$$

There holds

$$E\left[C_{ijkl}\varepsilon_{kl}\left(\chi_{(pq)}\right)\right]= \int_{-\infty}^{+\infty} C_{ijkl}^{0}\left(\chi_{(pq)k,l}\right)^{0} p_{R}\left(\mathbf{b}(x)\right)d\mathbf{b}$$

$$+ \int_{-\infty}^{+\infty} \Delta b^{r} C_{ijkl}^{;r} \Delta b^{u}\left(\chi_{(pq)k,l}\right)^{;u} p_{R}\left(\mathbf{b}(x)\right)d\mathbf{b}$$

$$+ \frac{1}{2}\int_{-\infty}^{+\infty} C_{ijkl}^{0} \Delta b^{u}\Delta b^{r}\left(\chi_{(pq)k,l}\right)^{;uv} p_{R}\left(\mathbf{b}(x)\right)d\mathbf{b}$$

$$= C_{ijkl}^{0}\left(\chi_{(pq)k,l}\right)^{0} + \left\{C_{ijkl}^{;r}\left(\chi_{(pq)k,l}\right)^{;s} + \tfrac{1}{2}C_{ijkl}^{0}\left(\chi_{(pq)k,l}\right)^{;rs}\right\}Cov\left(b^{r},b^{s}\right) \qquad (2.215)$$

Averaging both sides of this equation over the region Ω and including in the relation (2.213) together with spatially averaged expected values of the original elasticity tensor, the expected values of the homogenised elasticity tensor are obtained. Next, the covariances of the effective elasticity tensor components can be derived similarly as

$$Cov\left(C_{ijkl}^{(eff)};C_{mnpq}^{(eff)}\right)= Cov\left(C_{ijkl},C_{mnpq}\right)+ Cov\left(C_{ijkl},C_{mnuv}\chi_{(pq)u,v}\right)$$
$$+ Cov\left(C_{ijrs}\chi_{(kl)r,s},C_{mnpq}\right)+ Cov\left(C_{ijrs}\chi_{(kl)r,s},C_{mnuv}\chi_{(pq)u,v}\right) \qquad (2.216)$$

Finally, the covariances of the effective elasticity tensor components are calculated below. Covariance of the first component in (2.216) is derived as

$$Cov\left(C_{ijkl};C_{mnpq}\right)= \int_{-\infty}^{+\infty}\left(C_{ijkl} - E\left[C_{ijkl}\right]\right)\left(C_{mnpq} - E\left[C_{mnpq}\right]\right)p_{R}\left(\mathbf{b}(x)\right)d\mathbf{b}$$

$$= \int_{-\infty}^{+\infty}\left(C_{ijkl}^{0} + \Delta b_{r}C_{ijkl}^{;r} - C_{ijkl}^{0}\right)\left(C_{mnpq}^{0} + \Delta b_{s}C_{mnpq}^{;s} - C_{mnpq}^{0}\right)p_{R}\left(\mathbf{b}(x)\right)d\mathbf{b} \qquad (2.217)$$

$$= C_{ijkl}^{;r}C_{mnpq}^{;s}\int_{-\infty}^{+\infty}\Delta b_{r}\Delta b_{s}\, p_{R}\left(\mathbf{b}(x)\right)d\mathbf{b} = C_{ijkl}^{;r}C_{mnpq}^{;s}Cov\left(b^{r},b^{s}\right)$$

Next, the cross−covariances of the second component are calculated and there holds

$$Cov\left(C_{ijtw}\chi_{(kl)t,w};C_{mnuv}\chi_{(pq)u,v}\right)= \int_{-\infty}^{+\infty}\left(C_{ijtw}\chi_{(kl)t,w} - E\left[C_{ijtw}\chi_{(kl)t,w}\right]\right)$$
$$\times\left(C_{mnuv}\chi_{(pq)u,v} - E\left[C_{mnuv}\chi_{(pq)u,v}\right]\right)p_{R}\left(\mathbf{b}(x)\right)d\mathbf{b} \qquad (2.218)$$

which, by introducing the simplifying notation, becomes

$$
\begin{aligned}
&\int_{-\infty}^{+\infty}\!\Big(\mathbf{C}^0\chi^0 + \mathbf{C}^{,r}\Delta b_r\chi^0 + \mathbf{C}^0\chi^{,u}\Delta b_u + \mathbf{C}^{,r}\Delta b_r\chi^{,u}\Delta b_u + \tfrac{1}{2}\mathbf{C}^0\chi^{,uv}\Delta b_u\Delta b_v \\
&\quad -\Big\{\mathbf{C}^0\chi^0 + \big(\mathbf{C}^{,r}\chi^{,s} + \tfrac{1}{2}\mathbf{C}^0\chi^{,rs}\big)Cov\big(b^r,b^s\big)\Big\}\Big)p_R(\mathbf{b}(x))\,d\mathbf{b} \\
&\times \int_{-\infty}^{+\infty}\!\Big(\mathbf{D}^0\varphi^0 + \mathbf{D}^{,a}\Delta b_a\varphi^0 + \mathbf{D}^0\varphi^{,c}\Delta b_c + \mathbf{D}^{,a}\Delta b_a\varphi^{,c}\Delta b_c + \tfrac{1}{2}\mathbf{D}^0\varphi^{,cd}\Delta b_c\Delta b_d \\
&\quad -\Big\{\mathbf{D}^0\varphi^0 + \big(\mathbf{D}^{,a}\varphi^{,c} + \tfrac{1}{2}\mathbf{D}^0\varphi^{,ac}\big)Cov\big(b^a,b^c\big)\Big\}\Big)p_R(\mathbf{b}(x))\,d\mathbf{b}
\end{aligned}
\tag{2.219}
$$

Further, it is obtained that

$$
\begin{aligned}
&\int_{-\infty}^{+\infty}\!\Big(\mathbf{C}^0\chi^0 + \mathbf{C}^{,r}\Delta b_r\chi^0 + \mathbf{C}^0\chi^{,u}\Delta b_u + \mathbf{C}^{,r}\Delta b_r\chi^{,u}\Delta b_u + \tfrac{1}{2}\mathbf{C}^0\chi^{,uv}\Delta b_u\Delta b_v \\
&\quad -\Big\{\mathbf{C}^0\chi^0 + \big(\mathbf{C}^{,r}\chi^{,s} + \tfrac{1}{2}\mathbf{C}^0\chi^{,rs}\big)Cov\big(b^r,b^s\big)\Big\}\Big)p_R(\mathbf{b}(x))\,d\mathbf{b} \\
&\times \int_{-\infty}^{+\infty}\!\Big(\mathbf{D}^0\varphi^0 + \mathbf{D}^{,a}\Delta b_a\varphi^0 + \mathbf{D}^0\varphi^{,c}\Delta b_c + \mathbf{D}^{,a}\Delta b_a\varphi^{,c}\Delta b_c + \tfrac{1}{2}\mathbf{D}^0\varphi^{,cd}\Delta b_c\Delta b_d \\
&\quad -\Big\{\mathbf{D}^0\varphi^0 + \big(\mathbf{D}^{,a}\varphi^{,c} + \tfrac{1}{2}\mathbf{D}^0\varphi^{,ac}\big)Cov\big(b^a,b^c\big)\Big\}\Big)p_R(\mathbf{b}(x))\,d\mathbf{b} \\
&= \int_{-\infty}^{+\infty}\!\mathbf{C}^{,r}\Delta b_r\chi^0\mathbf{D}^{,a}\Delta b_a\varphi^0\, p_R(\mathbf{b}(x))\,d\mathbf{b} + \int_{-\infty}^{+\infty}\!\mathbf{C}^{,r}\Delta b_r\chi^0\mathbf{D}^0\varphi^{,c}\Delta b_c\, p_R(\mathbf{b}(x))\,d\mathbf{b} \\
&+ \int_{-\infty}^{+\infty}\!\mathbf{C}^0\chi^{,u}\Delta b_u\mathbf{D}^{,a}\Delta b_a\varphi^0\, p_R(\mathbf{b}(x))\,d\mathbf{b} + \int_{-\infty}^{+\infty}\!\mathbf{C}^0\chi^{,u}\Delta b_u\mathbf{D}^0\varphi^{,c}\Delta b_c\, p_R(\mathbf{b}(x))\,d\mathbf{b}
\end{aligned}
\tag{2.220}
$$

Integration over the probability domain gives

$$
\begin{aligned}
&\int_{-\infty}^{+\infty}\!\mathbf{C}^{,r}\Delta b_r\chi^0\mathbf{D}^{,a}\Delta b_a\varphi^0\, p_R(\mathbf{b}(x))\,d\mathbf{b} + \int_{-\infty}^{+\infty}\!\mathbf{C}^{,r}\Delta b_r\chi^0\mathbf{D}^0\varphi^{,c}\Delta b_c\, p_R(\mathbf{b}(x))\,d\mathbf{b} \\
&+ \int_{-\infty}^{+\infty}\!\mathbf{C}^0\chi^{,u}\Delta b_u\mathbf{D}^{,a}\Delta b_a\varphi^0\, p_R(\mathbf{b}(x))\,d\mathbf{b} + \int_{-\infty}^{+\infty}\!\mathbf{C}^0\chi^{,u}\Delta b_u\mathbf{D}^0\varphi^{,c}\Delta b_c\, p_R(\mathbf{b}(x))\,d\mathbf{b} \\
&= \Big\{\mathbf{C}^{,r}\mathbf{D}^{,s}\chi^0\varphi^0 + \mathbf{C}^{,r}\chi^0\mathbf{D}^0\varphi^{,s} + \mathbf{C}^0\chi^{,r}\mathbf{D}^{,s}\varphi^0 + \mathbf{C}^0\chi^{,r}\mathbf{D}^0\varphi^{,s}\Big\}Cov\big(b^r,b^s\big)
\end{aligned}
\tag{2.221}
$$

or, in a more explicit way, that

$$Cov\left(C_{ijtw}\chi_{(kl)t,w};C_{mnuv}\chi_{(pq)u,v}\right)$$
$$=\left\{C_{ijtw}^{,r}C_{mnuv}^{,s}\left(\chi_{(kl)t,w}\right)^{0}\left(\chi_{(pq)u,v}\right)^{0}+C_{ijtw}^{,r}C_{mnuv}^{0}\left(\chi_{(kl)t,w}\right)^{0}\left(\chi_{(pq)u,v}\right)^{,s}\right.$$
$$\left.+C_{ijtw}^{0}C_{mnuv}^{,r}\left(\chi_{(kl)t,w}\right)^{,s}\left(\chi_{(pq)u,v}\right)^{0}+C_{ijtw}^{0}C_{mnuv}^{0}\left(\chi_{(kl)t,w}\right)^{,r}\left(\chi_{(pq)u,v}\right)^{,s}\right\} \qquad (2.222)$$
$$\times Cov\left(b^{r},b^{s}\right)$$

Now, the third component is transformed as follows:

$$Cov\left(C_{ijkl};C_{mnuv}\chi_{(pq)u,v}\right)=Cov(\mathbf{C};\mathbf{D}\chi)$$
$$=\int_{-\infty}^{+\infty}\left(\mathbf{C}^{0}+\mathbf{C}^{,r}\Delta b_{r}-\mathbf{C}^{0}\right)\cdot p_{R}(\mathbf{b}(x))\,d\mathbf{b}$$
$$\times\int_{-\infty}^{+\infty}\left(\mathbf{D}^{0}\chi^{0}+\mathbf{D}^{,a}\Delta b_{a}\chi^{0}+\mathbf{D}^{0}\chi^{,c}\Delta b_{c}+\mathbf{D}^{,a}\Delta b_{a}\chi^{,c}\Delta b_{c}+\tfrac{1}{2}\mathbf{D}^{0}\chi^{,cd}\Delta b_{c}\Delta b_{d}\right. \qquad (2.223)$$
$$\left.-\left\{\mathbf{D}^{0}\chi^{0}+\left(\mathbf{D}^{,a}\chi^{,c}+\tfrac{1}{2}\mathbf{D}^{0}\chi^{,ac}\right)Cov\left(b^{a},b^{c}\right)\right\}\right)p_{R}(\mathbf{b}(x))\,d\mathbf{b}$$
$$=\int_{-\infty}^{+\infty}\mathbf{C}^{,r}\Delta b_{r}\mathbf{D}^{,a}\Delta b_{a}\chi^{0}p_{R}(\mathbf{b}(x))\,d\mathbf{b}+\int_{-\infty}^{+\infty}\mathbf{C}^{,r}\Delta b_{r}\mathbf{D}^{0}\chi^{,c}\Delta b_{c}\,p_{R}(\mathbf{b}(x))\,d\mathbf{b}$$
$$=\left\{\mathbf{C}^{,r}\mathbf{D}^{,s}\chi^{0}+\mathbf{C}^{,r}\mathbf{D}^{0}\chi^{,s}\right\}Cov\left(b^{r},b^{s}\right)$$

Introducing the symbolic summation notation for the tensor function considered above it can be written that

$$Cov\left(C_{ijkl};C_{mnuv}\chi_{(pq)u,v}\right)$$
$$=Cov(\mathbf{C};\mathbf{D}\chi)=\left\{\mathbf{C}^{,r}\mathbf{D}^{,s}\chi^{0}+\mathbf{C}^{,r}\mathbf{D}^{0}\chi^{,s}\right\}Cov\left(b^{r},b^{s}\right) \qquad (2.224)$$
$$=\left\{C_{ijkl}^{,r}C_{mnuv}^{,s}\left(\chi_{(pq)u,v}\right)^{0}+C_{ijkl}^{,r}C_{mnuv}^{0}\left(\chi_{(pq)u,v}\right)^{,s}\right\}Cov\left(b^{r},b^{s}\right)$$

By the analogous way, it is obtained

$$Cov\left(C_{ijtw}\chi_{(kl)t,w};C_{mnpq}\right)$$
$$=Cov(\mathbf{C}\chi;\mathbf{D})=\left\{\mathbf{C}^{,r}\chi^{0}\mathbf{D}^{,s}+\mathbf{C}^{0}\chi^{,r}\mathbf{D}^{,s}\right\}Cov\left(b^{r},b^{s}\right) \qquad (2.225)$$
$$=\left\{C_{ijtw}^{,r}\left(\chi_{(kl)t,w}\right)^{0}C_{mnpq}^{,s}+C_{ijtw}^{,r}\left(\chi_{(kl)t,w}\right)^{,s}C_{mnpq}^{0}\right\}Cov\left(b^{r},b^{s}\right)$$

The components of effective elasticity tensor covariances are found. Starting from the classical definition

$$Cov\left(C_{ijkl}^{(eff)}; C_{mnpq}^{(eff)}\right)$$

$$= Cov\left(C_{ijkl} + C_{ijtw}\chi_{(kl)t,w}; C_{mnpq} + C_{mnuv}\chi_{(pq)u,v}\right)$$

$$= \int\limits_{-\infty}^{+\infty}\left(C_{ijkl} + C_{ijtw}\chi_{(kl)t,w} - E[C_{ijkl}] - E[C_{ijtw}\chi_{(kl)t,w}]\right)$$

$$\times\left(C_{mnpq} + C_{mnuv}\chi_{(pq)u,v} - E[C_{mnpq}] - E[C_{mnuv}\chi_{(pq)u,v}]\right)p_R(\mathbf{b}(x))\,d\mathbf{b}$$

(2.226)

Transforming the respective integrands and using Fubini theorem applied to the integrals of random functions we obtain further

$$\int\limits_{-\infty}^{+\infty}\left(C_{ijkl} - E[C_{ijkl}]\right)\left(C_{mnpq} - E[C_{mnpq}]\right)p_R(\mathbf{b}(x))\,d\mathbf{b}$$

$$\times\int\limits_{-\infty}^{+\infty}\left(C_{ijkl} - E[C_{ijkl}]\right)\left(C_{mnuv}\chi_{(pq)u,v} - E[C_{mnuv}\chi_{(pq)u,v}]\right)p_R(\mathbf{b}(x))\,d\mathbf{b}$$

$$\times\int\limits_{-\infty}^{+\infty}\left(C_{ijtw}\chi_{(kl)t,w} - E[C_{ijtw}\chi_{(kl)t,w}]\right)\left(C_{mnpq} - E[C_{mnpq}]\right)p_R(\mathbf{b}(x))\,d\mathbf{b}$$

$$\times\int\limits_{-\infty}^{+\infty}\left(C_{ijtw}\chi_{(kl)t,w} - E[C_{ijtw}\chi_{(kl)t,w}]\right)\left(C_{mnuv}\chi_{(pq)u,v} - E[C_{mnuv}\chi_{(pq)u,v}]\right)p_R\,d\mathbf{b}$$

(2.227)

which, using the classical definition of the covariance, is equal to

$$Cov\left(C_{ijkl}, C_{mnpq}\right) + Cov\left(C_{ijkl}, C_{mnuv}\chi_{(pq)u,v}\right) +$$
$$+ Cov\left(C_{ijtw}\chi_{(kl)t,w}, C_{mnpq}\right) + Cov\left(C_{ijtw}\chi_{(kl)t,w}, C_{mnuv}\chi_{(pq)u,v}\right)$$

(2.228)

Introducing all the statements into the last one it can finally be written that

$$Cov\left(C_{ijkl}^{(eff)}; C_{mnpq}^{(eff)}\right)$$

$$= \left\{ C_{ijkl}^{,r}C_{mnpq}^{,s} + C_{ijtw}^{,r}\left(\chi_{(kl)t,w}\right)^0 C_{mnpq}^{,s} + C_{ijtw}^{,r}\left(\chi_{(kl)t,w}\right)^{,s} C_{mnpq}^0 \right.$$

$$+ C_{ijkl}^{,r}C_{mnuv}^{,s}\left(\chi_{(pq)u,v}\right)^0 + C_{ijkl}^{,r}C_{mnuv}^0\left(\chi_{(pq)u,v}\right)^{,s}$$

$$+ C_{ijtw}^{,r}C_{mnuv}^{,s}\left(\chi_{(kl)t,w}\right)^0\left(\chi_{(pq)u,v}\right)^0 + C_{ijtw}^{,r}C_{mnuv}^0\left(\chi_{(kl)t,w}\right)^0\left(\chi_{(pq)u,v}\right)^{,s}$$

$$\left. + C_{ijtw}^0 C_{mnuv}^{,r}\left(\chi_{(kl)t,w}\right)^{,s}\left(\chi_{(pq)u,v}\right)^0 + C_{ijtw}^0 C_{mnuv}^0\left(\chi_{(kl)t,w}\right)^{,r}\left(\chi_{(pq)u,v}\right)^{,s} \right\}$$

$$\times Cov\left(b^r, b^s\right)$$

(2.229)

It should be underlined here that the above equations give complete a description of the effective elasticity tensor components in the stochastic second moment and second order perturbation approach. Finally, let us note that many simplifications resulted here thanks to the assumption that the input random variables of the homogenisation problem are just the Young moduli of the fibre and matrix. If the Poisson ratios are treated as random, the second order derivatives of the constitutive tensor would generally differ from 0 and the stochastic finite element formulation of the homogenisation procedure would be essentially more complicated.

For the periodicity cell and its discretisation shown in Figure 2.128 elastic properties of the glass fibre and the matrix are adopted as follows: the Young moduli expected values $E[e_1] = 84$ GPa, $E[e_2] = 4.0$ GPa, while the deterministic Poisson ratios are taken as equal to $v_1 = 0.22$ in fibre and $v_2 = 0.34$ – in the matrix.

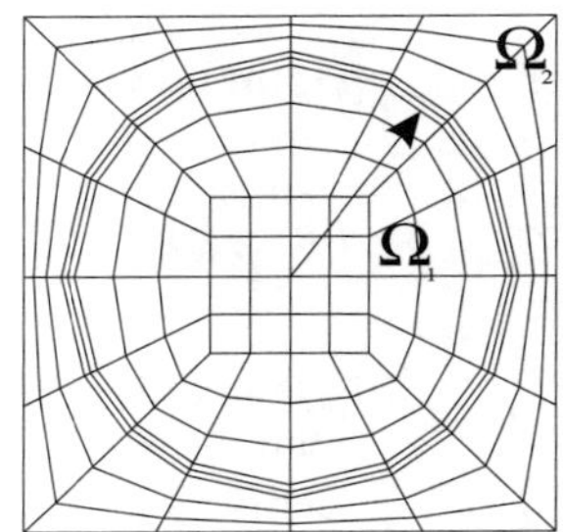

Figure 2.128. Periodicity cell tested

Five different sets of Young moduli coefficients of variation are analysed according to Table 2.21 – various values between 0.05 and 0.15 have been adopted to verify the influence of the component data randomness on the respective probabilistic moments of the homogenised elasticity tensor. The finite difference numerical technique has been employed to determine the relevant derivatives with respect to the input random variables adopted.

Table 2.21. The coefficient of variation of the input random variables

Test number	$\alpha(e_1)$	$\alpha(e_2)$
1	0.050	0.050
2	0.075	0.075
3	0.100	0.100
4	0.125	0.125
5	0.150	0.150

The cross-sectional fibre area equals to about a half of the total periodicity cell area. The results in the form of expected values and coefficients of variation of the homogenised tensor components obtained from four computational tests are shown in Table 2.22 and compared against the corresponding values obtained by using the MCS technique for the total number of random trials taken as 10^3.

Table 2.22. Coefficients of variation for the effective elasticity tensor

Test	$\alpha\left(C_{1111}^{(eff)}(\omega)\right)$		$\alpha\left(C_{1122}^{(eff)}(\omega)\right)$	
	SFEM	MCS	SFEM	MCS
1	0.0410	0.0516	0.7152	0.0517
2	0.0622	0.0777	0.1073	0.0777
3	0.0830	0.1037	0.1430	0.1037
4	0.1036	0.1297	0.1788	0.1297
5	0.1244	0.1557	0.2146	0.1557

It is seen that the results of the SFEM−based computations are slightly smaller than those resulting from the Monte Carlo simulations in the case of $\alpha\left(C_{1111}^{(eff)}(\omega)\right)$; the opposite trend is observed for $\alpha\left(C_{1122}^{(eff)}(\omega)\right)$. The differences between both models are acceptable for very small input coefficients of variation and above the value 0.1 (second order approach limitation) they enormously increase. It is also observed that the coefficients from the MCS analysis are equal with each other, while the SFEM returns different values for both effective tensor components. It follows the fact that the first partial derivatives of both components with respect to Young moduli of the fibre and matrix are different. These derivatives are included in the SFEM equations for the second order moments and, in the same time, they do not influence the MCS homogenisation model at all. Furthermore, a linear dependence between the results obtained and the input coefficients of variation of the components Young moduli is observed.

The main reason for numerical implementation of the SFEM equations for modelling of the homogenisation problem is a decisive decrease in computation time in comparison to that necessary by the MCS technique. It should be mentioned that the Monte Carlo sampling time can be approximated as a product of the following times:

(a) a single deterministic cell problem solution,
(b) the total number of homogenisation functions required (three functions $\chi_{(11)}$, $\chi_{(12)}$ and $\chi_{(22)}$ in this plane strain analysis),
(c) the total number of random trials performed.

There are some time consuming procedures in the MCS programs such as random numbers generation, post−processing estimation procedure and the subroutines for averaging the needed parameters within the RVE, which are not included, however their times are negligible in comparison with the routines pointed out before.

On the other hand, the time for Stochastic Finite Element Analysis can be approximated by multiplication of the following procedure times: (a) the SFE solution of the cell problem (with the same order of the cost considered as the deterministic analysis) and the total number of necessary homogenisation functions. Taking into account the remarks posed above, the difference in computational time between MCS and SFEM approaches to the homogenisation problem is of the order of about $(n-1)\tau$ provided that n is the total number of MCS samples and τ stands for the time of a deterministic problem solution. Observing this and considering negligible differences between the results of both these methods for smaller random dispersion of input variables, the stochastic second order and second moment computational analysis of composite materials should be preferred in most engineering problems. The only disadvantage is the complexity of the equations, which have to be implemented in the respective program as well as the bounds dealing with randomness of input variables (the coefficients of variation should be generally smaller than about 0.15).

2.3.4 Upper and Lower Bounds for Effective Characteristics

Let us consider the coefficients of the following linear second order elliptic problem [65]:

$$-div(\mathbf{C}^\varepsilon \varepsilon(\mathbf{u}^\varepsilon)) = \mathbf{f} \; ; \quad x \in \Omega \tag{2.230}$$

$$\varepsilon_{ij}(\mathbf{u}^\varepsilon) = \tfrac{1}{2}(u_{i,j}^\varepsilon + u_{j,i}^\varepsilon) \; ; \quad x \in \Omega \tag{2.231}$$

$$\mathbf{C}^\varepsilon = \psi^{\varepsilon(p)}(x)\, \mathbf{C}^{\varepsilon(p)} \tag{2.232}$$

with boundary conditions

$$\mathbf{u}^\varepsilon = 0 \; ; \quad x \in \partial\Omega \tag{2.233}$$

In the above equations $\mathbf{u}^\varepsilon$, $\varepsilon(\mathbf{u}^\varepsilon)$ and $\mathbf{f}$ denote the displacement field, strain tensor and vector of external loadings, respectively. As was presented in Sec. 2.3.3.2, the effective (homogenised) tensor $\mathbf{C}^0$ is such a tensor that replacing $\mathbf{C}^\varepsilon$ and $\mathbf{C}^0$ in the above system gives $\mathbf{u}^0$ as a solution, which is a weak limit of $\mathbf{u}^\varepsilon$ with scale parameter tends to 0. It should be mentioned that without any other assumptions on Ω microgeometry the bounded set of effective properties is generated. Moreover, it can be proved that there exist such tensors $\inf(C_{ijkl})$ and $\sup(C_{ijkl})$ that

$$\inf(C_{ijkl}) \le C_{ijkl}^0 \le \sup(C_{ijkl}) \tag{2.234}$$

It is well known that the theorem of minimum potential energy gives the upper bounds of the effective tensor, whereas the minimum complementary energy approximates the lower bounds. Thanks to the Eshelby formula the explicit equations are as follows:

$$\begin{cases} \sup \kappa = \left[\sum_{r=1}^{N} C_r (\kappa_u + \kappa_r)^{-1} \right]^{-1} - \kappa_u \\[2ex] \sup \mu = \left[\sum_{r=1}^{N} C_r (\mu_u + \mu_r)^{-1} \right]^{-1} - \mu_u \end{cases} \tag{2.235}$$

where κ_u, μ_u have the following form:

$$\begin{cases} \kappa_u = \tfrac{4}{3} \mu_{\max} \\[2ex] \mu_u = \tfrac{3}{2} \left(\dfrac{1}{\mu_{\max}} + \dfrac{10}{9\kappa_{\max} + 8\mu_{\max}} \right)^{-1} \end{cases} \tag{2.236}$$

Further, lower bounds for the elasticity tensor are obtained as

$$\begin{cases} \inf \kappa = \left[\sum_{r=1}^{N} C_r (\kappa_l + \kappa_r)^{-1} \right]^{-1} - \kappa_l \\[2ex] \inf \mu = \left[\sum_{r=1}^{N} C_r (\mu_l + \mu_r)^{-1} \right]^{-1} - \mu_l \end{cases} \tag{2.237}$$

where it holds that

$$\begin{cases} \kappa_l = \tfrac{4}{3} \mu_{\min} \\[2ex] \mu_l = \tfrac{3}{2} \left(\dfrac{1}{\mu_{\min}} + \dfrac{10}{9\kappa_{\min} + 8\mu_{\min}} \right)^{-1} \end{cases} \tag{2.328}$$

and n is a total number of composite constituents where $c_r, 1 \le r \le n$ denote their volume fractions. It should be noted that

$$\kappa = \frac{e}{3(1 - 2\upsilon)} \tag{2.239}$$

$$\mu = \frac{e}{2(1 + \upsilon)} \tag{2.240}$$

$$\lambda = \kappa - \tfrac{2}{3}\mu \tag{2.241}$$

$$C_{ijkl} = \delta_{ij}\delta_{kl}\lambda + (\delta_{ik}\delta_{jl} + \delta_{il}\delta_{jk})\mu \tag{2.242}$$

From the engineering point of view the most interesting is the effectiveness of such a characterisation of C_{ijkl}, which can be approximated as the difference between upper and lower estimates and, on the other hand, sensitivity of the effective tensor with respect to material characteristics of the constituents. The Monte Carlo simulation technique has been used to compute probabilistic moments of the effective elasticity tensor components for the periodic superconductor analysed before. The superconducting cable consists of fibres made of a superconductor placed around a thin−walled pipe (tube) covered with a jacket and insulating material. Experimental data describing elastic characteristics of the composite constituents are collected in Table 2.23.

Table 2.23. Probabilistic elastic characteristics of the superconductor components

Material	$E[e]$	$\sigma(e)$	$E[v]$	$\sigma(v)$
316LN	205 GPa	8 GPa	0.265	0.010
Incoloy 908 'annealed'	182 GPa	-	0.303	-
'cold worked'	184 GPa	-	0.299	-
Titanium	126 GPa	12 GPa	0.311	0.012
Insulation G10-CR	36 GPa	-	0.21	-

Because of negligible differences in the elastic properties of Incoloy (between the 'annealed' and 'cold worked' state) the 'annealed' state of the superconductor is considered further. All the results obtained in the computational experiments have been collected in Table 2.24 and Figures 2.129−2.137. Because of the fact that the expected values appeared to be rather insensitive to the total number of random trials in the Monte Carlo simulations, results of the relevant convergence tests have been omitted in the tables and presented further in the figures. The expected values considered have been collected in Table 2.24 for M=10,000 random trials.

Table 2.24. Effective elasticity tensor components and their expected values (in GPa)

Effective property type	Analysis type					
	Deterministic			probabilistic		
	$C^{(eff)}_{JJJJ}$	$C^{(eff)}_{JKKJ}$	$C^{(eff)}_{JKJK}$	$C^{(eff)}_{JJJJ}$	$C^{(eff)}_{JKKJ}$	$C^{(eff)}_{JKJK}$
sup-VR	189.56	81.83	53.86	189.94	82.30	53.82
Sup	178.44	76.07	51.18	178.57	76.37	51.10
Inf	156.99	62.70	47.14	156.68	62.61	47.03
Inf-VR	137.93	51.86	43.03	137.54	51.71	42.92

Effective properties collected in this chapter (*sup, inf* in Table 2.24) have been compared with the Voigt−Reuss ones (*sup-VR, inf-VR* in Table 2.24). Considering the results obtained, it should be noted that these first approximators are generally more restrictive than the Voigt−Reuss ones. Further, it can be observed that deterministic values are, with acceptable accuracy, equal to the corresponding expected values. Thus, for relatively small standard deviations of the input elastic characteristics, the randomness in the effective characteristics can be neglected.

Finally, it can be noted that more restrictive bounds can be used to determine the effective elasticity tensor in a more efficient way. Taking as a basis the arithmetic average of the upper and lower bounds, the difference between these bounds is in the range of 13% for $C_{JJJJ}^{(eff)}$ bound component, 19% for $C_{JKJK}^{(eff)}$ bound component and 8% for $C_{JKKJ}^{(eff)}$ bound component.

The following figures contain the results of the convergence analysis of the coefficient of variation, asymmetry and concentration with respect to increasing total number of Monte Carlo random trials. All these coefficients are presented for $C_{JJJJ}^{(eff)}$ bounds in Figures 2.129, 2.132 and 2.135, for $C_{JKJK}^{(eff)}$ bounds in Figures 2.130, 2.133 and 2.136 and for $C_{JKKJ}^{(eff)}$ in Figures 2.131, 2.134 and 2.137. On the horizontal axes of these figures the total number of Monte Carlo random trials M is marked, while the vertical is used for the coefficient of variation.

General observation here is that the $C_{JKJK}^{(eff)}$ bounds are the most sensitive with respect to the randomness of input elastic characteristics. These coefficients for $C_{JKJK}^{(eff)}$ bounds appeared to be the greatest and then we obtain the coefficients for $C_{JJJJ}^{(eff)}$ and $C_{JKKJ}^{(eff)}$, respectively. Next, it can be mentioned that the estimators of the coefficients of variation show fast convergence to their limits. Efficient approximation of final coefficients for various components of the tensor $C_{ijkl}^{(eff)}$ bounds is obtained for M equal to about 2,500 random trials. Generally, it is observed that the coefficients of variation of effective elasticity tensor fulfil the inequalities detected in case of the expected values. The greatest coefficients are obtained for Reuss bounds, next the upper and lower bounds proposed in this chapter, and the smallest for the Voigt lower bounds.

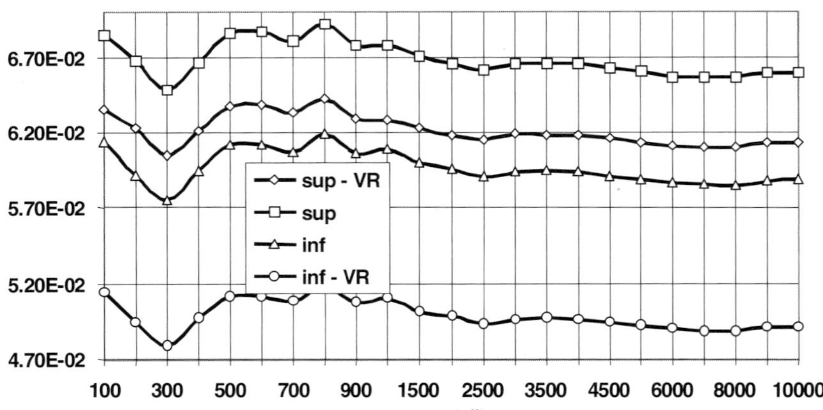

Figure 2.129. The coefficients of variation of $C_{JJJJ}^{(eff)}$ bounds

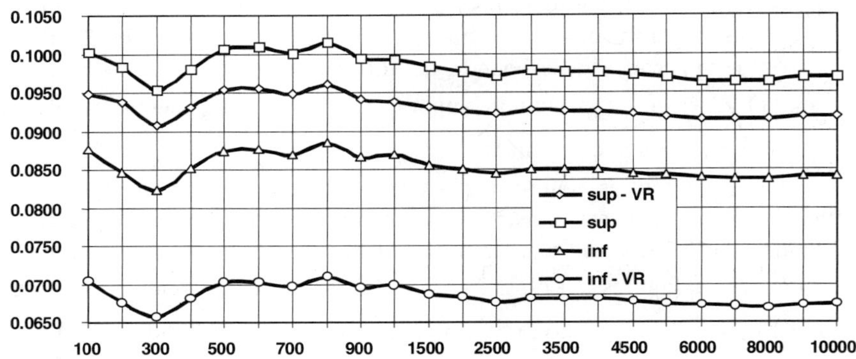

Figure 2.130. The coefficients of variation of $C_{JKJK}^{(eff)}$ bounds

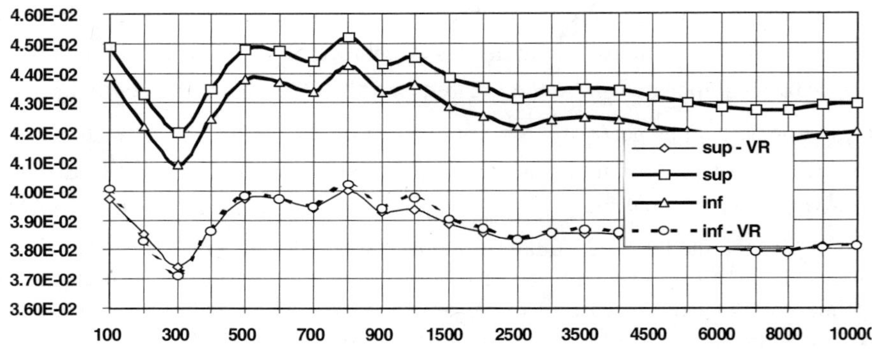

Figure 2.131. The coefficients of variation of $C_{JKKJ}^{(eff)}$ bounds

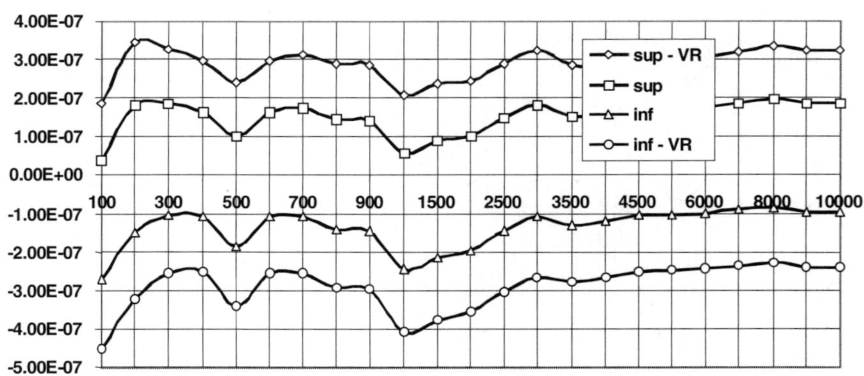

Figure 2.132. The coefficients of asymmetry of $C_{JJJJ}^{(eff)}$ bounds

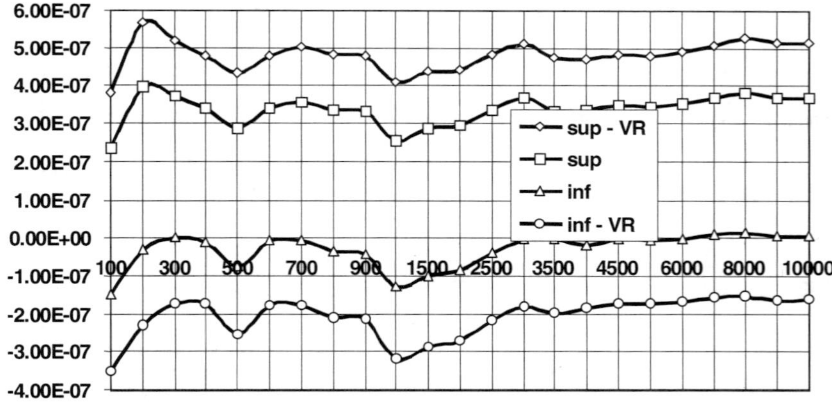

Figure 2.133. The coefficients of asymmetry of $C_{JKJK}^{(eff)}$ bounds

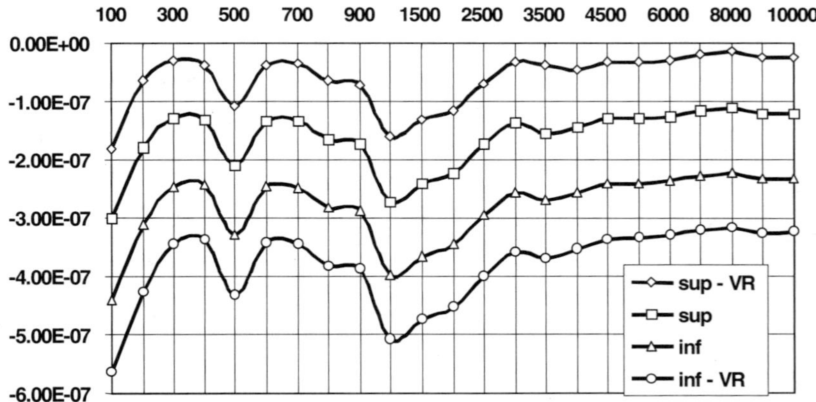

Figure 2.134. The coefficients of asymmetry of $C_{JKKJ}^{(eff)}$ bounds

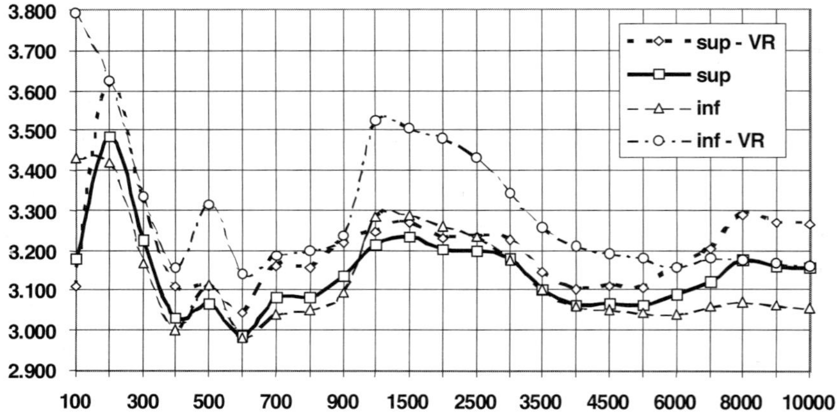

Figure 2.135. The coefficients of concentration of $C_{JJJJ}^{(eff)}$ bounds

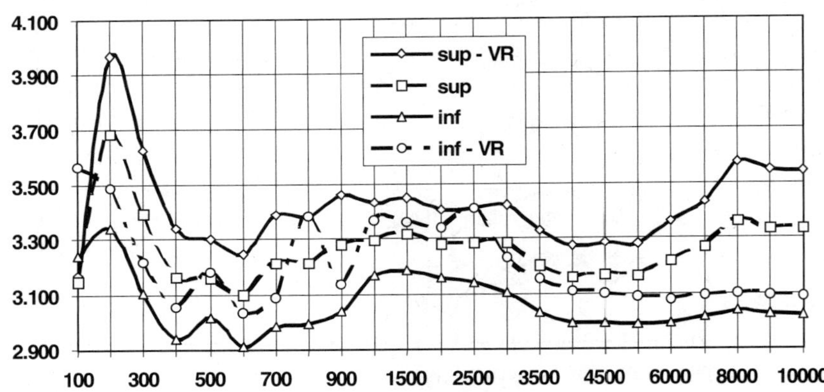

Figure 2.136. The coefficients of concentration of $C_{JKJK}^{(eff)}$ bounds

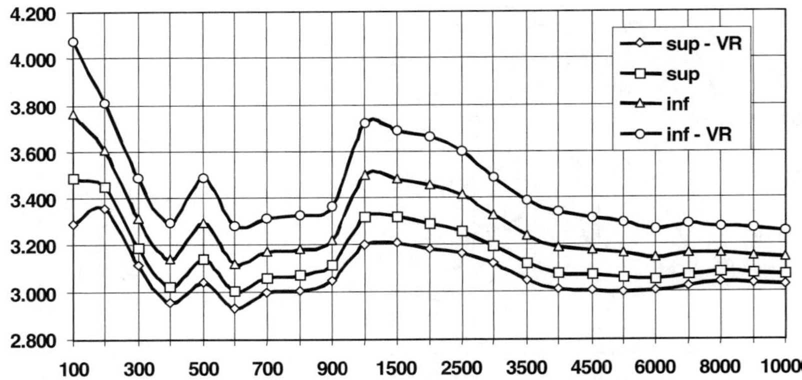

Figure 2.137. The coefficients of concentration of $C_{JKKJ}^{(eff)}$ bounds

Observing the results presented in Figures 2.132 and 2.134 it can be observed that all coefficients of asymmetry of $C_{ijkl}^{(eff)}$ verified tend to 0 with increasing total number of random trials. Comparing $C_{JJJJ}^{(eff)}$ and $C_{JKJK}^{(eff)}$ against $C_{JKKJ}^{(eff)}$ bounds it can be stated that the first two variables have minimum positive asymmetry, while the last have a negative one. It should be mentioned that for such probabilistic distributions with non−zero coefficients of asymmetry, the expected value is not equal to the most probable one.

Moreover, taking into account the convergence of coefficients of asymmetry it is seen that they are generally more slowly convergent than coefficients of variation estimators. *M* larger than 5,000 is required to compute these estimators with satisfactory accuracy. Analogous to the coefficients of variation, the hierarchy of the expected values of $C_{ijkl}^{(eff)}$, which has been discussed above, is fulfilled.

Figures 2.135–2.137 present the coefficients of concentration for different components of the effective elasticity tensor. The estimator convergence analysis proves that M equal to almost 10,000 is needed to compute these coefficients properly. The convergence of these estimators is more complex than the previous ones, but generally their values are greater than 3, which is characteristic for the Gaussian variables. Thus it can be stated that the $C_{ijkl}^{(eff)}$ probabilistic distributions obtained are more concentrated around their expected values than the Gaussian variables, but this difference is no greater than a maximum of 15% for the $C_{JKJK}^{(eff)}$ bounds.

Figures 2.138–2.140 illustrate the probability density functions of the upper and lower bounds for $C_{JJJJ}^{(eff)}$, $C_{JKJK}^{(eff)}$ and $C_{JKKJ}^{(eff)}$ components of the effective elasticity tensor. On the horizontal axes of these figures the values computed for these components are marked, while on the vertical axes the relevant probability density function (PDF) is given.

The PDFs for the tensor $C_{ijkl}^{(eff)}$ computed together with the additional coefficients of asymmetry and concentration β, γ show that these functions have distributions quite similar to the bell–shaped Gaussian distribution curve. Thus, in further analyses proposed in the conclusions, we assume that for the input random variables being elastic characteristics (Young moduli and Poisson ratios) being Gaussian uncorrelated random variables, the upper and lower bounds computed having also a Gaussian distribution, which essentially simplifies further estimation and related numerical analyses.

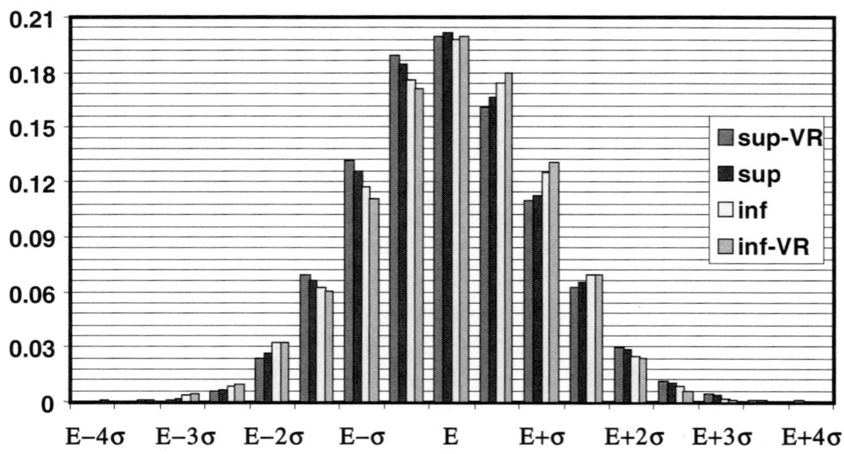

Figure 2.138. The probability densities of $C_{JJJJ}^{(eff)}$ bounds

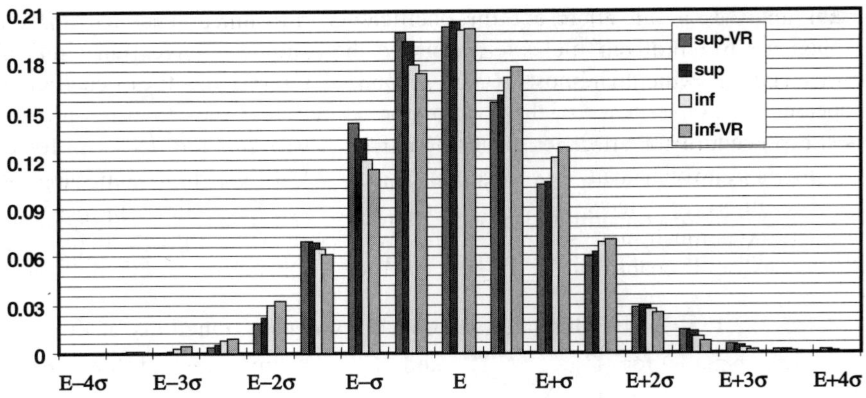

Figure 2.139. The probability densities of $C_{JKJK}^{(eff)}$ bounds

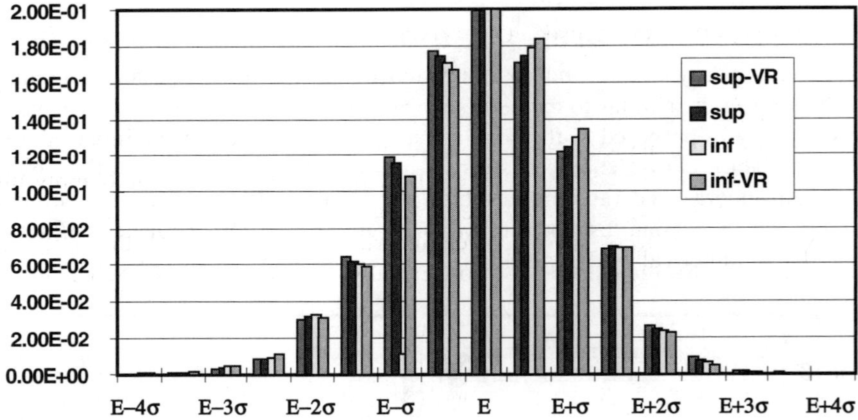

Figure 2.140. The probability densities of $C_{JKKJ}^{(eff)}$ bounds

The results of numerical tests performed lead us to the conclusion that the probabilistic upper and lower bounds of the effective elasticity tensor may be very efficient in the characterisation of superconducting composites with randomly defined elastic characteristics because of negligible relative differences between the upper and lower bounds. Considering the computational time cost they appear to be much more useful in engineering practice than other FEM–based direct methods.

Computational experiments carried out prove that the coefficients of variation of the bounds computed are in the range of the input random variables of the problem. Considering further analyses of homogenised superconducting coils, this fact confirms the need for the application of the SFEM in such computations, which is important for essential time savings in comparison with the simulation methods.

The probabilistic sensitivity of the effective elastic characteristics with respect to the probabilistic material parameters should be verified computationally in further analyses as an effect of regression test, for instance. Such an analysis enables us to find out these parameters of composite constituent elastic characteristics, which are the most influencing for global superconductor behaviour.

The procedure for effective elastic properties approximation seems to be the only method, which can be successfully applied to the homogenisation of stochastic interface defects. Such an approach will make the elastic properties of the interphases much more sensitive to the presence of structural defects than was in case of the Probabilistic Averaging Method. Considering this, the bounds presented should be implemented in numerical analysis of stochastic structural defects into the artificial composite interphases.

2.3.5 Effective Constitutive Relations for the Steel Reinforced Concrete Plates

The homogenisation method proposed for composite plates analysis is not based on any mathematical model. However it seems to be very effective for high contrast steel−reinforced concrete plates [160]. The next main reason to apply this model is that the composite plate need not be periodic in the applied approach, which perfectly reflects the civil engineering needs. To get the effective characterisation for the elasticity tensor, Eshelby theorem can be used since upper and lower bounds for this tensor are determined. However it is proved by comparison with collected experimental results, either lower and upper bounds are very effective in computational modelling of a real plate. Both of them can be used to calculate the zeroth, first and second order stiffness matrix and the resulting probabilistic moments of displacements and stresses for the composite plate during the SFEM analysis. It decisively simplifies the numerical analysis in comparison to the traditional FEM modelling of such structures (where reinforcement discretisation is complicated); more accurate results, especially in terms of thin periodic plate vibration analysis, are shown in [155]. Finally, it should be mentioned that the homogenised effective characteristics for composite shells can be derived analogously, following considerations presented in [227,338].

Numerical test deals with the homogenisation of steel−reinforced concrete plates characterised by the data collected in Table 2.25; the coefficients of variation randomized Young moduli are taken as 0.1 as in all previous experiments. The concrete rectangular plate with horizontal dimensions 0.90 m x 0.90 m and thickness 0.045 m, supported at its corners and loaded by the vertical concentrated force is examined and Table 2.26 contains the deterministic and probabilistic homogenisation output. It can be observed that, as in previous examples, the deterministic and expected values are close to each other, respectively, and the resulting coefficients of variation are obtained as smaller or equal to those taken for input random variables.

Table 2.25. Material data of the composite plate

Material properties	Steel	Concrete
Young modulus	200.0 GPa	28.6 GPa
Poisson ratio	0.30	0.15
Volume fraction	0.0367	0.9633
Yield stress	345.0 GPa	20.68 GPa

Table 2.26. Effective materials characteristics

Effective elasticity tensor components	Deterministic	Expected value	Variation
$E\left[\inf\left(C_{1111}\right)\right]$	42.53 GPa	42.52 GPa	0.0985
$E\left[\sup\left(C_{1111}\right)\right]$	44.84 GPa	44.84 GPa	0.0905
$E\left[\inf\left(C_{1212}\right)\right]$	13.13 GPa	13.12 GPa	0.0982
$E\left[\sup\left(C_{1212}\right)\right]$	13.88 GPa	13.88 GPa	0.0896
$E\left[\inf\left(C_{1122}\right)\right]$	16.27 GPa	16.28 GPa	0.0991
$E\left[\sup\left(C_{1122}\right)\right]$	17.09 GPa	17.09 GPa	0.0896

The most important observation is that the lower and upper bounds are almost equal for any of the effective elasticity tensor components. Thus it does not matter which of them are used in the approximation of the real composite structure. Hence, the very complicated discretisation process of this particular concrete structure type (ABAQUS) can be replaced with an analysis of the homogeneous plate with elasticity tensor components calculated as proposed above. After successful verification of other reinforced concrete plates with various combinations of input parameters, such formulas for the effective elasticity tensor could be incorporated in the finite element stiffness formation process to speed up the FEM modelling procedures for these structures.

The variability analysis for expected values and the coefficients of variation of the effective elasticity tensor is presented in Figures 2.141 and 2.142 as a function of Young moduli expectations of the steel and concrete. It is seen that the Young modulus of the concrete matrix is detected as a crucial parameter for both probabilistic moments. It is due to the fact that the matrix is the dominating component (in the volumetric context) while the equations for homogenised tensor are rewritten as functions of the volume ratios of the composite components.

Considering the above, the behaviour of a real composite is compared against the homogenised one, cf. Figure 2.143. It is seen that the central deflection increments for both models are almost equal in the elastic range and, further, some expressions for the nonlinear range should be proposed and verified.

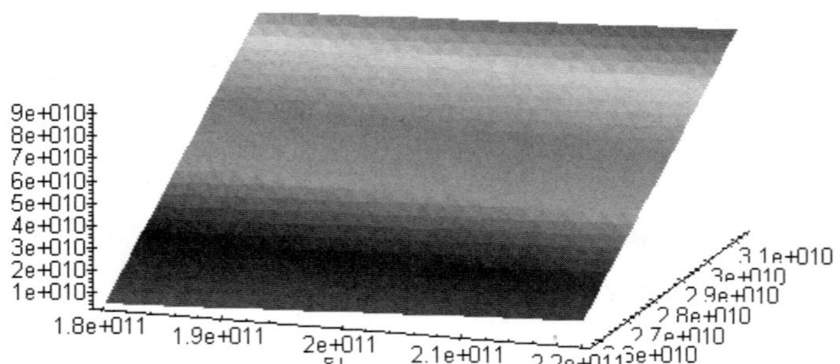

Figure 2.141. Expected value of upper bound for the component C_{1111}

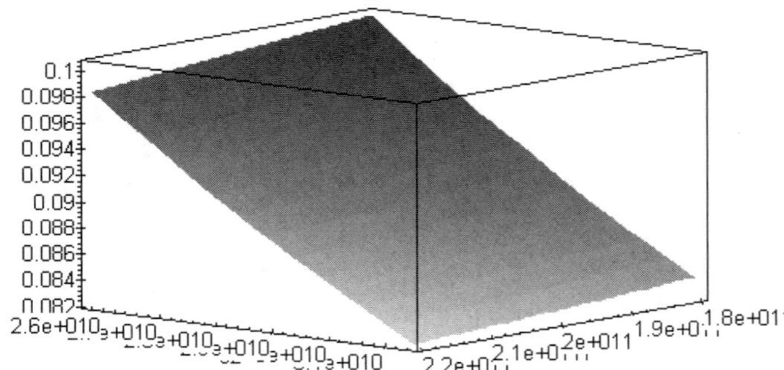

Figure 2.142. Coefficient of variation of upper bound for the component C_{1111}

A very broad discussion on theoretical and numerical modelling concepts in reinforced concrete structures have been presented in [22] − fracture analysis contained in this study can be incorporated into the SFEM using the approach described in [33]. Future analyses devoted to the application of homogenisation technique in reinforced plates modelling should focus on incorporation of the microcracks appearing in real matrices. It can be done using initial homogenisation of the cracks into the matrix [92,266,321] to find equivalent homogeneous medium; further homogenisation follows the above considerations.

Taking into account all the results of this test as well as the previous analyses on the homogeneous plates with random parameters, the application of the Stochastic Finite Element Method for the homogenised plate should approximate the probabilistic moments of displacements [63] in linear elastic range for the real plate very well. The expected values and variances of the effective elasticity tensor can be obtained for this purpose by using symbolic MAPLE computations analogous to those presented above.

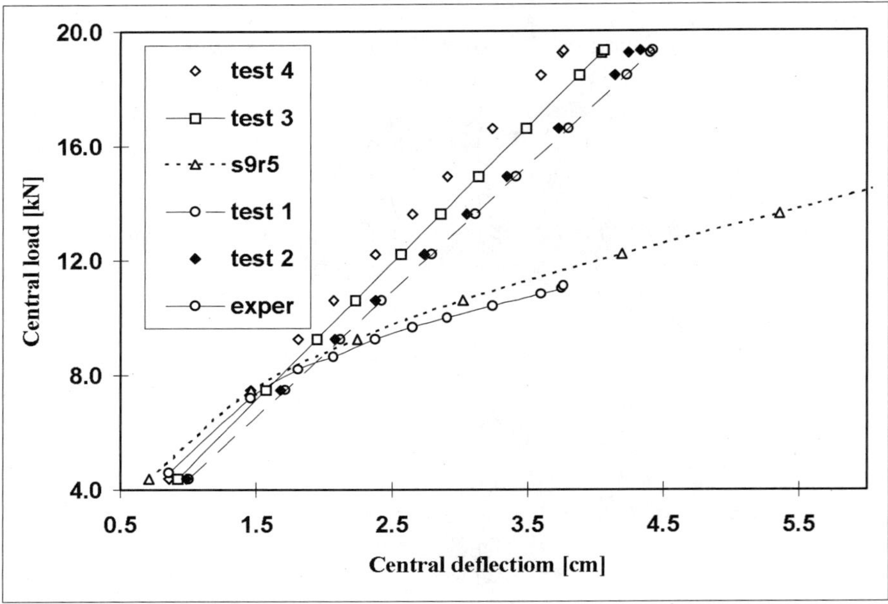

Figure 2.143. Vertical displacements of the composite plate centre

2.4 Conclusions

The main advantage of the homogenisation approach proposed is that any randomness in geometry or elasticity of the composite structures is replaced by a single effective random variable of the elasticity tensor components characterising such a structure. Hence, computational studies of engineering composites with different random variables using a homogeneous one with deterministically defined geometry and equivalent probability density function of the elastic properties can be carried out. It is observed that using an analytical expression for the homogenised elastic properties, the randomness in geometry for the periodicity cell can be introduced and can result in random fluctuations of the effective parameters only. Furthermore, even if the composite structure is not periodic, the results of homogenisation method application are satisfactory, i.e. the probabilistic response of the structure homogenised approximates very well the real composite model; analytical solution in the correlative approach for random quasi−periodic structures can be found in [278].

The basic value of the proposed homogenisation method is that the equations for the expected values and covariances of effective characteristics do not depend on the PDF type of the input random fields. However, in case of greater values of higher order probabilistic moments related to the first two as well as the lack of the

PDFs symmetry, a higher order version of the perturbation method is recommended. It is important since the probability density function of the input may not always be assumed properly, while in most experimental cases it is a subject of the statistical approximation only. Application of a stochastic higher order perturbation technique is relatively easy for closed form homogenisation equations considering the symbolic differentiation approach. It should be emphasised that, taking into account the capability of MAPLE links with FORTRAN routines, the program can be used in further SFEM computations as an intermediate procedure for symbolic homogenisation and sequential order perturbation derivation.

It should be underlined that the method proposed can find its application in stochastic reliability studies (SOSM approach) for various composite structures. This homogenisation technique makes it possible to reduce significantly the total number of degrees of freedom for such a structure, while the expected values and covariances of displacements and stresses enable one to estimate the second order second moment reliability (SORM) index or even third order reliability coefficients (W-SOTM). In the same time, both probabilistic methodologies have [171,175,180] and can find further applications in determination of effective heat conductivity coefficients in various models [216,294] including fibre-reinforced structures with some interfacial thermal resistance [303].

Due to the satisfactory accuracy of the homogenisation approach in modelling of composite structures, the model worked out can be treated as the first step for so-called self-homogenising finite elements, where the computer program automatically homogenises the entire structure using original material composite characteristics and finally calculates the displacements and stresses probabilistic moments for an equivalent homogeneous medium. On the other hand, the stochastic perturbation homogenisation procedure can be further modified for elastoplastic composite structures using Transformation Field Analysis (TFA) or Fast Fourier Transform (FFT) approaches. In the same time, the study of stochastic elastodynamic effective behaviour is recommended since the still growing range of composites has possible engineering applications.

2.5 Appendix

We prove, in the context of the composite model introduced in this chapter, that $\mathbf{u}(\mathbf{x},\mathbf{y})$ being a solution of problem (2.121) is constant in the region Ω. For this purpose, let us consider $\boldsymbol{u}(\boldsymbol{y})$ being a $\Omega-$periodic displacement function and the solution of the following boundary value problem:

$$\begin{cases} -\dfrac{\partial}{\partial y_j}\sigma_{ij} = F_i(\mathbf{y}),\ \mathbf{x}\in\Omega \\ \sigma_{ij} = C_{ijkl}(\mathbf{y})\varepsilon_{kl}(\mathbf{u}),\ \mathbf{x}\in\Omega_1\cup\Omega_2 \\ |\sigma_N| = g(\mathbf{y}),\ \mathbf{x}\in\Gamma_r, r=1,\dots,m \end{cases} \qquad (A2.1)$$

where $g(\mathbf{y})$ are given functions defined on Ω or Γ_r, $r=1,\dots,m$ with m being the total number of various interface boundaries. The variational formulation of (A2.1) may be stated as follows:

$$-\sum_{r=1}^{m}\int_{\Gamma_r}\sigma_{ij}n_j v_i d\Gamma + \sum_{a=1}^{n}\int_{\Omega_a}\sigma_{ij}\varepsilon_{ij}(\mathbf{v})d\Omega = \int_{\Omega}f_i v_i d\Omega \qquad (A2.2)$$

for $\mathbf{v}\in V$ being the following space:

$$V = \left\{ v_i ; v_i \in \left[H^1(\Omega_i)\right]^3 ; [v_i]=0\ on\ \Gamma_r ; r=1,\dots,m \right\} \qquad (A2.3)$$

while the corresponding components of the vector $\mathbf{v}$ are equal on the two opposite faces of Ω. Taking into account these conditions and neglecting body forces, we arrive at the well−known relation, cf. (2.121):

$$a(\mathbf{u},\mathbf{v}) = L(\mathbf{v}) \qquad (A2.4)$$

If $v_i = c$ is taken, which belongs to the set C of vectors constant on Ω, there holds for all $c\in C$

$$L(c) = 0 \qquad (A2.5)$$

Thus, if $g(\mathbf{y})$ from (A2.1) is such that $L(c)\neq 0$, there is no solution for the problem (A2.1). Next, let us introduce the space $S=V/C$ and let us denote by $\|.\|_k$ the norm in $\left[H^1(\Omega_k)\right]^3$ and by $.\|_k$ the norm in $L^2(\Omega_k)$. Let us observe that

(1)   V is a subspace of $\tilde{V} : \left[H^1(\Omega_1)\right]^3 \times \left[H^2(\Omega_2)\right]^3$;

(2) $\tilde{V}$ is a Hilbert space for the norm

$$\|\mathbf{v}\| = \left(\|v_1\|_1^2 + \|v_2\|_2^2\right)^{\frac{1}{2}} \qquad (A2.6)$$

(3)   There holds

$$\|\mathbf{v}\|_k^2 = \sum_{i,j}\left|\varepsilon_{ij}(\mathbf{v})\right|_k^2 + \sum_i \left|v_i\right|_k^2 \qquad (A2.7)$$

(4)   (A8) may be written equivalently as

$$\|\mathbf{v}\| = \left(\|v_1\|_1^2 + \sum_{i,j}\left|\varepsilon_{ij}(v_2)\right|_2^2\right)^{\frac{1}{2}}$$

(A2.8)

(5)   It can be proved that S is an Hilbert space for the norm

$$\|\mathbf{v}\| = \operatorname*{Inf}_{c\in C}\|\mathbf{v}+c\|$$

(A2.9)

(6)   The norm equivalent to (A2.9) on space S may be rewritten as

(A2.10)

$$N(\mathbf{v}) = \left(\sum_{i,j}\left|\varepsilon_{ij}(\mathbf{v})\right|_1^2 + \sum_{i,j}\left|\varepsilon_{ij}(\mathbf{v})\right|_2^2\right)^{\frac{1}{2}}$$

If we prove the statement (6) thus, due to the fact that $N(\mathbf{v})$ is continuous as well as coercive on space S and, further, applying the Lax–Millgram theorem we arrive at the conclusion that there exists a unique solution for (A2.1). To show this fact let us note that

$$\|\mathbf{v}\|^2 = \operatorname*{Inf}_{c\in C}\|\mathbf{v}+c\|^2 = N^2(\mathbf{v}) + \operatorname*{Inf}_{c\in C}\|v_1+c\|_1^2$$

(A2.11)

where

$$\|\mathbf{v}\|^2 \le N^2(\mathbf{v})$$

(A2.12)

There exists such a constant c_1 that for all $\mathbf{v}\in W$ that there holds

$$\|\mathbf{v}\|^2 \le c_1 N^2(\mathbf{v})$$

(A2.13)

Let us introduce the orthogonal projection operator O such that

$$O:\left(L^2(\Omega_1)\right)^3 \to C$$

(A2.14)

with respect to the scalar product corresponding to $\left|.\right|_1$. It yields

$$\|\mathbf{v}\|^2 = N^2(\mathbf{v}) + \left|v_1 - Ov_1\right|_1^2$$

(A2.15)

Equation (A2.15) is true if and only if for all $\mathbf{v}\in V$ there holds

$$\left|v_1 - Ov_1\right|_1^2 \le c_1 N^2(\mathbf{v})$$

(A2.16)

We assume that it is possible to improve $\mathbf{v}^n = \left(v_1^n, v_2^n\right)\in V$ for any positive n such that

$$\left|v_1^n - Ov_1^n\right|_1^2 = 1, \; N^2\left(\mathbf{v}^n\right) \le \frac{1}{n} \tag{A2.17}$$

Setting $\mathbf{w}^n = \left(v_1^n - Ov_1^n, v_2^n - Ov_2^n\right)$ we get for all $c \in C$ that

$$N^2\left(\mathbf{w}^n\right) \le \frac{1}{n}; \; \left|w_1^n\right|_1 = 1; \; \left(w_1^n, c\right) = 0 \tag{A2.18}$$

Then, $\left\{\mathbf{w}^n\right\}$ is bounded in V and there exists such a subsequence $\left\{\mathbf{w}^m = \left(w_1^m, w_2^m\right)\right\}$, which converges weakly to w_0 in V. Since that, w_1^μ converges strongly in $\left(L^2\left(\Omega_1\right)\right)^3$

$$\left|w_1^0\right|_1 = 1 \tag{A2.19}$$

Due to the lower semi−continuity there holds

$$N^2\left(\mathbf{w}^0\right) \le \underline{\lim} N^2\left(\mathbf{w}^m\right) = 0 \tag{A2.20}$$

Finally, it is obtained that

$$\varepsilon_{ij}\left(w_1^0\right) = 0; \; \varepsilon_{ij}\left(w_2^0\right) = 0 \tag{A2.21}$$

which gives as the result

$$w_1^0 = w_2^0 = c \in C \tag{A2.22}$$

3 Elastoplastic Problems

3.1 Introduction

There are numerous well−established techniques to calculate effective material characteristics for composite materials. In the case of composite components volume fractions only, one can use the closed form algebraic equations on upper and lower bounds or direct estimates for the effective material tensor components. Otherwise, the cell problems are formulated and solved using their Finite Element Method (FEM) or, alternatively, the Boundary Element Method (BEM) numerical implementations that enable direct computations of the effective characteristics. Recent advances in the area of computational methods in homogenisation of the nonlinear effective characterisation of heterogeneous materials and structures are reported in [4,85,86,107,112,136,250,325]. In the same time, stochastic analysis is still being developed to estimate or to compute probabilistic moments of homogenised material tensors.

Homogenisation of composite materials with elastoplastic constituents is presented below using the so−called Transformation Field Analysis (TFA) proposed by Dvorak and now applied to approximate the effective nonlinear behaviour of a three−component periodic composite. The self−consistent model and Mori−Tanaka theory, providing the estimation of the overall thermoelastic constants of composites on the basis of constituent properties and volume fractions, are partially incorporated in this model. Computational implementation of the method consists of the utilisation of the program ABAQUS to enable automatic homogenisation of n−component periodic composites in a general configuration of the components in the RVE. Numerical examples of the three−component periodic composite homogenisation make it possible to compare the nonlinear behaviour of a composite for its real and homogenised models in the case of the specific boundary problem defined for the cell. The next step in the development of this approach would be to determine the parameter sensitivity of the homogenised properties of the composite with respect to the material characteristics of the constituents as well as to some geometrical data defining the RVE. Statistical and stochastic simulation of probabilistic moments of the effective material tensors would be possible after such a sensitivity determination, taking into account the experimental knowledge of the statistical parameters of the composite constituents.

3.2 Homogenisation Method

The periodic n−component composite in the plane orthogonal to the fibre direction is considered where perfectly bonded components are assumed to be elastoplastic. Mechanical behaviour of the composite constituents is represented by time and temperature dependent constitutive relations under the assumption that for any time τ the total strains and stresses can be decomposed as

$$\varepsilon_r(\mathbf{y},\tau) = \varepsilon_r^{el}(\mathbf{y},\tau) + \varepsilon_r^{*}(\mathbf{y},\tau) \tag{3.1}$$

$$\sigma_r(\mathbf{y},\tau) = \sigma_r^{el}(\mathbf{y},\tau) + \sigma_r^{*}(\mathbf{y},\tau) \tag{3.2}$$

where ε_r^{el}, ε_r^{*} denote elastic strain resulting from a given displacement boundary condition applied on the region Ω_r and the eigenstrain in the same subregion, respectively; σ_r^{el}, σ_r^{*} stand for the elastic stress and eigenstress tensor components in Ω_r. The eigenstrain and eigenstress fields considered here as transformation field may be decomposed in the case of thermal and inelastic effects as

$$\varepsilon_r^{*}(\mathbf{y},\tau) = \mathbf{m}_r \alpha_T(\tau) + \varepsilon_r^{inel}(\mathbf{y},\tau) \tag{3.3}$$

$$\sigma_r^{*}(\mathbf{y},\tau) = \mathbf{l}_r \alpha_T(\tau) + \sigma_r^{rel}(\mathbf{y},\tau) \tag{3.4}$$

where m_r denotes the thermal strain tensor, l_r is the thermal stress tensor, while $\alpha_T(\tau)$ represents the linear thermal expansion coefficient. A procedure of determination of the effective thermal expansion coefficients for various composites has been described in [253,305,311]. Since ε_r^{inel} is the inelastic strain and σ_r^{rel} is the relaxation stress, (3.1) and (3.2) can be written as

$$\varepsilon_r(\mathbf{y},\tau) = \mathbf{M}_r \sigma_r(\mathbf{y},\tau) + \mathbf{m}_r \theta(t) + \varepsilon_r^{inel}(\mathbf{y},\tau) \tag{3.5}$$

$$\sigma_r(\mathbf{y},\tau) = \mathbf{C}_r \varepsilon_r(\mathbf{y},\tau) + \mathbf{l}_r \theta(t) + \sigma_r^{rel}(\mathbf{y},\tau) \tag{3.6}$$

where $\mathbf{C}_r$ and $\mathbf{M}_r$ are the elastic and compliance tensor components for the subregion Ω_r. Hence, it is possible to write the following relations between $\mathbf{m}_r, \mathbf{l}_r, \mathbf{M}_r, \mathbf{C}_r, \varepsilon_r^{inel}$ and σ_r^{rel} :

$$\mathbf{M}_r = \mathbf{C}_r^{-1} \tag{3.7}$$

$$\mathbf{m}_r = -\mathbf{M}_r \mathbf{l}_r \tag{3.8}$$

$$\mathbf{l}_r = -\mathbf{C}_r \mathbf{m}_r \tag{3.9}$$

$$\varepsilon_r^{inel}(\mathbf{y},\tau) = -\mathbf{M}_r \sigma_r^{rel}(\mathbf{y},\tau) \tag{3.10}$$

$$\sigma_r^{rel}(\mathbf{y},\tau) = -\mathbf{C}_r \varepsilon_r^{inel}(\mathbf{y},\tau) \tag{3.11}$$

Mechanical and thermal elastic influence functions are given by the following relations:

$$\varepsilon_r(\mathbf{y},\tau) = \mathbf{A}_r(\mathbf{y})\varepsilon(\tau) + \mathbf{D}_{rs}(\mathbf{y},\mathbf{y}')\varepsilon_s^*(\mathbf{y},\tau) \tag{3.12}$$

$$\sigma_r(\mathbf{y},\tau) = \mathbf{B}_r(\mathbf{y})\varepsilon(\tau) + \mathbf{F}_{rs}(\mathbf{y},\mathbf{y}')\sigma_s^*(\mathbf{y},\tau) \tag{3.13}$$

Matrices $\mathbf{B}_r(\mathbf{y})$ and $\mathbf{A}_r(\mathbf{y})$ in (3.12) and (3.13) denote stress and strain concentration factor tensors representing the volume averages of the corresponding functions over the periodicity cell, as is proposed in (3.14) to (3.17). To describe the overall homogenised response of volume Ω, the resulting strains and stresses are combined with their corresponding local components described by (3.3) to (3.6) as

$$\varepsilon(\tau) = \frac{1}{|\Omega|}\int_\Omega \varepsilon_r(\mathbf{y},\tau)d\Omega$$

$$= \frac{1}{|\Omega|}\int_\Omega [\varepsilon_r^{el}(\mathbf{y},\tau) + \varepsilon_r^*(\mathbf{y},\tau)]d\Omega = \varepsilon^{el}(\tau) + \varepsilon^*(\tau) \tag{3.14}$$

$$\sigma(\tau) = \frac{1}{|\Omega|}\int_V \sigma_r(\mathbf{y},\tau)d\Omega =$$

$$\frac{1}{|\Omega|}\int_\Omega [\sigma_r^{el}(\mathbf{y},\tau) + \sigma_r^*(\mathbf{y},\tau)]d\Omega = \sigma^{el}(\tau) + \sigma^*(\tau) \tag{3.15}$$

Then, local elastic fields may be written as

$$\varepsilon^{el}(\tau) = \frac{1}{|\Omega|}\int_\Omega [\mathbf{A}_r(\mathbf{y})\varepsilon(\tau) + \mathbf{a}_r(\mathbf{y})\alpha_T(\tau)]\, d\Omega \tag{3.16}$$

$$\sigma^{el}(\tau) = \frac{1}{|\Omega|}\int_\Omega [\mathbf{B}_r(\mathbf{y})\sigma(\tau) + \mathbf{b}_r(\mathbf{y})\alpha_T(\tau)]\, d\Omega \tag{3.17}$$

where $\mathbf{a}_r(\mathbf{y})$ and $\mathbf{b}_r(\mathbf{y})$ are the thermoelastic strain and stress concentration factors [86,94]. The strain transformation field $\varepsilon^*(\mathbf{y},\tau)$ defined in Ω results in the displacements on the unconstrained part of surface $\partial\Omega$, while the transformation stress $\sigma^*(\mathbf{y},\tau)$ generates surface tractions on Ω being constrained. The relation between the local and global transformation fields is proposed as

$$\varepsilon^*(\tau) = \frac{1}{|\Omega|}\int_\Omega [\varepsilon_r(\mathbf{y},\tau) - \mathbf{A}_r(\mathbf{y})\varepsilon(\tau)]\, d\Omega = \frac{1}{|\Omega|}\int_\Omega \mathbf{A}_r(\mathbf{y})\varepsilon^*(\mathbf{y},\tau)\, d\Omega \tag{3.18}$$

$$\sigma^*(\tau) = \frac{1}{|\Omega|}\int_\Omega [\sigma_r(\mathbf{y}) - \mathbf{B}_r(\mathbf{y})\sigma(\tau)]\, d\Omega = \frac{1}{|\Omega|}\int_V \mathbf{B}_r^T(\mathbf{y})\sigma^*(\mathbf{y},\tau)\, d\Omega \tag{3.19}$$

The elastic local strain $\varepsilon_r(\mathbf{y},\tau)$ and stress fields $\sigma_r(\mathbf{y},\tau)$ are found from a superposition of the elastic local fields $\varepsilon_r^{el}(\mathbf{y},\tau)$ and $\sigma_r^{el}(\mathbf{y},\tau)$ with local eigenstrains $\varepsilon_r^*(\mathbf{y},\tau)$ and eigenstresses $\sigma_r^*(\mathbf{y},\tau)$, respectively; the same model in the context of global scale is introduced analogously. These two different scales are linked using the formulation for local strain and stress fields in the following form:

$$\varepsilon_r(\mathbf{y'}) = \mathbf{A}_r(\mathbf{y'})\varepsilon + \mathbf{D}_{rs}(\mathbf{y},\mathbf{y'})\varepsilon_s^*(\mathbf{y'}) \tag{3.20}$$

$$\sigma_r(\mathbf{y}) = \mathbf{B}_r(\mathbf{y})\sigma + \mathbf{F}_{rs}(\mathbf{y},\mathbf{y'})\sigma_s^*(\mathbf{y'}) \tag{3.21}$$

$\mathbf{D}_{rs}(\mathbf{y},\mathbf{y'})$, $\mathbf{F}_{rs}(\mathbf{y},\mathbf{y'})$ are transformation strain and stress influence functions, which enable us to relate the strain and stress tensor components on the macroscale defined by $\mathbf{y}$ and the microscale identified by $\mathbf{y'}$. Solving the following boundary value problem on the RVE we get

$$div\sigma(\mathbf{y}) = \frac{\partial\sigma(\mathbf{y})}{\partial\mathbf{y}} = 0 \tag{3.22}$$

$$\varepsilon_r(\mathbf{y}) = \mathbf{M}_r\sigma_r(\mathbf{y}) + \varepsilon_r^*(\mathbf{y}) \tag{3.23}$$

$$\frac{1}{|\Omega|}\int_\Omega \varepsilon_r(\mathbf{y})d\Omega = \langle\varepsilon\rangle_\Omega \tag{3.24}$$

$$\mathbf{u}(\mathbf{y}) = \varepsilon\mathbf{y} + \mathbf{u}^*(\mathbf{y}) \tag{3.25}$$

where the local uniform strain field ε_r is found using the matrices $\mathbf{A}_r(\mathbf{y})$, $\mathbf{D}_{rs}(\mathbf{y},\mathbf{y'})$. Further, it is possible to determine the approximated expression of the averaged strain in the subvolume Ω_r given as

$$\varepsilon_r = \mathbf{A}_r\varepsilon + \sum_{r,s=1}^N \mathbf{D}_{rs}\varepsilon_r^* \tag{3.26}$$

Analogous to (3.26), the averaged stresses in the subregion Ω_r can be written in the form

$$\sigma_r = \mathbf{B}_r\sigma + \sum_{r,s=1}^N \mathbf{F}_{rs}\sigma_r^* \tag{3.27}$$

It is observed that $\mathbf{F}_r(\mathbf{y},\mathbf{y'})$ and $\mathbf{D}_r(\mathbf{y},\mathbf{y'})$ are eigenstress and eigenstrain influence functions, that reflect the effect on the scale $\mathbf{y}$ resulting from a transformation on the scale $\mathbf{y'}$ under overall uniform applied stress or strain. The additional influence functions are determined in terms of averages and piecewise constant approximations in the introduced subregions inside the RVE. Therefore, under overall strain $\varepsilon(t)=0$, the transformation concentration factor tensor $\mathbf{D}_{rs}$ gives the strain induced in the subvolume Ω_r by a unit uniform eigenstrain in Ω_s. Under overall stress $\sigma(t)=0$, the concentration factor tensor $\mathbf{F}_{rs}$ defines the stress in Ω_r resulting from the unit eigenstrain located in Ω_s. It can be shown that these tensors can be determined in the case of multiphase medium as

$$\mathbf{D}_{sr} = (\mathbf{I}-\mathbf{A}_r)(\mathbf{C}_r-\mathbf{C})^{-1}(\delta_{rs}\mathbf{I}-\mathbf{c}_s\mathbf{A}_s^T)\mathbf{C}_s \tag{3.28}$$

$$\mathbf{F}_{sr} = (\mathbf{I}-\mathbf{B}_r)(\mathbf{M}_r-\mathbf{M})^{-1}(\delta_{rs}\mathbf{I}-\mathbf{c}_s\mathbf{B}_s^T)\mathbf{M}_s \tag{3.29}$$

$(r,s=1,...,N$, without summation over repeated indices)

which for a two–component composite gives

$$\mathbf{D}_{p\alpha} = (\mathbf{I}-\mathbf{A}_p)(\mathbf{C}_\alpha-\mathbf{C}_\beta)^{-1}\mathbf{C}_\alpha$$

$$\mathbf{D}_{p\beta} = -(\mathbf{I}-\mathbf{A}_p)(\mathbf{C}_\alpha-\mathbf{C}_\beta)^{-1}\mathbf{C}_\beta \tag{3.30}$$

$$\mathbf{F}_{p\alpha} = (\mathbf{I}-\mathbf{B}_p)(\mathbf{M}_\alpha-\mathbf{M}_\beta)^{-1}\mathbf{M}_\alpha$$

$$\mathbf{F}_{p\beta} = -(\mathbf{I}-\mathbf{B}_p)(\mathbf{M}_\alpha-\mathbf{M}_\beta)^{-1}\mathbf{M}_\beta \tag{3.31}$$

$$\text{for } p=\alpha,\beta$$

This completes the description of the homogenisation method for a composite with elastoplastic coefficients by use of the Transformation Field Analysis (TFA). It should be underlined that, in comparison to the homogenisation model valid for the linear elastic range, the necessity of transformation matrix computations is crucial for the proposed nonlinear FEM analysis.

3.3 Finite Element Equations of Elastoplasticity

The following boundary value problem is now considered [206,210]:

$$\Delta\sigma_{kl,l} = 0 ; \ \mathbf{x}\in\Omega \tag{3.32}$$

$$\Delta\tilde{\sigma}_{kl} = C_{klmn}\Delta\varepsilon_{mn} ; \ \mathbf{x}\in\Omega \tag{3.33}$$

$$\Delta\varepsilon_{mn} = \tfrac{1}{2}[\Delta u_{k,l}+\Delta u_{l,k}+u_{i,k}\Delta u_{i,l}+\Delta u_{i,k}u_{i,l}+\Delta u_{i,k}\Delta u_{i,l}] ; \ \mathbf{x}\in\Omega \tag{3.34}$$

with the boundary conditions

$$\Delta\sigma_{\bar{k}l}n_l = \Delta t_{\bar{k}} ; \ \mathbf{x}\in\partial\Omega_\sigma , \ \bar{k}=1,2,3 \tag{3.35}$$

$$\Delta u_{\hat{k}} = \Delta\hat{u}_{\hat{k}} ; \ \mathbf{x}\in\partial\Omega_u , \ \hat{k}=1,2,3 \tag{3.36}$$

This problem is solved for displacements $u_k(\mathbf{x})$, strain $\varepsilon_{kl}(\mathbf{x})$ and stress $\sigma_{kl}(\mathbf{x})$ tensor components fulfilling equilibrium equations (3.32)–(3.36). Let us note that the stress tensor increments $\Delta\sigma_{kl}(\mathbf{x})$, $\Delta\tilde{\sigma}_{kl}(\mathbf{x})$ denote here the first and second Piola–Kirchhoff tensors

$$\Delta\sigma_{kl} = \Delta F_{km}\Delta\tilde{\sigma}_{ml} + F_{km}\Delta\tilde{\sigma}_{ml} + \Delta F_{km}\tilde{\sigma}_{ml} \ ; \ \mathbf{x}\in\Omega \qquad (3.37)$$

where

$$\Delta F_{km} = \Delta u_{k,m} \ ; \ \mathbf{x}\in\Omega \qquad (3.38)$$

To get the solution, the following functional defined on the displacement increments as Δu_k is introduced:

$$J(\Delta u_k) = \int_{\Omega}\left(\tfrac{1}{2}C_{klmn}\Delta\varepsilon_{kl}\Delta\varepsilon_{mn} + \tfrac{1}{2}\tilde{\sigma}_{kl}\Delta u_{i,k}\Delta u_{i,l}\right)d\Omega - \int_{\partial\Omega}\Delta\hat{t}_k\Delta u_k\,d(\partial\,\Omega) \qquad (3.39)$$

Let us note that this methodology is common for homogeneous materials as well as heterogeneous media. In case of composites, the last equation can be decomposed into the integrals valid for particular constituents and their boundaries and interfaces, separately.

Now, let us introduce the displacement increment function $\Delta u_k(\mathbf{x})$ being continuous and differentiable on Ω and, consequently, including all geometrically continuous and coherent subsets (finite elements) Ω_e, $e=1,...,E$ discretising the entire Ω. It is not assumed that $\Delta u_k(\mathbf{x})$ is differentiable on the interelement surfaces and boundaries $\partial\Omega_{ef}$ (for $e,f=1,...,E$, $e\neq f$). Next, let us consider the following approximation of $\Delta u_k(\mathbf{x})$ for $\mathbf{x}\in\Omega$:

$$\Delta u_k(\mathbf{x}) = \sum_{\zeta=1}^{N_e}\varphi_{\zeta k}(\mathbf{x})\Delta u_{\zeta}^{(N)} \qquad (3.40)$$

where $\varphi_{\zeta k}(\mathbf{x})$ are the shape functions for node k, $\Delta u_{\zeta}^{(N)}$ represents the degrees of freedom (DOF) vector, while N_e is the total number of the DOF in this node. Considering above, the displacements and strains gradients are rewritten as follows:

$$\Delta u_{k,l}(\mathbf{x}) = \varphi_{k,l}^{\zeta}(\mathbf{x})\Delta u_{\zeta}^{(N)} \qquad (3.41)$$

$$\Delta\bar{\varepsilon}_{kl}(\mathbf{x}) = [\overline{B}_{kl}^{(1)\zeta} + \overline{B}_{kl}^{(2)\zeta}]\Delta u_{\zeta}^{(N)} = \overline{B}_{kl}^{\zeta}\Delta u_{\zeta}^{(N)} \qquad (3.42)$$

$$\Delta\bar{\bar{\varepsilon}}_{kl}(\mathbf{x}) = \overline{\overline{B}}_{kl}^{\zeta\xi}\Delta u_{\zeta}^{(N)}\Delta u_{\xi}^{(N)} \qquad (3.43)$$

and finally

$$\Delta\varepsilon_{kl}(\mathbf{x}) = \Delta\bar{\varepsilon}_{kl}(\mathbf{x}) + \Delta\bar{\bar{\varepsilon}}_{kl}(\mathbf{x}) \tag{3.44}$$

The following notation is applied (3.42) and (3.43):

$$\bar{B}_{kl}^{(1)\zeta}(\mathbf{x}) = \varphi_{k,l}^{\zeta}(\mathbf{x}) \tag{3.45}$$

$$\bar{B}_{kl}^{(2)\zeta}(\mathbf{x}) = \varphi_{i,k}^{\zeta}(\mathbf{x})\varphi_{i,l}^{\xi}(\mathbf{x})u_{\xi}^{(N)} \tag{3.46}$$

$$\bar{\bar{B}}_{kl}^{\zeta\xi}(\mathbf{x}) = \tfrac{1}{2}\varphi_{i,k}^{\zeta}(\mathbf{x})\varphi_{i,l}^{\xi}(\mathbf{x}) \tag{3.47}$$

All these equations are substituted into the variational formulation of the problem (cf. (3.39)). There holds

$$\begin{aligned}
\tfrac{1}{2}C_{klmn}\Delta\varepsilon_{kl}\Delta\varepsilon_{mn} &= \tfrac{1}{2}C_{klmn}\left(\Delta\bar{\varepsilon}_{kl} + \Delta\bar{\bar{\varepsilon}}_{kl}\right)\left(\Delta\bar{\varepsilon}_{mn} + \Delta\bar{\bar{\varepsilon}}_{mn}\right) \\
&= \tfrac{1}{2}C_{klmn}\left(\Delta\bar{\varepsilon}_{kl}\Delta\bar{\varepsilon}_{mn} + \Delta\bar{\varepsilon}_{kl}\Delta\bar{\bar{\varepsilon}}_{mn} + \Delta\bar{\bar{\varepsilon}}_{kl}\Delta\bar{\varepsilon}_{mn} + \Delta\bar{\bar{\varepsilon}}_{kl}\Delta\bar{\bar{\varepsilon}}_{mn}\right) \\
&= \tfrac{1}{2}C_{klmn}\left(\bar{B}_{kl}^{\zeta}\Delta u_{\zeta}^{(N)}\bar{B}_{mn}^{\xi}\Delta u_{\xi}^{(N)} + \bar{B}_{kl}^{\zeta}\Delta u_{\zeta}^{(N)}\bar{B}_{mn}^{\mu\nu}\Delta u_{\mu}^{(N)}\Delta u_{\nu}^{(N)}\right. \\
&\left. + \bar{\bar{B}}_{kl}^{\zeta\xi}\Delta u_{\zeta}^{(N)}\Delta u_{\xi}^{(N)}\bar{B}_{mn}^{\mu}\Delta u_{\mu}^{(N)} + \bar{\bar{B}}_{kl}^{\zeta\xi}\Delta u_{\zeta}^{(N)}\Delta u_{\xi}^{(N)}\bar{\bar{B}}_{mn}^{\mu\nu}\Delta u_{\mu}^{(N)}\Delta u_{\nu}^{(N)}\right)
\end{aligned} \tag{3.48}$$

$$\tfrac{1}{2}\tilde{\sigma}_{kl}\Delta u_{i,k}\Delta u_{i,l} = \tfrac{1}{2}\tilde{\sigma}_{kl}\varphi_{i,k}^{\zeta}(\mathbf{x})\Delta u_{\zeta}^{(N)}\varphi_{i,l}^{\xi}(\mathbf{x})\Delta u_{\xi}^{(N)} \tag{3.49}$$

Next, the following notation is applied:

$$k_{\zeta\xi}^{(\sigma)e} = \int_{\Omega_e}\tilde{\sigma}_{kl}\varphi_{i,k}^{\zeta}(\mathbf{x})u_{\zeta}^{(N)}\varphi_{i,l}^{\xi}(\mathbf{x})d\Omega \tag{3.50}$$

$$k_{\zeta\xi}^{(con)e} = \int_{\Omega_e}\tfrac{1}{2}C_{klmn}\bar{B}_{kl}^{(1)\zeta}\bar{B}_{mn}^{(1)\xi}d\Omega \tag{3.51}$$

$$k_{\zeta\xi}^{(u)e} = \int_{\Omega_e}\tfrac{1}{2}C_{klmn}\left(\bar{B}_{kl}^{(1)\zeta}\bar{B}_{mn}^{(2)\xi} + \bar{B}_{kl}^{(2)\zeta}\bar{B}_{mn}^{(1)\xi} + \bar{B}_{kl}^{(2)\zeta}\bar{B}_{mn}^{(2)\xi}\right)d\Omega \tag{3.52}$$

where

$$k_{\zeta\xi}^{(1)e} = k_{\zeta\xi}^{(\sigma)e} + k_{\zeta\xi}^{(con)e} + k_{\zeta\xi}^{(u)e} \tag{3.53}$$

and for the second and third order terms

$$
\begin{aligned}
&k_{\zeta\xi}^{(2)e} \\
&= \int_{\Omega_e} \tfrac{3}{2} C_{klmn} \left(\overline{B}_{kl}^{\zeta} \Delta u_{\zeta}^{(N)} \overline{\overline{B}}_{mn}^{\mu\nu} \Delta u_{\mu}^{(N)} \Delta u_{\nu}^{(N)} + \overline{\overline{B}}_{kl}^{\zeta\xi} \Delta u_{\zeta}^{(N)} \Delta u_{\xi}^{(N)} \overline{B}_{mn}^{\mu} \Delta u_{\mu}^{(N)} \right) d\Omega
\end{aligned}
\tag{3.54}
$$

$$
k_{\zeta\xi}^{(3)e} = \int_{\Omega_e} 2 C_{klmn} \left(\overline{\overline{B}}_{kl}^{\zeta\xi} \Delta u_{\zeta}^{(N)} \Delta u_{\xi}^{(N)} \overline{\overline{B}}_{mn}^{\mu\nu} \Delta u_{\mu}^{(N)} \Delta u_{\nu}^{(N)} \right) d\Omega
\tag{3.55}
$$

Introducing $k_{\zeta\xi}^{(i)}$ for $i=1,2,3$ to the functional $J(\Delta u_k)$ in (3.39) and applying the transformation from the local to the global system by the use of the following formula, typical for the FEM implementation:

$$
\Delta u_{\zeta}^{(N)} = a_{\xi\alpha} \Delta q_{\alpha}
\tag{3.56}
$$

it is obtained that

$$
\begin{aligned}
J(\Delta q_{\alpha}) =\ & \tfrac{1}{2} K_{\alpha\beta}^{(1)} \Delta q_{\alpha} \Delta q_{\beta} + \tfrac{1}{3} K_{\alpha\beta\gamma}^{(2)} \Delta q_{\alpha} \Delta q_{\beta} \Delta q_{\gamma} \\
& + \tfrac{1}{4} K_{\alpha\beta\gamma\delta}^{(3)} \Delta q_{\alpha} \Delta q_{\beta} \Delta q_{\gamma} \Delta q_{\delta} - \Delta Q_{\alpha} \Delta q_{\alpha}
\end{aligned}
\tag{3.57}
$$

The stationarity of the functional $J(\Delta q_{\alpha})$ leads to the following algebraic equation:

$$
K_{\alpha\beta}^{(1)} \Delta q_{\beta} + K_{\alpha\beta\gamma}^{(2)} \Delta q_{\beta} \Delta q_{\gamma} + K_{\alpha\beta\gamma\delta}^{(3)} \Delta q_{\beta} \Delta q_{\gamma} \Delta q_{\delta} = \Delta Q_{\alpha}
\tag{3.58}
$$

being fulfilled for any configuration of Ω. The iterative numerical solution of this equation makes it possible, according to the specified boundary conditions, to compute the effective constitutive tensor components of the homogenised composite. It should be stressed that the first two components of the stiffness matrix are considered only in further numerical analysis (geometrical nonlinearity is omitted in the homogenisation process); a detailed description of the numerical integration and solution of (3.58) can be found in [12,72,271,276], for instance.

3.4 Numerical Analysis

As was mentioned above, the main goal of the TFA approach is to compute the transformation matrices $\mathbf{A}_r$, $\mathbf{D}_{rs}$ that are determined only once for the original geometry of the composite and assuming initially linear elastic characteristics of the constituents. There holds that

$$
\langle d\sigma_r \rangle_{\Omega_r} = \mathbf{C}_r^{\text{eff}} \langle d\varepsilon_r \rangle_{\Omega_r} = \mathbf{C}_r^{\text{eff}} \left\langle \left(d\varepsilon_r^{el} + d\varepsilon_r^{\text{inel}} \right) \right\rangle_{\Omega_r}
$$

$$\left\langle d\varepsilon_r \right\rangle_{\Omega_r} = \mathbf{A}_r d\varepsilon + \sum_{r,s=1}^{N} \mathbf{D}_{rs} \left\langle d\varepsilon_s^{inel} \right\rangle_{\Omega_s} \tag{3.59}$$

$$\left\langle d\boldsymbol{\sigma}_r^{el} \right\rangle_{\Omega_r} = \mathbf{C}_r^{el} \left\langle d\varepsilon_r^{el} \right\rangle_{\Omega_r} \quad \text{and} \quad \left\langle d\boldsymbol{\sigma}_r^{inel} \right\rangle_{\Omega_r} = \mathbf{C}_r \left\langle d\varepsilon_r^{inel} \right\rangle_{\Omega_r} \tag{3.60}$$

$$\mathbf{C}_r^{eff} = \mathbf{C}_r^{el} + \mathbf{C}_r \tag{3.61}$$

Further, using spatial averaging definitions, the averaged stress tensor components are calculated as follows:

$$\left\langle \varepsilon_r \right\rangle_{\Omega} = \varepsilon \quad \text{and} \quad \left\langle \sigma_r \right\rangle_{\Omega} = \sigma \tag{3.62}$$

Hence, the effective elasticity tensor components $\mathbf{C}^{eff}$ are derived for a given increment as

$$d\sigma = \mathbf{C}^{eff} d\varepsilon \tag{3.63}$$

$$\mathbf{C}^{eff} = \sum_{r=1}^{N} c_r \mathbf{C}_r^{el} + \sum_{r,s=1}^{N} c_r \left(\mathbf{D}_{rs} \varepsilon_s^{inel} \right)^{-1} : \sigma_r^{inel} \tag{3.64}$$

In the particular case of a two–component composite, the transformation and concentration matrices are obtained as, cf. (3.30) and (3.31)

$$\mathbf{D}_{11} = (\mathbf{I} - \mathbf{A}_1)(\mathbf{C}_1 - \mathbf{C}_2)^{-1} \mathbf{C}_1 \tag{3.65}$$

$$\mathbf{D}_{21} = (\mathbf{I} - \mathbf{A}_2)(\mathbf{C}_1 - \mathbf{C}_2)^{-1} \mathbf{C}_1 \tag{3.66}$$

$$\mathbf{D}_{12} = -(\mathbf{I} - \mathbf{A}_1)(\mathbf{C}_1 - \mathbf{C}_2)^{-1} \mathbf{C}_2 \tag{3.67}$$

$$\mathbf{D}_{22} = -(\mathbf{I} - \mathbf{A}_2)(\mathbf{C}_1 - \mathbf{C}_2)^{-1} \mathbf{C}_2 \tag{3.68}$$

$\mathbf{C}_1, \mathbf{C}_2$ denote here the components corresponding to elastic properties, while $\mathbf{A}_1, \mathbf{A}_2$ are mechanical concentration matrices. Finally, using (3.64) it is obtained that

$$\mathbf{C}^{eff} = \sum_{r=2}^{N} c_1 \mathbf{C}_1^{el} + c_2 \mathbf{C}_2^{el} + \sum c_1 \left(\mathbf{D}_{11} \varepsilon_1^{inel} \right)^{-1} : \sigma_1^{inel} + c_2 \left(\mathbf{D}_{12} \varepsilon_2^{inel} \right)^{-1} : \sigma_2^{inel}$$

$$+ \sum c_1 \left(\mathbf{D}_{21} \varepsilon_1^{inel} \right)^{-1} : \sigma_1^{inel} + c_2 \left(\mathbf{D}_{22} \varepsilon_2^{inel} \right)^{-1} : \sigma_2^{inel} \tag{3.69}$$

The FEM aspects of TFA computational implementation are discussed in detail in Section 3.4 below. Further, it should be noticed that there were some approaches in the elastoplastic approach to composites where, analogously to the linear elasticity homogenisation method, the approximation of the effective yield limit stresses of a composite is proposed as a quite simple closed form function

$$\sigma^{(eff)} = \sqrt{\sigma_1 \sigma_2}$$ (3.70)

or, in terms of the effective yield surface, in the following form:

$$\Phi(\sigma) = \max_{y \geq 0} \left\{ \left[c_2 + c_1 \left(\frac{\sigma_y^{(1)}}{\sigma_y^{(2)}} \right) y \right]^{-1} \sigma \left[m(y)\sigma - \left(\sigma_y^{(2)} \right)^2 \right] \right\}$$ (3.71)

where $m(y = \frac{\mu_1}{\mu_2}) = 3 \ \mu_2 V(\mu_1, \mu_2)$ and V is any estimate of the viscosity compliance tensor defined using the viscosities μ_1 and μ_2. A review of the most recent theories in this field can be found in [381], for instance.

The main aim of computational experiment presented is to determine the global nonlinear homogenised constitutive law for two component composites with elastoplastic components; the FEM based program ABAQUS [1] is used in all computational procedures. However the method presented can be implemented in any nonlinear FEM plane strain/stress code such as [60], for instance. The numerical experiments are carried out in the microstructural (RVE) level, and that is why the global response of the composite is predicted starting from the behaviour of the periodicity cell. The numerical micromechanical model consists of a three−component periodicity cell with horizontal and vertical symmetry axes and dimensions 3.0 cm (horizontal) × 2.13 cm (vertical) (cf. Figure 3.1 and 3.2). The composite is made of epoxy matrix and metal reinforcement with material properties of the components collected in Table 1. The void embedded into the steel casting simulates a lack of any matrix in the periodicity cell. Some nonzero material data are introduced to avoid numerical singularities during the homogenisation problem solution.

The 10−node biquadratic, quadrilateral hybrid linear pressure reduced integration plane strain finite elements with 4 integration Gaussian points are used to discretise the cell. Periodic boundary conditions are imposed to ensure periodic character of the entire structure behaviour. A suitable formulation of displacement boundary conditions has the following form:

$$u_i = \varepsilon_{ij}(y(P_2) - y(P_1))$$ (3.72)

where $u_i = \{u_1, u_2\}$ represents the displacement function components, ε_{ij} is the global total strain imposed on the periodicity cell, while $y(P_1)$ and $y(P_2)$ denote coordinates of the points lying on the opposite sides of the RVE.

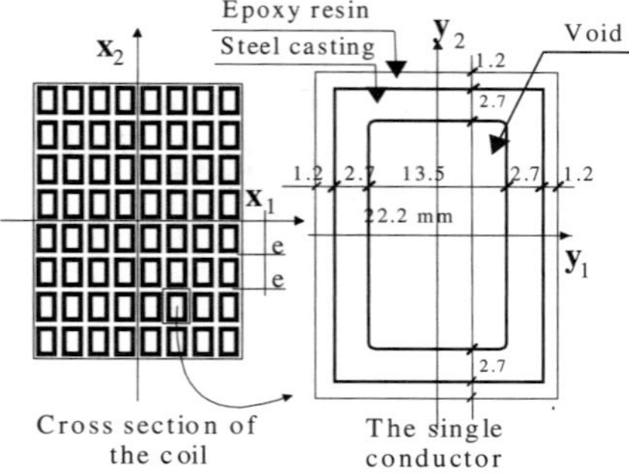

Figure 3.1. Cross section of a superconducting coil

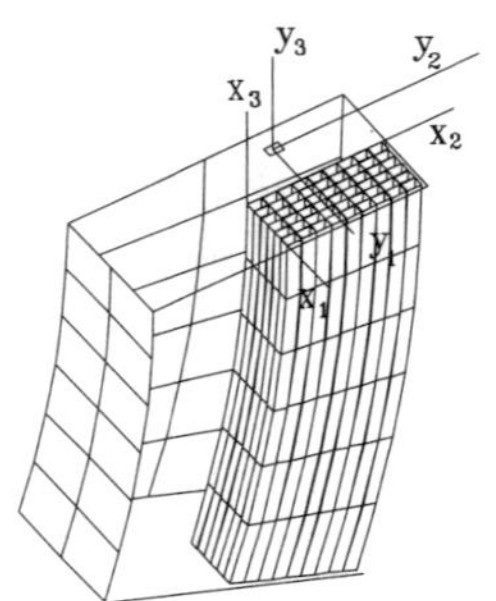

Figure 3.2. 3D view of the superconducting coil part

Table 3.1. Material characteristics of composite constituents

No	Material	Young modulus	Poisson ratio	Yield stress
1	Epoxy resin	7000.0	0.3	10.0
2	Metal	42000.0	0.2	22.0
3	Void	70.0	0.1	0.1

To calculate the effective tensor components, the boundary value problem given by
(3.22) − (3.25) is solved first, where the periodicity cell is discretised with 25 finite
elements of the type CGPE10R implemtented into the system ABAQUS. The
displacement boundary conditions are introduced at the edges of the RVE quarter
as is shown in Figures 3.3 and 3.4.

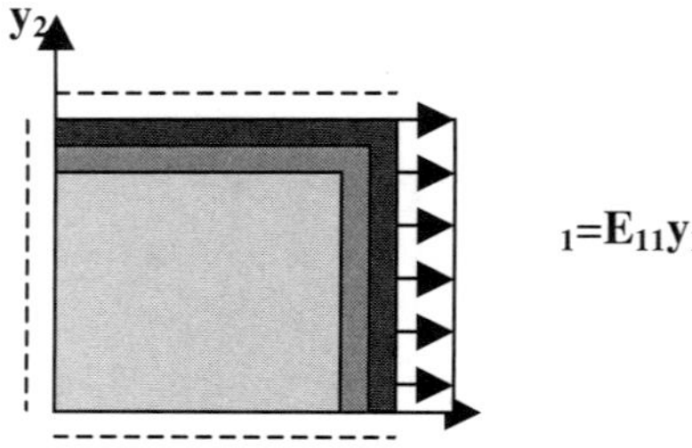

Figure 3.3. Boundary conditions for $E_{11} \neq 0$

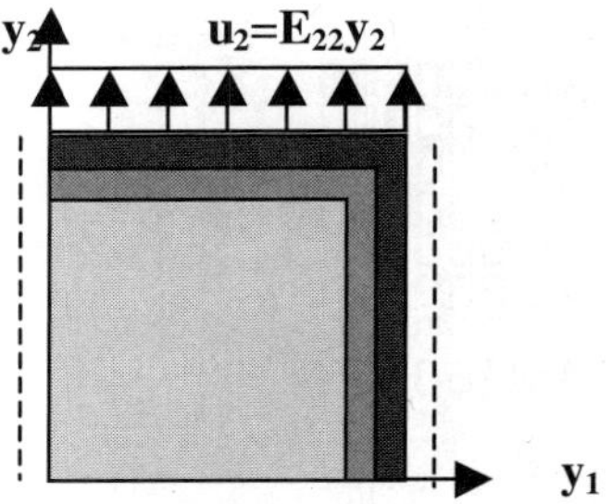

Figure 3.4. Boundary conditions for $E_{22} \neq 0$

Further, since the generalised plane strain is considered, the matrices computed have a rank $\alpha = 4$ and the total dimensions of the matrices $\mathbf{A}_r$ and $\mathbf{D}_{rs}$ are $[4 \times 4]$. The first step in the numerical analysis is to compute mechanical and transformation concentration matrices $\mathbf{A}_r$ and $\mathbf{D}_{rs}$, which is carried out according to the special purpose implementation in the computer system ABAQUS. Transformation matrices $\mathbf{A}_r$ and $\mathbf{D}_{rs}$ are evaluated as

(1) matrix $\mathbf{A}_r$ by means of the overall strain loading case $\varepsilon = \{\varepsilon_{ij}\} = [\varepsilon_{11}, \varepsilon_{22}, 2\varepsilon_{12}, \varepsilon_{33}]^T$ introduced using displacements

$$u_i = \varepsilon_{ij} y_j , \tag{3.73}$$

(2) matrix $\mathbf{D}_{rs}$ imposing the uniform eigenstrain in the subvolume V_r or V_s as the uniform stress; since it is not possible to introduce the eigenstrain directly in each subvolume in the program ABAQUS, the stress tensor components are calculated as

$$\sigma_r^* = -\mathbf{C}_r : \varepsilon_r^* , \text{ and } r{=}1,...,N \tag{3.74}$$

and imposed on each of the N subvolumes, where the elasticity tensor $\mathbf{C}_r$ is given by

$$\mathbf{C}_r = \frac{E_r}{(1-v_r)(1-2v_r)} \begin{bmatrix} 1-v_r & v_r & v_r & 0 \\ v_r & 1-v_r & v_r & 0 \\ v_r & v_r & 1-v_r & 0 \\ 0 & 0 & 0 & \dfrac{1-2v_r}{2} \end{bmatrix} \tag{3.75}$$

The accuracy of the homogenisation method applied for a given material model is verified by comparison with the results obtained for real heterogeneous composite under the same boundary conditions. For this purpose, the same boundary value problem is solved with four different loading cases. The elastoplastic static analysis consists of 25 incremental load steps (with a constant increment in each step) and is performed using the Radial Return algorithm for the perfect J_2 elastoplastic material. The results in the form of stress strain relations are shown in Figures 3.5 to 3.8, while the stress distribution in the periodicity cell can be compared in Figures 3.9 to 3.12.

Generally, it is observed that the elastic range is very well approximated by the TFA model results. However, the homogenised material seems to be a little stiffer than the heterogeneous one, especially in the nonlinear range in the direction y_1 of the RVE. At the same time, for the interrelation of shear strain and stress, the last incremental steps show almost linear behaviour and that is why practically there is no difference between heterogeneous and homogeneous material. To obtain more efficient effective elastoplastic properties, homogenisation method presented above should be corrected to include the increments of transformation matrices during the loading process.

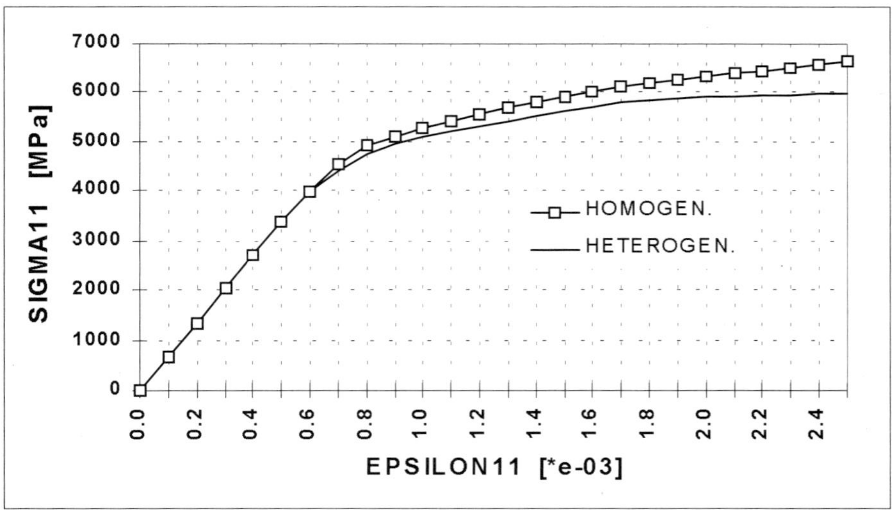

Figure 3.5. Constitutive σ_{11}-ε_{11} relation for homogenised and real composites

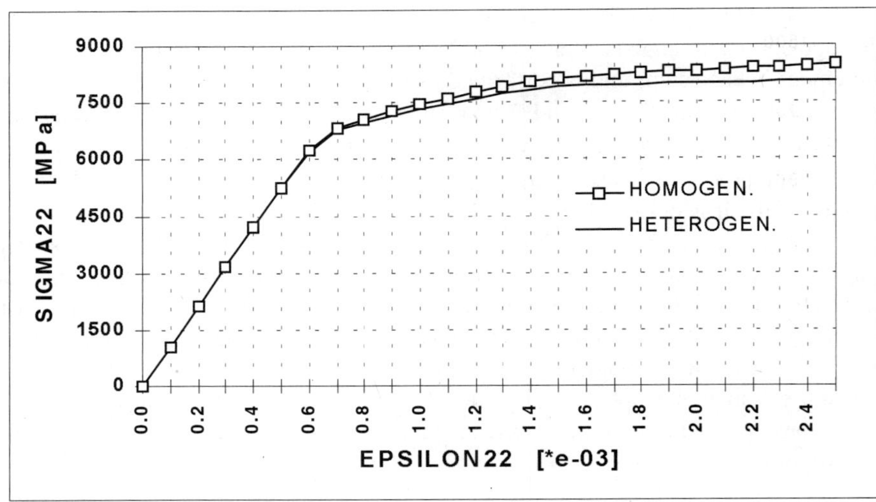

Figure 3.6. Constitutive σ_{22}-ε_{22} relation for homogenised and real composites

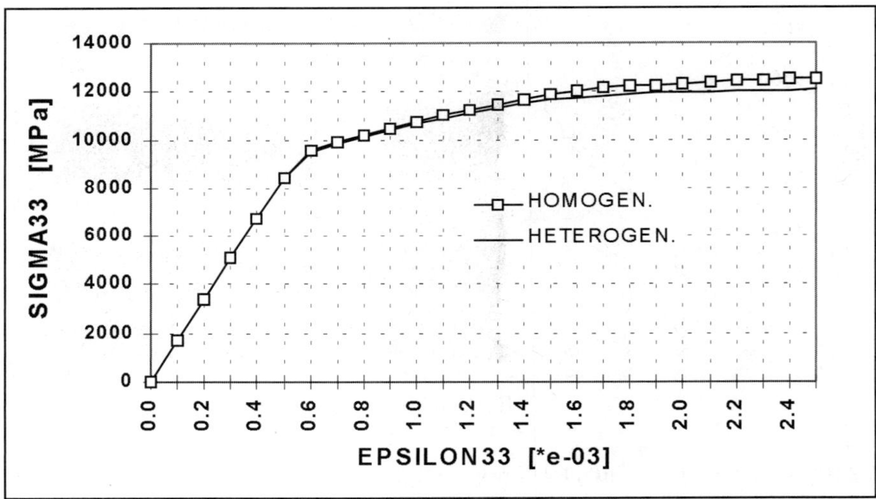

Figure 3.7. Constitutive σ_{33}-ε_{33} relation for homogenised and real composites

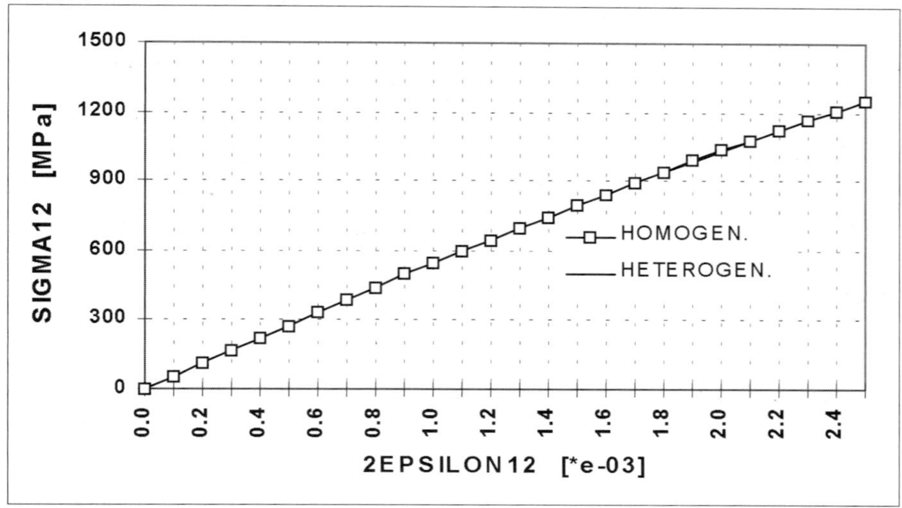

Figure 3.8. Constitutive σ_{12}-$2\varepsilon_{12}$ relation for homogenised and real composites

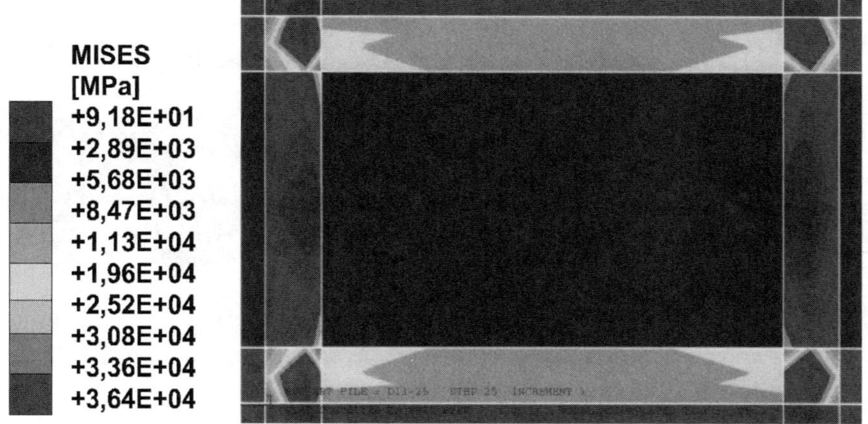

Figure 3.9. The equivalent stress σ_{11}^{eq} distribution in the RVE

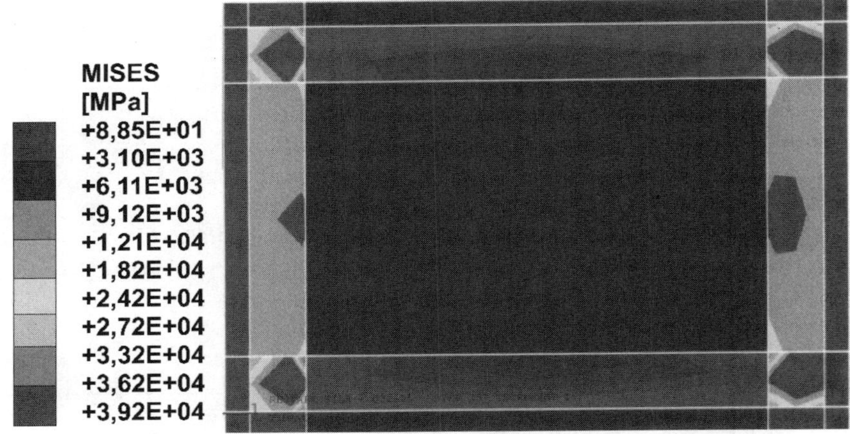

Figure 3.10. The equivalent stress σ_{22}^{eq} distribution in the RVE

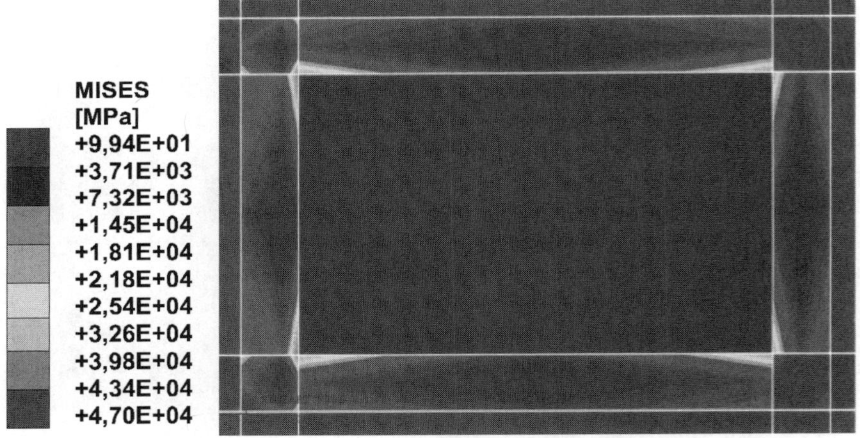

Figure 3.11. The equivalent stress σ_{12}^{eq} distribution in the RVE

Figure 3.12. The equivalent stress σ_{33}^{eq} distribution in the RVE

That is why the FEM mesh should employed the most precisely around all interfaces – its density along the external RVE edges does not need to be so precise. Comparing the stresses fields spatial variations with analogous results collected in Sec. 2.3.3.2 it is seen that maximum stresses variations are obtained along the interface in RVE. This observation does not depend on the homogenisation approach used as well as on its FEM solution, so it is common for various cell problem solutions.

Further, the effective properties of the homogenised material are computed starting from the properties of the composite constituents and the constitutive relation verified for all strain increments during the computational incremental analysis. We use the relation (3.70) and therefore

$$
C^{el} = \begin{bmatrix} k+\mu & k-\mu & k-\mu & 0 \\ k-\mu & k+\mu & k-\mu & 0 \\ k-\mu & k-\mu & k+\mu & 0 \\ 0 & 0 & 0 & \mu \end{bmatrix} = \begin{bmatrix} 15776.83 & 2601.7 & 2601.7 & 0 \\ 2601.7 & 15776.83 & 2601.7 & 0 \\ 2601.7 & 2601.7 & 15776.83 & 0 \\ 0 & 0 & 0 & 6587.57 \end{bmatrix}
$$

while the inelastic part of the effective constitutive tensor can be written as

$$
\mathbf{C}^{inel} = \sigma : \varepsilon^{-1} \rightarrow \mathbf{C}^{inel} = \begin{bmatrix} 3393.88 & 1505.92 & 2080.0 & 0 \\ 749.44 & 2653.36 & 1576.12 & 0 \\ 1631.4 & 1546.4 & 5017.2 & 0 \\ 0 & 0 & 0 & 499.600 \end{bmatrix},
$$

which completes the calculations of the effective elastoplastic characteristics of the composite considered. As it is shown here, the homogenisation technique presented can be very efficiently used in case of linear elastic constituents of the composite. It can be used instead of the previous method, where the symmetry conditions have been applied on the external edges of the RVE and some specific stress boundary conditions were applied on the bimaterial (or multimaterial) interfaces.

3.5 Some Comments on Probabilistic Effective Properties

Deterministic approaches to homogenisation of elastoplastic or viscoelastoplastic composites worked out recently are more complicated than the analysis presented above. However some authors presented simplified approximations for the effective yield stresses or yield conditions. It is known that for some special case where the volume fractions of the fibre–matrix constituents are equal or almost equal, the effective yield stresses can be described as

$$\Sigma^{(eff)} = \sqrt{\Sigma_1 \Sigma_2}$$

(3.75)

where Σ_1, Σ_2 denote the yield stresses for the two–component composite. This relation is used to show how to calculate the probabilistic moments in case, where yield stresses are characterized by their first two probabilistic moments. These parameters are defined for fibre and matrix using $E[\Sigma_1]$, $\sigma(\Sigma_1)$ and $E[\Sigma_2]$, $\sigma(\Sigma_2)$. Considering the above, then the first two probabilistic moments of the effective parameter can be calculated starting from

$$\left(\Sigma^{(eff)}\right)^2 = \Sigma_1 \Sigma_2$$

(3.76)

and by using the second order perturbation method, we get

$$E\left[\left(\Sigma^{(eff)}\right)^2\right]$$

$$= \int_{-\infty}^{+\infty} \left(\Sigma_2^0 + \varepsilon \Delta b_u \Sigma_2^{;u} + \tfrac{1}{2}\varepsilon^2 \Delta b_u \Delta b_v \Sigma_2^{;uv}\right) p_R(\mathbf{b}(x)) d\mathbf{b}$$

$$\times \int_{-\infty}^{+\infty} \left(\Sigma_1^0 + \varepsilon \Delta b_r \Sigma_1^{;r} + \tfrac{1}{2}\varepsilon^2 \Delta b_r \Delta b_s \Sigma_1^{;rs}\right) p_R(\mathbf{b}(x)) d\mathbf{b}$$

$$= E[\Sigma_1]E[\Sigma_2]$$

(3.77)

since the second order derivatives of the effective yield stresses are equal to 0. Then, omitting second order terms being equal to 0, the variance of effective yield stresses can be calculated as

$$Var\left(\left(\Sigma^{(eff)}\right)^2\right) = Var(\Sigma_1\Sigma_2) = \int_{-\infty}^{+\infty}\left(\Sigma_1\Sigma_2 - E[\Sigma_1\Sigma_2]\right)p_R(\mathbf{b}(x))\,d\mathbf{b}$$

$$= \int_{-\infty}^{+\infty}\left\{\left(\Sigma_1^0 + \varepsilon\Delta b_r\Sigma_1^{;r}\right)\left(\Sigma_2^0 + \varepsilon\Delta b_s\Sigma_2^{;s}\right) - \Sigma_1^0\Sigma_2^0\right\}p_R(\mathbf{b}(x))\,d\mathbf{b}$$

$$= Var^2(\Sigma_1) + Var^2(\Sigma_2)$$

(3.78)

which gives a combination of variances under the assumption of uncorrelation of the variables Σ_1 and Σ_2. Finally, the expected value and variance of the effective yield stress can be determined from the following equation system:

$$E\left[\left(\Sigma^{(eff)}\right)^2\right] = E^2\left[\Sigma^{(eff)}\right] + Var\left(\Sigma^{(eff)}\right)$$

(3.79)

and

$$Var\left(\left(\Sigma^{(eff)}\right)^2\right) = 2Var\left(\Sigma^{(eff)}\right)\left(2E^2\left[\Sigma^{(eff)}\right] + Var\left(\Sigma^{(eff)}\right)\right)$$

(3.80)

Probabilistic moments of effective yield stresses of a composite can be found as a solution of (3.77) and (3.78) in conjunction with the statements (3.79) and (3.80). For illustration, using a matrix built up with the following material data: $E[\Sigma_2] = 80\ MPa$, $\sigma(\Sigma_2) = 8.0\ MPa$ and the fibre as $E[\Sigma_1] = 4100\ MPa$, $\sigma(\Sigma_1) = 410\ MPa$, the effective plastic stress is obtained as the expected value $E[\Sigma^{(eff)}] = 552.969\ GPa$ and the standard deviation $\sigma\left(\Sigma^{(eff)}\right) = 40.367\ GPa$.

3.6 Conclusions

As is shown in the computational experiments, the homogenised material obtained as a result of the Transformation Field Analysis (TFA) is stiffer in the nonlinear range than the original composite. It is caused by the fact that the constitutive relations during the whole iteration procedure are based on constant constitutive tensors $\mathbf{C}_r$ with elastic properties. To obtain a better effective approximation of composite behaviour, these matrices should be divided into elastic and plastic parts, after yielding, by means of the consistent tangent matrices.

Since the Transformation Field Analysis makes it possible to characterise explicitly the effective elastoplastic behaviour starting from composite component material properties, it is possible to carry out the numerical sensitivity studies of homogenised composite properties with respect to its original material characteristics. Such computational studies make it possible to determine the most decisive material parameter for the overall elastoplastic behaviour of the composite, which may be important in the context of optimisation techniques applied in composite engineering design studies.

Due to the fact that most of the composite components material characteristics are obtained experimentally as statistical estimators, the next step to utilise the present approach is probabilistic implementation of the homogenisation problem. It will generally enable us to compute the respective probabilistic moments and coefficients of effective properties, starting from the expected values and standard deviations of composite component elastoplastic characteristics. As is known, it can be done using the Monte Carlo simulation technique, for instance. Further, it should be mentioned that such an implementation makes it possible to specify the stochastic sensitivity of composite effective characteristics to the randomness of the component material nonlinear behaviour.

3.7 Appendix

The two-component transversely isotropic RVE of volume Ω is subjected to a uniform overall strain increment $\Delta \mathbf{E}$ or stress $\Delta \Sigma$. A possible description of the local uniform strain and stress increment field is suggested as

$$\Delta \varepsilon_r = \mathbf{A}_r \Delta \mathbf{E} , \; r=1,2 \tag{A3.1}$$

$$\Delta \sigma_r = \mathbf{B}_r \Delta \Sigma , \; r=1,2 \tag{A3.2}$$

with further relations

$$c_1 \mathbf{A}_1 + c_2 \mathbf{A}_2 = \mathbf{I} \tag{A3.3}$$

$$c_1 \mathbf{B}_1 + c_2 \mathbf{B}_2 = \mathbf{I} \tag{A3.4}$$

Further, the local and overall increments are expressed as

$$\Delta \Sigma = c_1 \Delta \sigma_1 + c_2 \Delta \sigma_2 \quad \text{and} \quad \Delta \mathbf{E} = c_1 \Delta \varepsilon_1 + c_2 \Delta \varepsilon_2 \tag{A3.5}$$

with composite constituent volume fractions

$$c_1 = \frac{\Omega_1}{\Omega} , \; c_2 = \frac{\Omega_2}{\Omega} \text{ and } \Omega_1 \cup \Omega_2 = \Omega . \tag{A3.6}$$

The constitutive relations for composite constituents in elastoplastic range are defined as

$$\Delta\sigma_r = \mathbf{C}_r\Delta\varepsilon_r\,, \quad \Delta\varepsilon_r = \mathbf{M}_r\Delta\sigma_r \quad \text{and} \quad \mathbf{M}_r = \mathbf{C}_r^{-1} \tag{A3.7}$$

while the overall properties are

$$\Delta\Sigma = \mathbf{C}\Delta\mathbf{E}\,, \quad \Delta\mathbf{E} = \mathbf{M}\Delta\Sigma \quad \text{and} \quad \mathbf{M} = \mathbf{C}^{-1} \tag{A3.8}$$

The constitutive and compliance matrices are given as the relevant spatial averages over the RVE

$$\mathbf{C} = \sum_{r=1}^{2} c_r\mathbf{C}_r\mathbf{A}_r\,, \quad \mathbf{M} = \sum_{r=1}^{2} c_r\mathbf{M}_r\mathbf{B}_r \tag{A3.9}$$

The individual components of $\mathbf{B}_r$ and $\mathbf{A}_r$ may be found as solutions of (A3.3) and (A3–4). There holds that

$$\mathbf{A}_r = \begin{bmatrix} 0.5(k/k_r + \mu/\mu_r) & 0.5(k/k_r - \mu/\mu_r) & 0.5(1-1_r)/k_r & 0 \\ 0.5(k/k_r - \mu/\mu_r) & 0.5(k/k_r + \mu/\mu_r) & 0.5(1-1_r)/k_r & 0 \\ 0 & 0 & 1 & 0 \\ 0 & 0 & 0 & \mu/\mu_r \end{bmatrix} \tag{A3.10}$$

as well as

$$\mathbf{B}_r = \frac{1}{E_C}\begin{bmatrix} E_C & 0 & 0 & 0 \\ 0 & E_C & 0 & 0 \\ (1-c_r)a_r & (1-c_r)a_r & E_{rL} & 0 \\ 0 & 0 & 0 & E_C \end{bmatrix}, \quad \text{for } r=1,2 \tag{A3.11}$$

where
$$E_C = c_1 E_{(1)L} + c_2 E_{(2)L} \quad \text{and} \quad a_1 = \left(v_{(1)L}E_{(2)L} - v_{(2)L}E_{(1)L}\right) = -a_2 \tag{A3.12}$$

Next, the transformation and concentration matrices $\mathbf{D}_{rs}$, $\mathbf{F}_{rs}$ are calculated as

$$\mathbf{D}_{r1} = (\mathbf{I} - \mathbf{A}_r)(\mathbf{C}_1 - \mathbf{C}_2)^{-1}\mathbf{C}_1\,, \quad \mathbf{D}_{r2} = (\mathbf{I} - \mathbf{A}_r)(\mathbf{C}_1 - \mathbf{C}_2)^{-1}\mathbf{C}_2 \tag{A3.13}$$

$$\mathbf{F}_{r1} = (\mathbf{I} - \mathbf{B}_r)(\mathbf{M}_1 - \mathbf{M}_2)^{-1}\mathbf{M}_1\,, \quad \mathbf{F}_{r2} = (\mathbf{I} - \mathbf{A}_r)(\mathbf{M}_1 - \mathbf{M}_2)^{-1}\mathbf{M}_2 \tag{A3.14}$$

with $\mathbf{D}_{11}$, $\mathbf{D}_{12}$, $\mathbf{D}_{21}$, $\mathbf{D}_{22}$, $\mathbf{F}_{11}$, $\mathbf{F}_{12}$, $\mathbf{F}_{21}$, $\mathbf{F}_{22}$ to be calculated. The components of the matrix $\mathbf{D}_{11}$ are obtained as, for instance,

$$\mathbf{D}_{11} = \begin{bmatrix} D_{1111} & D_{1112} & D_{1113} & D_{1114} \\ D_{1121} & D_{1122} & D_{1123} & D_{1124} \\ D_{1131} & D_{1132} & D_{1133} & D_{1134} \\ D_{1141} & D_{1142} & D_{1143} & D_{1144} \end{bmatrix} \tag{A3.15}$$

then

$$D_{1111} = \left(1 - \frac{k}{2k_1} - \frac{\mu}{2\mu_1}\right)(2k_1 + l_1)$$
$$\times \left[\frac{2l_1(k_1 - k_2) + 2l_2(k_2 - k_1) + (k_1 - k_2)^2 + (\mu_1 - \mu_2)^2}{\left((k_1 - k_2)^2 + (\mu_1 - \mu_2)^2\right)(l_1 - l_2)}\right] \tag{A3.16}$$

$$D_{1112} = \left(1 - \frac{k}{2k_1} + \frac{\mu}{2\mu_1}\right)(2k_1 + l_1)$$
$$\times \left[\frac{2l_1(k_1 - k_2) + 2l_2(k_2 - k_1) + (k_1 - k_2)^2 + (\mu_2 - \mu_1)^2}{\left((k_1 - k_2)^2 + (\mu_2 - \mu_1)^2\right)(l_1 - l_2)}\right] \tag{A3.17}$$

$$D_{1113} = \left[1 - \frac{k_1}{2}(l - l_1)\right](2l_1 + n_1)\left[\frac{l_1 - l_2 + 2(n_1 - n_2)}{(l_1 - l_2)(n_1 - n_2)}\right] \tag{A3.18}$$

$$D_{1121} = \left(1 - \frac{k}{2k_1} + \frac{\mu}{2\mu_1}\right)(2k_1 + l_1)$$
$$\times \left[\frac{2l_1(k_1 - k_2) + 2l_2(k_2 - k_1) + (k_1 - k_2)^2 + (\mu_2 - \mu_1)^2}{\left((k_1 - k_2)^2 + (\mu_2 - \mu_1)^2\right)(l_1 - l_2)}\right] \tag{A3.19}$$

$$D_{1122} = \left(1 - \frac{k}{2k_1} - \frac{\mu}{2\mu_1}\right)(2k_1 + l_1)$$
$$\times \left[\frac{2l_1(k_1 - k_2) + 2l_2(k_2 - k_1) + (k_1 - k_2)^2 + (\mu_1 - \mu_2)^2}{\left((k_1 - k_2)^2 + (\mu_1 - \mu_2)^2\right)(l_1 - l_2)}\right], \tag{A3.20}$$

$$D_{1123} = \left[1 - \frac{k_1}{2}(l - l_1)\right](2l_1 + n_1)\left[\frac{l_1 - l_2 + 2(n_1 - n_2)}{(l_1 - l_2)(n_1 - n_2)}\right] \tag{A3.21}$$

$$D_{1144} = \mu_1\left(1 - \frac{\mu}{\mu_1}\right)\left(\frac{1}{\mu_1 - \mu_2}\right) \tag{A3.22}$$

The remaining coefficients of $\mathbf{D}_{11}$ are equal to 0.

4 Sensitivity Analysis for Some Composites

4.1 Deterministic Problems

As is known, the sensitivity analysis in engineering systems is employed to verify how input parameters of a specific engineering problem influence the analysed state functions (displacements, stresses and temperatures, for instance). The sensitivity coefficients [269], being the purpose of such an analysis, are computed using partial derivatives of the considered state function with respect to the particular input parameter(s). These derivatives can be obtained numerically starting from the fundamental algebraic equations system of the problem, for instance, or alternatively, by a simple derivation if only a closed form solution exists; some combined analytical–numerical methods are also known [99]. It is important to underline that this methodology is common for all discrete numerical techniques: Boundary Element Method (BEM) [51,206], Finite Difference Method (FDM) [90,206], FEM [7,21,387] as well as hybrid and meshless strategies [81].

From the computational point of view, there are the following numerical methods in structural design sensitivity analysis [75,76,103,134,207]: the Direct Differentiation Method (DDM), the Adjoint Variable Method (AVM) applied together with the Material Derivative Approach (MDA) or the Domain Parametrisation Approach (DPA) suitable for shape sensitivity studies. Considering these capabilities and, on the other hand, a very complex structure of composite materials, sensitivity analysis should be applied especially in design studies for such structures. Instead of a single (or two) parameters characterising the elastic response of a homogeneous structure, the total number of design parameters is obtained as a product of component numbers in a composite and the number of material and geometrical parameters for a single component. Even some extra state variables should be analysed to define interfacial behaviour, general interaction of the constituents and/or the lack of periodicity. Usually, to reduce the complexity of the original composite, the so–called effective homogenisation medium having the same strain (or complementary) energy is analysed.

This chapter is devoted to general computational sensitivity studies of the homogenisation method for some periodic composite materials with linear elastic and transversely isotropic constituents. The composite is first homogenised, the effective material tensor components are computed using the FEM–based additional computer program. Further, material parameters of the composite most decisive for its effective material properties are determined numerically. It should be underlined that the homogenisation method is generally an intermediate numerical tool applied to exclude the necessity of composite micro–scale discretisation and, in the same time, to reduce the total number of degrees of freedom of the entire model. On the other hand, there are many numerical homogenisation techniques. They can be divided generally into two essentially different approaches: stress averaging (the boundary stresses are introduced between the composite constituents plus displacement–type periodicity conditions) and strain approach (uniform extensions of the RVE boundaries in various directions plus periodicity conditions on the remaining cell edges). Considering this, different results of the homogenisation method in terms of the effective material tensors are obtained (as a result quite different sensitivity gradients must be computed in these two approaches). The sensitivity analysis introduces a new aspect of the homogenisation technique – it can be verified if the homogenised and original structures have the same or even analogous (in terms of their signs) sensitivity gradients. The composites can be optimised then by manipulating its material or geometrical design parameters [310] as well as by choosing various constituent materials with computationally determined shape for the new designed composite structure.

The sensitivity gradients are computed here by application of a homogenisation–oriented computer program MCCEFF according to the DDM implementation approach and presented as functions of the composite design parameters – Young moduli and Poisson ratios of the constituents. Since a finite difference scheme is used for the sensitivity gradient computations, numerical sensitivity of the final results to the increase of an arbitrarily introduced parameter must be verified. This numerical phenomenon makes it necessary to determine the most suitable interval of parameter increments for the particular effective elasticity tensor components.

The entire computational methodology is illustrated with two examples – 1D and 2D two component periodic composites. The closed form effective Young modulus is used in the first example, while the homogenisation function is to be computed in the second case. Both illustrations show that different components of the effective elasticity tensor show different sensitivities to particular mechanical properties of the original composite and, further, the illustrations make it possible to determine the most decisive elastic parameters for the homogenisation–based computational design studies. Quite similar sensitivity studies are carried out in the case of heat conductivity coefficient for 1D, 2D and 3D two component composites.

It should be noticed that sensitivity analysis can be used for validation of various homogenisation methods. In most cases an increase in Young moduli of composite components should result in a corresponding increase of the effective material tensor components; an opposite phenomenon can be observed for some specific cases, but usually in an extremely small range only. Therefore, if the sensitivity analysis shows that most of the gradients are negative, the homogenisation theory should be essentially corrected.

The applied effective modulus method is verified below using the examples of 1D distributed heterogeneities in the periodic two–component bar structure and of the fibre–reinforced periodic composite. As is demonstrated for plane composite structure, the sensitivity gradients of a homogenised elasticity tensor show some instabilities observed for an extremely small value of the perturbation parameter.

At the same time, for Poisson ratios values tending to their physical bounds, an uncontrolled increase of all sensitivity gradients is observed. That is why a continuation of this study is necessary in the context of computational error, to extend constitutive models of composite components as well as to evaluate geometrical and material sensitivity gradients for more complex heterogeneous structures, especially in the probabilistic context.

Another important topic studied here is the application of the parameter finite difference analysis to the sensitivity analysis of the uniform plane strain problem of the real composite. This is done under the assumption that the RVE of plane cross−section is uniformly extended in two perpendicular directions and the unit shear strain is applied on the RVE. Therefore, the sensitivity functional is proposed as the elastic strain energy stored in the cell, which is treated as some type of representative strain state of the composite under real conditions. To reflect the real conditions of the composite service more accurately, the particular strain component can be scaled over some multipliers to illustrate pure horizontal and/or vertical extension of the composite specimen. The sensitivity of this functional is taken as a measure of influence of various material parameters on the overall behaviour of the composite. According to the previous results, we observe the Poisson ratio of the matrix as a dominating material parameter for the fibre−reinforced periodic composites with the RVE specified below.

Finally, it should be mentioned that this sensitivity analysis is introduced and performed to validate the homogenisation theory itself. In the case when the external boundary conditions are known together with the micromorphology of a certain composite, the homogenisation theory makes it possible to determine the effective characteristics of this structure and, according to the sensitivity analysis the sensitivity gradients of both real and homogenised structures are computed. If these gradients have consistent signs and comparable values, the homogenisation algorithm proposed is useful in computational modelling; otherwise another method should be proposed. It can happen that some homogenisation theories (or even closed form equations) are valid for some specific boundary value problems and it can be verified in this way. Another promising field of application of such an analysis is optimization and/or identification of composite materials and structures.

Sensitivity gradients cannot be obtained analytically if the homogenisation function components are determined numerically in some cell problem solutions. Hence, two separate ways can be followed, the first one being purely computational finite difference based studies, where the gradients are obtained as differences of some slightly modified homogenisation tests. Alternatively, a semi−analytical method can be implemented where the spatial averages of the constitutive tensor components (independent from homogenisation functions) are differentiated symbolically and the remaining part resulting from homogenisation FEM tests is analysed using the finite differences; analogous opportunities are available for probabilistic (and next stochastic) analyses. Taking into account the consistency of the Monte Carlo simulation application and the computational time savings, full numerical differentiation is implemented. A semi−analytical approach can be implemented partially in some mathematical symbolic computation packages, where probabilistic moments can be derived according to the classical integral definitions, while the random fields of homogenising stresses averaged over the RVE are treated using the numerical differentiation approach.

The results of computations in the form of deterministic derivatives or their probabilistic equivalents can next be implemented in deterministic and/or probabilistic optimisation problems based on the gradient techniques. Such an analysis will enable us to optimise various composites [84,240,264,281,320] using their homogenised models – without the necessity of complicated multiscale problem discretisation and their further solution. The main benefits of the integrated computational approach to the composites are (a) the most effective choice of composite components (sensitivity to the expected values of material parameters), (b) selection of the best processing technology from the necessary accuracy point of view (standard deviation levels), (c) efficient durability control and analysis (sensitivity to the interface and structural defects parameters), etc. The proposed method is significantly more complicated than the previous approaches. However it makes the computational model of composite materials and their behaviour more realistic and focused on the engineering analyses.

4.1.1 Sensitivity Analysis Methods

The main aim of the structural design sensitivity analysis is to study the interrelation between the response (or state variables) of a structure determined from a solution for the boundary–value problem and design variables begin the input data for the solution process. Displacements, stresses, temperatures or velocities can be taken as the structural response measures, whereas such parameters as truss and beam cross–sectional areas, plate and shell thicknesses and material characteristics are usually chosen as design variables. Let us note that even for linear elastic problems the equilibrium equations may generally contain some nonlinear expressions for the state and design variables – this is the case of plate/shell thickness and/or truss lengths and, especially, material parameters in composites.

The sensitivity gradients are the main numerical tool to evaluate the design sensitivity of a structure with respect to some design parameter. For engineers a more interesting issue is the overall sensitivity of the structure examined under general loading conditions than particular state function gradients. The gradients of the structural response functionals with respect to design variables give a useful measure of structural response variation together with the change of a given design input.

The sensitivity analysis is especially applicable with common implementation with one of the well-established numerical methods of structural analysis, i.e. with the finite element formulation. To illustrate the main ideas let us consider the static structural response of a linear elastic system with N degrees of freedom defined by the functional [208]

$$\Im\left(h^d\right)= G\left[q_\alpha\left(h^d\right),h^d\right], \quad d=1,2,\ldots,D; \quad \alpha = 1,2,\ldots,N \tag{4.1}$$

where G is a given function of structural displacements vector (q_α) and design variables, h^d represents a D–dimensional vector of design variables; the displacement vector satisfies classical equilibrium equations, i.e.

$$K_{\alpha\beta}\left(h^d\right)q_\beta\left(h^d\right)= Q_\alpha\left(h^d\right) \tag{4.2}$$

The displacement vector is assumed to be an implicit function of design variables, because the stiffness matrix $K_{\alpha\beta}$ and the load vector Q_α are some functions of these variables.

Now, the SDS analysis is employed to determine the changes of the structural response functional with variations in design parameters, so the so–called sensitivity gradient $\partial\Im/\partial h^d$ is to be determined. The chain rule of differentiation applied to (4.1) returns here

$$\Im^{\cdot d} = G^{\cdot d} + G_{.\alpha}q_\alpha^{\cdot d} \tag{4.3}$$

where $(.)^{\cdot d}$ and $(.)_{.\alpha}$ denote first partial derivatives with respect to the dth design variable and the αth nodal displacement, respectively. The design variables h^d are introduced as the only arguments in the functions $\Im$, $K_{\alpha\beta}$, q_α, Q_α and, therefore, partial derivatives of these functions with respect to h^d are in fact equal to the corresponding total derivatives. Nevertheless, there holds $\partial G/\partial h^d = \Im^{\cdot d}$ in case of G. Since it is an explicitly given function of h^d and q_α, the derivatives $G^{\cdot d}$ and $G_{.\alpha}$ may be computed directly, while $q_\alpha^{\cdot d}$ is to be determined numerically.

The first technique for computing of the sensitivity gradients known as the direct differentiation method (DDM) extensively employed in structural optimisation reflects the following algorithm. Let us assume that $K_{\alpha\beta}\left(h^d\right)$ and $Q_\alpha\left(h^d\right)$ are continuously differentiable with respect to the design variables h^d; then, the vector $q_\beta\left(h^d\right)$ is also continuously differentiable. Differentiation of both sides of (4.2) with respect to h^d gives

$$K_{\alpha\beta}q_\beta^{\cdot d} = Q_\alpha^{\cdot d} - K_{\alpha\beta}^{\cdot d}q_\beta \tag{4.4}$$

Since the stiffness matrix $K_{\alpha\beta}$ is assumed to be nonsingular, (4.4) can be solved for $q_\beta^{\cdot d}$; it yields

$$\mathfrak{I}^{.d} = G^{.d} + G_{.\beta} K_{\alpha\beta}^{-1} \left(Q_\alpha^{.d} - K_{\alpha\gamma}^{.d} q_\gamma \right) \tag{4.5}$$

The alternative AVM strategy begins with the introduction of an adjoint variable vector λ_α, $\alpha=1,2,\ldots,N$ such that

$$\lambda_\alpha = G_{.\beta} K_{\alpha\beta}^{-1} \tag{4.6}$$

It yields the adjoint equations for λ_α in the form

$$K_{\alpha\beta} \lambda_\beta = G_{.\alpha} \tag{4.7}$$

and then, the sensitivity gradient coefficients may be obtained as

$$\mathfrak{I}^{.d} = G^{.d} + \lambda_\alpha \left(Q_\alpha^{.d} - K_{\alpha\beta}^{.d} q_\beta \right) \tag{4.8}$$

having solved the above equation for the adjoint variables λ_β. The main ideas of the DDM and AVM seem to be identical but in realistic engineering design problems their computer performance is considerably different. Since most of the functions are given explicitly in the problems considered, the DDM technique has found its application below.

The matrices of derivatives of practically any order of the global stiffness matrix with respect to design variables are obtained simply by adding derivatives of element stiffness expressed in the global coordinate system. It is done quite similarly to the assembling procedure for the global stiffness matrix. This process is usually essentially simplified, because almost all entries in the matrices of their derivatives with respect to the particular design variables are equal to 0 and then all arithmetic operations can be carried out at the element level.

Effective computation of stiffness derivatives with respect to design variables for finite elements is another issue to be taken into account in developments of any sensitivity-oriented software. Most up-to-date finite element codes engage numerical integration instead of using the closed form expressions in terms of design variables to generate the element stiffness matrices. For such numerically generated element matrices a differentiation process with respect to design variables can be performed through a sequence of computations (at least two solution for initial and for a slightly perturbed design parameter) used to generate these matrices, leading to implicit design derivative procedures.

The element matrices of the design derivatives can also be obtained by using a finite difference scheme, which is demonstrated for the eth element of the stiffness matrix

$$\frac{\partial K_{\alpha\beta}^{(e)}}{\partial h^d} \cong \frac{1}{\varepsilon} \left[K_{\alpha\beta}^{(e)} \left(h + 1_{(\bar{d})} \varepsilon \right) - K_{\alpha\beta}^{(e)} (h) \right], \; \bar{d} = 1,2,\ldots,D, \tag{4.9}$$

where $K_{\alpha\beta}^{(e)}$ is the eth element stiffness matrix, $h^{\bar{d}}$ is the $\bar{d}$ th component of the D−dimensional design variable vector h, ε represents a small perturbation and the D−dimensional vector $1_{(\bar{d})}$ is equal to 1 at the $\bar{d}$ th position and zeroes elsewhere.

Such a scheme is known as forward finite difference rule, however backward and central differences can be applied too. Backward differentiation uses the values of a function in actual (h) and previous point $(h\text{-}\varepsilon)$, while central difference is returned from arithmetic averaging of equations containing forward and backward differences.

4.1.2 Sensitivity of Homogenised Heat Conductivity

As is known, it is possible to obtain the effective heat conductivity tensor components by the application of some algebraic approximations for particular types of composite materials. However, numerical procedure is not very general in this case. The effective heat conductivity for a periodic fibre−reinforced composite in a 2D problem where the fibre has the round cross−section and the total composite volume is relatively large in comparison to the single inclusion can be approximated using the Cylinder Assemblage Model (CAM) for a fibre−reinforced plane structure. The Spherical Inclusion Model (SIM) [65] for spherical inclusions distributed periodically (3D composite). The heat conductivity coefficients of composite components k_1, k_2 are such that $k_1 > k_2$ (the same results hold true for electrical conductivity, magnetic permeability and the dielectric constant for composites, for instance).

A concept of the first test is to compare the effective heat conductivities obtained for the 1D, 2D (fibre) and 3D (particle-reinforced) composites in terms of various reinforcement volume ratios and the interrelation between heat conductivity coefficients for both components. The following equations are used:

- 1D composite

$$k^{(eff)} = \frac{|\Omega|}{\displaystyle\int_{\Omega} \frac{d\Omega}{k(y)}}$$

- 2D composite

$$k_{2D}^{(eff)} = k_2\left(1 + v_f\left(\frac{1-v_f}{2} + \frac{k_2}{k_1 - k_2}\right)^{-1}\right)$$

- 3D composite

$$k_{3D}^{(eff)} = k_2\left(1 + v_f\left(\frac{1-v_f}{3} + \frac{k_2}{k_1 - k_2}\right)^{-1}\right)$$

where v_f is the reinforcement volume fraction, while k_1, k_2 are heat conductivity coefficients of composite components such that $k_1 > k_2$.

Furthermore, the sensitivities of effective heat conductivity with respect to those characterising original composite components are determined: the computations are performed using the mathematical package MAPLE. All the results of the numerical experiments are presented in Figures 4.1–4.9: the effective heat conductivities for the 1D, 2D and 3D composites are plotted in Figures 4.1–4.3, their material sensitivities with respect to design variable k_1 in Figures 4.4–4.6, while sensitivity studies with respect to the parameter k_2 are presented in Figures 4.7–4.9.

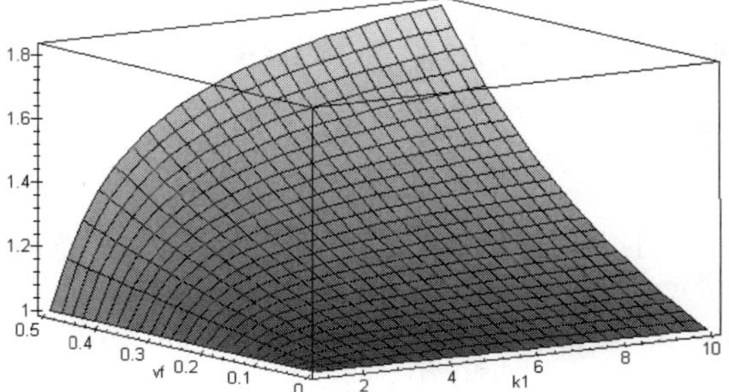

Figure 4.1. Effective heat conductivity for 1D composite

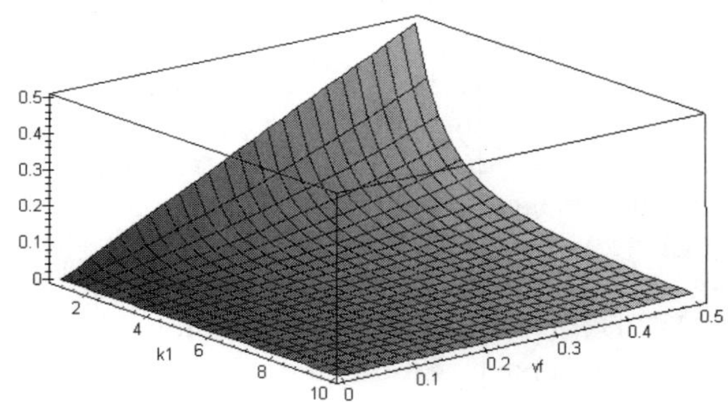

Figure 4.2. Material sensitivity of $k^{(eff)}$ in 1D problem to k_1

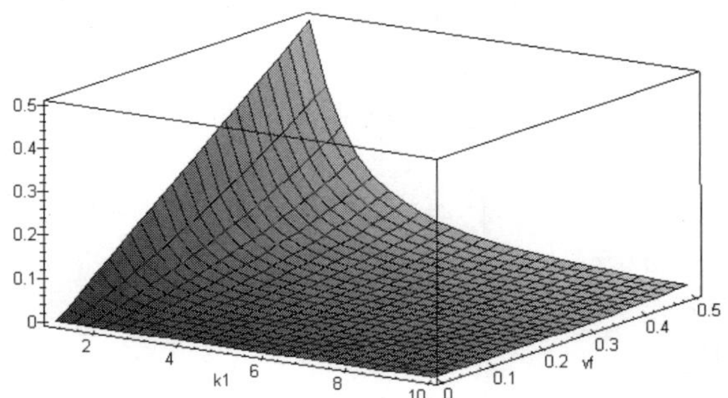

Figure 4.3. Sensitivity of $k^{(eff)}$ in 1D problem to v_f

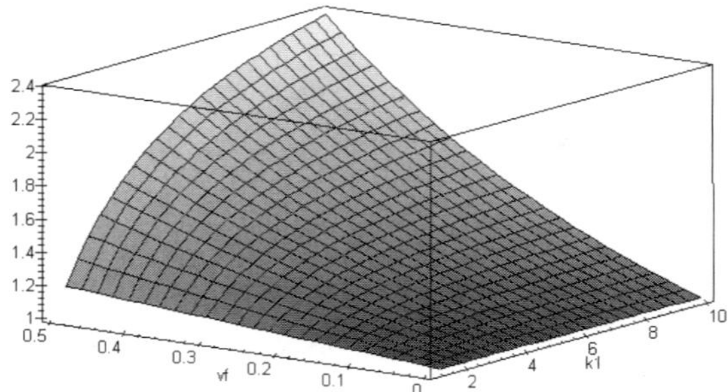

Figure 4.4. Effective heat conductivity for 2D composite

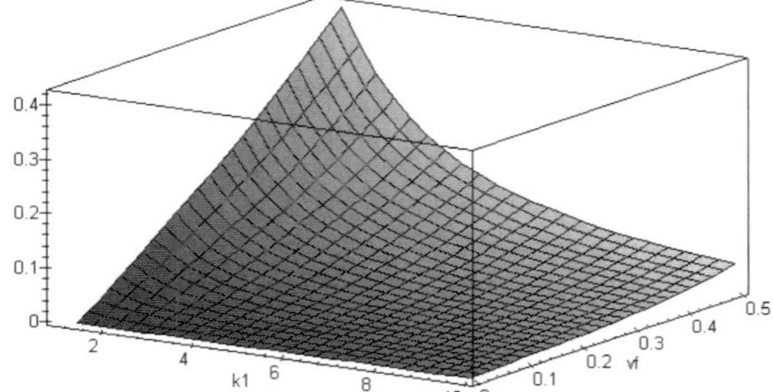

Figure 4.5. Material sensitivity of $k^{(eff)}$ in 2D problem to k_1

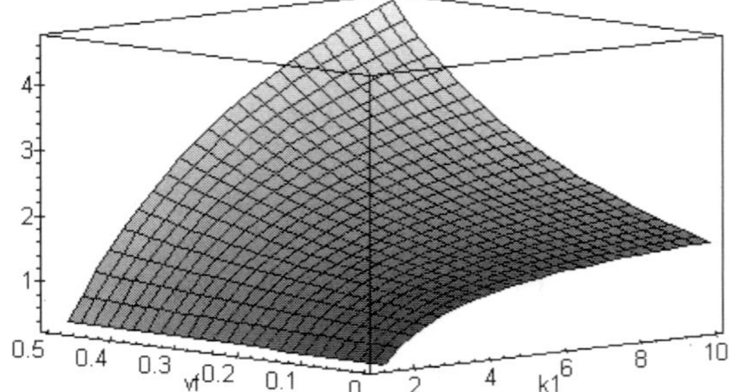

Figure 4.6. Sensitivity of $k^{(eff)}$ in 2D problem to v_{f}

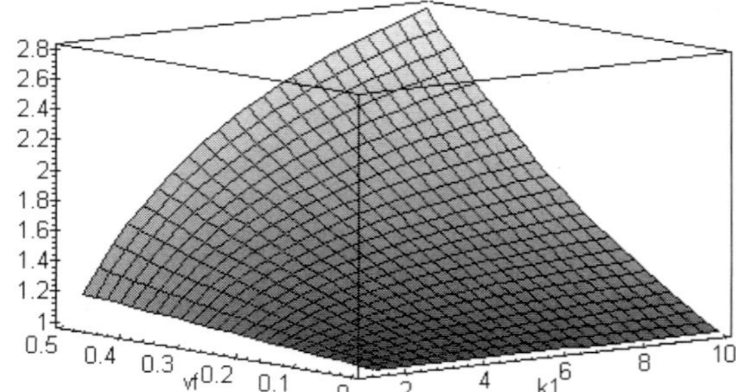

Figure 4.7. Effective heat conductivity for 3D composite

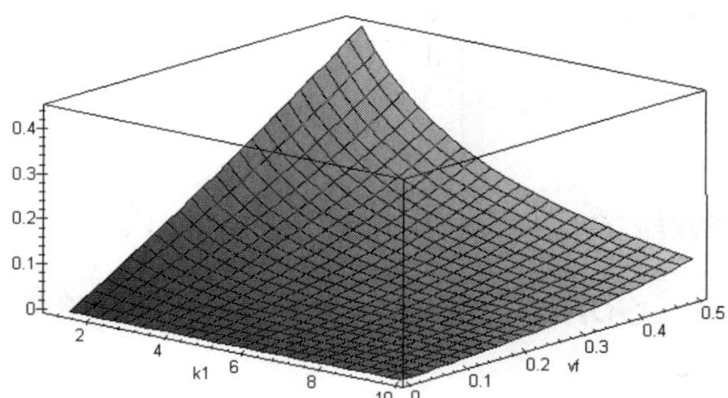

Figure 4.8. Material sensitivity of $k^{(eff)}$ in 3D problem to k_1

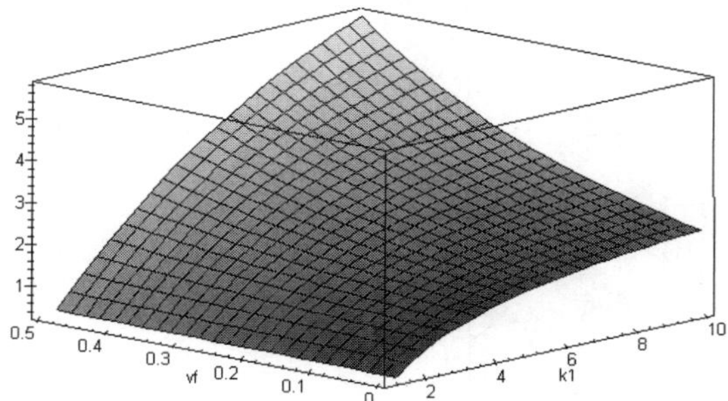

Figure 4.9. Sensitivity of $k^{(eff)}$ in 3D problem to v_f

Analysing numerical results it can be observed that the effective heat conductivity surface has an analogous shapes for 1D, 2D and 3D composites. However the values of this coefficient obtained for the same reinforcement ratio are largest for 3D composite with spherical inclusion, next largest for 2D fibre–reinforced composite, and smallest for the 1D case. Therefore, 3D composites seem to be most optimal – using the same volume of reinforcement, the highest value of the effective material property is obtained. According to engineering intuition, it is found that increasing both k_1 and v_f an increasing of final value of $k^{(eff)}$ is obtained. The results of sensitivity studies presented in Figures 4.3, 4.6 and 4.9 make it possible to observe the greatest sensitivity of composite effective characteristics with respect to both design parameters (k_1 and v_f) for extremely small values of the coefficient k_1 and the largest value of the reinforcement ratio. The sensitivity gradients of $k^{(eff)}$ with respect to v_f have almost constant value, while with respect to k_1 are efficiently nonlinear and reach the maximum for v_f=0.5 (cf. Figure 4.2 and 4.3, for instance). This result means that the effective conductivity value is most sensitive to the changes of k_1, if the reinforcement volume ratio is maximal, which is predictable result and it positively validates this homogenisation method.

The smallest sensitivity of $k^{(eff)}$ to the parameter k_1 can be noted for v_f tending to 0, while the inverse relation is observed with respect to the reinforcement volume fraction. The variability of the sensitivity surface for $k^{(eff)}$ with respect to the heat conductivity coefficient k_1 is almost the same for all composites. However in the case of sensitivity to v_f the 2D and 3D models are similar, while the 1D case is essentially different – it results from the relevant equations forms.

4.1.3 Sensitivity of Homogenised Young Modulus for Periodic Composite Bars

Let us consider periodic composite bar applied to the compressive/tensile stresses and the homogenised Young modulus of such a structure. For such a unidirectional n–component composite structure, one can readily obtain the sensitivity gradients of the effective parameter $e^{(eff)}$ with respect to the modulus of its jth component e_j as

$$\frac{\partial e^{(eff)}}{\partial e_j} = \frac{\prod_{i=1}^{j-1} e_i \prod_{j+1}^{n} e_i \left(\sum_{i=1}^{n} A_i l_i e_1 e_2 \ldots e_{i-1} e_{i+1} \ldots e_n \right)}{\left(\sum_{i=1}^{n} A_i l_i e_1 e_2 \ldots e_{i-1} e_{i+1} \ldots e_n \right)^2}$$

$$- \frac{\prod_{i=1}^{n} e_i \frac{1}{e_j} \left(\sum_{i=1}^{j-1} A_i l_i e_1 e_2 \ldots e_{i-1} e_{i+1} \ldots e_n + \sum_{j+1}^{n} A_i l_i e_1 e_2 \ldots e_{i-1} e_{i+1} \ldots e_n \right)}{\left(\sum_{i=1}^{n} A_i l_i e_1 e_2 \ldots e_{i-1} e_{i+1} \ldots e_n \right)^2} \tag{4.10}$$

The geometrical sensitivity with respect to the cross-sectional area A_j is determined as

$$\frac{\partial e^{(eff)}}{\partial A_j} = - \frac{\prod_{i=1}^{n} e_i \left(l_j e_1 e_2 \ldots e_{j-1} e_{j+1} \ldots e_n \right)}{\left(\sum_{i=1}^{n} A_i l_i e_1 e_2 \ldots e_{i-1} e_{i+1} \ldots e_n \right)^2} \tag{4.11}$$

Analogously, geometrical sensitivity with respect to the member length l_j is calculated from the following formula:

$$\frac{\partial e^{(eff)}}{\partial l_j} = - \frac{\prod_{i=1}^{n} e_i \left(A_j e_1 e_2 \ldots e_{j-1} e_{j+1} \ldots e_n \right)}{\left(\sum_{i=1}^{n} A_i l_i e_1 e_2 \ldots e_{i-1} e_{i+1} \ldots e_n \right)^2} \tag{4.12}$$

It should be underlined that the equations obtained above can be relatively easily inserted in the 1D implementations of the FEM formulation for elastostatics as well as heat conduction problems, both in deterministic and stochastic computation.

Now, the sensitivity gradients are derived first for a 1D two–component composite with the RVE presented in Figure 2.42. Considering the fact that composite materials are characterised by numerous parameters, it is essential to reduce this number by introduction of non–dimensional normalised parameters between the corresponding material and geometric characteristics of a composite. It is recommended to make the sensitivity analysis more focused with opportunity to compare the sensitivity gradients with each other.

Determination of the first sensitivity gradient, cf. (4.11), makes it possible to verify how the interrelation between cross–sectional area α of both components influences the final effective Young modulus of the composite. The next gradient is responsible for the sensitivity of the composite to the length of both components ratio γ, while the last one gives information about the influence of interrelation β of the Young moduli for composite components.

The general observation in this analysis is that an increase in analysed structural geometrical parameters results in a decrease of the effective parameter value (negative derivative sign) and vice versa. Analogously, it is observed that increasing any Young modulus of composite components, the increase of the effective homogenised parameter is obtained. Quantitative verification of the most decisive parameter depends on the interrelations between particular material and geometrical characteristics and should be analysed in detail in further studies. In case of the unidirectional composite, the shape sensitivity studies with respect to the interface location can be done analytically. All the sensitivities calculated above enable us to design, during engineering studies, the most suitable interrelations between particular components for unidirectional tensioned/compressed structural members. Considering the nature of the presented 1D homogenisation approach, it is clear that the sensitivity of the Young modulus holds true for the effective heat conductivity and other related coefficients.

The first and second order sensitivity gradients together with the mean value of the homogenised Young modulus have been computed and collected in the figures below. The following input data are adopted: e_2=2.0E9, the coefficient γ relating the lengths of composite components is arbitrarily taken as equal to 1. Other parameters are adopted in the following form: A_2=0.2 and l_2=10.0. The effective Young modulus is determined with respect to the reinforcement ratio as well as to the cross-sectional area ratio of the components and presented below.

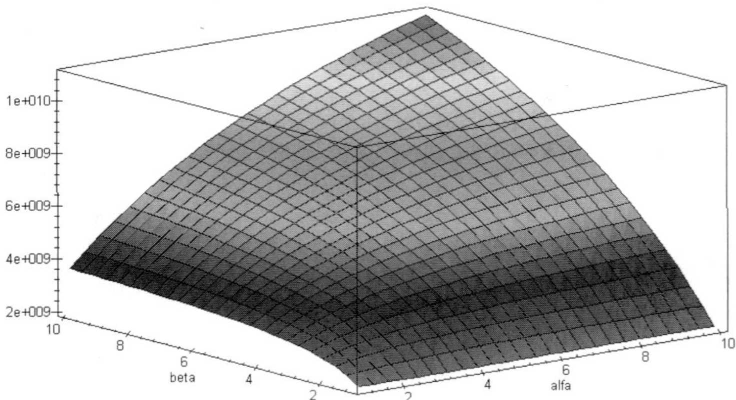

Figure 4.10. Parameter variability of the effective Young modulus

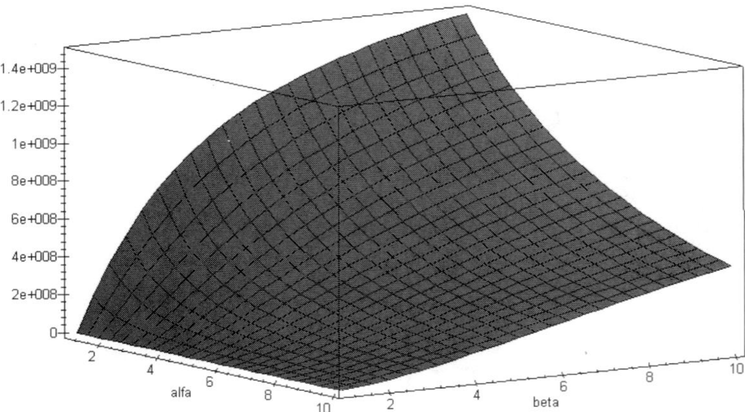

Figure 4.11. Parameter variability of $e^{(eff)}$ sensitivity gradient wrt parameter α

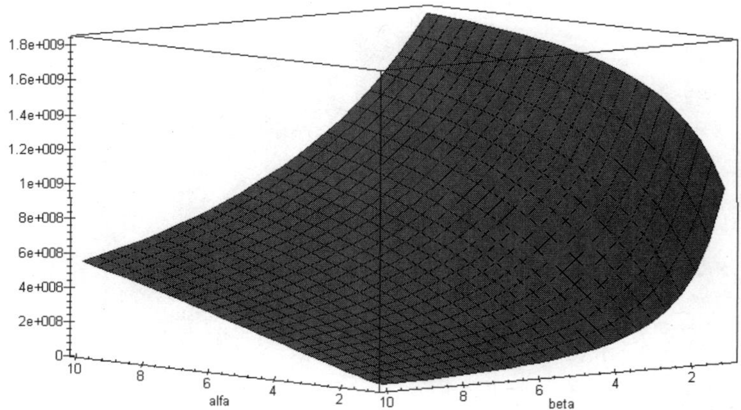

Figure 4.12. Parameter variability of $e^{(eff)}$ sensitivity gradient wrt parameter β

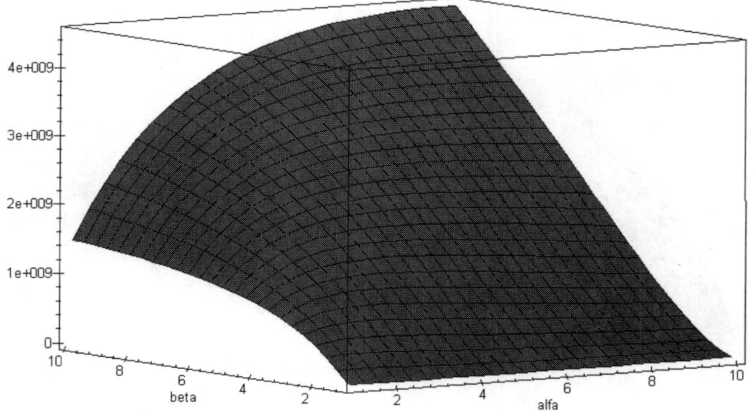

Figure 4.13. Parameter variability of $e^{(eff)}$ sensitivity gradient wrt parameter γ

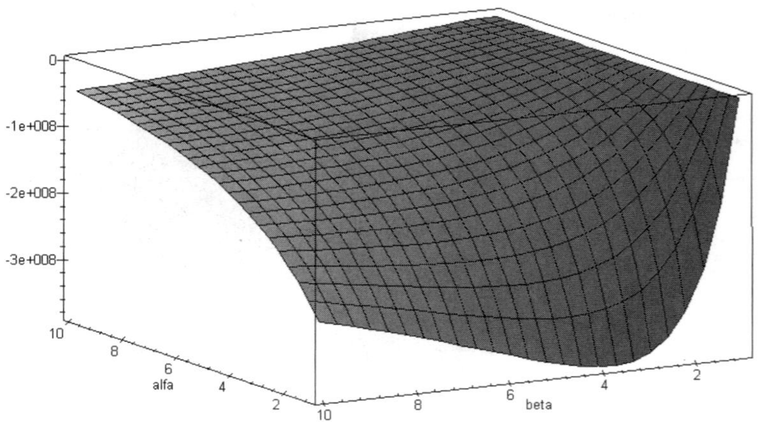

Figure 4.14. Second order sensitivity gradient of $e^{(eff)}$ wrt parameter α

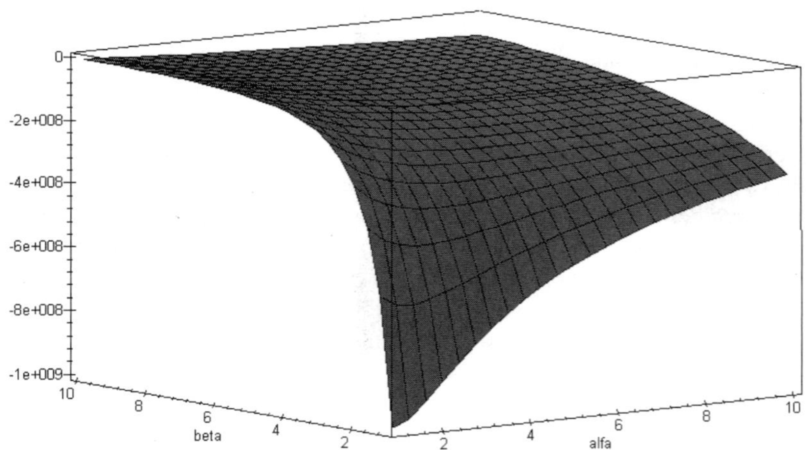

Figure 4.15. Second order sensitivity gradient of $e^{(eff)}$ wrt parameter β

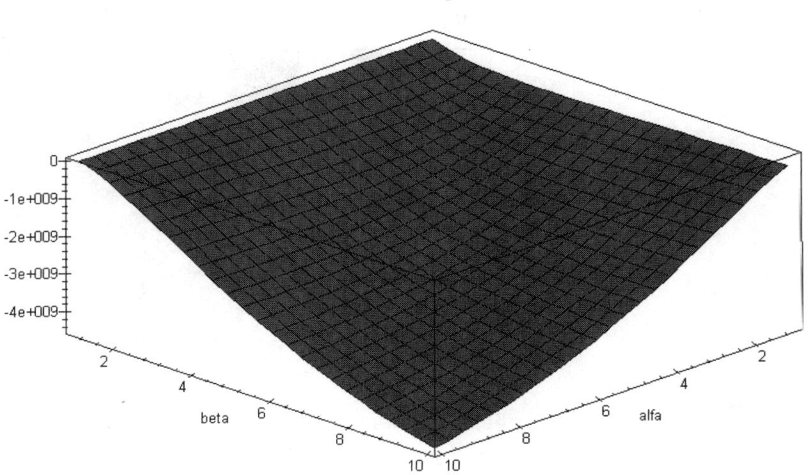

Figure 4.16. Second order sensitivity gradient of $e^{(eff)}$ wrt parameter γ

It is seen that in the case of both ratios equal to 1, the effective elasticity modulus is obtained as the value corresponding to a weaker material, which perfectly agrees with engineering intuition. Next, first and second order derivatives of the effective Young modulus of the composite with respect to the coefficients relating composite components are computed and analysed. It is typical that all the first order gradients are positive, while second order derivatives are less equal to 0. It reflects the fact that the overall effective Young modulus increase is obtained by the corresponding increase of any of these parameters. The second order sensitivity gradients computed and visualised above enable one to confirm the existence of an extremum of the first order derivatives presented before.

4.1.4 Material Sensitivity of Unidirectional Periodic Composites

The formulas describing the effective elasticity tensor components for the periodic composite with unidirectional distribution of the heterogeneities (see (2.103) − (2.107)) have been implemented in the symbolic computations package MAPLE to derive the appropriate sensitivity gradients [177]. The two−component composite shown schematically in Figure 4.17 was examined with the following input data for (a) weaker material e_2=4.0E9, ν_2=0.34, c_2=1-c_1 and (b) stronger material: e_1=4.0 E9 α, ν_1=0.34 β, c_1=0.5.

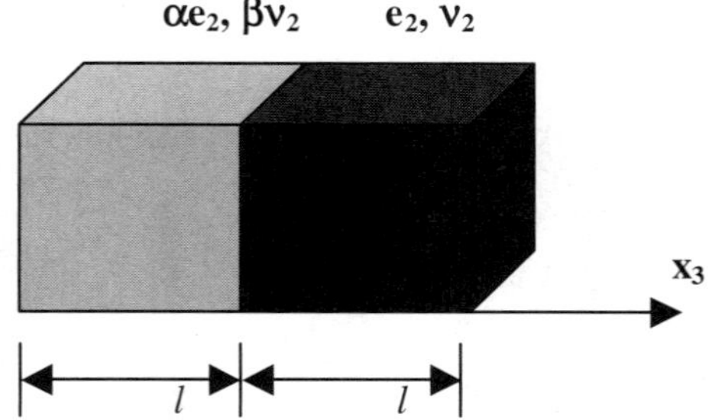

Figure 4.17. RVE of two–component composite bar

Design parameters α and β are introduced to make the visualisation of particular sensitivity gradients for some variations of the contrast between Young moduli and Poisson ratios of laminate layers. It will enable more successful optimisation of the composite in case of the homogenisation theory applications. The gradients collected on figures given below are normalised to make all the surfaces presented comparable to each other. First, quite obvious engineering interpretation of these results is that if particular gradient is less than 0 – an increase of design parameter accompanies a decrease of particular effective characteristic value. Otherwise (gradient greater than 0), an increase of the design parameter results in the appropriate increase of the homogenised quantity, while gradient comparable to 0 means that the given design parameter almost does not influence the overall effective characteristic. The figures plotted from the specially implemented MAPLE script present the sensitivity gradients of the homogenised elasticity tensor components – for $C^{(eff)}_{1111}$ (Figures 4.18–4.21), $C^{(eff)}_{3333}$ (Figures 4.22–4.25), $C^{(eff)}_{1133}$ (Figures 4.26–4.29), $C^{(eff)}_{1122}$ (Figures 4.30–4.33) and $C^{(eff)}_{1212}$ (Figures 4.34–4.37). Parameters α and β equivalent to the contrasts between stronger and weaker materials Young moduli and Poisson ratios are marked on the vertical axes of these figures, correspondingly.

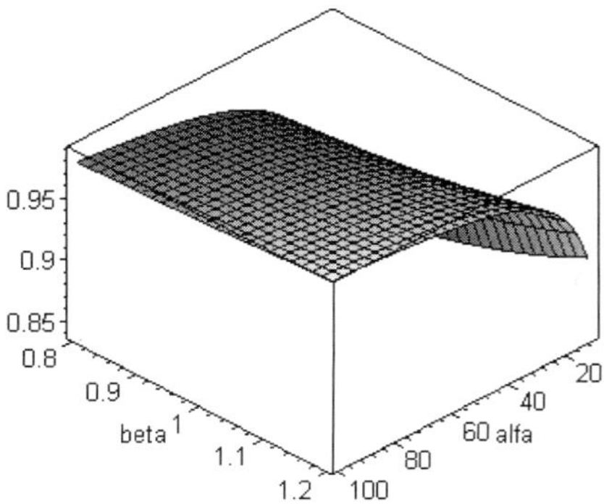

Figure 4.18. Sensitivity of $C^{(eff)}_{1111}$ wrt e_1

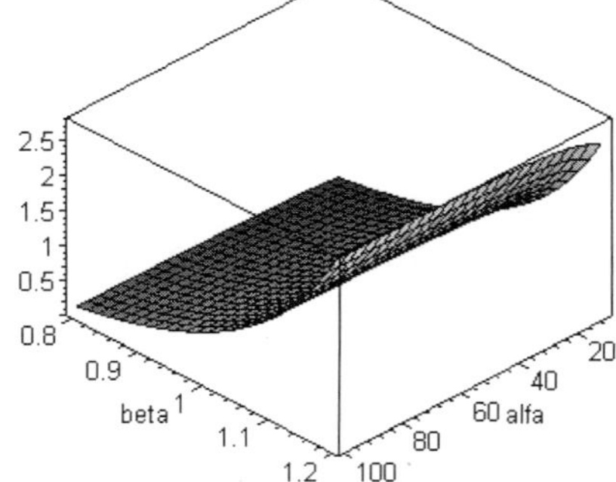

Figure 4.19. Sensitivity of $C^{(eff)}_{1111}$ wrt ν_1

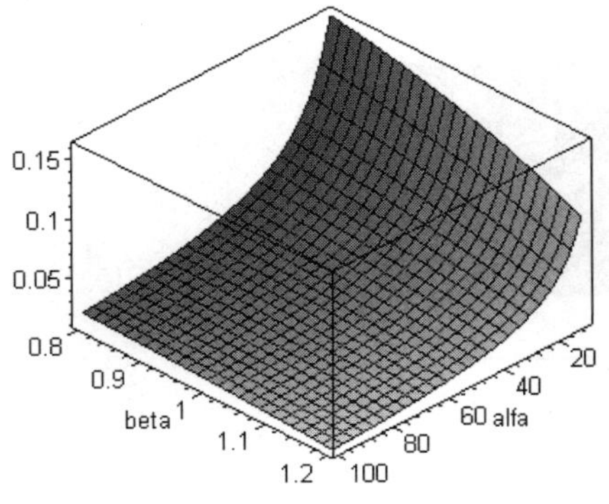

Figure 4.20. Sensitivity of $C^{(eff)}_{1111}$ wrt e_2

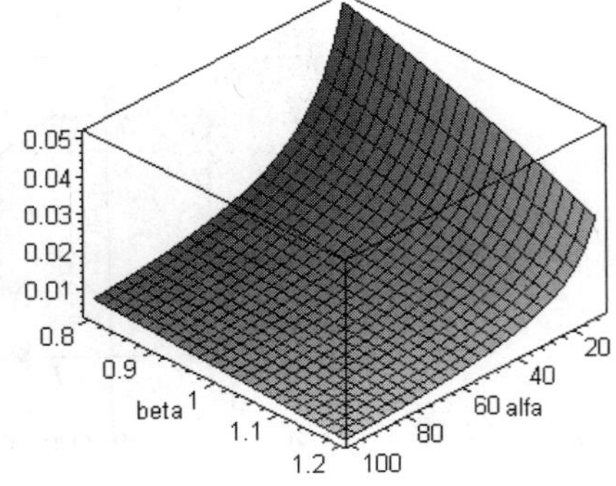

Figure 4.21. Sensitivity of $C^{(eff)}_{1111}$ wrt ν_2

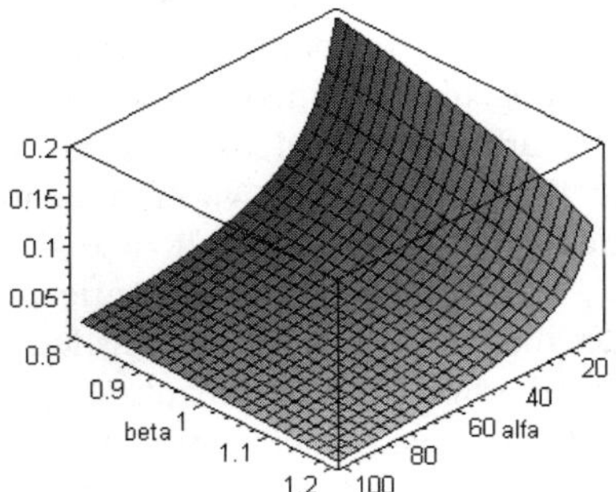

Figure 4.22. Sensitivity of $C^{(eff)}_{3333}$ wrt e_1

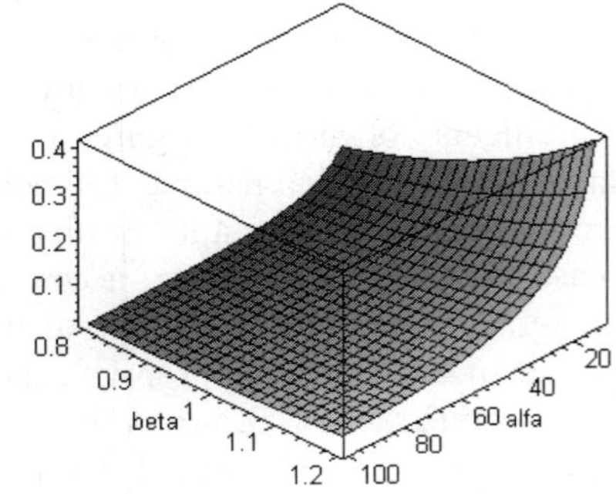

Figure 4.23. Sensitivity of $C^{(eff)}_{3333}$ wrt ν_1

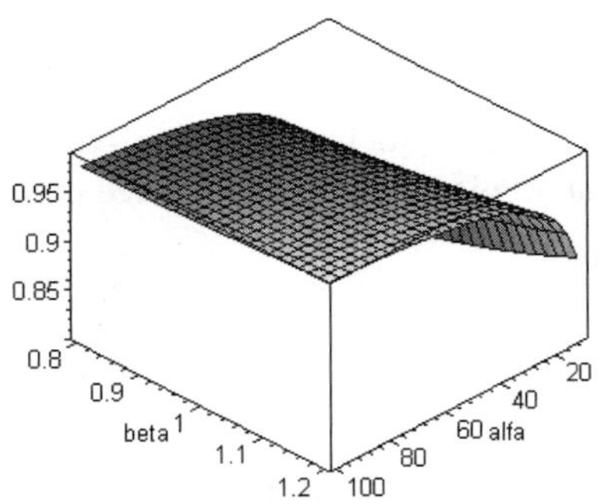

Figure 4.24. Sensitivity of $C^{(eff)}_{3333}$ wrt e_2

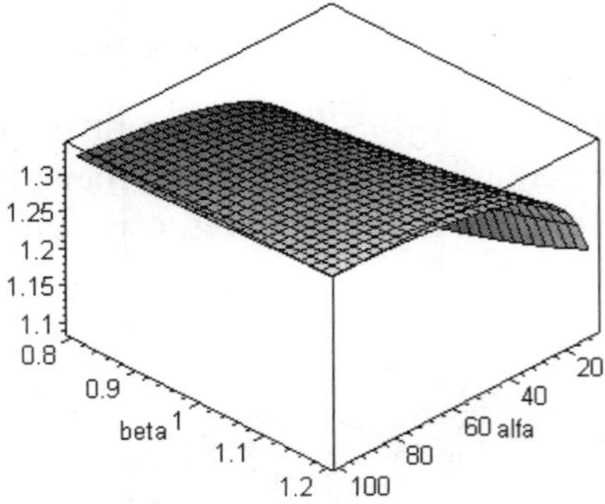

Figure 4.25. Sensitivity of $C^{(eff)}_{3333}$ wrt ν_2

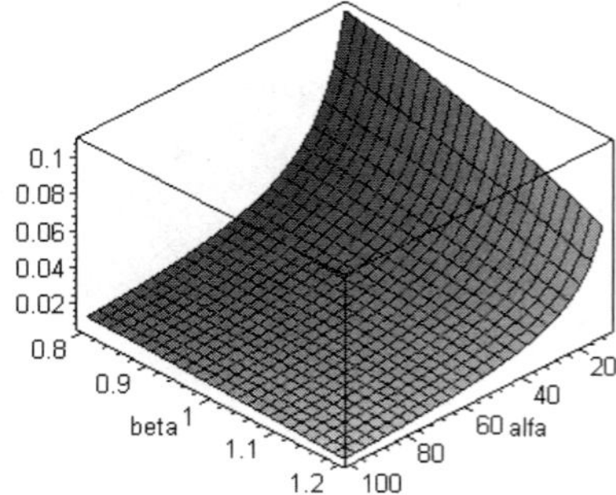

Figure 4.26. Sensitivity of $C^{(eff)}_{1133}$ wrt e_1

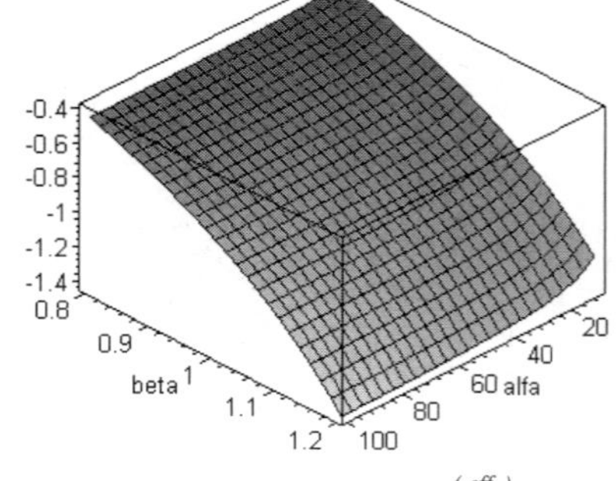

Figure 4.27. Sensitivity of $C^{(eff)}_{1133}$ wrt ν_1

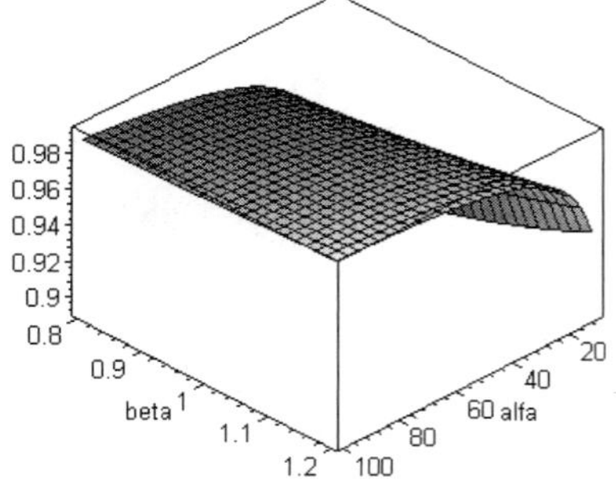

Figure 4.28. Sensitivity of $C^{(eff)}_{1133}$ wrt e_2

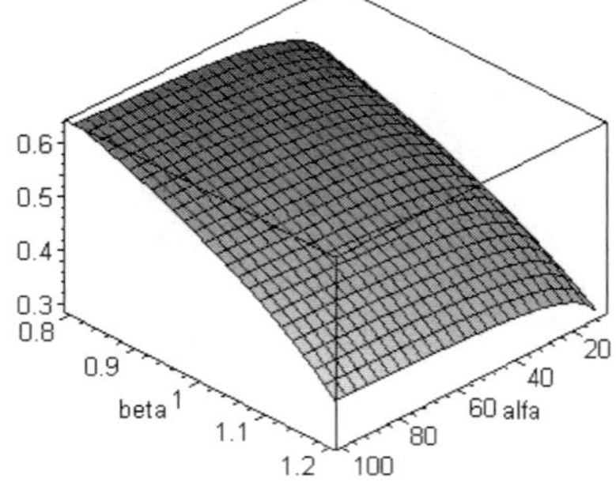

Figure 4.29. Sensitivity of $C^{(eff)}_{1133}$ wrt ν_2

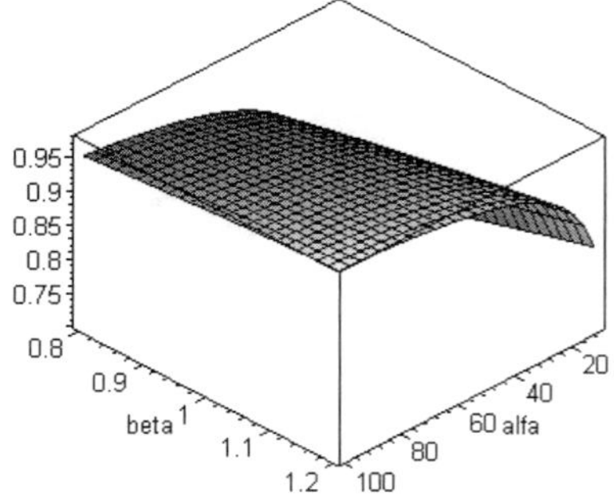

Figure 4.30. Sensitivity of $C^{(eff)}_{1122}$ wrt e_1

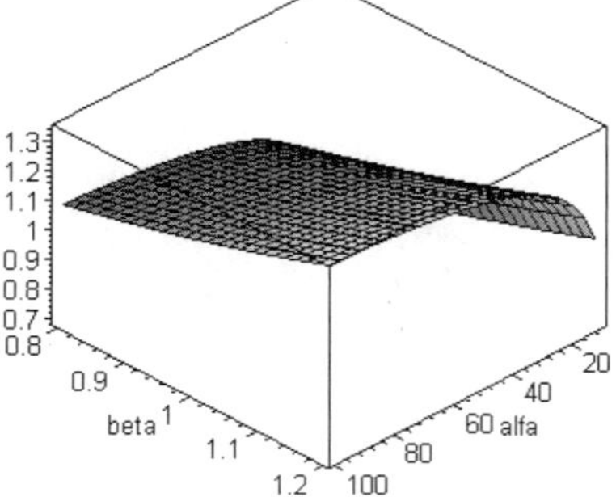

Figure 4.31. Sensitivity of $C^{(eff)}_{1122}$ wrt ν_1

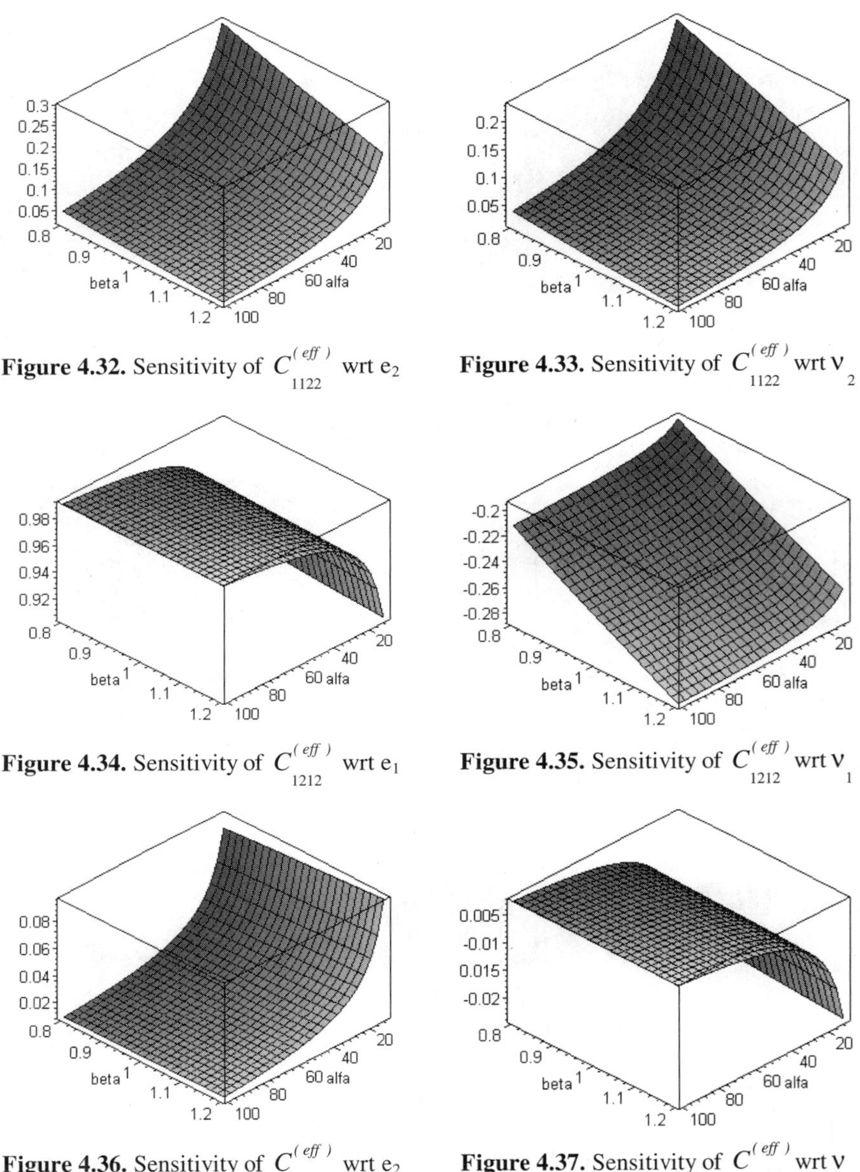

Figure 4.32. Sensitivity of $C^{(eff)}_{1122}$ wrt e_2

Figure 4.33. Sensitivity of $C^{(eff)}_{1122}$ wrt ν_2

Figure 4.34. Sensitivity of $C^{(eff)}_{1212}$ wrt e_1

Figure 4.35. Sensitivity of $C^{(eff)}_{1212}$ wrt ν_1

Figure 4.36. Sensitivity of $C^{(eff)}_{1212}$ wrt e_2

Figure 4.37. Sensitivity of $C^{(eff)}_{1212}$ wrt ν_2

How is demonstrated in all these figures, an increase of Young moduli of both stronger and weaker material result in the increase of all effective elasticity tensor components. Sensitivity gradients computed with respect to Poisson ratios of both composite components have mixed signs and all gradients essentially differ from 0. Taking into account particular variations and values of these results it can be observed that

(a) $C^{(eff)}_{1111}$ is sensitive at most wrt v_1, then to e_1 and e_2 and at least to v_2 and all gradients are positive;

(b) $C^{(eff)}_{3333}$ is most sensitive to v_2, then to e_2, v_1 and at least to e_1; all values are positive;

(c) $C^{(eff)}_{1133}$ is the most sensitive wrt e_2 and then to v_2, e_1 and at least to v_1 where the last parameter sensitivity analysis results in the negative gradient;

(d) $C^{(eff)}_{1122}$ (similarly to $C^{(eff)}_{1111}$) is most sensitive to v_1, then to e_1 and e_2 and at least to v_2 and all gradients have positive values;

(e) $C^{(eff)}_{1212}$ shows the greatest sensitivity wrt e_1, then to e_2 and finally to v_2 and v_1 where the last two variables give negative gradients.

Table 4.1. Sensitivity gradients for the unidirectional periodic composite

h	$\dfrac{\partial C^{(eff)}_{1111}}{\partial h}$	$\dfrac{\partial C^{(eff)}_{3333}}{\partial h}$	$\dfrac{\partial C^{(eff)}_{1133}}{\partial h}$	$\dfrac{\partial C^{(eff)}_{1122}}{\partial h}$	$\dfrac{\partial C^{(eff)}_{1212}}{\partial h}$	$G^{,h}$
e_1	0.9041	0.1138	0.0603	0.7696	0.9584	3.9570
e_2	0.0959	0.8862	0.9397	0.2304	0.0451	2.5430
v_1	-0.0476	0.0368	-0.2811	0.7849	-0.1728	-1.1099
v_2	0.0338	1.2018	0.6254	0.1891	-0.0105	1.8538

Furthermore, the sensitivity gradients of $G^{,h}$ with respect to all design parameters, i.e. Young moduli and Poisson ratios of both layers have been computed symbolically. They are found for equal values of components volume fractions in the RVE (50%) with the following material parameters: e_1=84.0 GPa, v_1=0,22 and e_2=4.0 GPa, v_2=0.34. All the gradients are collected in Tab. 4.1 – for particular components of the effective elasticity tensor and global composite structural response functional G. It is visible from these results that positive values of $G^{,h}$ are determined for e_1 and both material parameters of a weaker material, whereas negative – in case of stronger material Poisson coefficient. It should be mentioned that uniform strain field with $\varepsilon_{ij} = 1$ is applied at the RVE to define this functional.

Particular values of the quantities $G^{,h}$ lead to the conclusion that the entire composite is the most sensitive with respect to Young modulus of stronger material, then to the parameters e_2 and v_2 and at least – to the parameter v_1. Comparing these results with analogous results obtained for the fibre–reinforced composite and collected in Tab. 2 it is observed that quite similar values are obtained in both cases and, moreover, both composites show negative sensitivity to Poisson ratios of stronger material. The fibre–reinforced composite is however the most sensitive with respect to the Poisson ratio of a composite weaker component.

Finally, is can be noted that since the procedure presented for unidirectional composite contains the algebraic approximations of homogenised characteristics depending on volume fractions of the components, the sensitivity gradients can be easily recalculated to include the volume fractions of both (or greater number of) constituents.

4.1.5 Sensitivity of Homogenised Properties for Fibre-Reinforced Periodic Composites

Material sensitivity of the periodic fibre–reinforced plane composite is studied here according to the numerical homogenisation method employed in Chapter 2. The sensitivity coefficients for effective elasticity tensor components with respect to the design parameters vector represented by $\mathbf{h}$ can be calculated using formula (2.131) as [167,177]

$$\frac{dC_{ijpq}^{(eff)}}{d\mathbf{h}} = \frac{\partial}{\partial\mathbf{h}}\left\{\frac{1}{|\Omega|}\int_\Omega C_{ijpq}d\Omega\right\} + \frac{\partial}{\partial\mathbf{h}}\left\{\frac{1}{|\Omega|}\int_\Omega C_{ijkl}\varepsilon_{kl}\left(\chi_{(pq)}\right)d\Omega\right\} \tag{4.13}$$

which can be rewritten in the following form:

$$\begin{aligned}&\frac{dC_{ijpq}^{(eff)}}{d\mathbf{h}}\\ &= \frac{1}{|\Omega|}\int_\Omega \frac{\partial C_{ijpq}}{\partial\mathbf{h}}d\Omega + \frac{1}{|\Omega|}\int_\Omega \frac{\partial C_{ijkl}}{\partial\mathbf{h}}\varepsilon_{kl}\left(\chi_{(pq)}\right)d\Omega + \frac{1}{|\Omega|}\int_\Omega C_{ijkl}\frac{\partial\varepsilon_{kl}\left(\chi_{(pq)}\right)}{\partial\mathbf{h}}d\Omega\end{aligned} \tag{4.14}$$

It is necessary to underline that differentiation with respect to any design sensitivity parameter can be inserted under the integration sign over the RVE, only if geometrical sensitivity with respect to composite composite dimensions is not accounted. It is observed that if the input sensitivity parameters are not the arguments of the elasticity tensor C_{ijkl}, the formula (4.14) simplifies to

$$\frac{dC_{ijpq}^{(eff)}}{d\mathbf{h}} = \frac{1}{|\Omega|}\int_\Omega C_{ijkl}\frac{\partial\varepsilon_{kl}\left(\chi_{(pq)}\right)}{\partial\mathbf{h}}d\Omega \tag{4.15}$$

while the derivatives of the homogenisation functions $\chi_{(pq)}$ with respect to the components of vector h can be determined computationally by only. The first component of the sensitivity gradients in eqn (4.14) can be computed using analytical methods implemented in any symbolic computation packages. Furthermore, the sensitivity of $C_{ijpq}^{(eff)}$ components with respect to the fibre shape can be derived. However the final equations have a decisively more complicated form and they could be shown, only if the homogenisation function is derived analytically. Finally, the homogenised tensor derivatives are normalized as follows:

$$\left(\frac{dC_{ijpq}^{(eff)}}{d\mathbf{h}}\right)_{scaled} = \frac{\partial C_{ijpq}^{(eff)}}{\partial \mathbf{h}} \cdot \frac{h}{C_{ijpq}^{(eff)}(h)} \quad \text{(no summation over } i,j,p,q) \qquad (4.16)$$

which makes it possible to compare all the homogenised tensor sensitivity gradients with each other and to establish quantitatively the most decisive parameters.

The most interesting problem however is not to determine the sensitivity coefficients of the homogenised tensor with respect to particular composite parameters but to approximate the sensitivity of the entire structure to its some design parameters. That is why, following previous considerations, we need to establish some structural response functional being an implicit function of the homogenisation function of the original composite design parameters [75,76,207,208]. This functional must represent the overall elastic strain (or complementary) energy for such a plane strain problem defined on the RVE which, after some minor modifications only, can be valid for numerous engineering applications in the composites engineering.

Therefore, let us define the sensitivity functional as the strain energy of the homogenised composite under a combination of the uniform constant strains in horizontal and vertical directions as well as for the transverse strain ε_{xy} as is illustrated below. In this case, the sensitivity functional can be expressed as

$$\begin{aligned}
G &= \tfrac{1}{2}\int_\Omega \sigma_{ij}\varepsilon_{ij}\,d\Omega = \tfrac{1}{2}\int_\Omega \left(\sigma_{11}\varepsilon_{11} + \sigma_{12}\varepsilon_{12} + \sigma_{21}\varepsilon_{21} + \sigma_{22}\varepsilon_{22}\right)d\Omega \\
&= \tfrac{1}{2}\int_\Omega \left(\left\{C_{1111}^{(eff)}\varepsilon_{11} + C_{1122}^{(eff)}\varepsilon_{22}\right\}\varepsilon_{11} + \left\{C_{1212}^{(eff)}\varepsilon_{12} + C_{1221}^{(eff)}\varepsilon_{21}\right\}\varepsilon_{12}\right)d\Omega \\
&\quad + \tfrac{1}{2}\int_\Omega \left(\left\{C_{2121}^{(eff)}\varepsilon_{21} + C_{2112}^{(eff)}\varepsilon_{12}\right\}\varepsilon_{21} + \left\{C_{2211}^{(eff)}\varepsilon_{11} + C_{2222}^{(eff)}\varepsilon_{22}\right\}\varepsilon_{22}\right)d\Omega
\end{aligned} \qquad (4.17)$$

The strain state relevant to this functional can represent (a) uniaxial and/or biaxial compression/tension of the RVE; (b) shear (or torsion) of the composite specimen for $\varepsilon_{11}=0$ and $\varepsilon_{22}=0$ or (c) some combined strain state for the homogenised material.

Let us note that the difference between the vertical and horizontal strain tensor components is important in the case of an elliptical fibre and/or rectangular RVE where the extension of the cell give the unsymmetric strain field. Integrating over the RVE domain, recalling the assumed constant strain over this cell as well as a constant character of $C_{ijkl}^{(eff)}$ on Ω, one can get

$$G = \frac{l^2}{2}\left\{C_{1111}^{(eff)} + C_{1122}^{(eff)} + C_{1212}^{(eff)} + C_{1221}^{(eff)} + C_{2121}^{(eff)} + C_{2112}^{(eff)} + C_{2211}^{(eff)} + C_{2222}^{(eff)}\right\} \tag{4.18}$$

where l is a basic dimension of the RVE cell. Taking into account the elasticity tensor symmetry, the functional G can be expressed as

$$G = l^2\left\{C_{1111}^{(eff)} + C_{1122}^{(eff)} + 2C_{1212}^{(eff)}\right\} \tag{4.19}$$

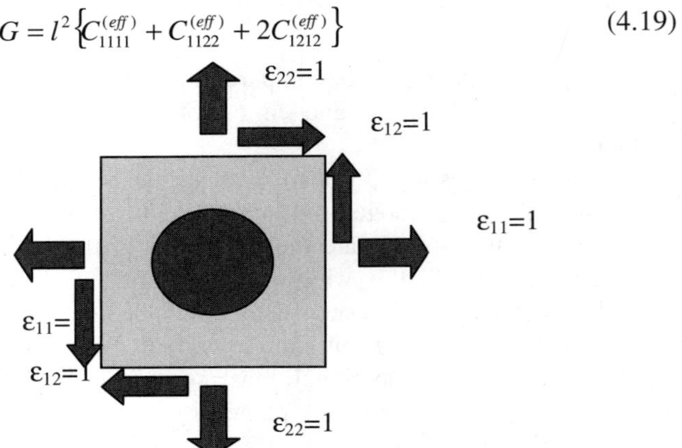

Figure 4.38. An idea of the structural response functional for the homogenised composite

Further, partial derivatives of G with respect to any component of the design parameters vector h can be calculated as

$$G^{,h} = \frac{\partial G}{\partial \mathbf{h}}$$

$$= \frac{\partial l^2}{\partial \mathbf{h}}\left\{C_{1111}^{(eff)} + C_{1122}^{(eff)} + 2C_{1212}^{(eff)}\right\} + l^2\left\{\frac{\partial C_{1111}^{(eff)}}{\partial \mathbf{h}} + \frac{\partial C_{1122}^{(eff)}}{\partial \mathbf{h}} + 2\frac{\partial C_{1212}^{(eff)}}{\partial \mathbf{h}}\right\} \tag{4.20}$$

The first component differs from 0 only if the design parameter vector contains the external diameter of the RVE. Otherwise, sensitivity gradients of this functional are determined as

$$G^{,h} = \frac{\partial G}{\partial \mathbf{h}} = l^2\left\{\frac{\partial C_{1111}^{(eff)}}{\partial \mathbf{h}} + \frac{\partial C_{1122}^{(eff)}}{\partial \mathbf{h}} + 2\frac{\partial C_{1212}^{(eff)}}{\partial \mathbf{h}}\right\} \tag{4.21}$$

Using this formula the most decisive design parameter for the homogenised composite in uniform plane strain can be determined having computed the effective elasticity tensor gradients from (4.13).

Finally, it is observed for the 1D heterogeneous structure with the constant cross–section A that the structural response functional G can be expressed as

$$G = \tfrac{1}{2}\int_\Omega \sigma\varepsilon d\Omega = \tfrac{A}{2}\int_l e^{(\mathit{eff})}\varepsilon\varepsilon dy = \frac{Ae^{(\mathit{eff})}}{2}\int_l \varepsilon^2 dy \tag{4.22}$$

which gives for the unit strain

$$G = \frac{Ae^{(\mathit{eff})}l}{2} \tag{4.23}$$

and then the sensitivity gradients of the functional G may be easily calculated by the chain rule as was proposed before.

The deterministic discretised homogenisation problem of elastic composites given by (4.14) is rewritten in the case of the DDM sensitivity studies as follows:

$$\frac{\partial K_{\alpha\beta}}{\partial \mathbf{h}} q_{(rs)\alpha} + K_{\alpha\beta}\frac{\partial q_{(rs)\alpha}}{\partial \mathbf{h}} = \frac{\partial Q_{(rs)\alpha}}{\partial \mathbf{h}} \tag{4.24}$$

where the sensitivity gradients of homogenisation function components are calculated as

$$\frac{\partial q_{(rs)\alpha}}{\partial \mathbf{h}} = K_{\alpha\beta}^{-1}\left(\frac{\partial Q_{(rs)\alpha}}{\partial \mathbf{h}} - \frac{\partial K_{\alpha\beta}}{\partial \mathbf{h}} q_{(rs)\alpha}\right) \tag{4.25}$$

If design variables are not the arguments of the RHS vector, it can be reduced to

$$\frac{\partial q_{(rs)\alpha}}{\partial \mathbf{h}} = -K_{\alpha\beta}^{-1}\frac{\partial K_{\alpha\beta}}{\partial \mathbf{h}} q_{(rs)\alpha} \tag{4.26}$$

The derivatives of the stiffness matrix components $\dfrac{\partial K_{\alpha\beta}}{\partial \mathbf{h}}$ can be computed explicitly during the stiffness process formation or, alternatively, thanks to the finite difference scheme (FDM) presented below. Therefore, sensitivity coefficients of the effective elasticity tensor components are calculated starting from the above equations as

$$\frac{\partial C^{(eff)}_{ijkl}}{\partial \mathbf{h}} = \frac{1}{|\Omega|}\int_\Omega \frac{\partial C_{ijkl}}{\partial \mathbf{h}}d\Omega + \frac{1}{|\Omega|}\int_\Omega \frac{\partial C_{ijpq}}{\partial \mathbf{h}}B_{kl\gamma}q_{(pq)\gamma}d\Omega$$

$$+ \frac{1}{|\Omega|}\int_\Omega C_{ijkl}B_{kl\gamma}\frac{\partial q_{(pq)\gamma}}{\partial \mathbf{h}}d\Omega \tag{4.27}$$

$$= \frac{1}{|\Omega|}\int_\Omega \frac{\partial C_{ijkl}}{\partial \mathbf{h}}d\Omega + \frac{1}{|\Omega|}\int_\Omega \frac{\partial C_{ijpq}}{\partial \mathbf{h}}B_{kl\gamma}K^{-1}_{\beta\gamma}Q_{(pq)\beta}d\Omega$$

$$- \frac{1}{|\Omega|}\int_\Omega C_{ijkl}B_{kl\gamma}K^{-1}_{\beta\gamma}\frac{\partial K_{\beta\gamma}}{\partial \mathbf{h}}Q_{(pq)\beta}d\Omega + \frac{1}{|\Omega|}\int_\Omega C_{ijkl}B_{kl\gamma}K^{-1}_{\beta\gamma}\frac{\partial Q_{(pq)\beta}}{\partial \mathbf{h}}d\Omega$$

for $\alpha, \beta, \gamma = 1,..., N$. If, for example, the sensitivity parameter is introduced as the Young modulus $h \equiv e_a$, then the elasticity tensor is rearranged as

$$C^{(a)}_{ijkl}(e(\mathbf{x});\nu(\mathbf{x})) = e_a(\mathbf{x})A^{(a)}_{ijkl}(\nu(\mathbf{x})) \tag{4.28}$$

and

$$\frac{\partial C^{(a)}_{ijkl}(e(\mathbf{x});\nu(\mathbf{x}))}{\partial e_a} = A^{(a)}_{ijkl}(\nu(\mathbf{x})) \tag{4.29}$$

while the finite element stiffness matrix component corresponding to ath material parameters can be expressed as

$$K^{(a)}_{\alpha\beta} = \int_{\Omega_a} C^{(a)}_{ijkl}B_{ij\alpha}B_{kl\beta}d\Omega = \int_{\Omega_a} e^{(a)}A^{(a)}_{ijkl}B_{ij\alpha}B_{kl\beta}d\Omega \tag{4.30}$$

As a result, the sensitivity of mth finite element stiffness matrix component with respect to the ath material Young modulus is computed as

$$\frac{\partial K^{(m)}_{\alpha\beta}}{\partial e^{(a)}} = \begin{cases} \int_{\Omega_a} A^{(a)}_{ijkl}B_{ij\alpha}B_{kl\beta}d\Omega; & \mathbf{x}^{(m)} \in \Omega_a \\ 0; & otherwise \end{cases} \tag{4.31}$$

Further, the sensitivity gradients of the RHS vector are obtained in a general form

$$\frac{\partial Q_{(pq)\alpha}}{\partial \mathbf{h}} = \frac{\partial \left(\left(C_{pq\alpha j}{}^{(a)}\right)\right)n_j}{\partial e_a} = \left[A_{pq\alpha j}{}^{(a)}\right]n_j \tag{4.32}$$

and the sensitivity of the effective elasticity tensor to Young modulus e_a is determined as

$$\frac{\partial C^{(eff)}_{ijkl}}{\partial e_a} = \frac{1}{|\Omega|}\int_\Omega A^{(a)}_{ijkl}\,d\Omega + \frac{1}{|\Omega|}\int_\Omega A^{(a)}_{ijkl}B_{i\gamma}K^{-1}_{\gamma\beta}Q_{(pq)\beta}\,d\Omega$$

$$+\frac{1}{|\Omega|}\int_\Omega C^{(a)}_{ijkl}B_{i\gamma}\left[\sum_{a=1}^{E}\int_{\Omega_a} A^{(a)}_{ijkl}B_{i\gamma}B_{kl\beta}\,d\Omega\right]Q_{(pq)\beta}\,d\Omega$$

$$+\frac{1}{|\Omega|}\int_\Omega C^{(a)}_{ijkl}B_{i\gamma}K^{-1}_{\beta\gamma}\left[A^{(a)}_{pq\beta j}\right]n_j\,d\Omega$$

(4.33)

Analogously, the sensitivity gradients of the effective elasticity tensor components for a composite with respect to the Poisson ratios can be calculated but since the elasticity tensor is a complex function of these ratios, the derivation is omitted here.

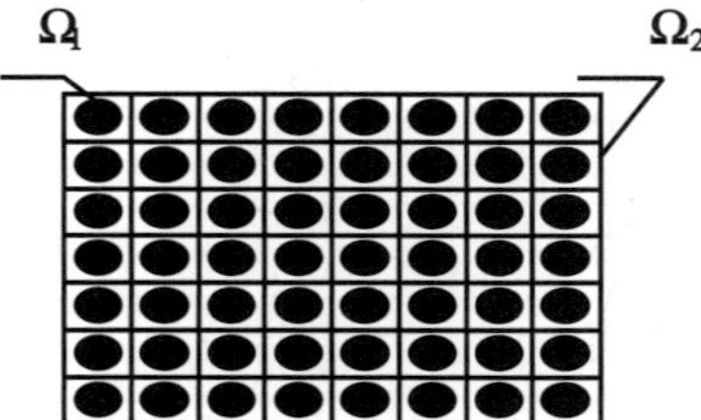

Figure 4.39. Periodic composite specimen

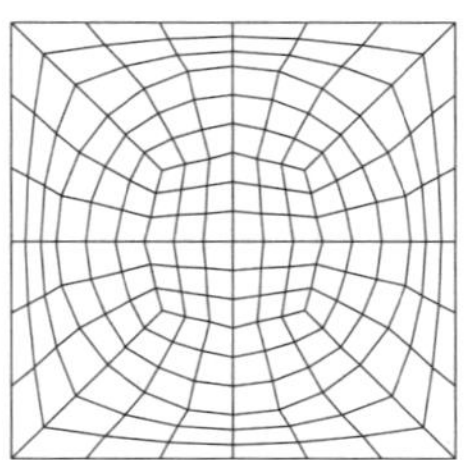

Figure 4.40. Mesh of the periodicity cell

Let us consider for illustration the composite with periodicity cell shown in Figure 4.39 – the fibre has a round cross–section and the entire RVE is rectangular. The analysed composite is assumed to be perfectly periodic with fibres distributed uniformly in the transverse cross–section, while the reinforcement ratio is equal to 50% of the total area of the RVE. Material characteristics for the computational analysis are taken as follows: e_1=84.0 GPa, e_2=4.0 GPa, v_1=0.34 and v_2=0.22; the FEM discretisation using 4-node linear plane strain elements is presented in Figure 4.40.

Computational sensitivity studies are carried out to determine the sensitivity gradients of the effective elasticity tensor components with respect to material parameters of the constituents, i.e. Young moduli and Poisson ratios of fibre and matrix. All computational tests are done by the use of the specially tailored computer program MCCEFF [167,173], designed and implemented for deterministic and stochastic computational homogenisation–based studies. The variability of the sensitivity gradients of the effective elasticity tensor components resulting from the perturbation parameter variations are presented in Figures 4.41 – 4.52: for the component $C_{1111}^{(eff)}$ (Figures 4.41–4.44), for the component $C_{1122}^{(eff)}$ (Figures 4.45–4.48), and for $C_{1122}^{(eff)}$ in Figures 4.49–4.52. The sensitivity gradients are marked on the horizontal axes for three different ranges of parameter increments shown on the vertical axes. These series correspond to the homogenised tensor increments in the range of promiles (O(-3)), percents (O(-2)) and tenths (O(-1)) of the verified parameter. The numerous experiments result from the fact that, as was expected and shown numerically, particular values of sensitivity gradients of the effective tensor components depend on the perturbation of a given material parameter employed as the design parameter.

Table 4.2. Sensitivity gradients of the effective elasticity tensor

h	$\dfrac{\partial C_{1111}^{(eff)}}{\partial h}$	$\dfrac{\partial C_{1122}^{(eff)}}{\partial h}$	$\dfrac{\partial C_{1212}^{(eff)}}{\partial h}$	$G^{.h}$
e_1	0.141	0.072	0.958	2.129
v_1	0.056	0.180	-0.173	-0.090
e_2	0.867	0.926	0.044	1.881
v_2	1.205	2.814	-0.011	3.987

As can be observed on all these graphs, the worst numerical stability of sensitivity gradients is obtained for the smallest perturbation order O(-3) and can result from the computational error of the homogenisation method itself. This numerical phenomenon can be studied in terms of the discretisation density of the RVE in the homogenisation analysis and with respect to the reinforcement ratio of the entire composite. Another phenomenon, resulting from physical aspects of the composite being visible especially in Figures 4.44, 4.48 and 4.52 in the case of the sensitivities of O(-1) order, is caused by the fact that the Poisson ratio of the matrix tends to its upper physical limit for this variable, which results in an uncontrolled increase of the components $C_{1111}^{(eff)}$ and $C_{1122}^{(eff)}$ sensitivity gradients. Because of that, greater values of $C_{ijkl}^{(eff)}$ derivatives with respect to Δv_2 do not exist.

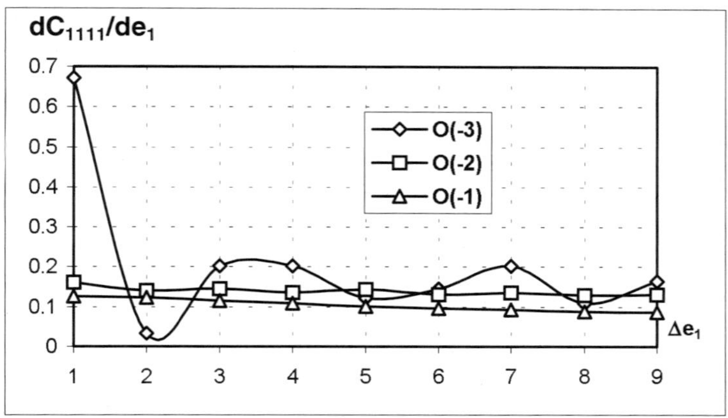

Figure 4.41. Sensitivity of $C_{1111}^{(eff)}$ wrt e_1

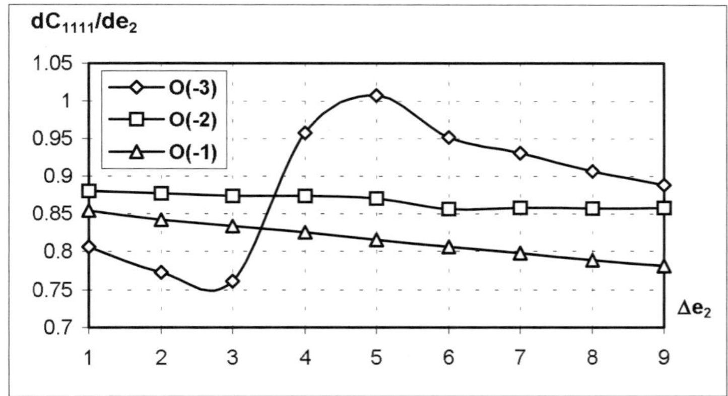

Figure 4.42. Sensitivity of $C_{1111}^{(eff)}$ wrt e_2

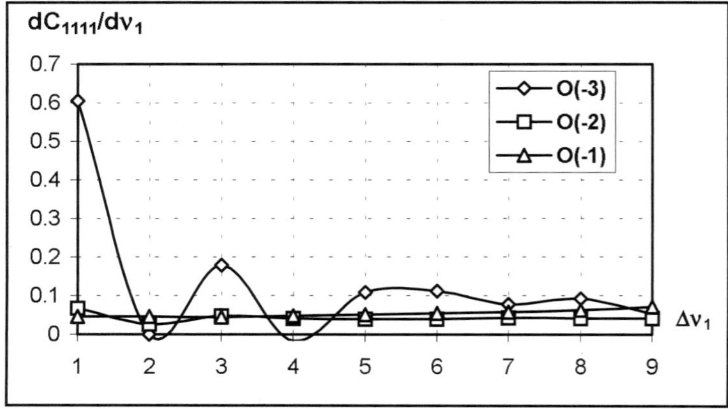

Figure 4.43. Sensitivity of $C_{1111}^{(eff)}$ wrt v_1

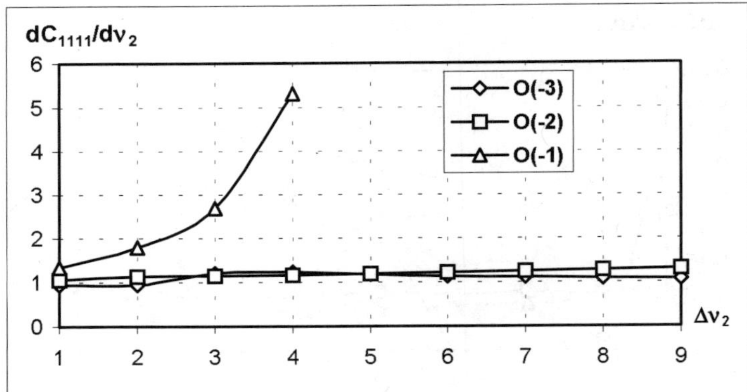

Figure 4.44. Sensitivity of $C_{1111}^{(eff)}$ wrt v_2

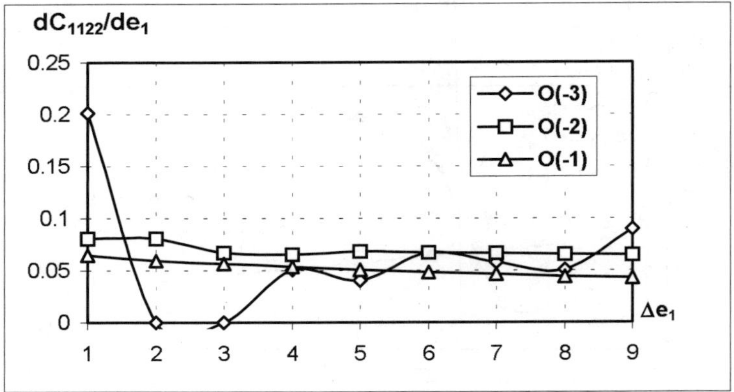

Figure 4.45. Sensitivity of $C_{1122}^{(eff)}$ wrt e_1

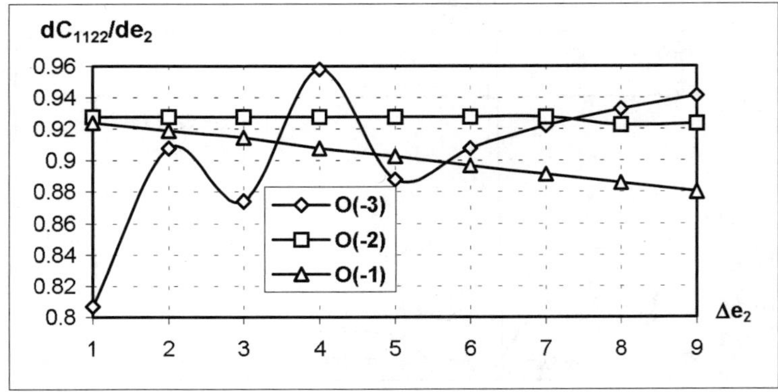

Figure 4.46. Sensitivity of $C_{1122}^{(eff)}$ wrt e_2

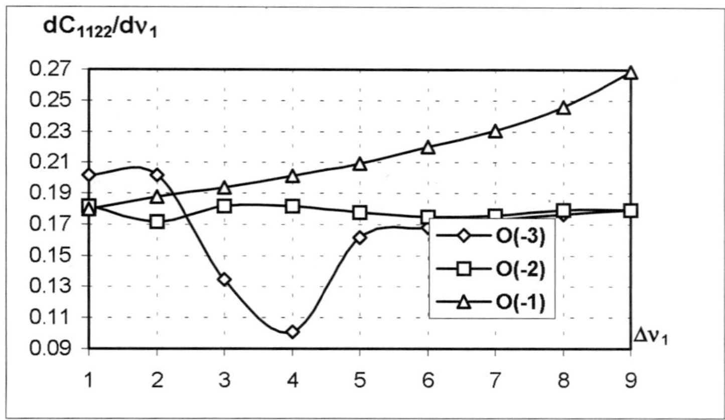

Figure 4.47. Sensitivity of $C_{1122}^{(eff)}$ wrt v_1

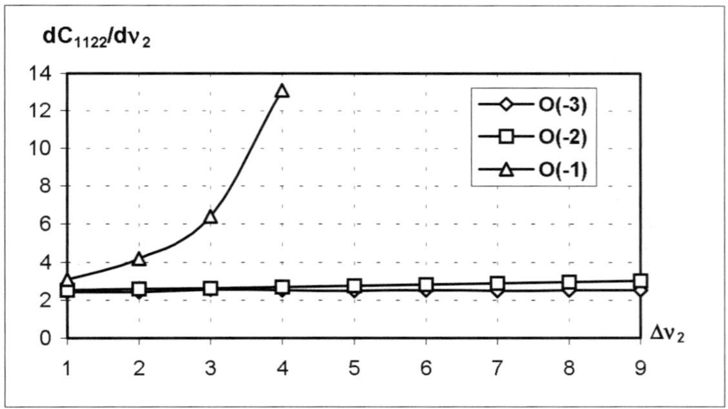

Figure 4.48. Sensitivity of $C_{1122}^{(eff)}$ wrt v_2

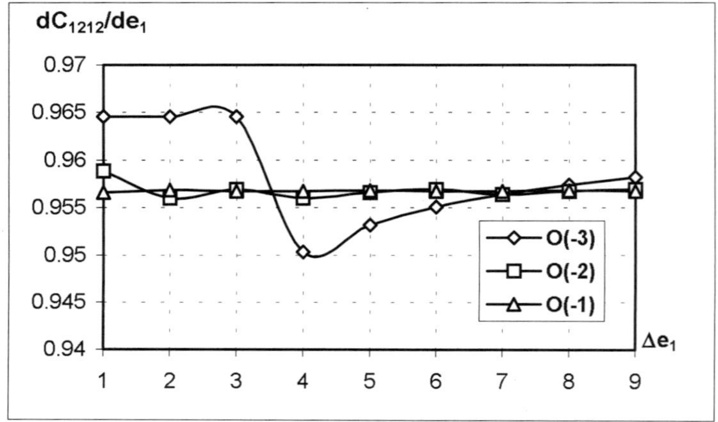

Figure 4.49. Sensitivity of $C_{1212}^{(eff)}$ wrt e_1

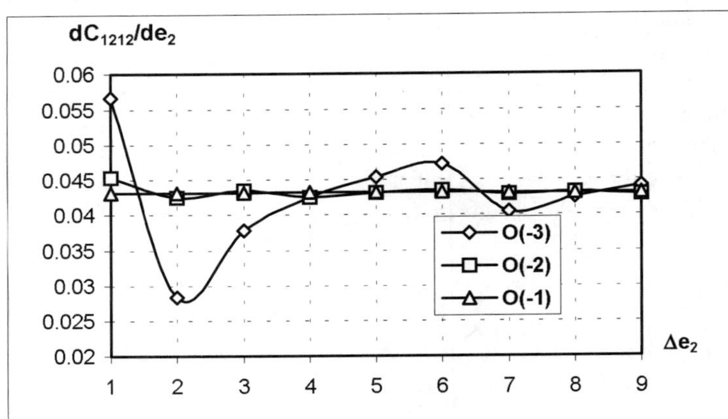

Figure 4.50. Sensitivity of $C_{1212}^{(eff)}$ wrt e_2

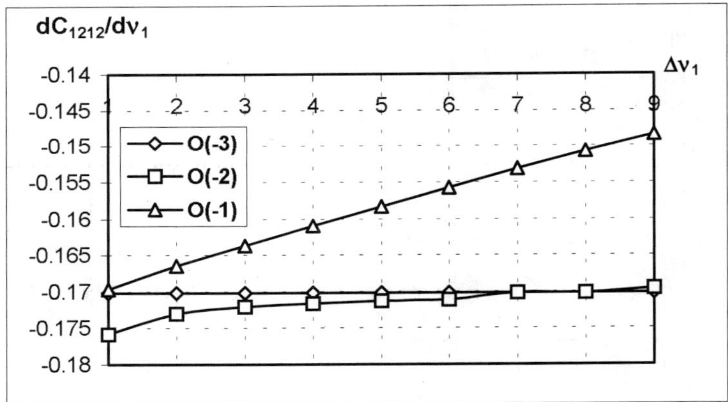

Figure 4.51. Sensitivity of $C_{1212}^{(eff)}$ wrt v_1

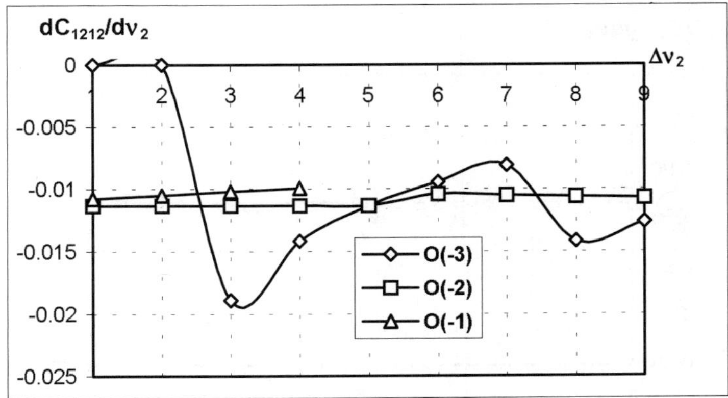

Figure 4.52. Sensitivity of $C_{1212}^{(eff)}$ wrt v_2

Next, comparing all the figures, it is seen that the best numerical stability is obtained for the computational series corresponding to O(-2). The sensitivity gradients in this range are almost constant for 1–10% of all input parameters perturbations. Starting from these results, a more detailed comparison, both from a computational and engineering point of view, can be carried out for various FEM mesh sizes and the interrelations of design parameter mean values. It should be noticed that, as for the 1D composite example, a negative sign of the sensitivity gradient is equivalent to a decrease of a particular $C_{ijkl}^{(eff)}$ component, accompanying the increase of this parameter value; a positive derivative corresponds to the opposite relation.

Observing the particular results of computational analysis it can be noticed that most sensitivity gradients are positive (except those shown in Figures 4.51 and 4.52), which basically means that an increase of most elastic characteristics of fibre–reinforced composite components results in corresponding increase of the overall effective elasticity tensor; the effective elasticity tensor components $C_{1111}^{(eff)}$ and $C_{1122}^{(eff)}$ are sensitive at most to Poisson ratios of the composite constituents. The component $C_{1212}^{(eff)}$ is sensitive to the Young modulus of the matrix (with a negative sign). Furthermore, it is seen that particular values of the presented gradients depend strongly on the interrelations between the main values of Young moduli and Poisson ratios of the entire structure. That is why the observed phenomena are the best illustration of the material sensitivity of glass–epoxy periodic composites only. The results collected in Table 4.2 can be compared against those obtained before for essentially different interrelations between the composite components – we observe in that study that for similar constituents the signs of particular gradients are exactly the same. However the values are qualitatively different. This is why such an analysis must be oriented to the particular composite; otherwise it should be carried out for the very particular engineering application of the analysed composite.

The sensitivity with respect to the reinforcement shape, local lack of periodicity as well as material parameters in the case of inelastic behaviour of the constituents may be verified numerically in further computational studies on homogenised properties of the composites. On the other hand, it seems to be reasonable to verify the computed sensitivity gradient to the interrelation between corresponding elastic properties of the components (the ratio of Young modulus in a fibre to Young modulus of a matrix, for instance). This study would validate if all groups of various composites with the same geometry of the RVE had the same or at least comparable sensitivity gradients.

Finally, the approximation of the determined gradients by some specific values is proposed and, considering all the remarks posed above, it is established as the arithmetic average of the gradients corresponding to 1% and 10% increments of the design parameters. These values are used to approximate the value of $G^{,h}$ computed on the basis of (4.21), which are scaled over the RVE total area. Using such a composite structure response functional, the Poisson ratio of a matrix and the next Young modulus of the fibre are detected as the most decisive design material parameters of this composite in the view of its homogenised elastic parameters.

4.2 Probabilistic Analysis

The main purpose of the sensitivity analysis is to verify numerically the influence of some particular input parameters on the analysed state functions. In the case of the homogenisation procedure, the sensitivity of the effective elasticity tensor can be verified in terms of material parameters of the constituents, reinforcement shape and its spatial distribution, the volume ratios of the components, etc.; material parameters of composite constituents are taken below as design parameters.

Analogous situation takes place in case of probabilistic analysis, however the total number of possible design parameters dramatically increases. It reflects the fact that each geometrical and material parameter is usually represented by its at least two probabilistic moments. Hence, all probabilistic moments of all input variables can be considered as design parameters. At the same time, the sensitivity gradients can be computed in addition to any probabilistic moment of the state function determined during a structural modeling. Therefore, we can determine the sensitivity gradients of expected values of displacement vector to the expected values and/or standard deviations of structural members thickness, length or elastic parameters. Similarly, the cross-correlation function or standard deviation of the resulting state variables can be the subject of the SDS analysis.

Using the definition of effective elasticity tensor, the sensitivity of the mth order probabilistic moment of this tensor with respect to the nth order central probabilistic moment of an input random design variable vector $\mathbf{h}$ can be formulated as

$$\frac{\partial\left(\mu_m\left(C_{ijkl}^{(eff)}(\mathbf{h};\omega)\right)\right)}{\partial\mu_n(\mathbf{h})} = \frac{\partial}{\partial\mu_n(\mathbf{h})}\left(\mu_m\left(\int_\Omega C_{ijkl}(\mathbf{h};\mathbf{x};\omega)d\Omega\right)\right) \\ + \frac{\partial}{\partial\mu_n(\mathbf{h})}\left(\mu_m\left(\int_\Omega \sigma_{ij}\left(\chi^{(kl)}(\mathbf{h};\mathbf{x};\omega)\right)d\Omega\right)\right) \tag{4.34}$$

The first component of the RHS summation can be derived analytically or symbolically, whereas the second one can be obtained numerically only by using the Finite or Boundary Element Method programs adopted for any probabilistic technique. If the effective material tensor is represented by the closed form function of the elastic properties of composite components, then it is possible to derive analytically probabilistic moments of a homogenised tensor. Alternatively, the Monte Carlo simulation technique may be used to randomise and estimate the sensitivity gradients of any order probabilistic moments with respect to the effective elasticity tensor components.

Numerical illustration is carried out by an application of the Monte Carlo simulation for a homogenisation cell problem have been performed using specially modified FEM code MCCEFF and its 4−node rectangular isoparametric plane strain finite elements. The results of computations are expected now as the partial derivatives of probabilistic moments of the effective elasticity tensor with respect to relevant probabilistic moments of elastic or geometrical characteristics of the fibre and/or matrix as well as the interface defects (Young modulus of the matrix is treated here as the design probabilistic variable).

Sensitivity gradients of up to the fourth order probabilistic characteristics of the homogenised elasticity tensor components with respect to $\mathbf{h} \equiv e_2(\omega)$ being uncorrelated Gaussian random variables are obtained for the following data: $E[e_1] = 84.0\ E6$, $E[e_2] = 4.0\ E6$, $Var(e_1) = 70.56\ E12$, $Var(e_2) = 16.0\ E10$; it corresponds to the coefficient of variation equal to 0.1 for both Young moduli. The interface defects (model with 'bubbles') are simulated numerically in such a way that a 10% elastic characteristic reduction in the interphase is obtained and they are compared against the results computed for a composite with perfectly bonded components (model with 'no bubbles'). The results of simulations are collected in Table 4.3 as sensitivity gradients of first two probabilistic moments of the homogenised elasticity tensor components with respect to expected value and the variance of the matrix Young modulus.

Table 5.3. Probabilistic sensitivity gradients of the homogenised elasticity tensor

Probabilistic moment	$\dfrac{\partial}{\partial E[e_2]}$		$\dfrac{\partial}{\partial Var(e_2)}$	
	'Bubbles'	'No bubbles'	'Bubbles'	'No bubbles'
$E[C_{1111}^{(eff)}]$	0.0819	0.0817	-0.0001	-0.0001
$\alpha(C_{1111}^{(eff)})$	-0.0748	-0.0747	0.0405	0.0405
$\beta(C_{1111}^{(eff)})$	-0.0076	-0.0127	-0.0005	-0.0064
$\gamma(C_{1111}^{(eff)})$	-0.0003	0.0005	-0.0004	0.0000
$E[C_{1122}^{(eff)}]$	0.0892	0.0893	-0.0001	-0.0001
$\alpha(C_{1122}^{(eff)})$	-0.0815	-0.0815	0.0438	0.0438
$\beta(C_{1122}^{(eff)})$	0.0082	0.0059	0.0012	0.0013
$\gamma(C_{1122}^{(eff)})$	0.0003	0.0002	0.0002	0.0003
$E[C_{1212}^{(eff)}]$	0.0043	0.0043	-0.00006	0.0000
$\alpha(C_{1212}^{(eff)})$	-0.0043	-0.0043	0.0020	0.0020
$\beta(C_{1212}^{(eff)})$	0.0000	-0.0001	0.0003	-0.00002
$\gamma(C_{1212}^{(eff)})$	0.0000	0.0000	0.0000	0.0000

The results collected above show that the most sensitive probabilistic moment of the homogenised tensor component with respect to $E[e_2]$ is the expected value of $E[C_{1122}^{(\mathit{eff})}]$, which is slightly greater than the result computed for $E[C_{1111}^{(\mathit{eff})}]$. An analogous relation is observed in addition to the coefficient of variation $\alpha\left(C_{1122}^{(\mathit{eff})}\right)$ and $Var(e_2)$. However sensitivity gradients determined with respect to the expected value are significantly greater than those obtained for the variance, which partially reflects the input coefficient of the variation of the matrix Young modulus. The smallest sensitivity of random variable $C_{1212}^{(\mathit{eff})}(\omega)$ with respect to the random input $e_2(\omega)$ is observed also, while the fourth order coefficients of concentration are in practice neither sensitive to $E[e_2]$ nor to $Var(e_2)$, which follows the Gaussian type of both input and output probabilistic distributions in homogenisation problems. The most sensitive statistical estimator is the coefficient of asymmetry – some differences are observed between the models with and without interface bubbles, where some sign changes are also noticeable.

Comparing the probabilistic gradients computed for the composite with and without the interphase some small variations between these two models are observed. These variations however can increase together with further weakening of the interphase and detailed computer simulation can verify this tendency.

Further computations are necessary to study the variability of the obtained results with respect to the chosen increment during the numerical differentiation process; the proposed value of 10% has been detected as the most effective in previous computations. An increase in effectiveness of the numerical procedure can be achieved by implementation of a semi–analytical homogenisation procedure, where the sensitivity gradients of spatially averaged effective elasticity tensor components are determined symbolically using the system MAPLE, for instance, and an averaged stress tensor is differentiated numerically using the finite difference scheme.

4.3 Conclusions

The sensitivity analysis of homogenised material tensors, proposed and carried out in this Chapter, makes it possible to consider the influence of particular material parameters of the composite components on the overall effective properties of a composite. Thanks to such an analysis, a composite designer can generally determine the most decisive material characteristics of the constituents (Poisson ratios of fibre and matrix for a 2D composite, for instance) and then, modifying their values during the design process, can optimise the composite structure for the effective parameters given *a priori*. The sensitivity equations for homogenisation of linear elastic composites can be extended to an analogous analysis for effective properties of composites with viscoelastoplastic components, both in a deterministic and probabilistic context. The proposed methodology has a general character, however further examination of various composites (beams, plates, 2D and 3D structures) should give different results.

Particular computational studies, performed in terms of perturbation parameter ε applied in sensitivity gradients analysis, show that the best numerical stability is obtained for $\varepsilon \approx O(-2)$. Smaller order taken in numerical analysis causes significantly greater deviations of the final result, while the lack of physical sense of the problem is obtained for $\varepsilon \approx O(-1)$, where the Poisson ratio is taken as a design parameter. Further computations should be carried out to determine the RVE mesh and the finite element type influence on the final result.

Considering the assumption that the scale factor between the periodicity cell and the entire composite structure tends to 0 and, on the other hand, that this quantity in real composites is small and positive, but differs from 0, the sensitivity of the effective characteristics for this parameter is to be calculated next using the so-called homogenisation micro–macro analysis, for instance. To make such an analysis, the scale parameter must be inserted in equations describing the effective quantities and then, the influence of the relation between the micro– and macrostructure must be shown.

Sensitivity analysis of the 1D, 2D and 3D homogenisation of effective heat conductivity carried out in this chapter may be applied for any linear potential field problem – irrotational and incompressible fluid flow, film lubrication, acoustic vibration as well as for electric conduction, electrostatic field and electromagnetic waves. To use these results for homogenisation of other engineering problems, the well–known field analogies may be applied to transform the effective Young modulus to related physical field parameters.

Proposed methodology of the sensitivity gradient computations for homogenised tensor components of periodic random composites is exact in the probabilistic sense because of the application of the Monte Carlo simulation technique. The use of the MCS technique preserves the existence and uniqueness of the classical homogenisation problem solution – it exists for each realisation separately. Therefore, thanks to the statistical estimation implementation, a mathematically correct probabilistic characterisation of the homogenised tensor components is obtained. The numerical weakness of the finite difference apparatus implemented in the homogenisation–oriented FEM code should be eliminated in further simulations by the verification of the sensitivity gradient values with respect to variations of the input parameter increments. As is documented by some previous computations, lack of numerical stability of such sensitivity computations is observed for physical parameters tending to their physical bounds.

An analogous procedure can be applied to determine the sensitivity gradients of homogenised characteristics of other composites, i.e. 1D periodic beams, plates, shells, periodic 3D structures with particles and fibres of various shapes as well as multi–component engineering structures as superconducting devices [168] studied before. A linear elastic model in sensitivity analysis may be extended to inelastic homogenised characteristics [118,230,307] as well as on stochastic optimisation of composites through the homogenisation method.

5 Fracture and Fatigue Models for Composites

5.1 Introduction

The effective fatigue model for engineering composites analysis is decisive for a precise estimation of the overall life of this structure and satisfactory reliability analysis of such materials. Various theoretical, experimental and computational criteria must be satisfied in the same time to obtain such a model [37,172,246,298]. These criteria may include material properties of composite constituents [226,258], composite type [229] (ductile or brittle components), spatial distribution, length (continuity) as well as size effect of the reinforcing fibres [219,220,335], frequency effects [350], load amplitude type [48] (constant or not), micromechanical phenomena [110,217,279], etc. First of all, a very precise, experimentally based deterministic idea of fatigue life cycle estimation has to be proposed. It should be adequate for the composite components, the technology applied and numerical methodology implemented. Monitoring of most engineering composites and preventing the fatigue failure is very complicated and usually demands very modern technology [360]. It is widely known that the interface conditions and phenomena can be decisive factors for both static fracture and fatigue resistance of laminates, fibre– and particle–reinforced composites. Analytical models even in the case of linear elasticity models are complicated [369], therefore numerical analysis is very popular in this area. Engineering FEM software makes it possible to simulate delamination processes [362] and fatigue damage [62,277] in fibre–reinforced composites as well as time–dependent interlaminar debonding processes [69], for instance.

The application of the well–known Palmgren–Miner or Paris–Erdogan laws is not always recommended as the most effective method in spite of their simplicity or wide technological usage. The choice of fatigue theory should be accompanied with a corresponding sensitivity analysis, where physical and material input parameters included into the fatigue life cycle equation are treated as design variables. Due to the sensitivity gradients determination, the most decisive parameters should be considered, while the remaining ones, considering further stochastic analysis complexity, may be omitted. The sensitivity gradients can be determined analytically using symbolic computation packages (MAPLE, MATLAB, MATHEMATICA, etc.) or may result from discrete FEM computations, for instance. A related problem is to decide if the local concept of composite fatigue is to be applied (critical element concept, for instance), where local fatigue damage causes global structural changes of the composite reliability. This results in computational FEM or Boundary Element Method (BEM) based analyses of the whole composite in its real configuration, including the microgeometry and all interface phenomena into it. Alternatively, the homogenisation method can be applied, where the complementary energy or potential energy of the entire system is the only measure of composite fatigue. Then, the global discretisation of the original structure is used instead and the equivalent, homogeneous medium is simulated numerically.

Next, an appropriate analytical or computational stochastic analysis method corresponding to the level of randomness of input parameters is considered. The Monte Carlo simulation based analysis, stochastic second or third order perturbation method or, alternatively, stochastic spectral analysis can be taken into account. The first method does not have any restrictions on input random variable probabilistic moment interrelations. However, time consuming computations can be expected. Numerical analysis using the second approach implementation is very fast, but not sufficiently effective for larger than 10% variations of input random parameters, while the last approach has some limitations on convergence of the output parameters and fields. The choice between the methods proposed is implied by the availability of the experimental techniques, considering the input randomness level. On the other hand this choice is determined by relevant reliability criteria for composites. Furthermore, having collected most of the deterministic fatigue concepts for composites, corresponding stochastic equations can be obtained automatically using analytical derivation or computer simulation techniques.

Combination of deterministic models and stochastic methods requires another engineering decision about the choice of the randomness type to be analysed. It is known from recent references in this area that (i) random variables, (ii) random fields as well as (iii) stochastic processes can be considered as the input of the entire fatigue analysis. According to the state−of−the−art research, the first two types of randomness can be considered together with FEM or BEM based computational simulation, while the stochastic processes can be used in terms of direct simulation of the fatigue process when the analytical solution is known. Some approximate methods of combining discrete modelling with stochastic degradation of homogeneous materials are available in reliability modeling; however without any application in engineering composites area until now.

Various fatigue models worked out for composites can be classified in different ways: using the scale of the model application (local or global) or considering the main goal of the analysis (fatigue cycle number, its stiffness reduction, its crack growth or damage function determination), the analysis type (deterministic, probabilistic or stochastic) as well as the composite material type (ceramic, polymer−based, metal matrix and so forth).

Considering various scales of engineering composites and fatigue phenomena related to them, the local and, alternatively, global approaches are considered. Local and microlocal models represented by the critical element concept [299], assume that there exists so−called critical element in the entire composite structure that controls the total fatigue damage (as well as subcrtitical elements, too), and then the local damage is governing the reliability of the whole composite structure.

This assumption results in the fact that the whole composite, together with microstructural defects increasing during fatigue processes, should be discretised for the FEM or BEM simulation. Taking into account the application of the probabilistic analysis, the model implies the randomness in microgeometry of the composite, which is extremely difficult in computational simulation, as is shown below. Some special purpose algorithms are introduced to replace the randomness in composite interface geometry with the stochasticity of material thermoelastic properties.

Alternatively, a homogenisation method is proposed for more efficient fracture and fatigue phenomena analysis [223] that originated from analysis of linear periodic elastic composites without defects. The main idea is to find the medium equivalent to the original composite in terms of complementary energy, or potential energy, equal for both media. The final goal of the homogenisation procedure is to find the effective material characteristics defining the equivalent homogeneous medium. The effective constitutive relations can be found for the composite with elastic, elastoplastic or even viscoelastoplastic components with and/or without microstructural defects. The general assumption of the model means, however, that every local phenomenon can be averaged in some sense in the entire composite volume and that the global, not local, phenomena result in the overall composite fatigue.

5.2 Existing Techniques Overview

Taking into account the results of fatigue analysis, four essentially different approaches can be observed: (i) direct determination of the fatigue cycle number N, (ii) fatigue stiffness reduction where mechanical properties of the composite are decreased in the function of N, (iii) observation of the crack length growth a as a function of fatigue cycle number (as da/dN, taking into account the physical nature of fatigue phenomenon) or, alternatively, (iv) estimation of the damage function in terms of dD/dN. A damage function is usually proposed as follows:

(1) $D=0$ with cycle number $n=0$;

(2) $D=1$, where failure occurs;

(3) $D = \sum_{i=1}^{n} \Delta D_i$, where ΔD_i is the amount of damage accumulation during fatigue at stress level r_i. Generally, the function D can be represented as

$$D = D(n, r, f, T, M, ...) \tag{5.1}$$

where n indexes a number of the current fatigue cycle, r is the applied stress level, f denotes applied stress frequency, T is temperature, while M denotes the moisture content. Then, contrary to the crack length growth analysis, the damage function can be proposed each time in a different form as a function of various structural parameters.

Let us note that direct determination of fatigue cycle number makes it possible to derive, without any further computational simulations, the life of the structure till the failure, while the stiffness reduction approach is frequently used together with the FEM or BEM structural analyses. The crack length growth and damage function approach are used together with the structural analysis FEM programs, usually to compute the stress intensity factors. However final direct or symbolic integration of crack length or damage function is necessary to complete the entire fatigue life computations.

Considering the mathematical nature of the fatigue life cycle estimation, the deterministic approach can be applied, where all input parameters are defined uniquely by their mean values. Otherwise, the whole variety of probabilistic approaches can be introduced where fatigue structural life is described as a simple random variable with structural parameters defined deterministically and random external loads. The cumulative fatigue damage can be treated as a random process, where all design parameters are modelled as stochastic parameters. However, in all probabilistic approaches sufficient statistical information about all input parameters is necessary, which is especially complicated in the last approach where random processes are considered due to the statistical input in some constant periods of time (using the same technology to assure the same randomness level).

The analysis of fatigue life cycle number begins with direct estimation of this parameter by a simple power function (A5.1) consisting of stress amplitude as well as some material constant(s). Alternatively, an exponential–logarithmic equation can be proposed (A5.2), where temperature, strength and residual stresses are inserted. Both of them have a deterministic form and can be randomised using any of the methods described below. The weak point is the homogeneous character of the material being analysed; to use these criteria for composites, the effective parameters should be calculated first. In contrary to theoretical models, the experimentally based probabilistic law can be proposed where parameters of the Weibull distribution of static strength are inserted (A5.3); it is important to underline that this law does not have its deterministic origin.

More complicated from the viewpoint of engineering practice are the stiffness reduction models (cf. A5.4–A5.7), where structural material characteristics are reduced together with a successive fatigue cycle number increase. The stiffness reduction model is used in FEM or BEM dynamical modelling to recalculate the component stiffness in each cycle. It is done using a linear model for stiffness reduction, cf. (A5.5), as well as some power laws (see (A5.4), for example) determined on the basis of mechanical properties reduction rewritten for homogeneous media only. An alternative power law presented as (A5.7) consists of the time of rupture, creep and fatigue, measured in hours. Considering the random analysis aspects, a probabilistic treatment of material properties seems to be much more justified.

Deterministic fatigue crack growth analysis presented by (A5.8) − (A5.29) can be classified taking into account the physical basis of this law formation, such as energy approaches (A5.8) − (A5.11), crack opening displacement (COD) based approaches (A5.12), (A5.15) − (A5.17), (A5.19) and (A5.20), continuous dislocation formalism (A5.13), skipband decohesion (A5.18), nucleation rate process models (A5.14) and (A5.15), dislocation approaches (A5.23) and (A5.24), monotonic yield strength dependence (A5.25) and (A5.31) as well as another mixed laws (A5.26) − (A5.30) and (A5.32) − (A5.35). Description of the derivative *da/dN* enables further integration and determination of the critical crack length. The second classification method is based on a verification of the validity of a particular theory in terms of elastic (A5.8) − (A5.20), (A5.26) − (A5.30), (A5.32) − (A5.34) or elastoplastic (A5.22) − (A5.25) and (A5.31) mechanism of material fracture. Most of them are used for composites, even though they are defined for homogeneous media, except for the Ratwani−Kan and Wang−Crossman models (A5.21) and (A5.22), where composite material characteristics are inserted. All of the homogeneous models contain stress intensity factor ΔK in various powers (from 2 to n), while composite−oriented theories are based on delamination length parameter. The structure of these equations enables one to include statistical information about any material or geometrical parameters and, next, to use a simulation or perturbation technique to determine expected values and variances of the critical crack length, which are very useful in stochastic reliability analysis.

An essentially different methodology is proposed for the statistical analysis [9,35,130,288,333,349,359] and in the stochastic case [241,244,373], where the crack size and/or components material parameters, their spatial distribution may be treated as random processes (cf. eqns (A5.36) − (A5.44)). Then, various representations and types of random fields and stochastic processes are used, such as stationary and nonstationary Gaussian white noise, homogeneous Poisson counting process [204] as well as Markovian [304], birth and death or renewal processes. However all of them are formulated for a globally homogeneous material. These methods are intuitively more efficient in real fatigue process modelling than deterministic ones, but they require definitely a more advanced mathematical apparatus. Further, randomised versions of deterministic models can be applied together with structural analysis programs, while stochastic characters of a random process cannot be included without any modification in the FEM or the BEM computer routines. An alternative option for stochastic models of fatigue is experimentally based formulation of fatigue law, where measurements of various material parameters are taken in constant time periods. Then, statistical information about expected values and higher order probabilistic characteristics histories is obtained, which allows approximation of the entire fatigue process. Such a method, used previously for homogeneous structural elements, is very efficient in stochastic reliability prognosis and then random fatigue process can be included in SFEM computations. Let us observe that formulations analogous to the ones presented above can be used for ductile fracture of composites where initiation, coalescence and closing of microvoids are observed under periodic or quasiperiodic external loads.

A wide variety of fatigue damage function models is collected at the end of the appendix. The basic rules are based on the numbers of cycles to failure ((A5.45) − (A5.48), (A5.54) − (A5.57), (A5.63) − (A5.65) and (A5.67)) illustrated with classical and modified Palmgren–Miner approach, for instance. This variable is most frequently treated as a random variable or a random process in stochastic modelling. Another group consists of mechanical models, where stress (A5.50) – (A5.53) or strain (A5.66) – (A5.67) limits are used instead of global life cycle number. Such models reflect the actual state of a composite during the fatigue process better and are more appropriate for the needs of computational probabilistic structural analysis. The combination of both approaches is proposed by Morrow in (A5.66) for constant stress amplitude and for different cycles by (A5.67). The overall fatigue analysis is then more complicated. However the most realistic model is obtained. Accidentally, Fong model is used, where damage function is represented by an exponential function of damage trend k, which is a compromise between counting fatigue cycles and mechanical tensor measurements.

The very important problem is to distinguish the scale of application of the proposed model, especially in the context of determination of a fatigue crack length. The models valid for long cracks do not account for the phenomena appearing at the microscale of the composite specimen. On the contrary, cf. (A5.33), the microstructural parameter d is introduced, which makes it possible to include material parameters in the microscale in the equation describing the fatigue crack growth.

All the models for the damage function can be extended on random variables theoretically, by perturbation methodology, or computationally, using the relevant MCS approach. The essential minor point observed in most of the formulae described above is a general lack of microstructural analysis. The two approaches analysed above can model cracks in real laminates, while other types of composites must be analysed using fatigue laws for homogeneous materials. This approach is not a very realistic one, since fatigue resistance of fibres, matrices, interfaces and interphases is essentially different. Considering the delamination phenomena during periodic stress changes, an analogous fatigue approach for fibre–matrix interface decohesion should be worked out. The probabilistic structural analysis of such a model can be made using SFEM computations or by a homogenisation. However a closed-form fatigue law should be completed first.

As is known, there exist a whole variety of effective probabilistic methods in engineering. The usage of any of these approaches depends on the following factors: (a) type of random variables (normal, lognormal or Weibull, for instance), (b) probabilistic information on the input random variables, fields or processes (in the form of moments or probability density function (PDF)), (c) interrelations between particular probabilistic characteristics of the input (of higher to the first order, especially), (d) method of solution of corresponding deterministic problem and (e) available computational time as well as (f) applied reliability criteria.

If the closed form solution is available or can be derived symbolically using computational algebra, then the probability density function (PDF) of the output can be found starting from analogous information about the input PDF. It can be done generally from definition – using integration methods, or, alternatively, by the characteristic function derivation. The following PDF are used in this case:

lognormal for stress and strain tensors, lognormal and Gaussian distributions for elastic properties as well as for the geometry of fatigue specimen. Weibull density function is used to simulate external loads (shifted Rayleigh PDF, alternatively), yield strength as well as the fracture toughness, while the initial crack length is analysed using a shifted exponential probability density function.

As is known [313], one of the following computational methods can be used in probabilistic fatigue modelling: Monte Carlo simulation technique, stochastic (second or higher order) perturbation analysis as well as some spectral techniques (Karhunen–Loeve or polynomial chaos decompositions). Alternatively, Hermitte–Gauss quadratures (HGQ) or various sampling methods (Latin Hypercube Sampling – LHS, for instance) in conjunction with one of the latters may be used. Computational experience shows that simulation and sampling techniques are or can be implemented as exact methods. However their time cost is very high. Perturbation-based approaches have their limitations on higher order probabilistic moments, but they are very fast. The efficiency of spectral methods depends on the order of decomposition being used, but computational time is close to that offered by the perturbation approach. Unfortunately, there is no available full comparison of all these techniques – comparison of MCS and SFEM can be found in [208], HGQ with SFEM in [237] and stochastic spectral FEM with MCS in [113,114]. A lot of numerical experiments have been conducted in this area, including cumulative damage analysis of composites by the MCS approach (Ma *et al.* [243]) and simulation of stochastic processes given by (A5.30) – (A5.38). However, the problem of an appropriate conjunction of stochastic processes and structural analysis using FEM or BEM techniques has not been solved yet.

Let us analyse the application of the perturbation technique to damage function D extension, where it is a function of random parameter vector b. Using a stochastic Taylor expansion it is obtained that

$$D(\mathbf{b}) = D^0\left(\mathbf{b}^0\right) + \varepsilon \Delta b^r D^{,r}\left(\mathbf{b}^0\right) + \tfrac{1}{2}\varepsilon^2 \Delta b^r \Delta b^s D^{,rs}\left(\mathbf{b}^0\right) \qquad (5.2)$$

Then, according to the classical definition, the expected value of this function can be derived as

$$E[D(\mathbf{b})] = \int\limits_{-\infty}^{+\infty} D(\mathbf{b})\, p(\mathbf{b})\, d\mathbf{b}$$

$$= \int\limits_{-\infty}^{+\infty}\left(D^0\left(\mathbf{b}^0\right) + \varepsilon \Delta b^r D^{,r}\left(\mathbf{b}^0\right) + \tfrac{1}{2}\varepsilon^2 \Delta b^r \Delta b^s D^{,rs}\left(\mathbf{b}^0\right)\right) p(\mathbf{b})\, d\mathbf{b} \qquad (5.3)$$

$$= D^0\left(\mathbf{b}^0\right) + \tfrac{1}{2} D^{,rs}\left(\mathbf{b}^0\right) Cov\left(b^r, b^s\right)$$

while variance is

$$Var(D) = \left(\frac{\partial D}{\partial \mathbf{b}}\right)^2 Var(\mathbf{b}) \qquad (5.4)$$

Since this function is usually used for damage control, which in the deterministic case is written as $D \leq 1$, an analogous stochastic formulation should be proposed. It can be done using some deterministic function being a combination of damage function probabilistic moments as follows:

$$D(\mathbf{b}) \leq g\big(\mu_k[D(\mathbf{b})]\big) \leq 1 \tag{5.5}$$

where $\mu_k(D(\mathbf{b}))$ denote some function of up to kth order probabilistic moments. Usually, it is carried out using a stochastic 'envelope' function being the upper bound for the entire probability density function as, for instance

$$g\big(\mu_k[D(\mathbf{b})]\big) = E[D(\mathbf{b})] - 3\sqrt{D(\mathbf{b})} \tag{5.6}$$

This formula holds true for Gaussian random deviates only. It should be underlined that this approximation should be modified in the case of other random variables, using the definition that the value of damage function should be smaller than 1 with probability almost equal to 1; the lower bound can be found or proposed analogously. In the case of classical Palmgren−Miner rule (A5.45), with fatigue life cycle number N treated as an input random variable,

$$D = \frac{n}{N}, \quad N \equiv b \tag{5.7}$$

the expected value is derived as follows [215]:

$$E[D] = D^0 + \tfrac{1}{2} D^{,NN} Var(N) = \frac{n}{N^0} + \frac{n}{\left(N^0\right)^3} Var(N) \tag{5.8}$$

and the variance in the form of

$$Var(D) = \left(D^{,N}\right)^2 Var(N) = \frac{n^2}{\left(N^0\right)^4} Var(N) \tag{5.9}$$

It is observed that the methodology can also be applied to randomise all of the functions D listed in the appendix to this chapter with respect to any single or any vector of composite input random parameters. In contrast to the classical derivation of the probabilistic moments from their definitions, there is no need to make detailed assumptions on input PDF to calculate expected values and variances for the inversed random variables in this approach.

Let us determine for illustration the number of fatigue cycles of cumulative damage of a crack at the weld subjected to cyclic random loading with the specified expected value and standard deviation (or another second order probabilistic characteristics) of $\Delta\sigma$. Let us assume that the crack in a weld is growing according to the Paris–Erdogan law, cf. (A5.26), described by the equation

$$\frac{da}{dN} = C\left(Y\Delta\sigma\sqrt{\pi}\right)^m a^{\frac{m}{2}}$$
(5.10)

and that $Y\neq Y(a)$. Then

$$\int \frac{da}{a^{\frac{m}{2}}} = \int C\left(Y\Delta\sigma\sqrt{\pi}\right)^m dN$$
(5.11)

which gives by integration that

$$\frac{1}{-\frac{m}{2}+1} a^{-\frac{m}{2}+1} = C\left(Y\Delta\sigma\sqrt{\pi}\right)^m N + D , \quad D\in\Re$$
(5.12)

Taking for $N=0$ the initial condition $a=a_i$, it is obtained that

$$\left(\frac{a}{a_i}\right)^k = \frac{1}{1-\beta N}$$
(5.13)

for

$$\kappa = \tfrac{m}{2}-1 , \quad \beta = \kappa a_i^\kappa C\left(Y\Delta\sigma\sqrt{\pi}\right)^m$$
(5.14)

Therefore, the number of cycles to failure is given by

$$N_f = \frac{1}{\beta}$$
(5.15)

The following equation is used to determine the probabilistic moments of the number of cycles for a crack to grow from the initial length a_i to its final length a_f:

$$\Delta N = \int_{a_i}^{a_f} \frac{1}{C(\Delta K)^m} da$$
(5.16)

Substituting for ΔK one obtains

$$\Delta N = \frac{1}{C(\Delta\sigma)^m \pi^{\frac{m}{2}}} \int_{a_i}^{a_f} \frac{1}{Y^m a^{\frac{m}{2}}} da$$
(5.17)

By the use of a stochastic second order perturbation technique we determine the expected value of ΔN as

$$E[\Delta N] = \Delta N(\Delta \sigma^0) + \frac{1}{2} \frac{\partial^2 (\Delta N(\Delta \sigma^0))}{\partial (\Delta \sigma^0)^2} Var(\Delta \sigma)$$

(5.18)

and the variance of number of cycles as

$$Var(\Delta N) = \left(\frac{\partial (\Delta N(\Delta \sigma^0))}{\partial (\Delta \sigma^0)} \right)^2 Var(\Delta \sigma)$$

(5.19)

Adopting $m=2$ it is calculated using (5.17) and (5.18) that

$$E[\Delta N] = \frac{1}{CY^2 \pi} \ln\left(\frac{a_f}{a_i} \right) \left[\frac{1}{E^2[\Delta \sigma]} + \frac{6}{E^4[\Delta \sigma]} Var(\Delta \sigma) \right]$$

(5.20)

and

$$Var(\Delta N) = \frac{4}{C^2 Y^4 \pi^2} \ln^2\left(\frac{a_f}{a_i} \right) \frac{\alpha(\Delta \sigma)}{E^4[\Delta \sigma]}$$

(5.21)

The following data are adopted in probabilistic symbolic computations: $E[\Delta \sigma] = \sigma_{max} - \sigma_{min} = 10.0 \, MPa$, $a_i=25$ mm and obtained experimentally $C=1.64 \times 10^{-10}$, $Y=1.15$. The visualisation of the first two probabilistic moments of fatigue cycle number is done using the symbolic computation program MAPLE as functions of the coefficient of variation $\alpha(\Delta \sigma)$ and the final crack length a_f. The results of the analysis in the form of deterministic values, corresponding expected values and standard deviations are presented below with the design parameters marked on the horizontal axes.

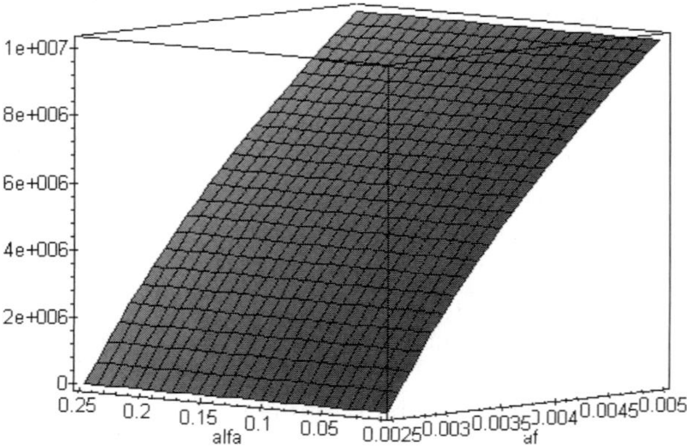

Figure 5.1. Deterministic values of fatigue cycles (dN)

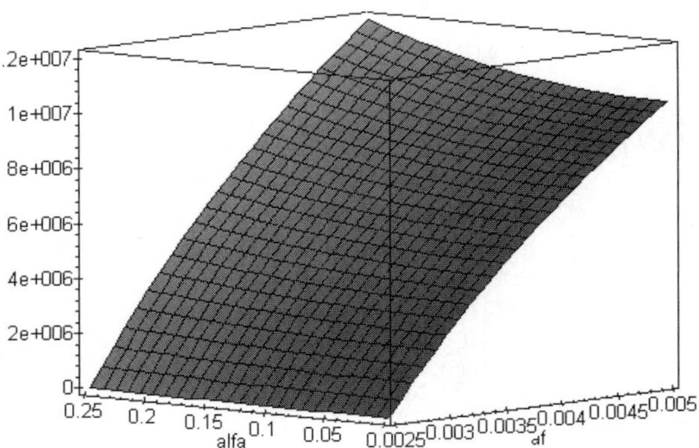

Figure 5.2. Expected values of fatigue cycles number (*EdN*)

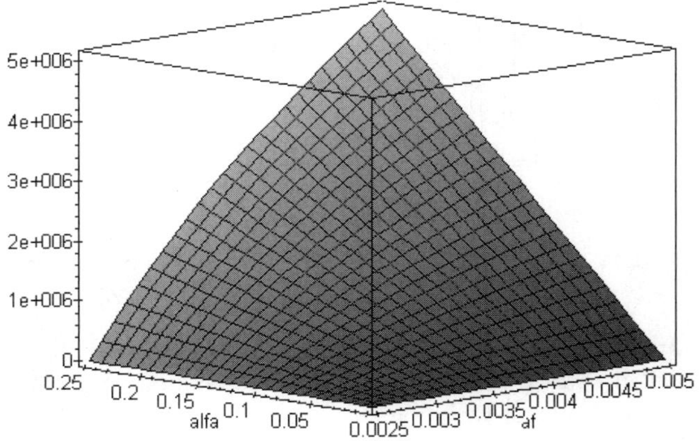

Figure 5.3. Standard deviations of fatigue cycles number (σdN)

Especially interesting here is a comparison between deterministic analysis and expected values obtained for analogous input data. It is seen that the expectations are essentially greater than the deterministic output, which results from (5.17), for instance. The difference increases nonlinearly together with an increase in the coefficient of variation of the stress amplitude $\Delta\sigma$. In the case of $\alpha(\Delta\sigma)=25\%$ this difference is equal to about 20% of the relevant deterministic values. This result can be used as the safety factor which could be proposed for deterministic analysis as $S=1.2$ for an analogous range of random variability of the stress amplitude. Furthermore, it is seen that the final crack length is remarkably more decisive for fatigue cycle number (even in a random case) than the coefficient of variation of the stress amplitude.

As shown in Figure 5.3, the variability of the examined standard deviation of $\Delta\sigma$ is essentially different from that typical for deterministic and expected values. The influences of final crack length and input coefficient of variation are almost the same for 25% increases of both parameters. Considering the above it can be concluded that the influence of the random character in fatigue cycle number is important in higher than first order probabilistic moments computations. It is clear that the presented symbolic computation methodology can be next exploited in the determination of stochastic sensitivity gradients of probabilistic moments of the fatigue cycle number with respect to particular random characteristics of the chosen input variables appearing in the fatigue life cycles formula. In particular, it will enable us to compare the sensitivity of various fatigue models with respect to the same parameters in which the sensitivity gradients are the most reasonable and realistic. The situation would be definitely more complicated if the variation of stress amplitude together with fatigue cycle number is analysed. Random fluctuations of $\Delta\sigma$ in time should be taken into account in this case and, therefore, $\Delta\sigma(\omega)=\Delta\sigma(\omega,t)$ is to be considered as a resulting nonstationary random process.

5.3 Computational Issues

Since the deterministic equations are valid for the Monte Carlo simulation analysis as well, then the essential theoretical differences are observed in the case of perturbation based analysis. The corresponding fatigue–oriented SFEM model begins with the new description of the material properties, where the stiffness reduction approach can result in the following equations for the Young modulus, Poisson ratio and material density as well as spring stiffness for interface modelling

$$e(n) = e_0\big(1 - D(n)\big), \ v(n) = v_0\big(1 - D(n)\big)$$
$$\rho(n) = \rho_0\big(1 - D(n)\big), \ k(n) = k_0\big(1 - D(n)\big) \tag{5.27}$$

Therefore, the first two probabilistic moments for the Young modulus can be represented as

$$E[e(n)] = E\big[e_0\big]\big(1 - E[D(n)]\big) \tag{5.28}$$
$$Var(e(n)) = Var\big(e_0\big(1 - D(n)\big)\big) = Var\big(e_0\big)Var\big(1 - D(n)\big) \tag{5.29}$$

and up to the second order perturbation equations are rewritten in the incremental formulation as follows:
- zeroth order

$$M^0_{\alpha\beta}(n)\Delta\ddot{q}^0_\beta(n) + C^0_{\alpha\beta}(n)\Delta\dot{q}^0_\beta(n) + K^0_{\alpha\beta}(n)\Delta q^0_\beta(n) = \Delta Q^0_\alpha(n) \tag{5.30}$$

- first order

$$M^0_{\alpha\beta}(n)\Delta\ddot{q}^{;r}_\beta(n) + C^0_{\alpha\beta}(n)\Delta\dot{q}^{;r}_\beta(n) + K^0_{\alpha\beta}(n)\Delta q^{;r}_\beta(n) = \Delta Q^{;r}_\alpha(n) +$$
$$- \left(M^{;r}_{\alpha\beta}(n)\Delta\ddot{q}^0_\beta(n) + C^{;r}_{\alpha\beta}(n)\Delta\dot{q}^0_\beta(n) + K^{;r}_{\alpha\beta}(n)\Delta q^0_\beta(n) \right) \qquad (5.31)$$

- second order

$$M^0_{\alpha\beta}(n)\Delta\ddot{q}^{(2)}_\beta(n) + C^0_{\alpha\beta}(n)\Delta\dot{q}^{(2)}_\beta(n) + K^0_{\alpha\beta}(n)\Delta q^{(2)}_\beta(n)$$
$$= \left\{ \Delta Q^{;rs}_\alpha(n) - \left(M^{;rs}_{\alpha\beta}(n)\Delta\ddot{q}^0_\beta(n) + C^{;rs}_{\alpha\beta}(n)\Delta\dot{q}^0_\beta(n) + K^{;rs}_{\alpha\beta}(n)\Delta q^0_\beta(n) \right) \right. \qquad (5.32)$$
$$\left. - \left(M^{;r}_{\alpha\beta}(n)\Delta\ddot{q}^{;s}_\beta(n) + C^{;r}_{\alpha\beta}(n)\Delta\dot{q}^{;s}_\beta(n) + K^{;r}_{\alpha\beta}(n)\Delta q^{;s}_\beta(n) \right) \right\}$$
$$Cov\left(b^r(n), b^s(n) \right)$$

where the stiffness matrix perturbation orders are defined as

$$K^{(.)}_{\alpha\beta}(n) = {}^{(con)}K^{(.)}_{\alpha\beta}(n) + {}^{(\sigma)}K^{(.)}_{\alpha\beta}(n) =$$
$$= \int_\Omega C^{(.)}_{ijkl}(n) B_{ij\alpha} B_{kl\beta}\, d\Omega + \int_\Omega \sigma_{ij}(n-1)\varphi_{k\alpha,i}\varphi_{k\beta,j}\, d\Omega \qquad (5.33)$$

so the dynamical structural response is given in the form

$$\Delta\ddot{q}^{(.)}_\beta = \ddot{q}^{(.)}_\beta(n+1) - \ddot{q}^{(.)}_\beta(n) \qquad (5.34)$$

The situation is more complicated when the crack phenomenon is considered apart from the material stochasticity and nonlinearity. In such a situation so−called direct methods are used or special purpose enriched finite elements with crack tip modelling can be applied alternatively. In the latter case, the displacements near the crack tip can be defined as

$$\begin{cases} u = K_I f_u + K_{II} g_u \\ v = K_I f_v + K_{II} g_v \end{cases} \qquad (5.35)$$

while the near field component f_u can be rewritten as

$$f_u = \frac{1}{4G}\sqrt{\frac{r}{2\pi}}$$
$$\left\{ \cos\phi \left[(2\gamma-1)\cos\frac{\theta}{2} - \cos\frac{3\theta}{2} \right] - \sin\phi \left[(2\gamma+1)\sin\frac{\theta}{2} - \sin\frac{3\theta}{2} \right] \right\} \qquad (5.36)$$

where ϕ denotes the orientation angle of a crack, which is measured from the positive x axis, r and θ are polar coordinates with origin at the crack tip and measured from the crack angle, G is shear modulus, while γ denotes $\gamma = 3 - 4v$ for plane strain problems or is equal to $\gamma = \dfrac{3-v}{1+v}$ for the plane stress analyses. The corresponding SFEM equations for displacements near the crack tip are rewritten using (5.36), while the stress intensity factors are computed using BEM or FEM techniques or, alternatively, are derived mathematically starting from stress equilibrium and displacement compatibility equations. The numerical results of SFEM analysis for composites with and/or without interface and volumetric microdefects are presented in [193,194], while in the case of the cracked medium they can be found in [33].

Alternatively, the structural microdefects are modelled by spherical voids during the ductile type fatigue fracture. Let us assume that the total number of the microdefects is equal to M_a, their radius is denoted by R_a in the composite component indexed with a. Adopting further that both of them are functions of the fatigue cycle, the modified elasticity tensor components can be calculated using stiffness reduction of the Young modulus and Poisson ratio as follows:

$$
C_{ijkl(a)}^{(eff)}(n) = \left(1 - \frac{\pi M_a(n)R_a^2(n)}{\Omega_a}\right) E_a(n)
$$

$$
\times \left\{ \frac{\left(1 - \dfrac{\pi M_a(n)R_a^2(n)}{\Omega_a}\right) v_a(n)}{\left[\left(1 + \left(1 - \dfrac{\pi M_a(n)R_a^2(n)}{\Omega_a}\right) v_a(n)\right)\left(1 - 2\left(1 - \dfrac{\pi M_a(n)R_a^2(n)}{\Omega_a}\right) v_a(n)\right)\right]} \right.
$$

$$
\left. \times \delta_{ij}\delta_{kl} + \frac{\left(\delta_{ik}\delta_{jl} + \delta_{il}\delta_{jk}\right)}{2\left(1 + \left(1 - \dfrac{\pi M_a(n)R_a^2(n)}{\Omega_a}\right) v_a(n)\right)} \right\}
\tag{5.37}
$$

The use of more advanced deterministic theories is known from the literature. However equivalent stochastic models are not available now. Similarly to a solid model with deterministic and stochastic microvoids, the stiffness reduction approach for cracked media can be applied as well [267]. The following material data are adopted for $n=0$: Young modulus $E_m=2.1$ E11, Poisson ratio $v_m=0.3$, expected value of microvoids radius $E[r]=0.1$ and standard deviation of microvoids radius $\sigma(r)=0.01$, expected value of microvoids total number $E[M]=1$ and variance of microvoids total number $Var(M)=0$. The Young modulus is taken with $\pm 10\%$ deviations from the mean value the microvoid ratio variability is included in the interval $[0,1.0]$. Therefore an adequate visualisation of the component $C_{1111}^{(eff)}$ can be obtained, cf. Figure 5.4.

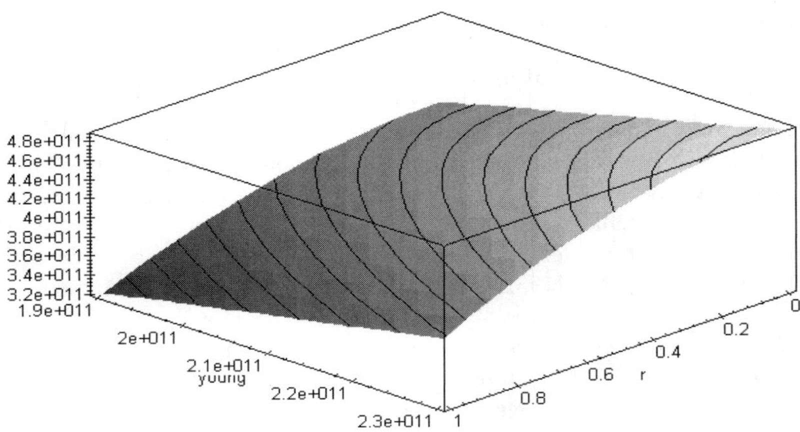

Figure 5.4. Parameter variability of $C_{1111}^{(eff)}$ for damaged homogeneous solid

Analysing the effective tensor surface, the expected linear dependence of this tensor on the Young modulus is observed as well as the nonlinear dependence on the microvoid mean radius (greater sensitivity to geometrical parameters of the structural defects). If only the statistical information about the input parameters is available, then the elasticity tensor can be rewritten using its first two probabilistic moments and introduced directly in SFEM analysis. If stochastic analysis in the elastoplastic range is necessary, the corresponding extension of the models presented in [355] can be applied. The microvoid volumetric ratio parameter is to be replaced with the two−parameter approach shown above and the probabilistic moments of these parameters are to be inserted as a function of the fatigue cycle.

As was mentioned before, the main goal of the homogenisation procedure is to find effective material properties of the homogeneous material, equivalent to the original composite. The most simplified method is to use the spatial average as the homogenised property and it is still used in terms of effective mass density, which can be rewritten for the nth cycle of fatigue analysis as

$$\rho^{(eff)}(n) = \left\langle \rho(n) \right\rangle_{\Omega} \tag{5.38}$$

Analogous homogenisation rule is applied in the case of heat capacity in transient heat transfer analysis and related thermoelastic or thermoelastoplastic coupled analyses of composites. The homogenisation of the elasticity tensor components is definitely more complicated and is usually carried out as

$$C_{ijkl}^{(eff)}(n) = \left\langle C_{ijkl}^{(a)}(n) \right\rangle_{\Omega} + \left\langle \sigma_{ij}\left(\chi_{kl}(n)\right) \right\rangle_{\Omega}, \quad \text{for } i,j,k,l=1,2,3 \tag{5.39}$$

where $\chi_{kl}(n)$ are the homogenisation function depending on the fatigue cycle.

The entire procedure can be applied for the fatigue analysis by rewriting the material properties of the composite components in terms of the current fatigue cycle number. Then, homogenising the constitutive law for each cycle, the whole composite fatigue can be modelled in a global scale, without the necessity for a very precise microscale discretisation or computational substructuring; an analogous analysis can be carried out for composite materials with cracks [336,337], for instance. It should be underlined that the described homogenisation procedure is sensitive to the RVE determination from the entire composite and to the scale parameter relating this element, dimensions to the dimensions of the entire composite. The formula for effective elasticity tensor is rewritten under the assumption that this parameter tends to 0, which is a very unrealistic model.

Furthermore, the homogenisation procedure can be established for random composites, too, only if the randomness does not influence the periodic character of the composite (especially during the fatigue process). Then, either MCS [191] as well as SFEM [192] can be utilised for this purpose. Therefore, starting from probabilistic characteristics of the composite properties, the expected values, variances (or standard deviations) as well as higher order moments (in the statistical estimation only) can be computed.

A very important issue from the technological point of view is the presence of the interface defects (usually with stochastic nature) appearing and growing between the composite components. Various computational models are proposed in this case in terms of special purpose spring finite elements or, alternatively, using the interphase as a new, separate material between the original composite components. This new material can be constructed from the original semicircular defects with random parameters, smeared (averaged probabilistically) over the entire interphase region according to the stochastic model introduced in Chapter 2; the composite with such an introduced interphase is then homogenised. To utilise the model for fatigue life cycle analysis, the geometrical and physical properties of the composite should be described in terms of the fatigue cycle number and then homogenised cycle by cycle for the needs of computational simulation of the composite.

5.3.1 Delamination of Two-Component Curved Laminates

Let us consider a two–component elastic transversely isotropic material in two–dimensional space Ω defined by the polar coordinate system $y=\{R,\Theta\}$ (cf. Figures 5.40–5.43). It is necessary to introduce the following relations:
(a) the gap between two surfaces

$$g(R,\Theta)= u_R^{(2)}(R,\Theta)-u_R^{(1)}(R,\Theta) \tag{5.40}$$

(b) the relative tangential slip of two surfaces

$$s(R,\Theta)= u_\Theta^{(2)}(R,\Theta)-u_\Theta^{(1)}(R,\Theta) \tag{5.41}$$

(c) the normal traction

$$\sigma_R(R,\Theta)= \sigma_R^{(2)}(R,\Theta)-\sigma_R^{(1)}(R,\Theta) \tag{5.42}$$

(d) the shear traction

$$\sigma_{R\Theta}(R,\Theta)= \sigma_{R\Theta}^{(2)}(R,\Theta)-\sigma_{R\Theta}^{(1)}(R,\Theta), \quad \Gamma_c =\left\{\Gamma_c : R = R_0 ; \Theta \in \langle 0,\infty\rangle\right\} \tag{5.43}$$

where R_0 is the radius of the interface curvature. Since (5.40) – (5.43) are referred to the composite interface (cracked or joined) Γ_c ($R=R_0$=const) only, then their radial dependence is neglected. The equilibrium problem of linear elasticity is given by the following equations system [95]:
  - equilibrium equations

$$\frac{\partial \sigma_R}{\partial R}+\frac{1}{R}\frac{\partial \sigma_{R\Theta}}{\partial R}+\frac{1}{R}(\sigma_R -\sigma_\Theta)+b_R =0 \tag{5.44}$$

$$\frac{\partial \sigma_{\Theta R}}{\partial R}+\frac{1}{R}\frac{\partial \sigma_\Theta}{\partial \Theta}+\frac{2}{R}\sigma_{R\Theta} +b_\Theta =0 \tag{5.45}$$

where b_R and b_Θ denote the body force components;
  - strain–displacement relations

$$\varepsilon_R = \frac{\partial u_R}{\partial R}, \quad \varepsilon_\Theta = \frac{1}{R}\frac{\partial u_\Theta}{\partial \Theta}+\frac{u_R}{R}, \quad \varepsilon_{R\Theta} =\frac{1}{2}\left(\frac{1}{R}\frac{\partial u_R}{\partial \Theta}+\frac{\partial u_\Theta}{\partial R}-\frac{u_\Theta}{R}\right) \tag{5.46}$$

- constitutive relations

$$\begin{Bmatrix} \sigma_R \\ \sigma_\Theta \\ \sigma_{R\Theta} \end{Bmatrix} = \begin{bmatrix} C_{1111} & C_{1112} & C_{1113} \\ C_{2221} & C_{2222} & C_{2223} \\ C_{3331} & C_{3332} & C_{3333} \end{bmatrix} \begin{Bmatrix} \varepsilon_R \\ \varepsilon_\Theta \\ \varepsilon_{R\Theta} \end{Bmatrix} \tag{5.47}$$

The following boundary conditions are employed:

$$u_R = \hat{u}_R \quad \text{and} \quad u_\Theta = \hat{u}_\Theta \ \text{on} \ \Gamma_u \tag{5.48}$$

$$t_R = \hat{t}_R \quad \text{and} \quad t_\Theta = \hat{t}_\Theta \ \text{on} \ \Gamma_\sigma \tag{5.49}$$

$$g(\Theta) = 0 \,; \ s(\Theta) = 0 \Rightarrow \sigma_R(\Theta) = 0 \,; \ \sigma_{R\Theta}(\Theta) = 0 \ \text{on} \ \Gamma_c \tag{5.50}$$

$$g(\Theta) = 0 \,; \ s(\Theta) \neq 0 \Rightarrow \sigma_R(\Theta) < 0 \,; \ \left| \sigma_{R\Theta}(\Theta) \right| = \mu \left| \sigma_R(\Theta) \right| \ \text{on} \ \Gamma_c \tag{5.51}$$

$$g(\Theta) > 0 \,; \ s(\Theta) = 0 \ \text{or} \ s(\Theta) \neq 0 \quad \sigma_R(\Theta) = 0 \,, \ \sigma_{R\Theta}(\Theta) = 0 \ \text{on} \ \Gamma_c \tag{5.52}$$

$$g(\Theta) < 0 \,; \ s(\Theta) \neq 0 \quad \sigma_R(\Theta) < 0 \,; \ \left| \sigma_{R\Theta}(\Theta) \right| = \mu \left| \sigma_R(\Theta) \right| \ \text{on} \ \Gamma_c \tag{5.53}$$

$$sign\left(\sigma_R(\Theta) \right) = sign\left(s(\Theta) \right) \ \text{on} \ \Gamma_c \tag{5.54}$$

where μ denotes the constant friction coefficient. Then, the near–tip stress field is described in the polar coordinate system as $\{x\}=\{r,\theta\}$ (cf. Figure 5.6).

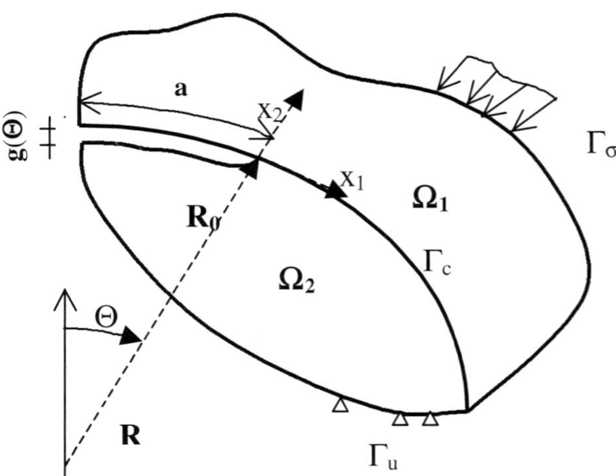

Figure 5.5. Two–component curved laminate structure

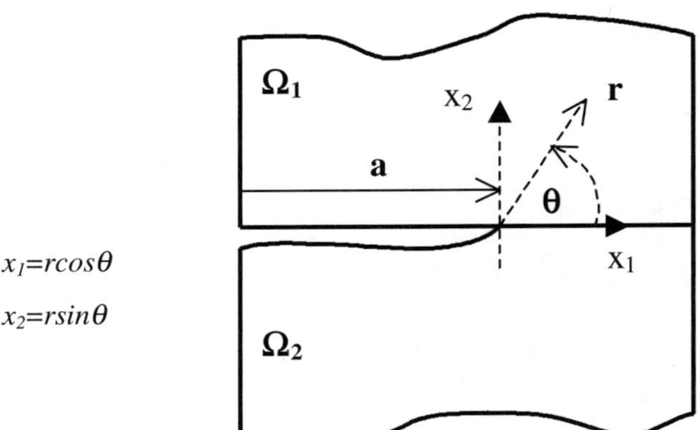

Figure 5.6. Near−tip field

It is assumed that both crack surfaces are modelled as perfectly smooth – there are no neither meso− nor micro−asperities on this surface in the context of the FEM contact model presented by [49,371,382], however application of the Boundary Element Method is also known, see [374]. Considering future particle-reinforced composites delamination simulations, the 3D contact algorithms must be employed [284,322]. The asymptotic nature of the elastic fields near a transition in the boundary conditions (crack tip) is expressed by the analytic functions and therefore, the description of a near−tip stress for an interface crack between two different transversely isotropic in a plane stress problem and the traction−free crack surfaces is given as follows [301]:

$$\sigma_{ij} = \mathrm{Re}\left[Kr^{i\in}\right](2\pi r)^{-0.5}\Sigma_{ij}^{I}(\theta,\in)+\mathrm{Im}\left[Kr^{i\in}\right](2\pi r)^{-0.5}\Sigma_{ij}^{II}(\theta,\in), \qquad (5.55)$$

where i,j=1,2, $\Sigma_{ij}^{I}(\theta,\in)$, $\Sigma_{ij}^{II}(\theta,\in)$ are the angular functions derived using the Muskhelishvili potentials; $r^{i\in}$ describes here the oscillatory stress singularity given as

$$r^{i\in} = \cos(\in \ln r)+i\sin(\in \ln r) \qquad (5.56)$$

The angular functions correspond to the normal and in−plane shear tractions, respectively, on interface ahead crack tip ($x_1>0$; $\theta=0$) at a distance r given by [140,222]:

$$\sigma_{22} +i\sigma_{12} = K(2\pi r)^{-0.5}r^{i\in} \quad \text{or}$$

$$\sigma_{22} = \mathrm{Re}\left[Kr^{i\in}\right](2\pi r)^{-0.5} \quad \text{and} \quad \sigma_{12} = \mathrm{Im}\left[Kr^{i\in}\right](2\pi r)^{-0.5} \qquad (5.57)$$

Moreover, the functions $\Sigma_{ij}^{I}(\theta,\in)$, $\Sigma_{ij}^{II}(\theta,\in)$ are related to the elastic properties of the bimaterial specimen using the oscillatory index $\in$ given by

$$\epsilon = \frac{1}{2\pi} \ln\left(\frac{\kappa_1/G_1 + 1/G_2}{\kappa_2/G_2 + 1/G_1}\right) \tag{5.58}$$

where the Kolosov constant κ_n is given as [158,259]

$$\kappa_n = \frac{3-v_n}{1+v_n} \quad \text{for the plane stress} \tag{5.59}$$

$$\kappa_n = 3 - 4v_n \quad \text{for the plane strain; } n=1,2. \tag{5.60}$$

where v_n and G_n denote the Poisson ratio and shear modulus of the nth component, respectively. Next, the elastic Dundur mismatch parameters are defined by

$$\alpha = \frac{G_1(\kappa_2+1) - G_2(\kappa_1+1)}{G_1(\kappa_2+1) + G_2(\kappa_1+1)} \quad \text{and} \quad \beta = \frac{G_1(\kappa_2-1) - G_2(\kappa_1-1)}{G_1(\kappa_2+1) + G_2(\kappa_1+1)} \tag{5.61}$$

Then, it is possible to rewrite (5.58) in the following way:

$$\epsilon = \frac{1}{2\pi} \ln\left(\frac{1-\beta}{1+\beta}\right) \tag{5.62}$$

The fracture modes I and II [54] of the SIF in the case of an interface crack between dissimilar isotropic materials are now coupled together into the single complex SIF $K=K_1+iK_2$ uniquely characterising the singular stress field; K_1 and K_2 are the functions of a distance r from the tip and may be denoted as follows:

$$K_1(r) = \text{Re}\left(Kr^{i\epsilon}\right) \text{ and } K_2(r) = \text{Im}\left(Kr^{i\epsilon}\right) \tag{5.63}$$

The associated relative crack surfaces displacements $(\Delta u_i = u_i(r,\theta=\pi) - u_i(r, \theta=-\pi))$ at a distance r behind the tip ($x_1<0$; $\theta=\pm\pi$) are described in the following way:

$$\Delta u_1 + i\Delta u_2 = \left(\frac{1-v_1}{G_1} + \frac{1-v_2}{G_2}\right)\frac{Kr^{i\epsilon}}{(1+2i\epsilon)\cosh(\pi\epsilon)}\left(\frac{r}{2\pi}\right)^{-0.5} \tag{5.64}$$

Finally, the ERR for the crack propagation along the interface may be given as

$$ERR = \left(\frac{1-v_1}{G_1} + \frac{1-v_2}{G_2}\right)\frac{K\bar{K}}{4\cosh^2(\pi\epsilon)}, \tag{5.65}$$

where $\bar{K}=K_1-iK_2$ is the conjugate complex SIF. It finally gives

$$ERR = \left(\frac{1-v_1}{G_1} + \frac{1-v_2}{G_2}\right)\frac{\left(K_1(r)\right)^2 + \left(K_2(r)\right)^2}{4\cosh^2(\pi \in)} , \qquad (5.66)$$

which makes it possible to calculate the material interface toughness starting from the local stress field under critical load.

The main goal of the computational experiments is to simulate the delamination process of a two–component layered composite subjected to shear loading in the shear device. It is predicted that near tip behaviour and frictional stresses along the crack surfaces are the main parameters governing the ERR. Therefore, the FEM–based numerical modelling of the delamination process is applied to get the accurate information about the following data: near–tip displacement and stress field description, normal stress distribution along the crack surfaces as well as the relation between the ERR and interface crack length.

A two–component curved composite beam is analysed numerically under the following assumptions: (i) the interlaminar adhesive layer has zero thickness (no contribution to ERR), (ii) near–tip stress field is analysed in the same way as the straight crack, (iii) the crack propagates along the interface only (kinking of a crack out of the interface is not considered), (iv) kinematic friction along the crack surfaces is accounted for, (v) friction between supporting jigs and the specimen is neglected (liquid lubrication of the tested material surface is assumed). The material components are homogeneous isotropic and linear elastic (cf. Table 5.1); geometrical data are given in Table 5.2 and Figure 5.7.

A FEM geometrical model is made from the three types of finite elements: 8–node plane stress quadrilateral with out–of–plane thickness (5.0e-3 m) and 4 integration points PLANE82 (structural solid), 3–node contact surface element with 2 integration points CONTA172 and 3–node target surface elements TARGE169. The last two element types simulate two essentially different kinds of material contact behaviour: flexible–to–flexible (between crack surfaces) and rigid–to–flexible (between the rigid curved device jigs and the curved specimen sides). For the present purposes, surface contact elements are more preferred than point–to–surface contact elements considering the curved geometry of a specimen and the requirements of a precise and detailed contact description as well as faster computational processing (smaller total number of contact finite elements). Moreover, target elements (CONTA172) simulating rigid curved jigs are modelled as longer than specimen curved sides ($\Theta_T+1°$) to prevent loss of the contact at the model edges during the loading process. Crack tip vicinity is modelled by the ring of 16 8–noded finite elements (cf. Figure 5.9) introduced around 6–node triangular elements (PLANE82). The required square–root singularity on the element sides is achieved by the motion of the midside nodes of crack tip elements into the quarter points. Size of the crack tip element ring is 5.0 E-06 m in the radial direction, which corresponds to 0.02% of the component thickness (the characteristic length of a composite specimen). The very dense discretisation (cf. Figures 5.8 and 5.9) makes it possible to analyse the near–tip stress zone where the singular stresses dominate (in so–called K–dominance zone) with the length about 3% of the component thickness.

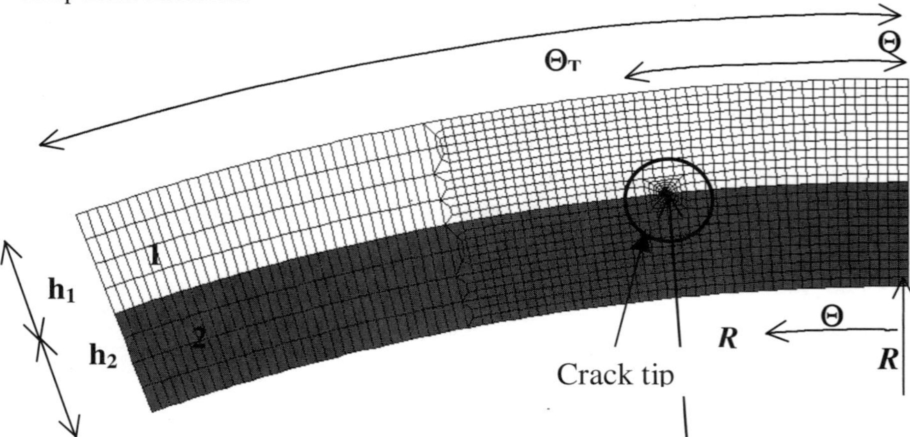

Figure 5.7. Composite beam geometry

Table 5.1. Material input data for FEM analysis

Material No.	Radial elastic modulus e_R [GPa]	Angular elastic modulus e_Θ [GPa]	Shear modulus $G_{R\Theta}$ [GPa]	Poisson ratio $v_{R\Theta}$
1	5.0	5.0	2.5	0.2
2	10.0	10.0	5	0.3

Table 5.2. Geometrical input data for FEM analysis

Component thickness [m]		Total angle Θ_T [deg]	Interface plane radius R_0 [m]	Crack propagation range Θ_a [deg]
h_1	h_2			
0.0025	0.0025	20	0.0525	<6-14>

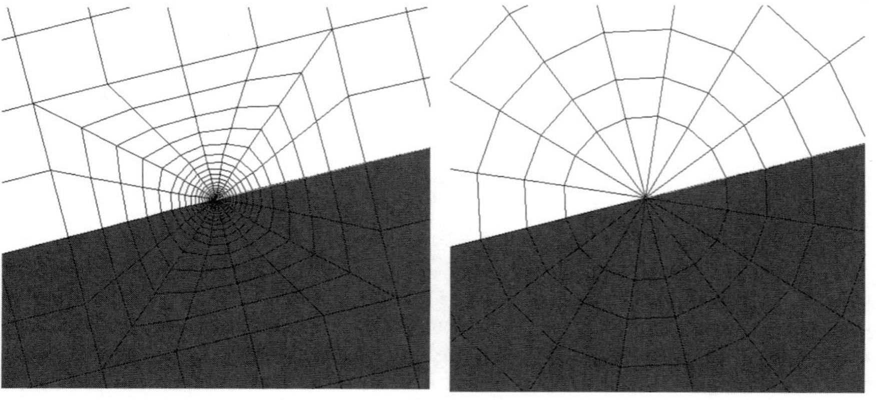

Figure 5.8. Crack tip zone discretisation   **Figure 5.9.** Crack tip mesh

The propagation of a crack is modelled computationally by the change of the crack tip position under constant radius value $R=R_0$. Thus, if the crack length increases during its propagation, the total number of elements and nodes increases as well as is comprised in the range between 2,302 and 2,878 finite elements (from 6,225 to 7,825 nodes).

The incremental nonlinear analyses (according to contact and friction) with two different boundary conditions (BC) and material configurations (cf. Table 5.3 and Figures 5.10 and 5.11) are performed to determine the influence of different load and material combinations on the contact between crack surfaces. The external loading is of static shear type and is applied in the form of displacement increments; the weaker component is loaded first.

Table 5.3. Geometrical boundary conditions for the composite beam

Case 1	Case 2
(i) $\Theta=\Theta_T$ and $R\in\langle R_0, R_0+h_1\rangle$: $u_R=0$	**(i)** $\Theta=\Theta_T$ and $R\in\langle R_0-h_2, R_0\rangle$: $u_R=0$
(ii) $\Theta\in\langle -1°, \Theta_T\rangle$, $R=R_0+h_1$: $u_R=u_\Theta=0$	**(ii)** $\Theta\in\langle -1° \Theta_T \rangle$, $R=R_0-h_2$: $u_R=u_\Theta=0$
(iii) $\Theta\in\langle 0°,\Theta_T+1°\rangle$, $R=R_0-h_2$: $u_R=0$; $u_\Theta=\tilde{u}$	**(iii)** $\Theta\in\langle 0°, \Theta_T+1°\rangle$, $R=R_0+h_1$: $u_R=0$; $u_\Theta=\tilde{u}$
(iv) $\Theta=0°$ and $R\in\langle R_0-h_2, R_0\rangle$: $u_\Theta=\tilde{u}$	**(iv)** $\Theta=0°$ and $R\in\langle R_0, R=R_0+h_1\rangle$: $u_\Theta=\tilde{u}$

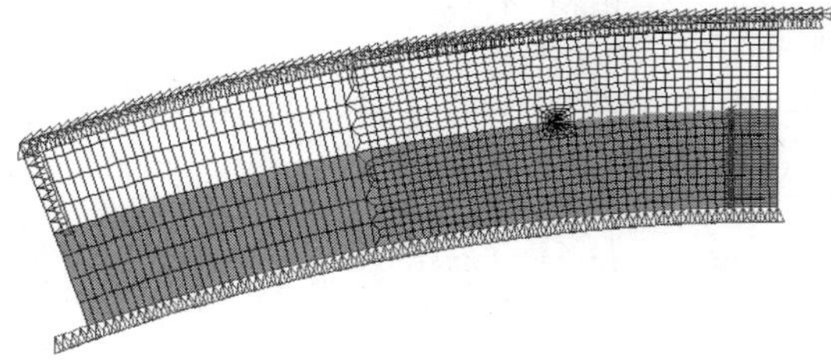

Figure 5.10. Model BC (case 1)

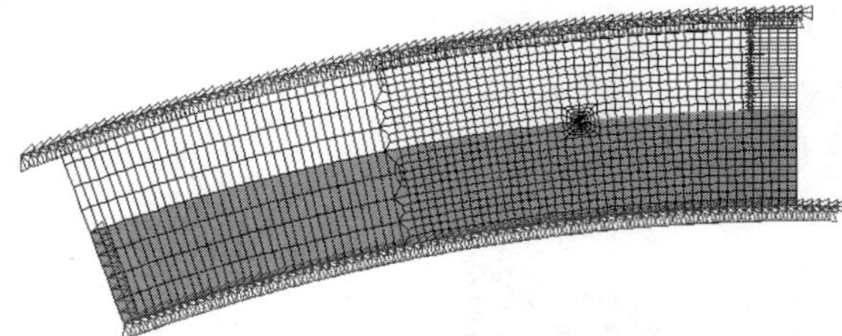

Figure 5.11. Model BC (case 2)

Frontal solver implemented of the ANSYS is used to solve the problem using the full Newton−Raphson iteration technique (stiffness matrix updated each time) together with the additional convergence enhancement tools: predictor−corrector and the line search options. The standard unilateral contact is modelled (pressure is equal to zero during separation) as well as one−pass contact (asymmetric contact) to obtain the equilibrium solution of contact tractions by means of the augmented Lagrangian method (iterative series of contact stiffness are updated for the contact stresses computation). Moreover, the unsymmetrical tangent stiffness matrix is used to derive of contact tractions what improved solution convergence in comparison with the symmetric tangent stiffness matrix approximation algorithm.

The cracks are closed on almost the entire length under applied shear loading, which results in sliding and sticking behaviour of the composite. Nevertheless, it is observed that in the case of weaker material loading, the area around the crack tip is opened which makes it possible to use the LEFM oscillatory theory of interfacial cracks in the ERR calculation. The length of the opened crack tip zones remains constant during crack propagation (about 1−2% of the total crack length, cf. Figure 5.12), while the crack opening maximum values are different for various crack lengths. It may be due to the change of the crack tip loading direction during crack propagation process. Moreover, the zero value of a crack opening shown in Figure 5.12 corresponds to sliding contact behaviour of the composite, which takes place in 98−99% of the crack length measured from the specimen edges; the asymptotic behaviour of stress is shown in Figure 5.13. The values of stresses depend asymptotically on the very high stress values up to values about 5 orders smaller and which are never equal to zero. Further, the oscillatory stress singularity is slightly influenced by the increasing friction coefficient μ and for extremal case (μ=1.0) the stress exponent is equal to λ=0.494.

Next, asymptotic behaviour of stress in the case of a completely closed crack (loading of stiffer component) is analysed in Figure 5.22. The extremal values of stresses (crack tip stress values) are considerably influenced by the friction coefficient increase and differ by about one order for μ=1. In this case the exponent λ depends on the friction coefficient μ and the interface fracture mechanics idea for the opened crack is no longer applicable. However, it is possible to calculate numerically the ERR for a closed crack with friction by means of the technique proposed in [292] using the FEM analysis [24], but here only the opened crack model is analysed. As can be expected, the stress tensor components around the crack for the test without the friction are essentially greater than those typical for the composite contact problem with a non−zero friction coefficient. It reflects the fact that some part of the internal strain energy is dissipated by the friction phenomenon in the second case, cf. Figures 5.14−5.21.

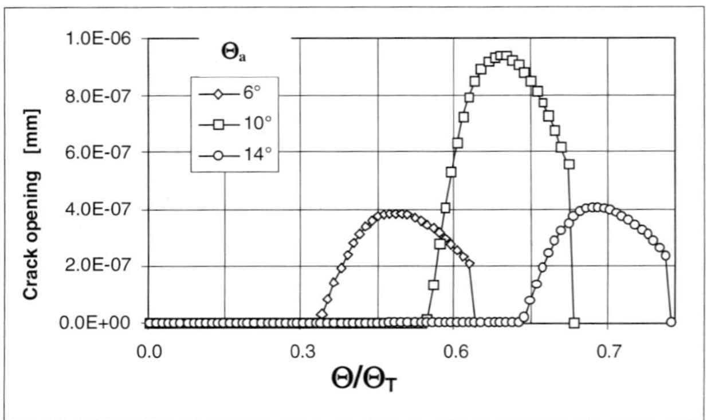

Figure 5.12. Crack opening displacement (case 1; μ=0.5)

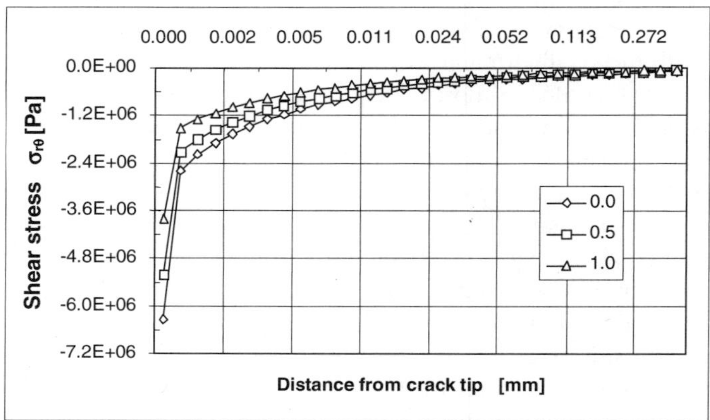

Shear stress $\sigma_{r\theta}$ [Pa]

Figure 5.13. Near-tip stress dependence on μ (opened crack tip)

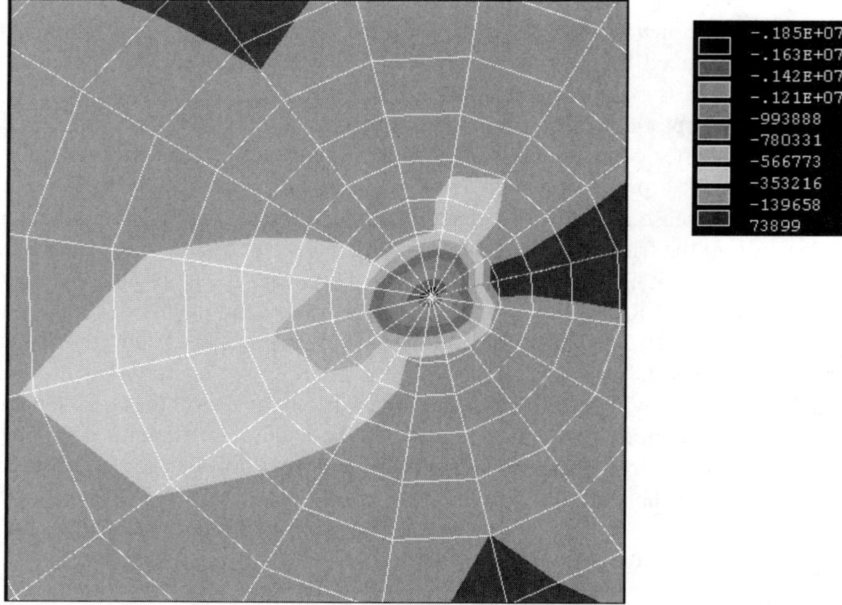

Figure 5.14. Near−tip stresses $\sigma_{r\theta}$ [Pa] (μ=0.5)

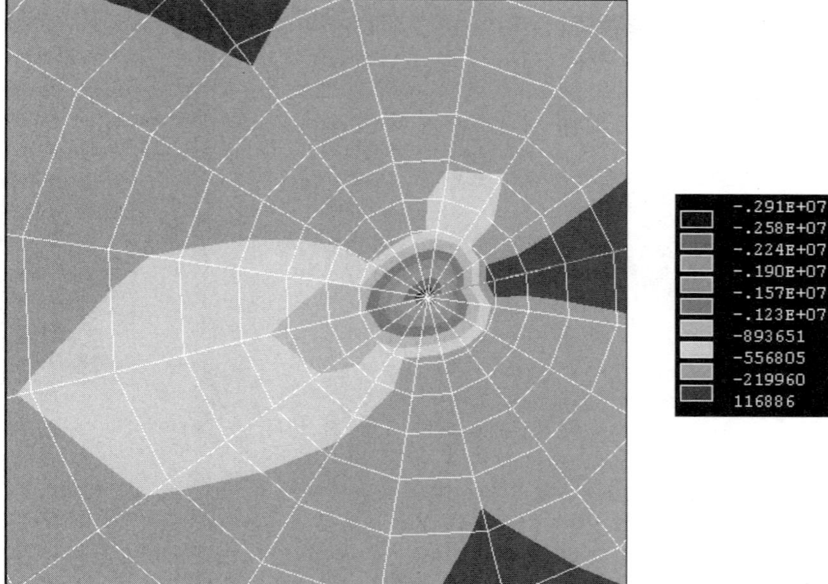

Figure 5.15. Near−tip stresses $\sigma_{r\theta}$ [Pa](μ=0)

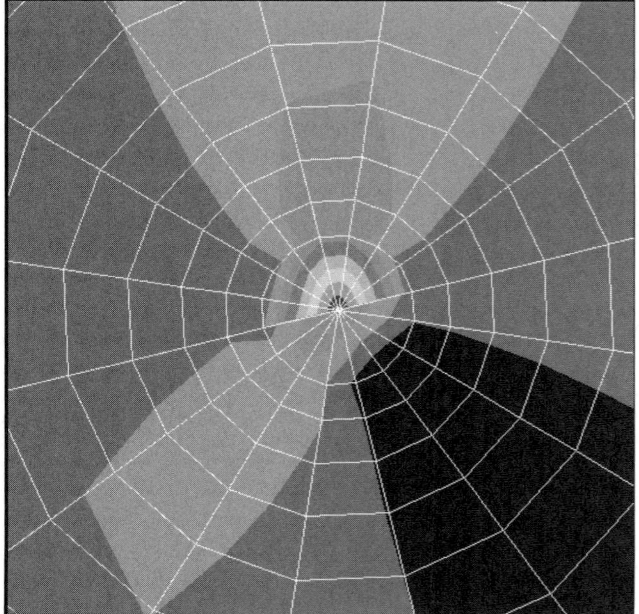

Figure 5.16. Near−tip stresses σ_r [Pa] (μ=0.5)

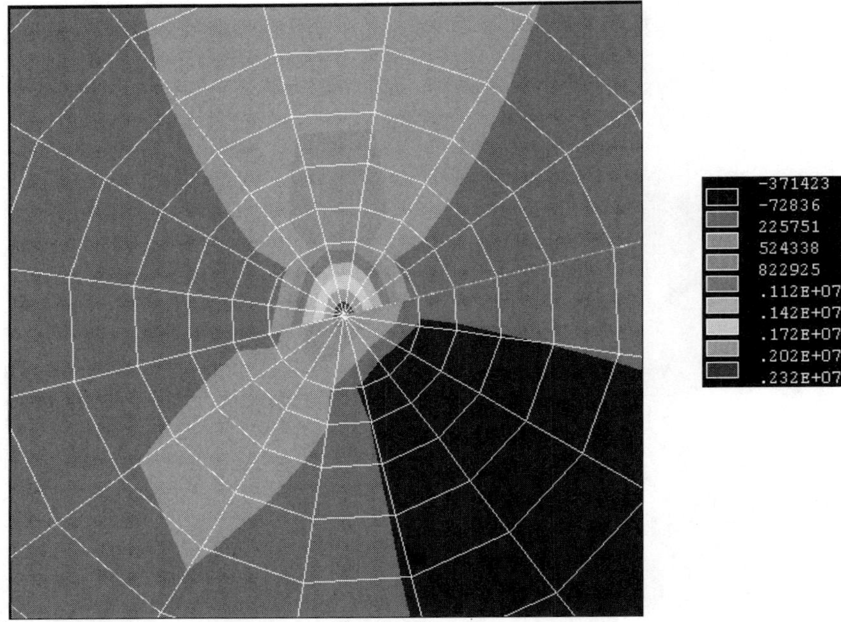

Figure 5.17. Near−tip stresses σ_r [Pa] ($\mu=0$)

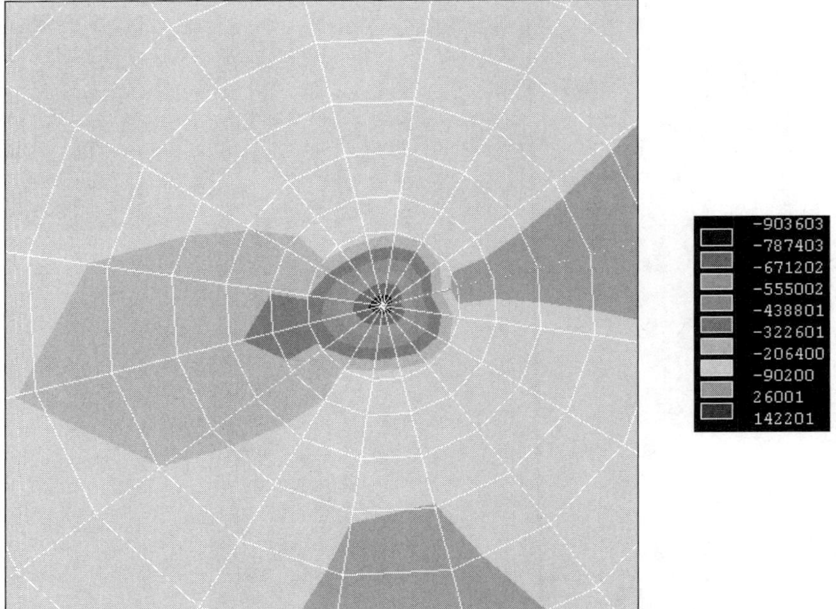

Figure 5.18. Near−tip stresses $\sigma_{r\theta}$ [Pa] ($\mu=0.5$)

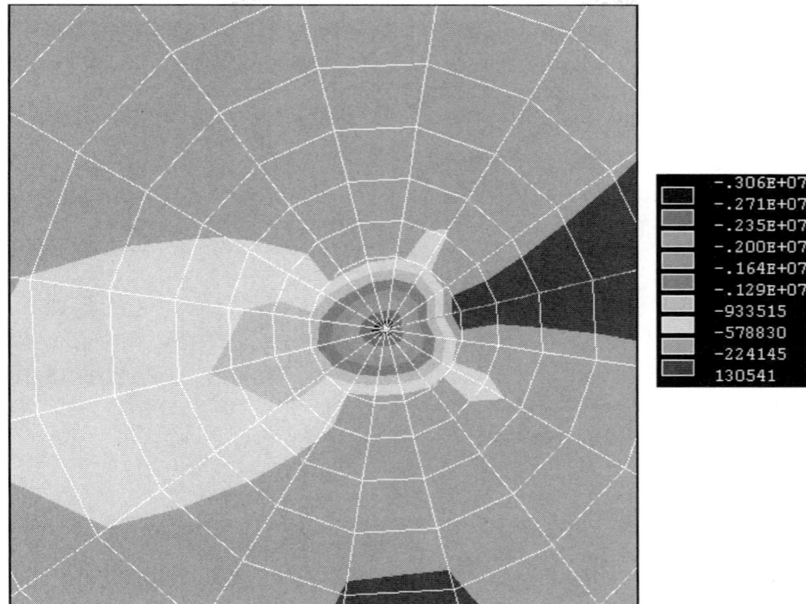

Figure 5.19. Near−tip stresses $\sigma_{r\theta}$ [Pa] (μ=0)

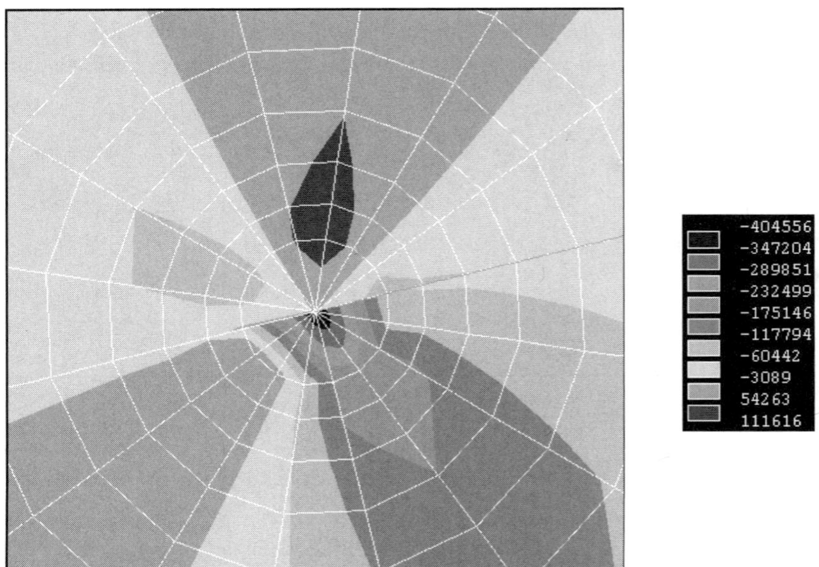

Figure 5.20. Near−tip stresses σ_r [Pa] (with μ=0.5)

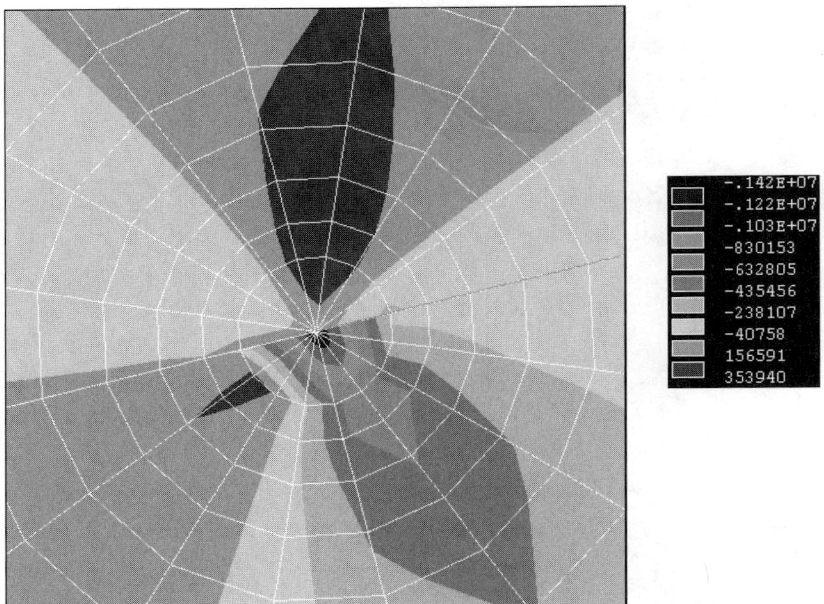

Figure 5.21. Near–tip stresses σ_r [Pa] ($\mu=0$)

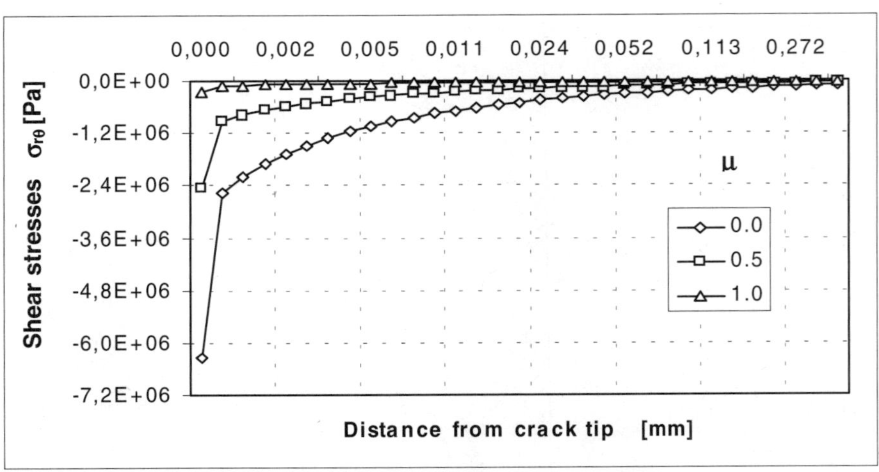

Figure 5.22. Near–tip stress dependence on μ (closed crack tip)

Further, normal stress distribution can be analysed along the crack length $\Theta \in \langle 0, \Theta_a \rangle$ (cf. Figures 5.23 and 5.24) from the model edge (zero crack length) to the crack tip for various crack lengths ($\Theta_a=6°$, $10°$ and $14°$). The uniform distribution of normal stresses σ_R, especially for longer cracks ($\Theta_a>9°$), which is obtained in conjunction with constant value of μ, results in a uniform frictional stress σ_Θ along the crack surfaces according to Coulomb law. The part of the crack surface with quasi-uniform normal stress distribution increases together with the crack length increase as follows: 3.44E-3 m (6°), 7.33E-3m (10°), 8.93E-3 m (14°) for Figure 5.21 and 3.89E-3 m (6°), 5.04E-3m (10°), 8.93E-3 (14°) for Figure 6.22. It is reasonable because of the greater non-uniform deformation of the composite edges (due to BC) decreases with respect to the entire crack length during its propagation.

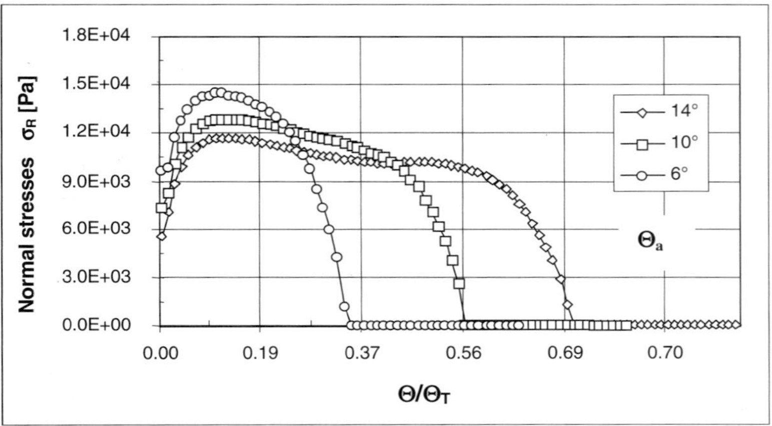

Figure 5.23. Normal stress distribution along the crack surface (case 1; μ=0.5)

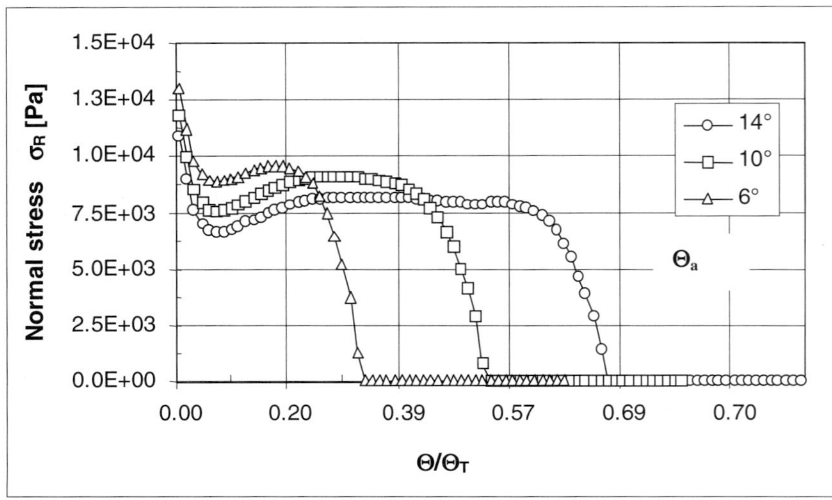

Figure 5.24. Normal stress distribution along the crack surface (case 2; μ=0.5)

The variable ERR is a function of the interface crack length and is computed for two different friction coefficients (μ=0 and μ=0.5). As is expected, a large decrease in ERR value follows the friction coefficient increase (cf. Figure 5.25).

For the shortest crack length (a_{min}=5.5E-3 m) the ERR takes value 1.54E-3 kJ/m^2 (μ=0) and 1.03E-3 kJ/m^2 (μ=0.5), while for the longest crack length (a_{max}=1.282E-2 m) the ERR value is equal to 8.71E-4 kJ/m^2 (μ=0) and 3.48E-4 kJ/m^2 (μ=0.5). Therefore, during the crack propagation from 5.5E-3 to 1.282E-2 m, the total amount of energy dissipated due to friction is comprised between 33 and 60% of the ERR value in the frictionless case. Moreover, the crack extension is stable (ERR decreases together with an increase of interface crack length), which means that a higher load should be applied to keep the growth of a crack. However, a friction phenomenon has a stabilising effect on the fracture process, which speeds up the ERR decrease together with crack length in comparison to a frictionless behaviour. Then, the quasi-stationary tendency of the ERR is observed for a certain crack length (a>1.1E-2 m) in frictional (from 3.9E-4 to 3.48E-4 kJ/m^2) and frictionless (from 9.06E-4 to 8.71E-4 kJ/m^2) cases. The stationary region of ERR may imply uniform crack tip load which would make it possible to determine experimentally the force responsible for delamination only; further analysis indicates the mode mixing of the fracture process. The shear mode prevails over the tensile mode of the ERR but the shear/tensile mode ratio (ERR2/ERR1) increases from 2.78 (a_{min}) to 2.88 (a_{max}) for μ=0 and decreases from 2.55 (a_{min}) to 2.45 (a_{max}) for μ=0.5. Although the friction influences both contributions of the ERR (ERR1 and ERR2), the ERR shear mode ERR2 is more reduced by the frictional stresses along the crack surfaces due to its direction during interface crack extension than the ERR tensile mode ERR1.

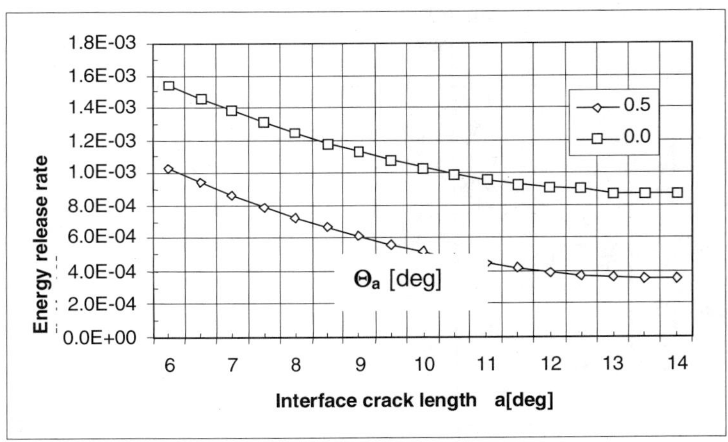

Figure 5.25. Energy release rate comparison

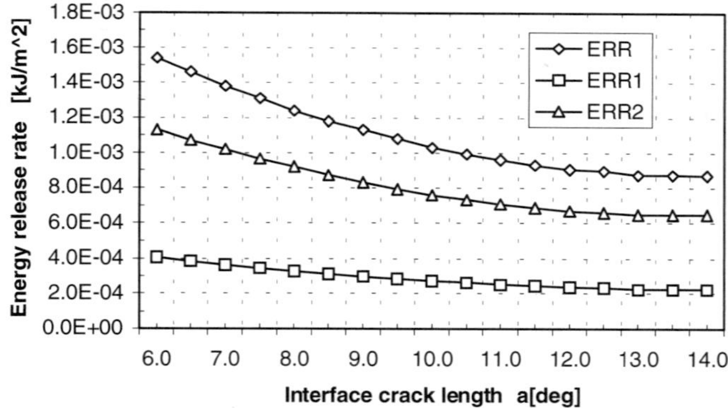

Figure 5.26. ERR contributions ($\mu=0$)

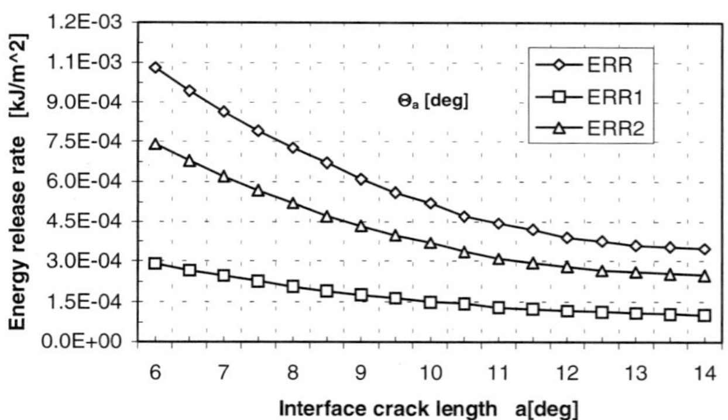

Figure 5.27. ERR contributions ($\mu=0.5$)

The computations are performed on a single processor machine (700 MHz) with 256 MB (RAM) memory; the computer processing time (CP) and cumulative number of iterations (CNI) are presented in Table 5.4 as functions of various crack lengths for the 20th loadstep of the displacement increment.

Table 5.4. Computational experiments technical data

Crack length a [deg]	Element number	CP [s]		CNI	
		$\mu=0$	$\mu=0.5$	$\mu=0$	$\mu=0.5$
6	2302	12413.580	11809.241	292	263
10	2590	12958.353	15646.158	270	327
14	2878	10506.438	10735.686	230	231

It is observed that the CP time is not affected by the friction coefficient and the finite element number, but depends on some certain crack tip positions as the result of the curved model geometry. Thus, it is possible to point out that the critical crack length maximising CP and CNI exists and is equal to about $\Theta_a=10°$.

5.3.2 Fatigue Analysis of a Composite Pipe Joint

A deterministic computational model of fatigue crack–like damage propagation in the composite pipe joint is introduced here and examined numerically using the FEM program ANSYS. The studies dealing with the other pressure vessels like longitudinally cracked pipes can be found in [142,218]. The model is built upon the following assumptions cf. [97,157,205,211,376]: (a) material components are linear elastic, (b) possible defect nucleation and growth is located within the adhesive layer and is caused by the high stress concentrations, (c) no initial manufacturing flaws, pre–cracks or other defects are assumed in the original adhesive layer (before the beginning of the fatigue loading process), (d) there are no microdefects forming and next coalescence during composite tension (typical for metallic materials) apart from crack formation and propagation, (e) the cyclic load has constant amplitude and (f) fatigue crack–like damage propagation is stable.

The stresses along the adhesive layer length are not uniform and their gradients arise at joint edges, which results from extension of the specimen layers in the opposite directions (composite pipe and coupling), cf. Figure 5.28. Then it is assumed that defects start to grow longitudinally along the adhesive layer and uniformly over all pipe circumference, under applied tensile load σ^{app}, when the resulting average shear stress $\langle\tau_{ad}\rangle$ over some distance d from the high stress concentration region is equal to or greater than the shear static strength τ_{ad}^u in adhesive layer. This criterion is expressed by the following equation:

$$\left\langle \tau_{ad} \right\rangle = \frac{1}{d}\int_0^d \tau_{ad}\,dX_A \geq \tau_{ad}^u \tag{5.69}$$

The formula (5.69) is called the average stress criterion after it was applied to notched strength prediction of laminated composites under uniaxial tension; a graphical representation of this criterion is shown schematically in Figure 5.28. The distance d is called the characteristic length and can stand for the damage accumulated or a nonlinear process zone. It is expressed here in terms of the critical fracture mechanics parameter as the critical Stress Intensity Factor (K_{IIc}) and shear strength of the adhesive layer as

$$d = \frac{1}{2\pi}\left(\frac{K_{IIc}}{\tau_{ad}^u}\right)^2 \tag{5.70}$$

Since (5.70) is based on the assumption of the square−root stress singularity in the front of the sharp crack tip, it does not precisely represent the stress distribution in the tubular adhesive layer in the stress concentration region. However, this characteristic length serves to estimate upper bound on the finite element size at the crack−like damage tip.

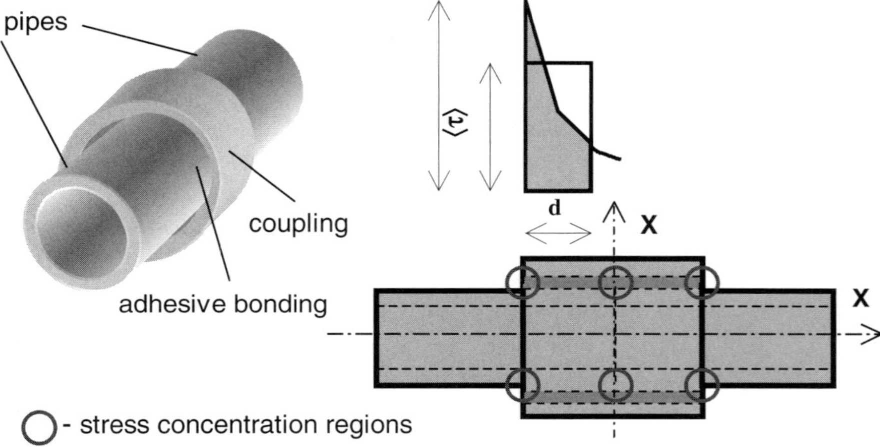

Figure 5.28. Pipe−to−pipe adhesive connection: 3D and 2D views

Then, it is postulated that after the crack−like defect had nucleated, it steadily propagates along the adhesive layer as the main single crack leading to an average stress increase over the distance d along with the number of load cycles N as

$$\langle \tau_{ad}(N) \rangle = \frac{1}{d}\int_0^d \tau_{ad}(N)dX_A \Rightarrow \frac{1}{d}\int_0^d \frac{\tau_{ad}(N)}{1-D(N)}dX_A \tag{5.71}$$

where $D(N)$ denotes the classical scalar damage variable, which may be written in terms of the nucleated and propagating main crack a as follows:

$$D(N) = \frac{a(N)}{l_a} \tag{5.72}$$

The defect propagation terminates according to the condition

$$D(N) = 1 \Leftrightarrow a(N) = l_a \tag{5.73}$$

which corresponds to the loss of stiffness for all those finite elements in the adhesive layer that are placed on the crack propagation path.

The boundary differential equation system, which describes fatigue defect propagation along the adhesive layer of a composite pipe joint may be defined over the pipe element of length $dl_a(N)=dX_A-da(N)$ as follows:

(i) equilibrium and damage equations

$$dF_p = dF_{ad}(N) \text{ and } dF_c = dF_{ad}(N) \tag{5.74}$$

$$d\sigma_p \frac{\pi}{4}\left(D_{op}^2 - D_{ip}^2\right) = \tau_{ad}(N)\pi\, D_{op} dl_a(N) \tag{5.75}$$

$$d\sigma_c \frac{\pi}{4}\left(D_{oc}^2 - D_{ic}^2\right) = \tau_{ad}(N)\pi\, D_{ic} dl_a(N) \tag{5.76}$$

(ii) constitutive relations

$$\frac{dw_p}{dl(N)} = \frac{\sigma_A}{E_p} \text{ and } \frac{dw_c}{dl(N)} = \frac{\sigma_A}{E_c} \tag{5.77}$$

$$\tau_{ad}(N) = \frac{G_{ad}\left(\gamma_p - \gamma_c\right)}{t_{ad}} \tag{5.78}$$

(iii) boundary conditions

$$\left.\frac{dw_p}{dl(N)}\right|_{X_A=L} = \frac{\sigma^{app}}{E_p} \text{ and } \left.\frac{dw_c}{dl(N)}\right|_{X_A=0} = \frac{\sigma^{app}}{E_c} \tag{5.79}$$

$$\left.\frac{dw_p}{dl(N)}\right|_{X_A=0} = 0 \text{ and } \left.\frac{dw_c}{dl(N)}\right|_{X_A=L} = 0 \tag{5.80}$$

where F_p, F_{ad}, F_c represent internal axial forces in a pipe, adhesive layer and coupling, respectively, internal axial stresses in the pipe, adhesive and coupling are denoted by σ_p, τ_{ad} and σ_c. Let us assume that E_p, E_c and G_{ad} are the axial modulus of the pipe, elastic modulus of the connecting layer and the adhesive shear modulus; w_p and w_c denote pipe and coupling axial displacements. This problem is now solved numerically for the pipe and coupling shear strains γ_p, γ_c and adhesive shear stresses $\tau_{ad}(N)$.

The main purpose of further computational studies is a prediction of crack damage propagation rate per a cycle in the composite pipe joint subjected to the pure tension fatigue load with the load time variations shown in Figure 5.29 (each load cycle is divided into two time intervals of 6 months). The cycle asymmetry ratio R is equal to 0, while the load amplitude is equal to the applied maximum load (σ_{max}^{app}). Since quasistatic fatigue load is applied, no frequency effect is therefore considered here.

Let us note that the axis symmetry of the composite pipe joint results in simplification of the entire computational model and essentially speeds up the analysis process – only half of the composite pipe joint in the axial direction is considered only. The final computational model geometrical data to the FEM displacement–based commercial program ANSYS [2] are shown in Figure 5.30. The pipe and coupling component are made up of E–glass/epoxy composite (50% fibre volume fraction) and the adhesive layer (rubber toughened epoxy). All material properties of the composite pipe joint components are listed in Table 5.5.

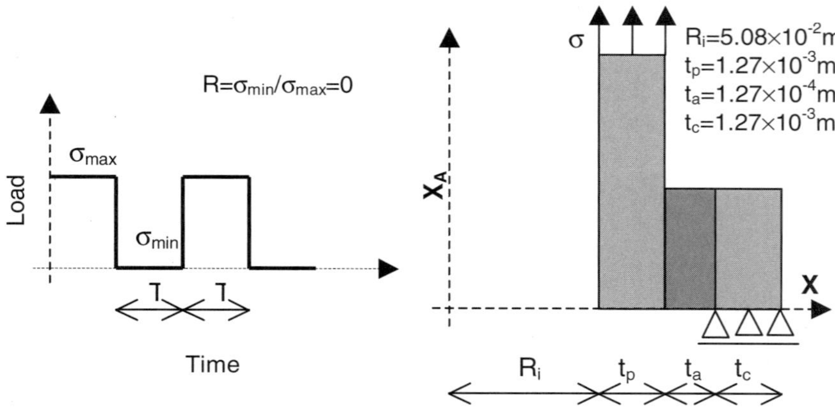

Figure 5.29. Applied fatigue load   **Figure 5.30.** Computational model

The axisymmetric FEM analysis is carried out using four node finite elements PLANE42 of three translational degrees of freedom (DOF) (u,v,w) at each node. The model mesh is made to obtain greater density in high stress concentration regions (at both edges of the adhesive layer) – in this region the finite element size was equal to the process zone d given by (5.70). During loading process, the average value of the shear stress component computed by ANSYS in the finite element is compared to the static shear strength (σ_{ad}^{u}) of the adhesive layer. After this value had been exceeded within a finite element, then finite element stiffness was multiplied by the reduction factor equal to 1×10^{-6}, and the element was deactivated, until analysis was terminated.

Table 5.5. Material properties of the model

Property	Rubber toughened epoxy (joint)	E-glass/epoxy
Longitudinal modulus [GPa]	3.05	45
Transverse modulus [GPa]	3.05	12
Shear modulus [GPa]	1.13	5.5
Poisson ratio	0.35	0.28
Shear strength [MPa]	54	70
Tensile strength [MPa]	82	1020
Fracture toughness G_{Ic} [kJ/m^2]	3.4	-
Fracture toughness G_{IIc} [kJ/m^2]	3.55	-

Supposing that the shear mode of failure is dominating in the problem, several different failure modes may occur in composite pipe joints subjected to the tensile static load. That is why the distribution of stresses within the pipe, adhesive layer and coupling was analysed first to find out whether the shear stresses are the most decisive stress components for failure initiation within the adhesive joint or not. For the pipe joint geometry considered (cf. Figure 5.30), the computations predicted the bonding failure is dominated by the shear stresses, while other stress components (orthogonal and parallel) values were at least one order smaller. These results excluded other modes of failure for this specific model and load amplitude $\sigma_{max}^{app} = 270$ MPa and, finally, confirmed appli cability of failure criterion (5.69).

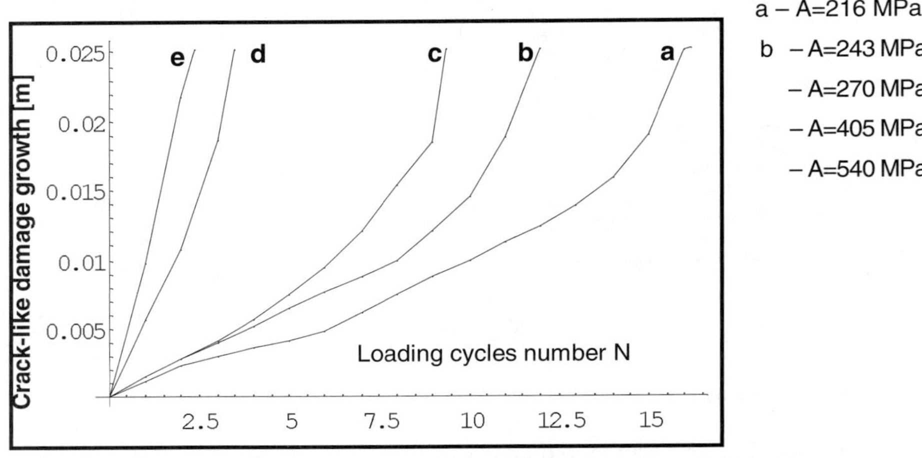

Figure 5.31. Crack−like damage growth under various amplitude fatigue loading

$$\sigma_m = 0.5\left(\sigma_{max} + \sigma_{min}\right) \text{ and } \left(\frac{da}{dN}\right)_m = \frac{1}{n}\sum_{i=1}^{n}\left(\frac{da}{dN}\right)_i \tag{5.81}$$

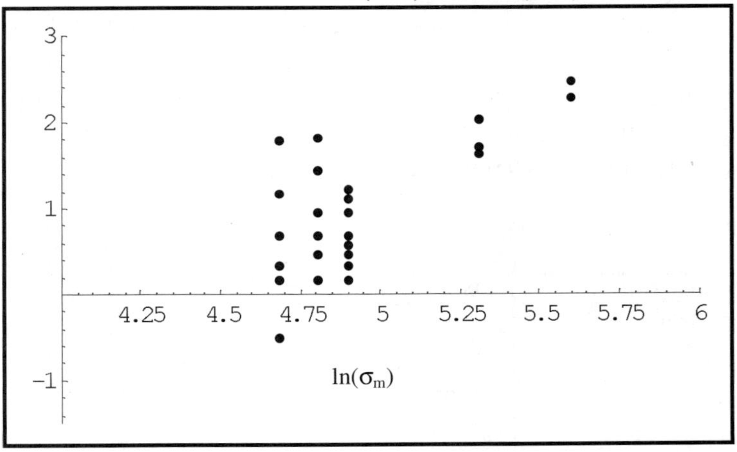

Figure 5.32. Crack-like damage growth per cycle

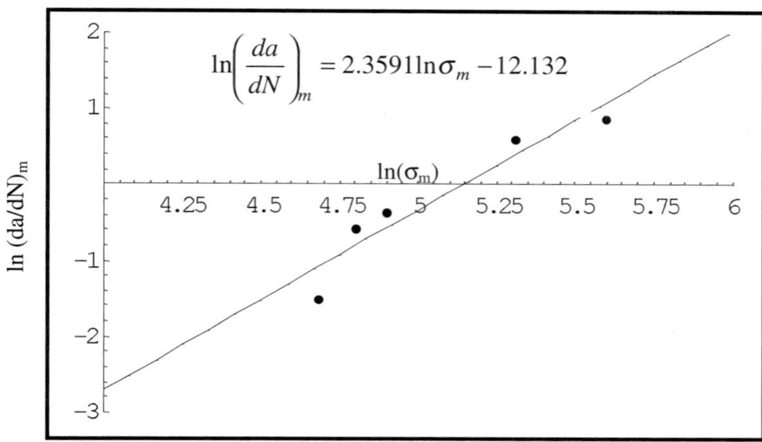

Figure 5.33. Fatigue constants estimation

The crack−like damage evolution in the adhesive layer is presented for five different load amplitudes $A = \sigma_{max}^{app} = 216$, 243, 270, 406 and 540 MPa as a function of load cycles. Those load amplitude values correspond to $4 \times \tau_{ad}^{u}$, $4.5 \times \tau_{ad}^{u}$, $5 \times \tau_{ad}^{u}$, $7.5 \times \tau_{ad}^{u}$ and $10 \times \tau_{ad}^{u}$, respectively. They were chosen to find out the load amplitude effect on a composite pipe joint. Since below an applied load amplitude $A=216$ MPa no damage nucleation was observed, then this load value may be assigned to the load threshold, A_{th}. The tendency of longitudinal crack−like damage propagation was obtained from the computer analysis as the difference between crack−like damage tip at Nth and (N-1)th cycle. The crack−like damage tip position was chosen to be the centroid of the finite element with reduced stiffness. Since the crack−like damage growth occurred from two opposite sides of the joint, thus two extreme longitudinal positions of the crack damage tips were considered and summed up to give a single crack−like damage value, as shown in Fig. 5.31. It is shown that an increase of amplitude resulted in a decrease of the load cycles were required for the final failure.

Then, the results from Figure 5.31 were used to calculate a mean crack−like damage propagation rate [mm/cycle] as a function of the applied mean fatigue−like load, calculated from (6.81) with the results shown in Figures 5.32 and 5.33.

A relation between the mean crack−like damage propagation rate and the applied mean stress is presented in Figure 5.33. The logarithmic form was taken in order to obtain coefficients $\alpha = 2.3591$ and $\beta = -12.132$ of the function $\ln(a) = \alpha \ln(b) + \beta$. The final relation between the mean crack damage propagation rate and the applied mean stress is given by the following equation:

$$\left(\frac{da}{dN}\right)_m = e^{\left[1\times10^{-4}\left(-121320.0+23591.0\ln\left(\sigma_m\right)\right)\right]} \qquad (5.82)$$

The usage of (5.82) makes it possible to estimate the mean crack damage propagation rate under applied mean fatigue load, although it should be compared with other computational approaches to the problem or the relevant experimental results. For composite containing different material properties, it would be necessary to repeat all numerical procedures carried out here, because α, β are load− and material−dependent constants.

In order to present stress distribution during crack−like damage propagation, shear stresses are plotted for different load cycles in Figures 5.34−5.38. These stresses were determined as a function of the joint length in the middle of the adhesive layer thickness. The crack−like damage tips on both sides of a joint are denoted by 'A' and 'B'. It is shown that shear stresses at the crack−like damage tips increase along with load cycle number, as was expected. It is caused by the fact that the load transfer area from pipe to coupling decreases. The crack−like damage propagation is initially the same for both tips 'A' and 'B' and supported by similar shear stress magnitudes. Then, the shear stress magnitude changes and it is different at opposite crack damage tips. It probably results from the non−uniform extension of the crack damage across the remained adhesive layer. It is necessary to mention that the lower part of the pipe overlapped coupling before the failure, which does not demonstrate a realistic situation, where pipe and coupling would slide over each other.

The tendency of fatigue crack propagation was also inspected under different failure conditions utilising the concept of the average stress criterion. That is why the average orthogonal and parallel stresses were compared with relevant strength values for different amplitudes of the applied load. Computations revealed that it would be necessary to modify failure criterion, given by (5.69) to predict fatigue life as a combination of the average shear stress with average longitudinal tensile stress in case when applied load amplitude is higher than $\sigma_{max}>406$ MPa.

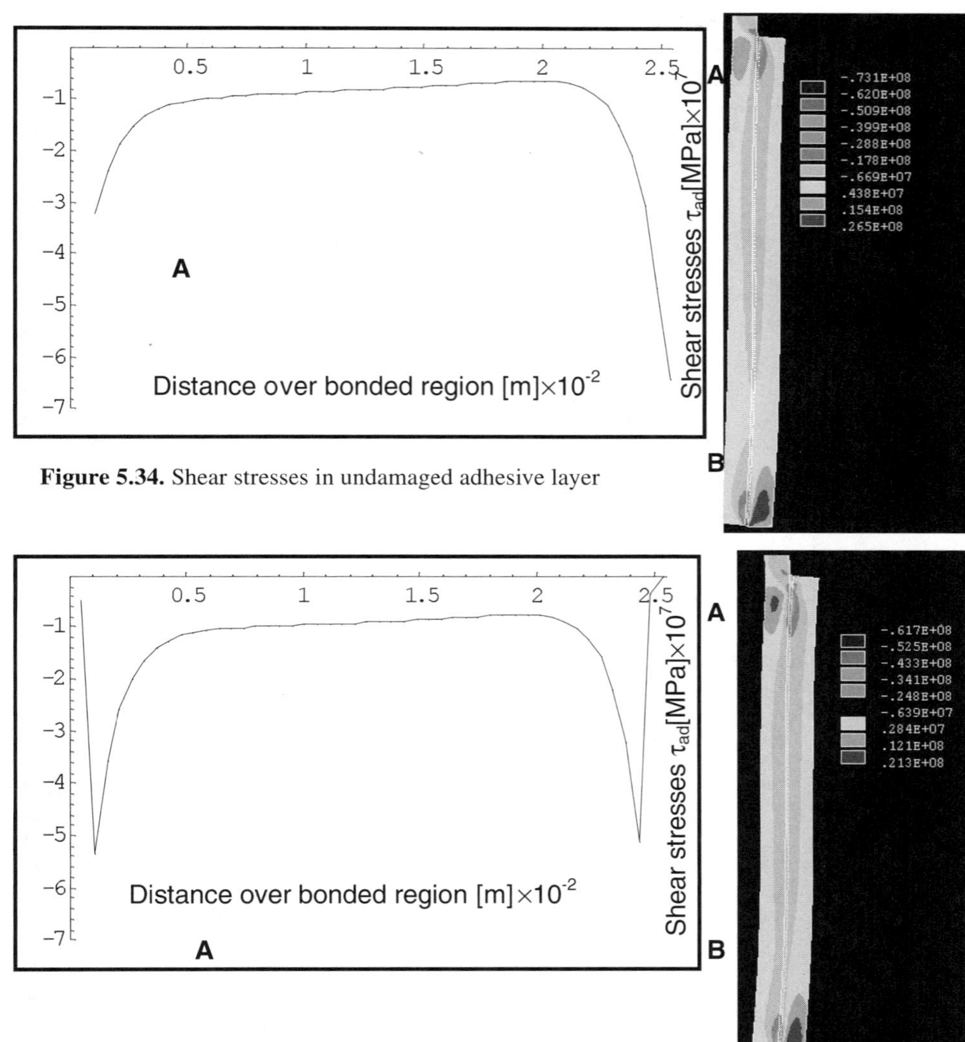

Figure 5.34. Shear stresses in undamaged adhesive layer

Figure 5.35. Shear stresses in adhesive layer after 1 cycle (1 year)

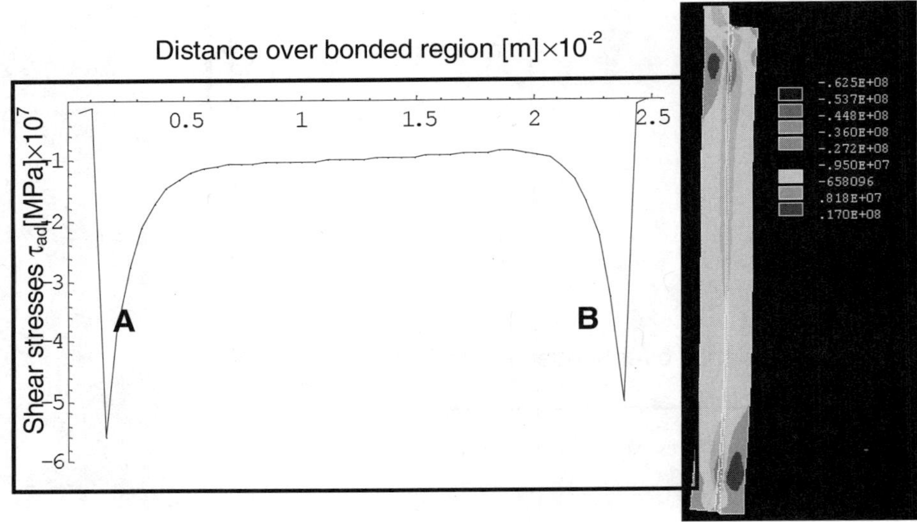

Figure 5.36. Shear stresses in adhesive layer after 2 cycles

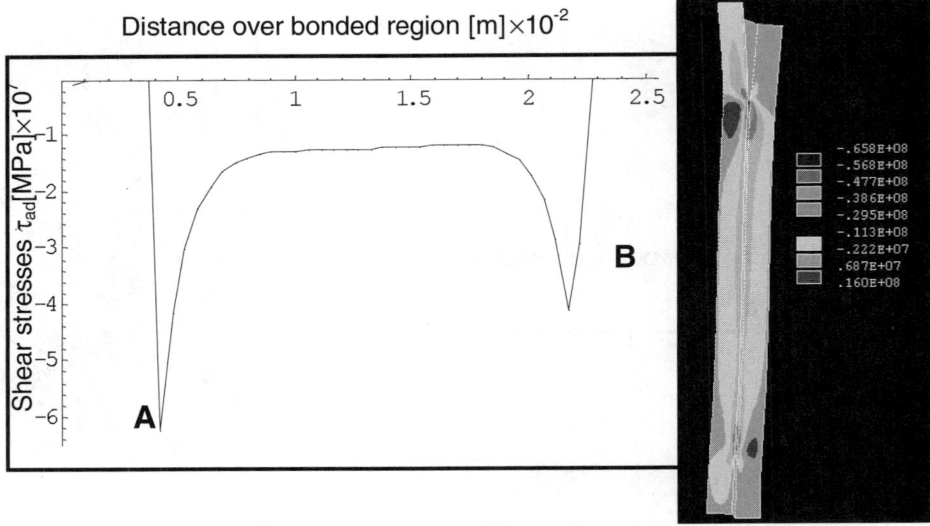

Figure 5.37. Shear stresses in adhesive layer after 5 cycles

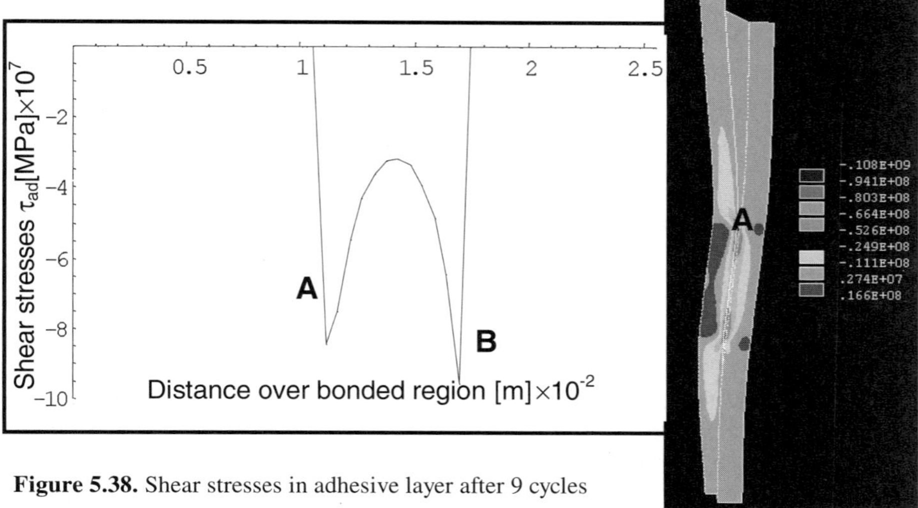

Figure 5.38. Shear stresses in adhesive layer after 9 cycles

Computations presented above are performed using 2,606 finite elements (254 in the adhesive layer); some numerical examples have been undertaken in order to estimate the total finite element number effect on the results. It was assumed that finite element number in the adhesive layer may only influence results by only. Thus the vertical mesh division effect was studied first with 400, 800, 1200, 1600, 2,000 and 4,000 finite elements, respectively. The results became independent from the decreasing finite element size (cf. Figure 5.39), while the critical finite element size for which results did not change was equal to $l_e \approx 0.0001$ m. It corresponds to about 250 vertical mesh divisions of the considered adhesive layer length.

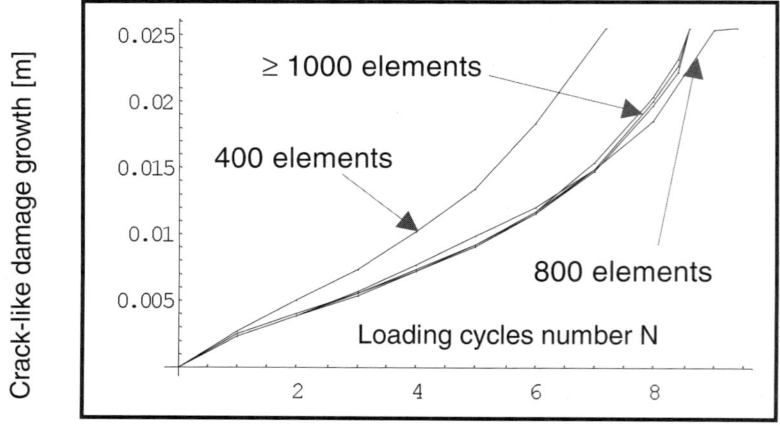

Figure 5.39. Fatigue life sensitivity to the finite elements number in adhesive layer

Numerical results presented in Figure 5.40 show that the finite element size simulating characteristic length d should be much smaller than those approximated by (5.70) and should be equal to $d{\approx}0.0007$ m. Similar comparative study was carried out for different horizontal divisions and they demonstrated a rather small mesh effect on fatigue life prediction, which oscillated in that case between 8.4 and 8.6 load cycles number (cf. Figure 5.39).

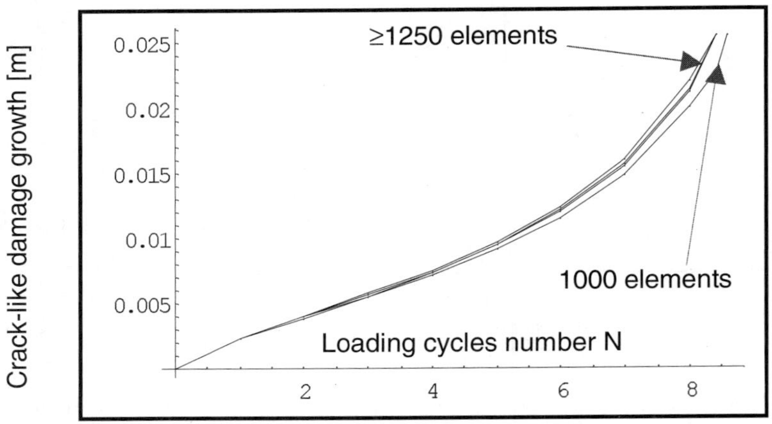

Figure 5.40. Fatigue life sensitivity to the finite elements number in adhesive layer

For the geometry of the model considered here, its finite element mesh of the adhesive layer should be designed using 5×250 elements (horizontal × vertical) in order to avoid a finite element mesh effect on the life prediction. Finally, it is suggested to solve numerically the problem by finite elements possessing a greater number of nodal degrees of freedom (nodal translations and rotations) such as shell finite elements, for instance, to improve the accuracy of the computational model.

The numerical approach proposed here enables efficient estimation of fatigue crack damage evolution rate in the composite pipe joint subjected to varying tensile load. This approach may be especially convenient in fatigue life prediction for the structures with high stress concentration regions, where internal stresses even under applied fatigue loading may be high enough to overcome material or component static strength. Qualitative numerical comparison of the fatigue crack damage evolution rate can be elaborated by the FEM displacement−based using cohesive zone fracture mechanics tools. In this case the damage of adhesive layer can be represented by a single crack model and crack evolution can be numerically determined e.g. through common spring finite elements, interface finite elements or solid finite elements with embedded discontinuity defined using the condition for a critical energy release rate growth.

5.3.3 Thermomechanical Fatigue of Curved Composite Beams

A two-component composite material with volume Ω is considered in the plane stress in an initially unstressed, undeformed and uncracked state, where its two constituents (Ω_1, Ω_2) are linear elastic and transversely isotropic materials; the effective elasticity tensor of the composite domain Ω is uniquely defined by their deterministic Young moduli and Poisson ratios. The problem is focused as before on the composite interface where a pre–crack of length a_o is introduced. Both crack surfaces are assumed to be perfectly smooth – there are neither meso– nor micro–asperities on their surfaces in the context of a contact model. The constant amplitude fatigue load σ_{ij} is applied with the coefficient of a cycle asymmetry $R = \sigma_{ij}^{min} / \sigma_{ij}^{max}$. The stress field under applied general transverse load at the crack tip is described by (5.57).

Now, let us analyse the fatigue phenomenon for such an interface [77,109,291,295], which results from thermo–mechanical external load cycles applied at the composite specimen [93,96]. Analogous to the classical Paris–Erdogan equation used to describe the fatigue crack growth rate in metals, its modified version is used

$$\frac{da}{dN} = c(\Delta G)^q \tag{5.83}$$

where c and q are some material constants determined experimentally. The energy release rate (ERR) range is described here as follows:

$$\Delta G = G^{max} - G^{min} \tag{5.84}$$

with G^{max} and G^{min} calculated for a certain applied load σ_{ij}^{max} and σ_{ij}^{min}, correspondingly. A quite similar fatigue analysis may also be applied in the case of thermal cycling or coupled thermomechanical fatigue analysis. However, it is necessary to apply the following equation to calculate the ERR range during periodic temperature variations:

$$\Delta G = G(T^{min}) - G(T^{max}) \tag{5.85}$$

where T^{min} and T^{max} are minimum and maximum temperatures for a given thermal cycle. The modified Paris–Erdogan equation (5.83) is used to estimate the number of fatigue cycles required for the steady state crack growth from an initial detectable precrack a_o to its critical length a_{cr}. It is assumed that once the critical crack length is reached, the crack grows continuously leading to the material failure by a delamination; this assumption determines the entire mechanism of a fatigue fracture of this particulate composite. Since that, the following fracture criterion is proposed:

$$\lim_{da \to 0} \frac{dG}{da} = \frac{G_{i+1} - G_i}{a_{i+1} - a_i} > 0 \Rightarrow a_{i+1} = a_{cr} \text{ and } G_{i+1} \geq 1.05 G_i \tag{5.86}$$

The 5% factor is used in (5.86) to prevent instabilities of crack propagation and which is based on some computational observations presented later. On the other hand, if the ERR is less than the threshold value G_{th}, then no crack growth is observed.

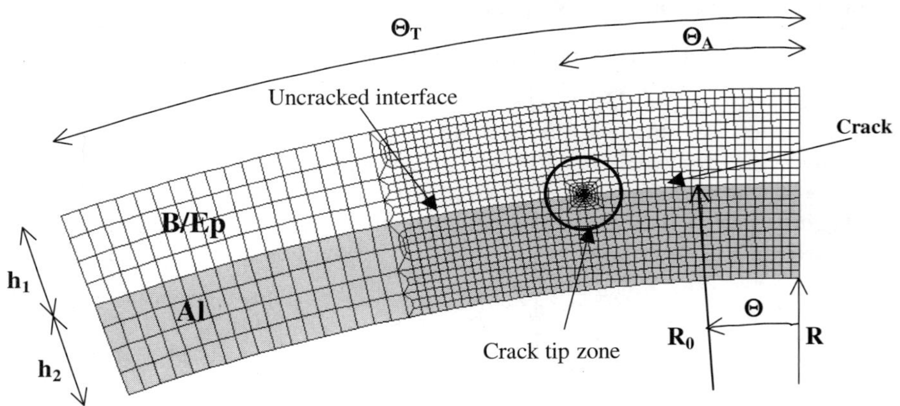

Figure 5.41. Composite FEM model

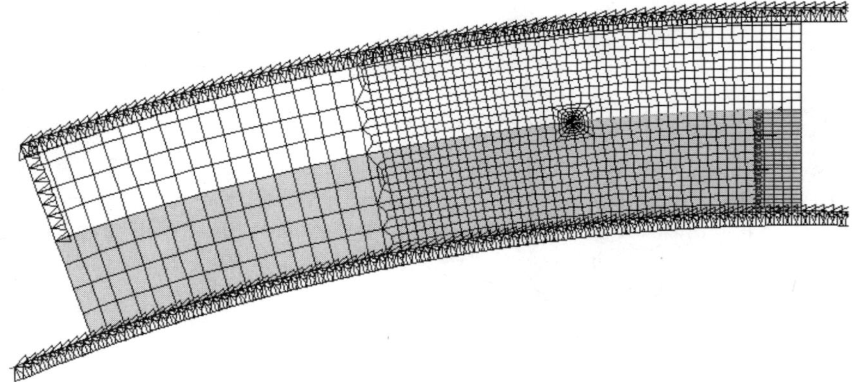

Figure 5.42. Mechanical boundary conditions

Moreover, it is possible to describe micro−crack density by the damage function as $D=a/a_{cr}$. In this case this function may be used to calculate the effective stress tensor for a cracked body as follows:

$$\sigma_{ij}^{eff(cr)} = \frac{\sigma_{ij}^{eff}}{1-D} = \frac{\sigma_{ij}\left(a_{cr}-a\right)}{a_{cr}}$$

(5.87)

where σ_{ij}^{eff} denotes the effective stress tensor of the initially perfectly bonded composite under applied load, which can be calculated by the classical homogenisation or mechanics of composite materials theory. Then, the effective stress tensor of a cracked body is estimated from (5.87) and is compared to the effective strength of a two−component curved composite.

The main purpose of computation is to estimate the number of load cycles required to composite fatigue failure by delamination as a function of the friction coefficient. The composite thermal cycling is simulated numerically to observe fatigue crack growth under non−mechanical loading. The analysis consists of the following steps in order to evaluate these parameters: (i) determination of the near-tip stress distribution under applied load (FEM analysis); (ii) evaluation of total ERR (and its contributions) as a function of the interface crack length and the friction coefficient; (iii) calculation of ERR range and (iv) determination of fatigue cycles to failure.

The composite FEM model for computer analysis is presented in Figures 5.41 and 5.42 − two linearly elastic transverse isotropic homogeneous components with the geometry parameters and material properties collected in Tables 5.6 and 5.7 are analysed.

Table 5.6. ANSYS geometrical input data

Component thickness [m]	h_1	0.0025
	h_2	0.0025
Total angle Θ_T [deg]; a_T [m]		20; 1.83×10^{-2}
Interface plane radius R_o [m]		5.25×10^{-2}
Pre-crack Θ_T [deg]; a_o [m]		6; 5.5×10^{-3}

Table 5.7. ANSYS input material properties

Property	Boron/epoxy	Aluminium 7075-T6
Density [kg/m^3]	2000	2810
Young modulus [GPa]	207	70.8
Poisson ratio	0.21	0.33
Shear modulus [GPa]	4.8	26.9
CTE [1/°C] $\times10^{-6}$	4.5	23.4
Conductivity [W/m°C]	14.7	130
Heat capacity [J/kg°C]	1150	960

The composite specimen is discretised in the FEM analysis using from 2,172 to 2,908 finite elements and from 5,863 to 7,879 nodal points to simulate the interface crack propagation. The crack length change is equal to 0.5deg (0.9×10^{-4}m). The very dense model discretisation around the crack tip needs a large effort for the singular near–tip stresses behaviour simulation. The elements used for model discretisation are 8–node plane stress solid elements PLANE82 (mechanical analysis) with 4 integration points and 8–node thermal solid elements PLANE77 (thermal analysis) with 9 integration points. Two–dimensional (2D) contact (CONTA171) and target (TARGE169) finite elements are used to simulate the contact with friction between crack surfaces and frictionless contact between external supports and model edges; the contact finite elements have 3 nodes and 2 integration points, while target finite elements are defined using 3 nodes. The numerical problem to be solved is geometrically nonlinear taking into account elastic contact with friction or frictionless elastic contact – that is why an incremental analysis is applied. The contact traction computation is possible thanks to the augmented Lagrangian technique with contact stiffness matrix symmetrisation. This technique as a combination of the two main constraint methods (penalty and Lagrange multiplier) is chosen in conjunction with predictor–corrector and the line–search numerical options to ensure satisfactory solution convergence.

The applied fatigue load is chosen as a compressive shear of 1.75 kN (138 MPa) with the cycle asymmetry factor $R=0.017$. It is observed that the shear contribution to the total ERR prevails ($\Delta G_2{\approx}\Delta G_T$) over tensile mode under the given fatigue load. Since the shear mode dominates, the ERR is taken from the range $\Delta G = G_2^{\max} - G_2^{\min}$ only and its dependence on the friction coefficient is shown in Figure 5.43. The values of ERR range vary together with the coefficient of friction from 147 ($a_o=5.49\times10^{-3}$ m) to 183 J/m^2 ($a_1=1.28\times10^{-2}$ m) for $\mu=0$ and from 108.4 (a_o) to 103.4 J/m^2 (a_1) for $\mu=0.15$. The energy dissipated due to friction results in a reduction of the ERR and alters the tendency of crack propagation – it stabilises the fracture process.

That is why the critical crack lengths corresponding to the lowest values of friction coefficients are equal to $a_{cr}=5.2$ mm for $\mu=0.0$ and $\mu=0.01$, which are smaller than those obtained for $\mu>0.01$ and equal to $a_{cr}=7.4$ mm. Thus, the number of cycles to composite failure by delamination is based on the critical crack length criterion and is calculated from (5.83). The parameter $q=10$ and the ERR threshold $\Delta G_{th}=100$ J/m^2 are applied together with the parameter $c=1\times10^{-26}$. The results of the composite life prediction are shown in Figure 5.44 – we observe there that the friction coefficient increases strongly and decreases the crack growth rate per cycle which finally leads to composite fatigue life improvement, under the assumption that interface delamination does not bring about other damage processes such as wear, for instance. Finally, the number of fatigue cycles to composite failure are estimated to be $N_f=61,865$ cycles for $\mu=0$ and $N_f=5.067040\times10^6$ cycles for $\mu=0.14$.

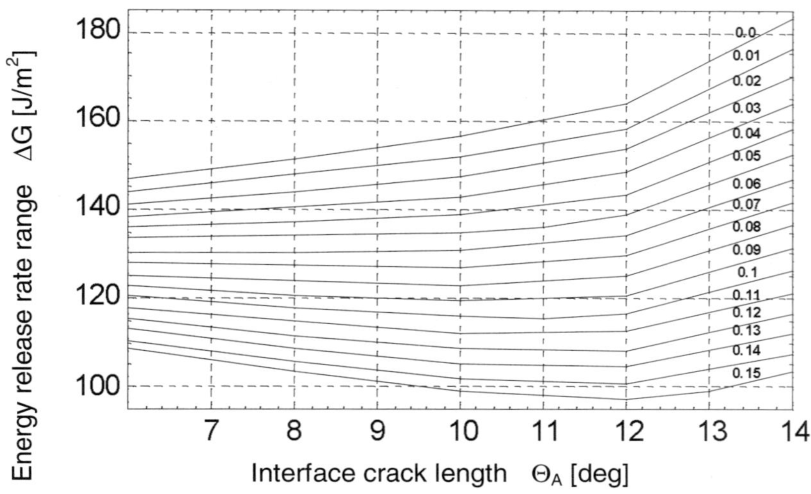

Figure 5.43. Energy release rate range during fatigue crack growth

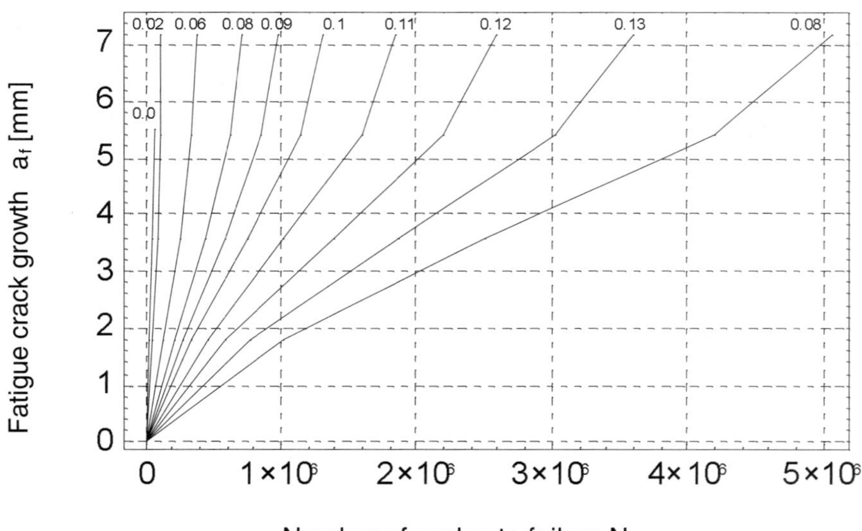

Figure 5.44. Composite mechanical fatigue life

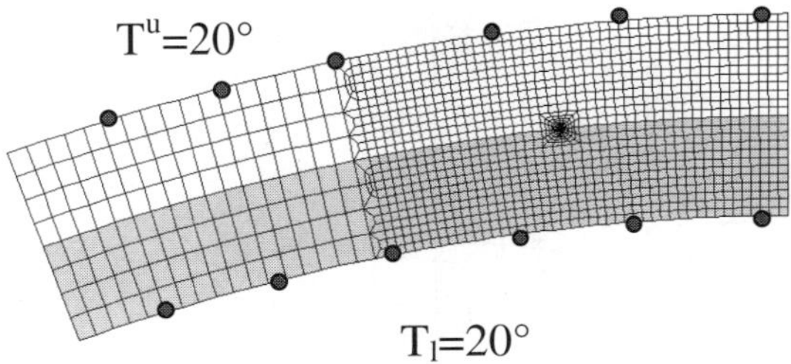

Figure 5.45. Initial temperature conditions

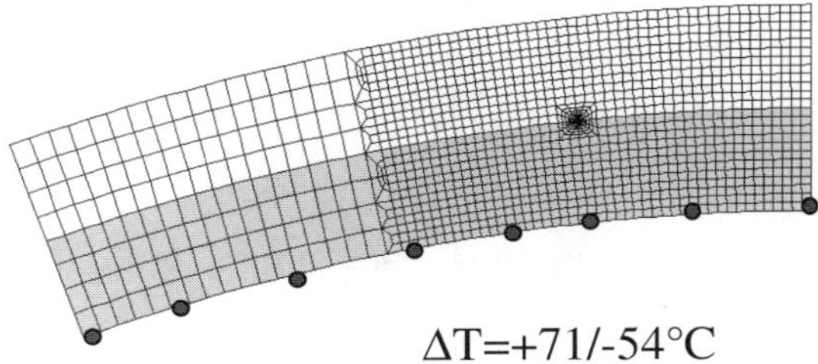

Figure 5.46. Thermal cycling

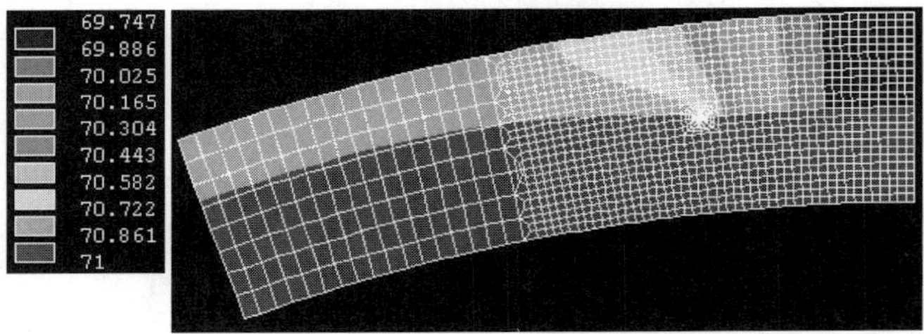

Figure 5.47. Temperature distribution (1st cycle, T=+71°C)

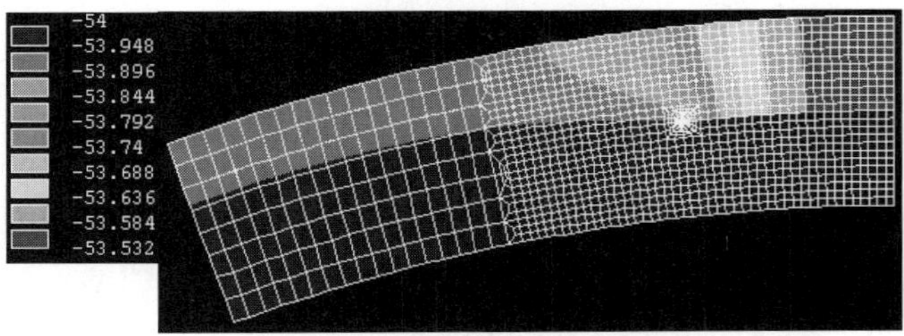

Figure 5.48. Temperature distribution (1st cycle, T=-54°C)

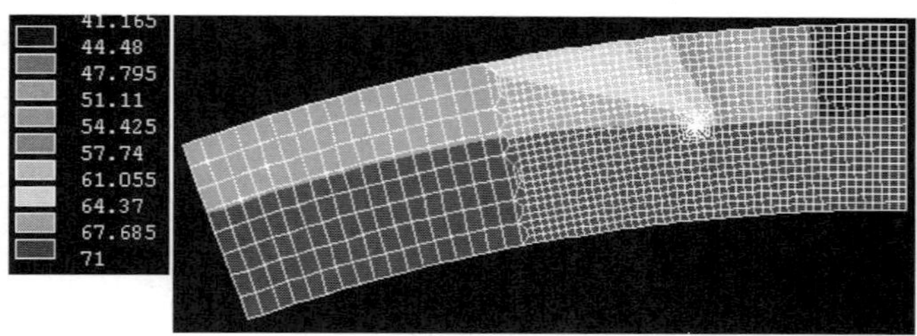

Figure 5.49. Temperature distribution (25th cycle, T=+71°C)

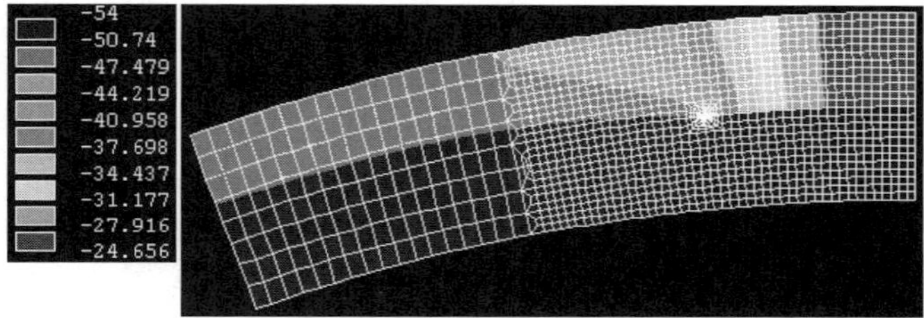

Figure 5.50. Temperature distribution (25th cycle, T=-54°C)

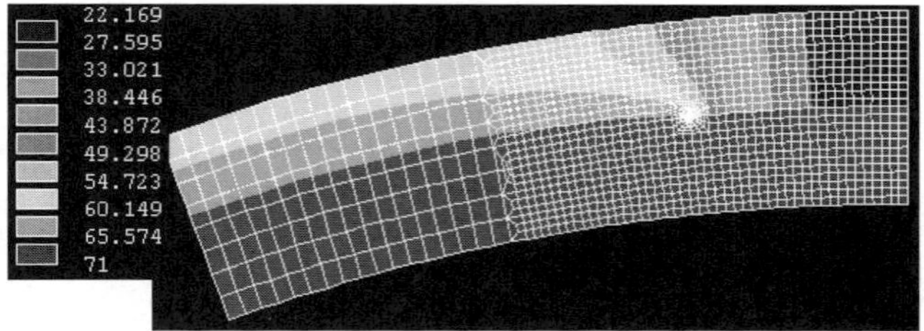

Figure 5.51. Temperature distribution (50th cycle, T=+71°C)

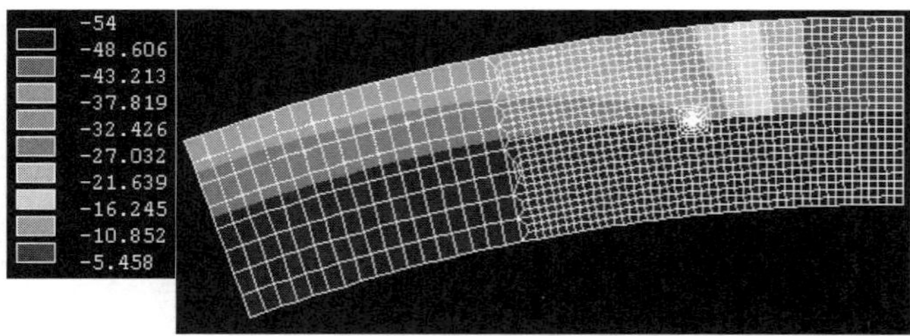

Figure 5.52. Temperature distribution (50th cycle, T=-54°C)

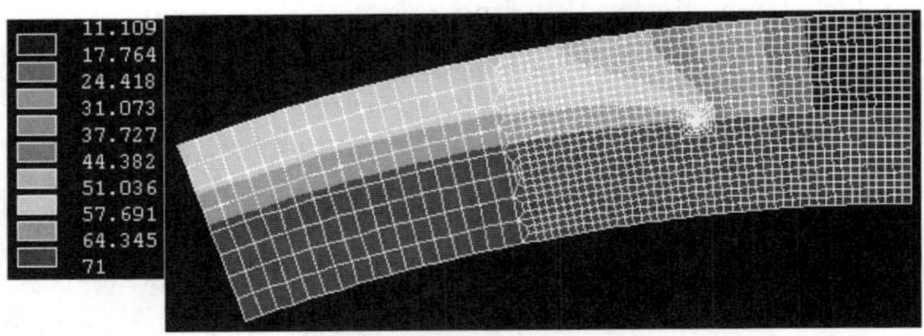

Figure 5.53. Temperature distribution (75th cycle, T=+71°C)

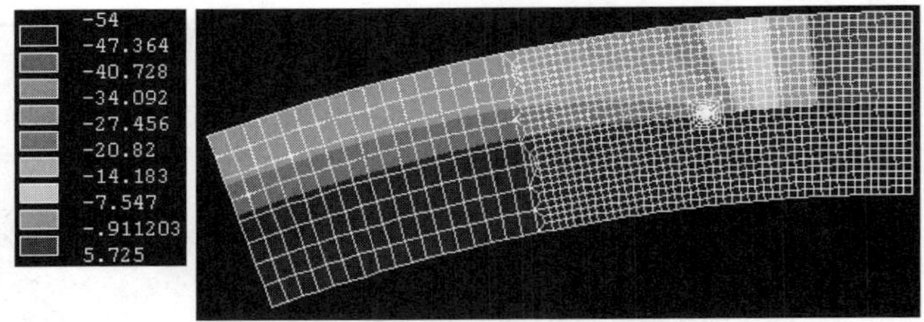

Figure 5.54. Temperature distribution (75th cycle, T=-54°C)

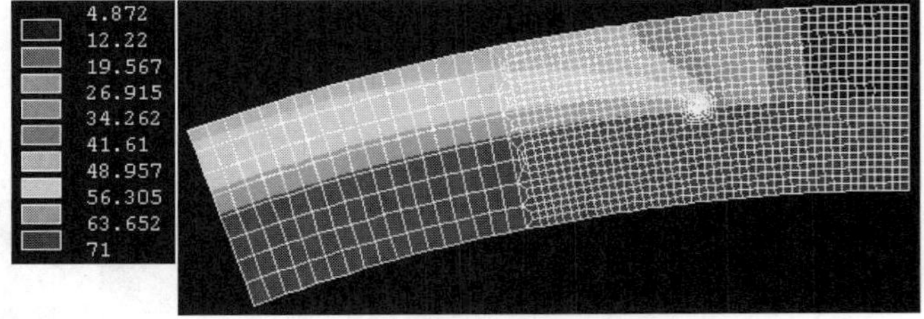

Figure 5.55. Temperature distribution (100th cycle, T=+71°C)

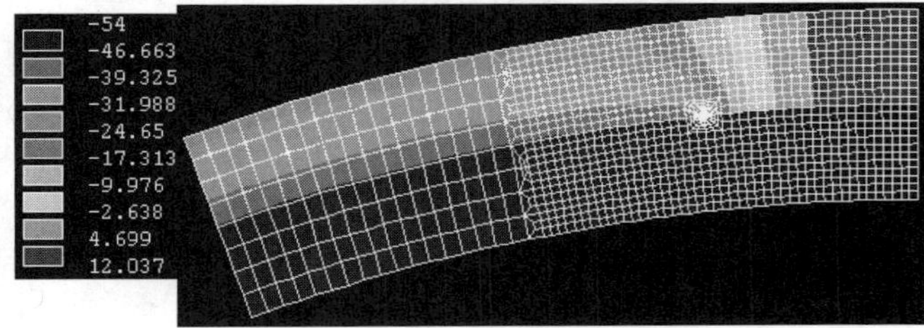

Figure 5.56. Temperature distribution (100th cycle, T=-54°C)

Computational thermal cycling is carried out for the composite specimen in the temperature range T^{max}=+71 and T^{min}=-54 – thermal boundary conditions are presented in Figures 5.44 and 5.45. First of all, the stationary thermal analysis is worked out to introduce the initial conditions for temperature distribution (cf.

Figure 5.44). Then, thermal cycling is carried out thanks to the non−stationary thermal analysis implemented in the program ANSYS. The number of simulated fatigue cycles is taken as 10 E5 cycles for +71/-54°C and corresponds to the total time t=252,000 s, where Δt=1260 s is used in numerical analysis as an incremental time step. As is shown later, the fatigue crack growth after 100 cycles is very small and equal to Δa=3×10^{-6} m. Therefore, the analysis is carried out for the initial crack length a_o=5.5×10^{-3} m. Initially, the temperature has almost the same value in all composite regions. Then, the difference in temperature increases together with the number of thermal cycles, and even temperatures with opposite signs are observed in opposite composite regions (upper and lower component). It is predicted that the near-tip stress range can be reduced if the temperatures of opposite signs appear on either side of the composite interface.

The temperature distribution over the laminate cross−section is presented for 25, 50, 75 and 100 cycles in Figures 5.49−5.56 for two temperatures mentioned above. Comparing Figures 5.47, 5.49, 5.51 and 5.53 illustrating the temperature distributions for greater initial temperature at the bottom of a laminate, it is seen that the minimal temperature is decreasing together with an increase of the fatigue cycle number (a composite is frozen during a fatigue analysis). Quite the opposite observation can be made for T=-54°C (cf. Figures 5.48, 5.50, 5.52, 5.54 and 5.56). The maximal temperature increases from T=-53.5°C (for 1st cycle) to about 12°C which means that the composite is heated during the delamination process. For both initial temperatures at the bottom of a structure, spatial distributions of temperature gradients are exactly the same.

The results of non−stationary analysis give an input for a mechanical analysis carried out for a composite model subjected to zero external mechanical loads. However, the composite is circumferentially fixed by the target finite elements and on the left side of the upper component by the supports. This coupling makes it possible to analyse the thermal stresses in a composite and further, to determine the ERR range. As was noticed before, the near−tip stress range is reduced during thermal cycling.

The equivalent thermal stresses $\sigma^{(th-eqv)}$ distributions around the crack tip are shown in Figures 5.57−5.62 for an initial crack length (a_o) at the upper and lower limit temperatures (+74°C and −54°C). Thermal stresses range is reduced from $\Delta\sigma^{(th-eqv)}$=-1000 MPa (after the 1st cycle) to -930 MPa (after 100 cycles).

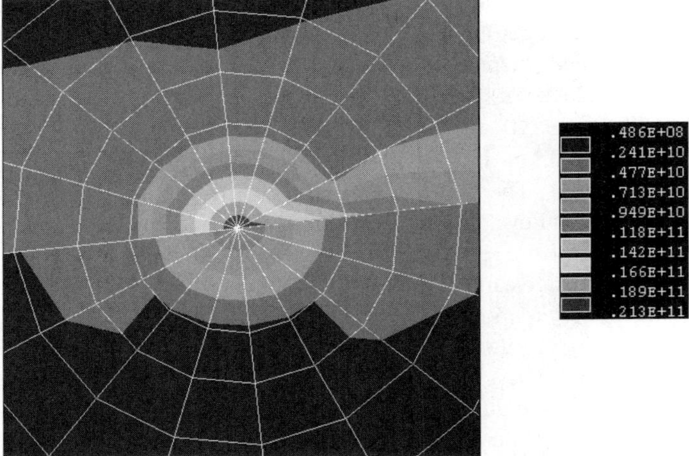

Figure 5.57. Thermal equivalent stress $\sigma^{(th\text{-}eqv)}$ [Pa] (1st cycle; +71°C)

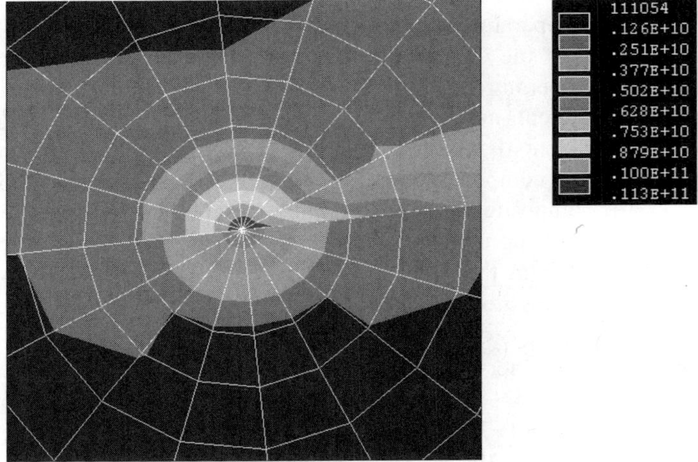

Figure 5.58. Thermal equivalent stress $\sigma^{(th\text{-}eqv)}$ [Pa] (1st cycle; -54°C)

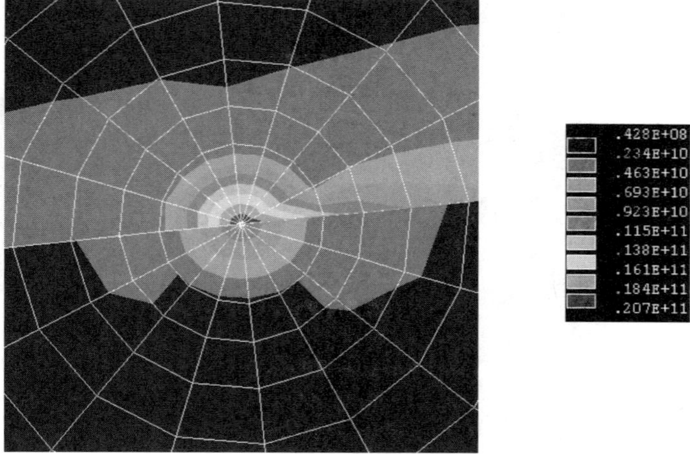

Figure 5.59. Thermal equivalent stress $\sigma^{(th\text{-}eqv)}$ [Pa] (50th cycle; +71°C)

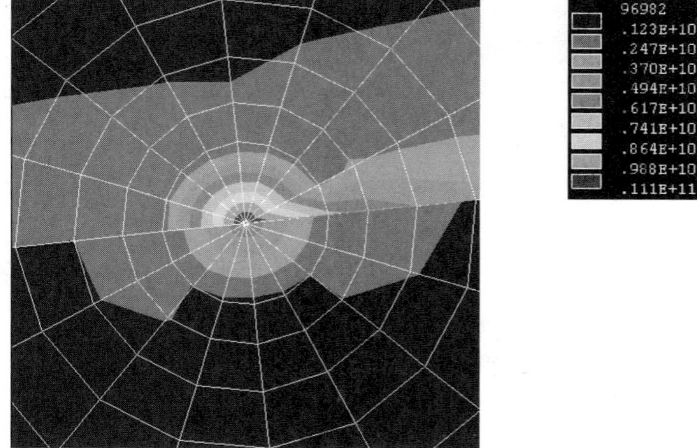

Figure 5.60. Thermal equivalent stress $\sigma^{(th\text{-}eqv)}$ [Pa] (50th cycle; -54°C)

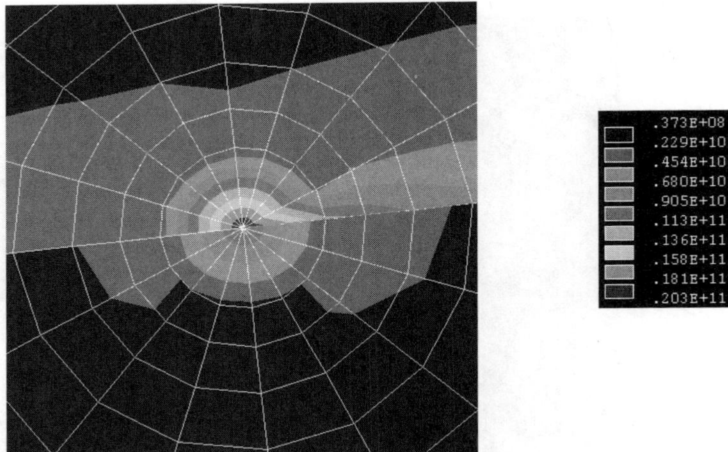

Figure 5.61. Thermal equivalent stress $\sigma^{(th\text{-}eqv)}$ [Pa] (100th cycle; +71°C)

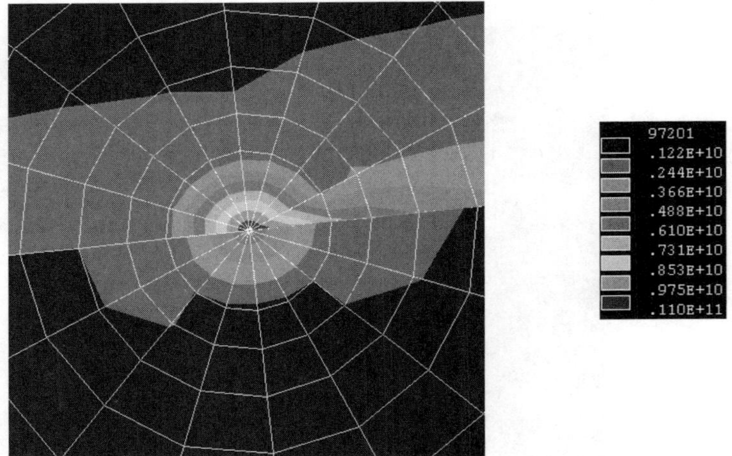

Figure 5.62. Thermal equivalent stress $\sigma^{(th\text{-}eqv)}$ [Pa] (100th cycle; -54°C)

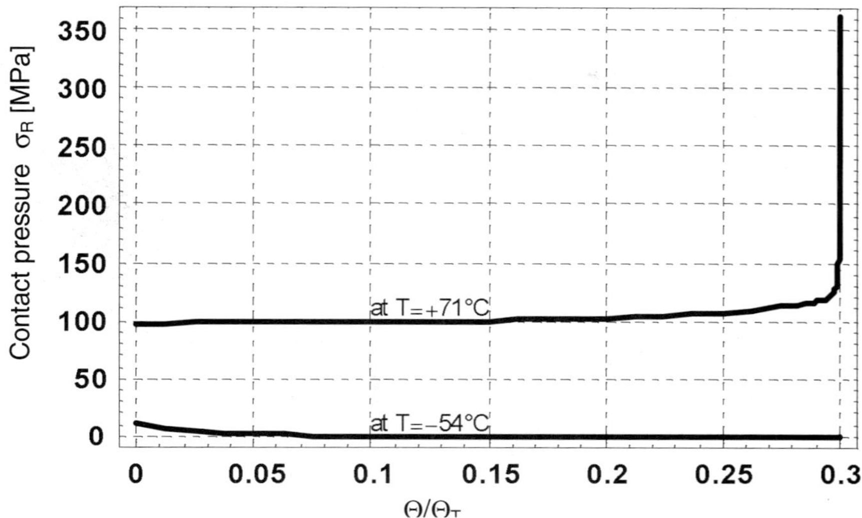

Figure 5.63. Contact pressure along the normalized crack length

Comparing these figures it can be noticed that thermal equivalent stresses are generally greater for greater initial temperature. Further, as can be expected, the maximum value of these stresses for both temperatures decreases together with an increase of the fatigue cycle number. Next, it is observed that material deformation at the upper temperature limit (+71°C) led to crack surface contact the over total crack length, while at the lower temperature limit (-54°C) the crack surfaces are opened along almost the entire crack length with the closed region near the specimen edge a_{cl}=1.14×10^{-3} m; it can be observed in Figure 5.63, where the contact pressure distribution along crack surfaces is presented after the 1st and 100th thermal cycle; a region characterised by the contact pressures σ_R=0 MPa corresponds to the crack opening. The normalised crack length equal to 0.3 is referred to the crack tip position, where the contact pressure at T=+74°C is reduced by about 10% after 100 thermal cycles.

The computed range of ERR is presented as a function of the interface crack length for a constant friction coefficient μ=0.0 in Figure 5.64. The total ERR range as a function of the interface crack length has a decreasing tendency. Mode I of the ERR range prevails, contrary to the ERR range contributions obtained from mechanical cycling, and is comprised of between 93.4% (a_o=5.4×10^{-3} m) and 95.2% (a=7.2×10^{-3} m) of a total ERR range, while the fatigue crack is arrested at a_{arr}=6.3×10^{-3} m according to the assumption of fatigue crack growth threshold ΔG_{th}=100 J/m^2.

Finally, the ERR range determination makes it possible to estimate the number of thermal cycles necessary to hold up the fatigue crack growth. The same fatigue constants as in the case mechanical fatigue are used to calculate the fatigue cycle number. The number of thermal cycles to increase the crack length from a_o to a_{rr} is equal to $N_{arr}=1.012155\times10^6$ cycles (cf. Figure 5.65). As fatigue crack is arrested and supported by the decreasing ERR range (see Figure 5.64), no criterion of composite failure is possible to take into account the crack propagation instability. That is why it would be feasible to use (5.87) to estimate the fatigue damage accumulation influence on the overall composite properties, replacing a_{cr} by a_{arr}.

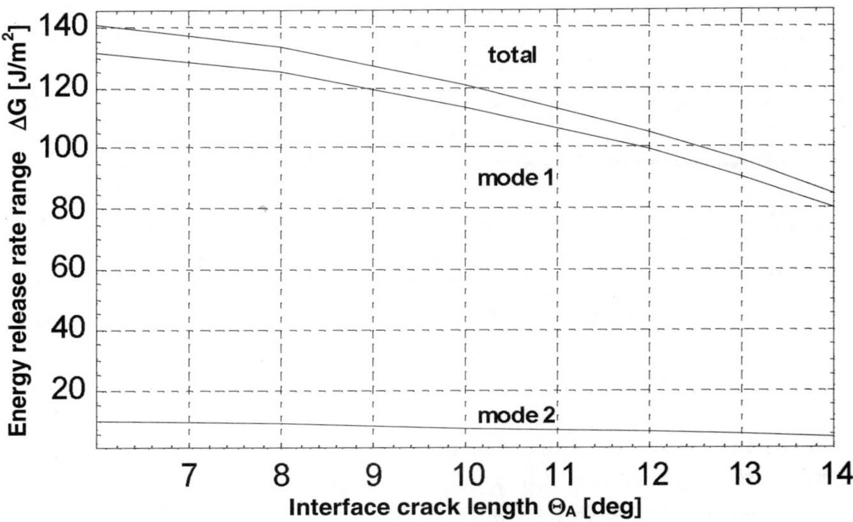

Figure 5.64. Energy release rate range

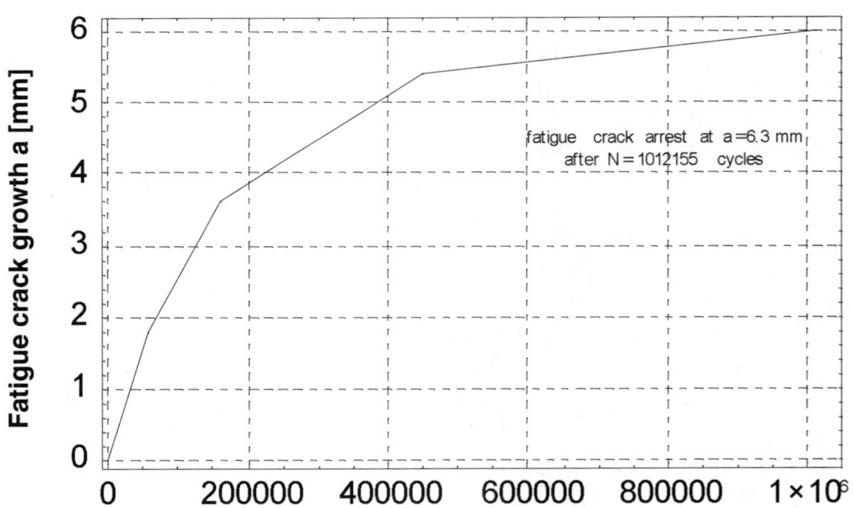

Figure 5.65. Number of cycles to fatigue crack arrest

5.4 Perturbation–based Fracture Criteria

Contrary to the traditional fracture criteria used in both deterministic analysis and stochastic Monte Carlo simulations, the probabilistic fracture criteria consist of probabilistic moments of material strength as well as the corresponding moments of external load and its direction angle. The second order perturbation technique is applied below to rewrite the Tsai–Hill criterion in terms of expected values and standard deviations of all quantities discussed.

It is expected that a failure criterion is a function of material strength and the stress (or strain) applied at the specimen. While for isotropic and homogeneous materials such a condition should not be relatively complicated, in the case of composites, the total strength is a function of composite type, the principal directions of the structure and the stress applied as well as the angle relating this stress to the direction introduced. One of the most popular in composite engineering are Tsai–Hill and Tsai–Wu failure criteria, which may be rewritten as follows:

$$\frac{\cos^4\theta}{X^2}+\left(\frac{1}{S^2}-\frac{1}{X^2}\right)\cos^2\theta\sin^2\theta+\frac{\sin^4\theta}{Y^2}=\frac{1}{\sigma^2} \tag{5.88}$$

X, Y, S denote composite strengths in the three principal directions (longitudinal, transverse and shear, respectively) while σ and θ denote externally applied axial stress and the angle between this stress and principal direction X [352];

$$\frac{\cos^4\theta}{X_t X_c}-\frac{\sin^2\theta\cos^2\theta}{\sqrt{X_t X_c Y_t Y_c}}+\frac{\sin^4\theta}{Y_t Y_c}+\frac{\sin^2\theta\cos^2\theta}{S^2}$$

$$+\left(\frac{1}{X_t}-\frac{1}{X_c}\right)\cos^2\theta+\left(\frac{1}{X_t}-\frac{1}{X_c}\right)\sin^2\theta=\frac{1}{\sigma^2} \tag{5.89}$$

where X_t, X_c, Y_t, Y_c and S are axial strength in tension, compression and transverse strength in tension and compression as well as shear strength, as previously. Equations (5.88) and (5.89) can be rewritten for the needs of probabilistic analysis after some basic algebraic transformations, while all of the quantities appearing in these equations may be treated as random variables.

Let us consider a fracture criterion for a composite being a function of material properties and stress tensor components to introduce the perturbation–based fracture analysis

$$f(\sigma;\mathbf{X})=0 \tag{5.90}$$

In terms of random loads and/or probabilistically given composite properties it can be said that (6.90) is verified with probability almost equal to 1. Since the fact that the character of the final probability density function (PDF) is unknown then denoting by $\mu_k(f(\sigma,\mathbf{X}))$ the kth order probabilistic moment of the failure function $f(\sigma;\mathbf{X})$, there holds

$$f(\sigma,\mathbf{X}) \geq F(\mu_k(f(\sigma,\mathbf{X}))) = 0 \tag{5.91}$$

where $F(\mu_k(f(\sigma,\mathbf{X})))$ is some deterministic function of probabilistic moments $\mu_k(f(\sigma,\mathbf{X}))$. The function can be evaluated starting from integration over the probability domain method, the characteristic function differentiation approach, Monte Carlo simulation technique or, alternatively, stochastic perturbation theory. Using the classical second order version of the perturbation technique, zeroth, first and second order equations for Tsai–Hill criteria in the form of

$$f(\sigma;\mathbf{X}) = 1 - \left(\frac{\sigma_1^2}{X^2} - \frac{\sigma_1\sigma_2}{X^2} + \frac{\sigma_1^2}{Y^2} + \frac{\tau_{12}^2}{S^2} \right) \tag{5.92}$$

can be written as
- zeroth order equation:

$$f^0(\sigma;\mathbf{X}) = 1 - \left(\frac{\sigma_1^2}{X^2} - \frac{\sigma_1\sigma_2}{X^2} + \frac{\sigma_1^2}{Y^2} + \frac{\tau_{12}^2}{S^2} \right) \tag{5.93}$$

- first order equations:

$$f^{,r}(\sigma;\mathbf{X}) = -\left(\left(\frac{1}{X^2}\right)^{,r} \sigma_1^2 + \frac{1}{X^2}\left(\sigma_1^2\right)^{,r} - \left(\frac{1}{X^2}\right)^{,r} \sigma_1\sigma_2 - \frac{1}{X^2}(\sigma_1\sigma_2)^{,r} \right.$$
$$\left. + \left(\frac{1}{Y^2}\right)^{,r} \sigma_1^2 + \frac{1}{Y^2}\left(\sigma_1^2\right)^{,r} + \left(\frac{1}{S^2}\right)^{,r} \tau_{12}^2 + \frac{1}{S^2}\left(\tau_{12}^2\right)^{,r} \right) \tag{5.94}$$

- second order equation:

$$f^{(2)}(\sigma;\mathbf{X}) =$$
$$-\left\{ \left(\frac{1}{X^2}\right)^{,rs} \sigma_1^2 + 2\left(\frac{1}{X^2}\right)^{,r}\left(\sigma_1^2\right)^{,s} + \frac{1}{X^2}\left(\sigma_1^2\right)^{,rs} - \left(\frac{1}{X^2}\right)^{,rs} \sigma_1\sigma_2 \right.$$
$$-2\left(\frac{1}{X^2}\right)^{,r}(\sigma_1\sigma_2)^{,s} - \frac{1}{X^2}(\sigma_1\sigma_2)^{,rs} + \left(\frac{1}{Y^2}\right)^{,rs} \sigma_1^2 + 2\left(\frac{1}{Y^2}\right)^{,r}\left(\sigma_1^2\right)^{,s}$$
$$\left. + \frac{1}{Y^2}\left(\sigma_1^2\right)^{,rs} + \left(\frac{1}{S^2}\right)^{,rs} \tau_{12}^2 + 2\left(\frac{1}{S^2}\right)^{,r}\left(\tau_{12}^2\right)^{,s} + \frac{1}{S^2}\left(\tau_{12}^2\right)^{,rs} \right\} Cov(b^r, b^s) \tag{5.95}$$

Analogous zeroth, first and second order equations can be obtained from Tsai–Wu deterministic criteria, cf. (5.89). Then, the first two probabilistic moments for the failure function $f(\sigma; \mathbf{X})$ can be calculated in the form of expected values

$$E[f(\sigma; \mathbf{X})] = f^{(0)}(\sigma; \mathbf{X}) + \tfrac{1}{2} f^{(2)}(\sigma; \mathbf{X}) \tag{5.96}$$

and variances

$$Var(f(\sigma; \mathbf{X})) = \left(f^{,r}(\sigma; \mathbf{X})\right)^2 Var(\mathbf{b}) \tag{5.97}$$

using the relations derived above.

Starting from the first two probabilistic moments, various forms of the function $F\left(\mu_k(f(\sigma, \mathbf{X}))\right)$ can be proposed which depend generally on the probability density function of input random variables as well as the output PDF of the failure function $f(\sigma; \mathbf{X})$. The following form of F is proposed below:

$$f(\sigma; \mathbf{X}) \geq E[f(\sigma; \mathbf{X})] - 3\sqrt{Var(f(\sigma; \mathbf{X}))} \tag{5.98}$$

which gives the most accurate result for Gaussian deviates and all symmetric PDF with the same first two probabilistic moments and the fourth order coefficient of concentration greater than 3. By the 'desired result' it is understood that inequality (5.98) holds true with probability almost equal to 1. Let us note that such a function is called an envelope function in stochastic theories of structural reliability.

The probabilistic failure criteria presented above have been examined in terms of the angle θ and axial stress σ being input random variables for the following material properties of the composite X=5.0 GPa, Y=6.0 GPa, S=4.0 GPa for Tsai–Hill criterion and X_t=5.0 GPa, X_c=5.5 GPa, Y_t=6.0 GPa, Y_c=6.6 GPa, S=4.0 GPa in the case of Tsai–Wu model. The variability of the expected values of the input parameters is taken as $E[\theta]$ =0,....,45 and $E[\sigma]$=2.0 GPa,...,6.0 GPa, while their standard deviations are in the range of 10% of the corresponding expected values. All computations are done by the use of the symbolic computation mathematical package MAPLE – zeroth, first and second order failure surfaces are obtained and starting from them the expected values, standard deviations and 'envelope' failure surfaces are plotted. Figures 5.66–5.69 and 5.70–5.73 presented below contain deterministic, probabilistic envelopes, expected values and standard deviations of Tsai–Hill and Tsai–Wu failure surfaces. It is seen that the character of standard deviations for both criteria plots is essentially different from the other surfaces.

Analysing the results plotted in Figures 5.66–5.73 it should be underlined that deterministic surfaces are quite close to their expected values (see (5.96)). It is caused by the fact that the coefficient of variation of both input random variables is relatively small. Further, it is observed that the 'envelope' failure surfaces for both Tsai–Hill and Tsai–Wu criteria have generally the same character as the corresponding deterministic and expected values. However, essentially smaller values generally confirm its usefulness in the probabilistic analysis of composite failure and should be studied further in detail. Especially valuable would be the application of the methodology proposed in the case of full statistical information on composite strength properties and the external load applied.

Finally, it should be underlined that the symbolic approach to stochastic perturbation analysis makes possible any finite order computations of probabilistic moments of the output. Due to this fact, precise numerical studies on model convergence for different perturbation orders, various PDF of input variables and their probabilistic parameters should be carried out.

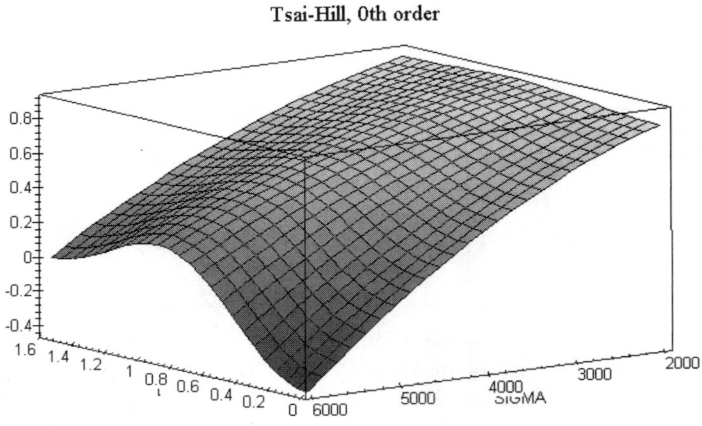

Figure 5.66. Tsai–Hill deterministic failure surface

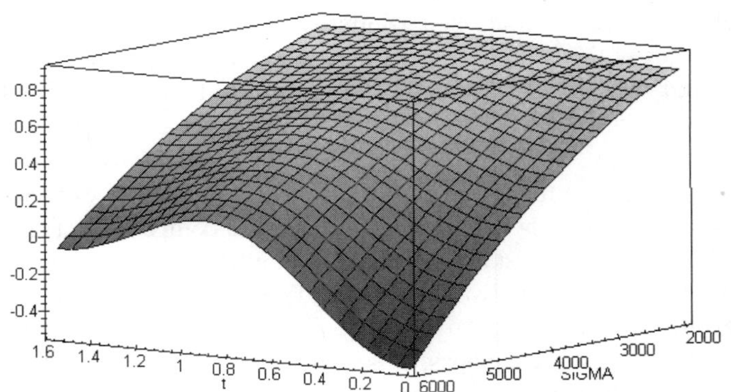

Figure 5.67. Tsai–Hill 'envelope' failure surface

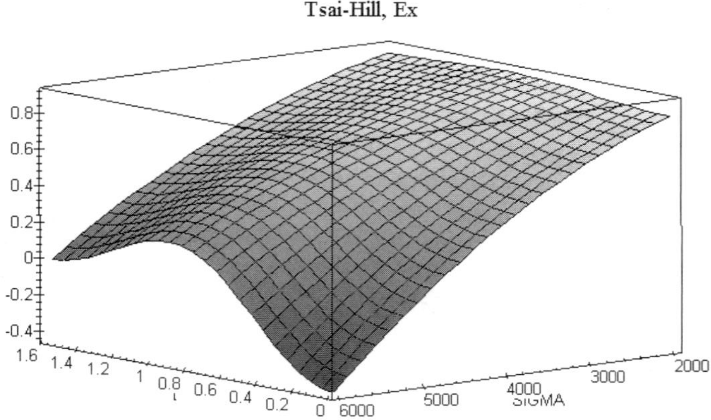

Figure 5.68. Expected values for Tsai−Hill failure surface

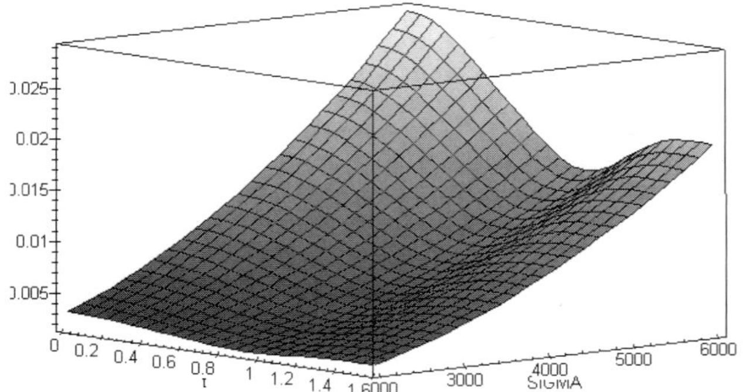

Figure 5.69. Standard deviations of Tsai−Hill failure surface

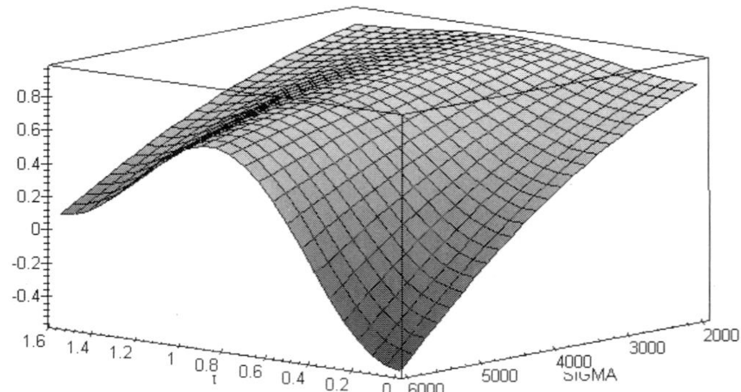

Figure 5.70. Tsai−Wu deterministic failure surface

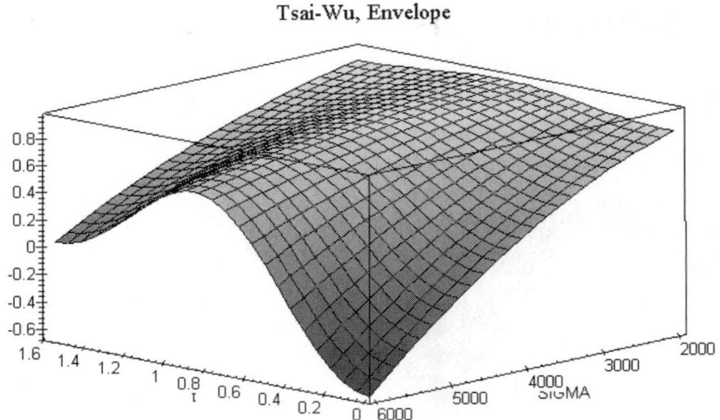

Figure 5.71. Tsai−Wu 'envelope' failure surface

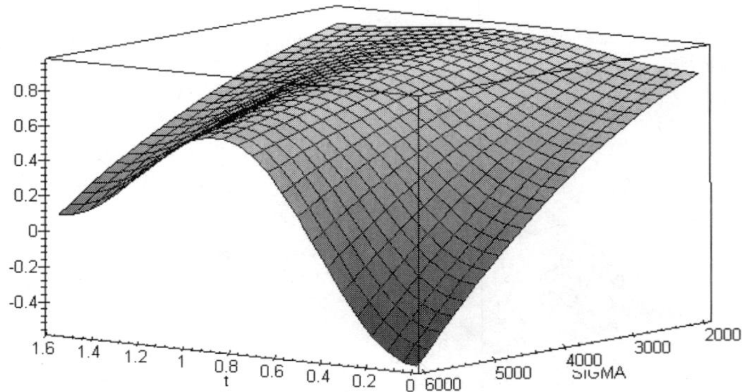

Figure 5.72. Expected values for Tsai−Wu failure surface

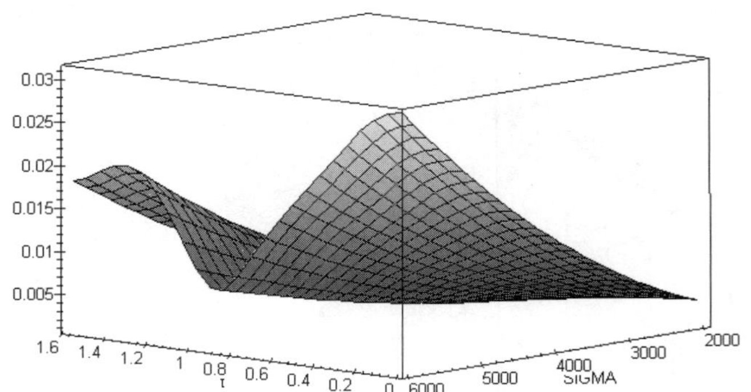

Figure 5.73. Standard deviations of Tsai −Wu failure surface

5.5 Concluding Remarks

The whole variety of mathematical and computational tools shown above makes it possible to analyse efficiently ordinary and cumulative deterministic and stochastic fatigue processes in different composite materials. Some local and global models are mentioned and the deterministic or stochastic techniques together with the approaches which enable randomisation of classical deterministic techniques to obtain at least the first two probabilistic moments of the structural response. For this purpose, most established composite oriented fatigue theories are classified and listed here. Next, the application of the perturbation–based SFEM has been demonstrated for various aspects of the fatigue process computational modelling to the W–SOTM reliability analysis, starting from direct FEM simulation in conjunction with fracture phenomena.

An alternative computational technique (MCS) is shown using the example of homogenisation analysis for a fibre–reinforced composite with stochastic interface defects simulating interface fatigue. Most of the computational illustrations presented above show, which is intuitively clear, that the expected values of structural functions decrease together with fatigue process progress. In the same time, the second order probabilistic characteristics (standard deviations, variances or coefficients of variation) increase together with the increase of fatigue cycle number, which means that the random uncertainty measure is increasing during the entire process.

The probabilistic modelling of composite materials fatigue processes summarised and proposed in this chapter is still an open question due to the fact that the area of composite material applications as well as the relevant technologies is still extending and because of the developments of the stochastic mechanics itself. The stochastic second or higher order perturbation theory for various problems shown above is very fast in randomisation of composite fatigue theories and in computational modelling. However, it is not sufficiently efficient in numerical simulation of engineering systems with increasing standard deviations of input structural parameters. The simulation methods based on the MCS approach are computationally exact, but not very effective in simple approximation of the probabilistic moments of the composite state functions, their failure criteria and the additional reliability index. Further usage of stochastic differential equations computer solvers [149] in conjunction with the FEM is recommended to include full stochastic nature of crack initiation and detection into the model.

An essential minor point of the up–to–date fatigue analysis methods (both deterministic and stochastic) is the lack of microstructural effects in the final formulae; some work is done for laminated structures. However interface phenomena in fibre–reinforced composites and stochastic microstructural problems in all composites are not included in the analysis until now. Finally, the lack of systematic sensitivity analysis of various models is observed, which makes it impossible to find a reasonable compromise between complexity of the fatigue analysis approach, probabilistic treatment of various phenomena resulting in cumulative damage and applied mathematical and numerical techniques. Such a sensitivity analysis should be carried out treating the expected values and higher order probabilistic moments of structural composite parameters as design variables, which seems to be necessary considering the application of random variables and fields in this area.

5.6 Appendix

Various fatigue models are collected below to give the overview of the capabilities of this analysis for both homogeneous and heterogeneous structures; they are listed according to the subject classification presented in this chapter.

A. Fatigue cycles number analysis – determine N:

1. Madsen (power law function) [244]:

$$N = KS^{-m} \tag{A5.1}$$

S is stress amplitude, K,m are some material constants;

2. Boyce and Chamis [42]:

$$N = 10\exp\left\{\log N_{MF} - \left[(\log N_{MF} - \log N_{MO})\right]\left[\frac{S}{S_0\left[\frac{T_F - T}{T_F - T_0}\right]^n\left[\frac{S_F - \sigma}{S_F - \sigma_0}\right]^m}\right]\right\}^{\frac{1}{q}} \tag{A5.2}$$

N_{MF} – final cycle, N_{MO} – reference cycle, S – fatigue strength, S_0 – reference fatigue strength, T_F – final temperature, T_0 – reference temperature, T – current temperature, σ – current mean stress, σ_0 – reference (residual) stress, n,q – empirical parameters;

3. Caprino, D'Amore and Facciolo [53]:

$$N = \beta\sqrt{1 + \frac{1}{\alpha(1-R)}\left\{\frac{\gamma}{\sigma_{\max}}\left|\ln(1 - f_n(N))\right|^{\frac{1}{\sigma}} - 1\right\}} \tag{A5.3}$$

$f_n(N)$ – probability of failure; γ,σ – scale parameter (characteristic strength) and the shape parameter of the Weibull distribution of the static strength; R – given stress ratio; $\sigma_{\max}$ – maximum stress level; α, β – constants from experiments.

B. Stiffness reduction models:

1. Whitworth [365]:

$$\frac{dE(N)}{dN} = -\frac{Df^a(E_0,S)}{aE^{a-1}(N)} \tag{A5.4}$$

$E(n)$ – residual modulus after n fatigue cycles, E_0 – initial modulus, $N=n/N^*$ – ratio of applied cycles to fatigue life; S,a,D – some constants, $f(E_0,S)$ – some function of E_0,S, i.e. $C\left[E_0 - \dfrac{S}{e_f}\right]$, e_f – constant depending on overall strain at failure;

2. Hansen [127]:

$$\frac{E}{E_0} = 1 - \beta, \; \beta = A\int_0^N\left(\frac{\varepsilon_e}{\varepsilon_0}\right)^n dN \tag{A5.5}$$

A – some constant, ε_e – effective strain level, ε_0 – damage strain where

$$\dot{\beta} = \frac{d\beta}{dN} = A\left(\frac{\varepsilon_e}{\varepsilon_0}\right)^n \tag{A5.6}$$

3. Bast and Boyce (creep component for the stiffness reduction) [20]:

$$\frac{S}{S_0} = \left[\frac{t_u - t_0}{t_u - t}\right]^{-\nu} \cong \left[\frac{10^6 - 0.25}{10^6 - t}\right]^{-\nu} \tag{A5.7}$$

t_u – ultimate strength of creep hours when rupture strength is very small, t_0 – reference number of creep hours where rupture strength is very large, t – current number of creep hours, ν – empirical material constant for the creep effect.

C. Fatigue crack growth analysis ($\dfrac{da}{dN}$) - *deterministic methods* (Yokobori [379]):

1. Liu (energy approach)

$$\alpha_1\left(\frac{\Delta K}{\sigma_{sy}}\right)^2 \tag{A5.8}$$

2. Paris (energy approach)

$$\frac{\alpha_2}{l_2}\left(\frac{\Delta K}{\sigma_{sy}}\right)^4 \tag{A5.9}$$

3. Raju (energy approach)

$$\alpha_3\frac{\Delta K^4}{\sigma_{sy}^2\left(K_{Ic}^2 - K_{max}^2\right)}; K_{max} \ll K_1 \tag{A5.10}$$

$$\alpha_3 \frac{\Delta K^4}{\sigma_{sy}^2 K_{Ic}^2}$$

4. Cherepanov (energy approach)

$$\alpha_4 \frac{\Delta K^4 E}{\sigma_{sy}^3 K_{Ic}^2} \qquad (A5.11)$$

5. Rice (crack opening displacement – COD)

$$\frac{\alpha_5}{l_5}\left(\frac{\Delta K}{\sigma_{sy}}\right)^4 \qquad (A5.12)$$

6. Weertman (continuous dislocation formalism)

$$\alpha_6 \frac{\Delta K^4}{\gamma E \sigma_{sy}^2} \qquad (A5.13)$$

7. Weertman, Mura and Lin (continuous dislocation formalism)

$$\alpha_7 \frac{\Delta K^4}{\gamma_p \mu \sigma_a^2} \qquad (A5.14)$$

8. Lardner (COD)

$$\alpha_8 \frac{\Delta K^2}{E \sigma_{sy}} \qquad (A5.15)$$

9. Schwalbe (COD)

$$\alpha_9 \frac{\Delta K^2}{E \sigma_{sy}} \qquad (A5.16)$$

10. Pook and Frost (COD)

$$\alpha_{10}\left(\frac{\Delta K}{E}\right)^2 \qquad (A5.17)$$

11. Tomkins (skipband decohesion)

$$\frac{\pi}{8}\left(\frac{\Delta K}{\sigma_{sy}}\right)^2\left(\frac{\Delta\sigma}{\sigma_0}\right)^{\frac{1}{\beta}} \qquad (A5.18)$$

12. McEvily (semi-experimental approach with COD)

$$\frac{(\Delta K - \Delta K_{TH})^2}{\sigma_{sy} E}\, f(\Delta K, K_{1c}, K_{max}) \qquad (A5.19)$$

13. Donahue *et al.* (COD)

$$\alpha_{11} \frac{\Delta K^2}{\mu \sigma_a} \qquad (A5.20)$$

14. Yokobori I (nucleation rate process approach)

$$\alpha_{12}\left(\frac{\Delta K}{\gamma E}\right)^{1/2\xi kT} \qquad (A5.21)$$

15. Yokobori II (nucleation rate process approach)

$$\alpha_{13}\left(\frac{\Delta K}{\sqrt{s}\,\sigma_{cy}}\right)^{2\beta/(1+\beta)}\left(\frac{b\sigma_{cy}^2}{\gamma E}\right)^{1/2\xi kT} \qquad (A5.22)$$

16. Yokobori III (dislocation approach)

$$\alpha_{14}\left(\frac{\Delta K}{\sqrt{s}\,E}\right)^{\frac{(m+1)^2}{m+2}} \qquad (A5.23)$$

17. Yokobori IV (dislocation approach)

$$\alpha_{15}\left(\frac{\Delta K}{\sqrt{s}\,\gamma}\right)^{\frac{2\beta}{1+\beta}\frac{(m+1)^2}{m+2}+\frac{1}{1+\beta}}\left(\frac{\sigma_{cy}}{E}\right)^{\frac{(m+1)^2}{m+2}} \qquad (A5.24)$$

where α_i, $i=1,15$ denote some experimentally determined material constants;

18. Yokobori V (monotonic
yield strength dependence)

$$B\left(\frac{\Delta K}{\sqrt{s}\sigma_c}\right)^n \tag{A5.25}$$

19. Paris–Erdogan [280]:

$$C(\Delta K)^m = C\left[Y(a)\Delta\sigma\sqrt{\pi a}\right]^m \tag{A5.26}$$

$Y(a)$ – geometry factor, $\Delta\sigma$ – stress range, C, m – some material constants;
20. Ratwani–Kan [296]:

$$C\left(\tau_{zma} - \tau_{zmi} - \tau_{th}\right)^{n_1} b^{m_1} \tag{A5.27}$$

τ_{zmi} – minimum interlaminar shear stresses, τ_{zma} – maximum interlaminar shear stresses, τ_{th} – interlaminar threshold shear stress range, C, n_1, m_1 – material constants, b – delamination length;
21. Wang–Crossman:

$$\alpha\left[\frac{C_e(a)t\sigma_m^2}{G_c E^2}\right] \tag{A5.28}$$

G_c – critical strain energy rate; σ_m – applied load, E – elastic modulus, a – delamination width; t – ply thickness;
22. Forman $et\ al.$ [101]:

$$\frac{C(\Delta K)^m}{(1-R)K_C - \Delta K} \ ; \qquad R = \frac{K_{min}}{K_{max}} \tag{A5.29}$$

where C, m are the material constants with $m \approx 3$ for steels and m$\approx$3-4 for alluminium alloys;
23. Donahue $et\ al.$ [82] for $\Delta K \rightarrow \Delta K_{th}$ obtained

$$C\left[\Delta K - \Delta K_{th}\right]^m \tag{A5.30}$$

24. McEvily and Groeger [247]

$$\frac{A}{E\sigma_Y}\left[\Delta K - \Delta K_{th}\right]^2\left[1 + \frac{\Delta K}{K_{IC} - K_{max}}\right] \tag{A5.31}$$

where σ_Y denotes the yield stresses of the specimen, A is an environment sensitive material parameter and K_{IC} is a plane strain fracture toughness.

25. Experimentally based law for combined mode I and mode II loadings proposed by Roberts and Kibler [302], where crack growth is obtained as

$$C(\Delta K_e)^m, \qquad K_e = \left(K_I^4 + 8K_{II}^4\right)^{\frac{1}{4}} \tag{A5.32}$$

26. Hobson [137] proposed one of the first quantitative models to describe short fatigue crack growth in terms of a microstructural parameter d assumed as a material characteristic

$$Ca^\alpha (d-a)^{1-\alpha}; a \le d \tag{A5.34}$$

where α, C are empirical constants (C depends on both material and loading parameters – Young modulus, yield stress and the applied cyclic stress);

27. Kitagawa–Takahashi curve: the LEFM (linear elastic fracture mechanics) approach determining the condition describing the stress level ΔK_{th} when the cracks can grow

$$\Delta K_{th} = Y\Delta\sigma\sqrt{\pi a} \tag{A5.35}$$

Let us recall that the LEFM approach is invalid when the small–scale yielding conditions are exceeded $\Delta\sigma \ge \frac{2}{3}\sigma_{cy}$ where σ_{cy} is the cyclic yield stress;

28. Priddle law [290]:

$$C\left(\frac{\Delta K}{K_F - K_{max}}\right)^2 \tag{A5.36}$$

C – growth resistance, K_F – critical value for the stress intensity factor;

D. Fatigue crack growth analysis – determination of $\dfrac{da}{dN}$ (*some stochastic methods*)

$$\frac{da(t)}{dt} = Q(\Delta K, K_{max}, S, A, R)X(t) = Q(a)(\mu + Y(t)) \tag{A5.37}$$

$a(t)$ – random crack size, Q – some nonnegative function, ΔK – stress intensity factor range, K_{max} – maximum stress intensity factor, $X(t)$ – nonnegative random process, $Y(t)$ – random process with 0 mean;

1. Ditlevsen and Sobczyk [80]:

$$\frac{da(t)}{dt} = a^p X(t, \gamma)$$
(A5.38)

p = 1,3/2,2 (experimental), $X(t)$ – Gaussian white noise, process with finite correlation time;

2. Lin and Yang [234]:

$$X(t, \gamma) = \sum_{k=1}^{N(t)} Z_k w(t, \tau_k)$$
(A5.39)

$N(t)$ – homogeneous Poisson counting process, τ_k – arrival time of kth pulse, Z_k – random amplitude of kth pulse with the following synergistic sine hyperbolic functional form:

$$\left[\frac{da(n)}{dn}\right]' = 10^{C_1 \sinh|C_2(\log \Delta K + C_3)| + C_4}$$
(A5.40)

$a(n)$ – half crack length, C_i – some parameters
– randomized form:

$$\frac{da(n)}{dn} = X(n)\left[\frac{da(n)}{dn}\right]'$$
(A5.41)

3. Spencer *et al.* [327]:

$$\frac{da(t)}{dt} = Q(a)10^Z , \quad \frac{dZ}{dt} = -\xi Z + G(t) , \quad a(0) = a_0, \quad Z(0) = Z_0$$
(A5.42)

where $G(t)$ – stationary Gaussian white noise, $Z(t)$ – nonstationary random process; the Pontriagin–Vitt equation is used

$$-nT^{n-1} = Q(a_0)10^{Z_0}\frac{\partial T^n}{\partial a_0} - \xi z_0 \frac{\partial T^n}{\partial z_0} + \pi S_0 \frac{\partial^2 T^n}{\partial z_0^2} , \quad n=1,2,\dots$$
(A5.43)

with the boundary conditions:

$$T^0(a_0, z_0) \equiv 1, \quad T^n(a_0, z_0) \to 0 : z_0 \to \infty$$
(A5.44)

$$\forall z_0 : T^n(a_c, z_0) = 0 , \quad \frac{\partial T^n(a_0, z_0)}{\partial z_0} \to 0 : z_0 \to -\infty$$
(A5.45)

Fatigue damage function based model – calculation of $\dfrac{dD}{dN}$:

1. Palmgren–Miner model [299]:

$$D = \frac{n}{N} \tag{A5.46}$$

n – number of fatigue cycle, N – number of cycles to failure;

2. Modified Palmgren–Miner model:

$$D = \left(\frac{n}{N} \right)^{C} \tag{A5.47}$$

C – constant independent of applied stress; some probabilistic aspects of this model can be found in [254];

3. Shanley model:

$$D = CS^{kb} n \tag{A5.48}$$

S – applied stress, C,K – constants, b – slope of central position of S–N curve;

4. Marco–Starkey model:

$$D = \left(\frac{n}{N} \right)^{C_i} \tag{A5.49}$$

$C_i>1$ – stress dependent constant;

5. Henry model:

$$D = \frac{\left(S_t - S_t' \right)}{S_t} \tag{A5.50}$$

S_t – fatigue of virgin specimen, S_t' – fatigue limit after damage;

6. Corten–Dolan model:

$$D = mcn^{\alpha} \tag{A5.51}$$

m – number of damage nuclei, c,a – function of stress condition; α – some constant;

7. Gatt model:

$$D = \left(S_t - S_t'\right)^{\alpha}$$ (A5.52)

8. Marin model:

$$S^k N = C$$ (A5.53)

9. Manson model:
– for crack initiation:

$$D = \frac{n}{N_I}$$ (A5.54)

– for crack propagation:

$$D = \frac{n}{N_P}$$ (A5.55)

10. Owen–Howe model:

$$D = B\left(\frac{n}{N}\right) - C\left(\frac{n}{N}\right)^2,$$ (A5.56)

B,C – some constants;
11. Srivatsavan–Subramanyan model:

$$D = \frac{\log N_t - \log N}{\log N_t - \log n}$$ (A5.57)

12. Lemaitre–Plumtree model:

$$D = 1 - \left(1 - \frac{n}{N}\right)^a$$ (A5.58)

$a = \dfrac{1}{p+1}$ strain controlled; $a = \dfrac{1}{c+p+1}$ stress controlled; p,c–material constants

13. Fong model [100]:

$$D = \frac{\exp(kx) - 1}{\exp(k) - 1}$$ (A5.59)

where k represents damage trend;
14. Cole model:

$$A_D = A - C$$ (A5.60)

A_D – attenuation due to damage, A – total attenuation, C – attenuation of virgin specimen;

15. Fitzgerald–Wang model:

$$D = 1 - \frac{E}{E^*} \qquad (A5.61)$$

E – modulus at a fatigue cycle; E^* – reference modulus;

16. Wool model:

$$-\frac{dD}{dt} = kD^a \qquad (A5.62)$$

17. Chou model:

$$D = \Delta F(n) \qquad (A5.63)$$

18. Hwang–Han model I [143]:

$$D = \frac{F_0 - F(n)}{F_0 - F_f} = \left(\frac{n}{N}\right)^c \qquad (A5.64)$$

F_0 – undamaged, F_f – damaged modulus;

19. Hwang–Han model II [144]:

$$D = \frac{\varepsilon(n)}{\varepsilon_f} = \frac{r}{1 - Kn^c} \qquad (A5.65)$$

ε_f – failure strain;

20. Hwang–Han model III:

$$D = \frac{\varepsilon(n) - \varepsilon_0}{\varepsilon_f - \varepsilon_0} = \frac{r}{1 - r} \frac{n^c}{B - n^c} \qquad (A5.66)$$

21. Morrow approach [257]:

$$D_i = \frac{n_i}{N_i} \left(\frac{S_i}{S_m}\right)^d \qquad (A5.67)$$

S_i – amplitude of stress causing fatigue damage, S_m – maximum stress amplitude, n_i – number of stress peak at level S_l, d – plastic work interaction exponent, N_i – number of stress peak to the failure if S_i =const.;
22. Morrow approach with different cycles:

$$D(t) = \sum_{i=1}^{n(t)} \frac{1}{N_i} \left(\frac{S_i}{S_m(t)} \right)^d \qquad (A5.68)$$

6 Reliability Analysis

6.1 Introductory Remarks

A very natural application of SFEM and the other probabilistic analytical and numerical methods [313] is the reliability assessment for both homogeneous [45,256,354] and heterogeneous structures [87,102,231,262]. The starting point of the analysis is to assume the limit state function in terms of any structural state parameters – displacements, stresses, temperatures or strains (as well as some combination of them in the coupled problems). Then, starting from statistical input on the structural parameter, probabilistic structural analysis is carried out and, finally, starting from the limit state function, the reliability index is computed. The reliability index should have the same properties as the classical Kolmogoroff probability and, in the same time, the damage function.

Following the stochastic structural analyses, First Order Reliability Method (FORM) and Second Order Reliability Method (SORM) are most frequently used [87,114,115,209]. The methods do not provide satisfactory results for non-symmetric PDF of the input and output in the same time and that is why the higher order moments are proposed. Considering numerous applications of the Weibull PDF in the composite material area, the corresponding Second Order Third Moment (W–SOTM) approach proposed for homogeneous media is described below. To illustrate this approach, let us denote the limit state function as g. The expected values, variances and skewnesses of this function are calculated or computed first using up to the second orders of this function, the limit state function derivatives with respect to the input random variables vector b as well as using its probabilistic moments (σ_i as a standard deviation). There holds

$$E[g] = g + \tfrac{1}{2}\sum_{i=1}^{n}\left(\frac{\partial^2 g}{\partial b_i^2}\right)\sigma_i^2 \tag{6.1}$$

$$\sigma^2(g) = \{g\}^2 + \sum_{i=1}^{n}\left[\left(\frac{\partial g}{\partial b_i}\right)^2 + g\frac{\partial^2 g}{\partial b_i^2}\right]\sigma_i^2 + \sum_{i=1}^{n}\left[\frac{\partial g}{\partial b_i}\frac{\partial^2 g}{\partial b_i^2}\right]S_i\sigma_i^2 - E^2[g] \tag{6.2}$$

$$S(g) = \left\{ \{g\}^3 + \frac{3}{2}\sum_{i=1}^{n}\left[2g\left(\frac{\partial g}{\partial b_i}\right)^2 + g^2\frac{\partial^2 g}{\partial b_i^2}\right]\sigma_i^2 \right.$$

$$\left. + \sum_{i=1}^{n}\left[\left(\frac{\partial g}{\partial b_i}\right)^3 + 3g\frac{\partial g}{\partial b_i}\frac{\partial^2 g}{\partial b_i^2}\right]S_i\sigma_i^3 - E^3[g] - 3E[g]\sigma(g)\right\}\frac{1}{\sigma^3(g)} \tag{6.3}$$

These formulae can be derived using the classical perturbation approach described previously. Next, parameters $\bar{x}$, β, λ of the Weibull distribution [8] are obtained as a solution of the following system of equations:

$$E[g] = \lambda\Gamma\left(1+\frac{1}{\beta}\right) + \bar{x} \tag{6.4}$$

$$\sigma(g) = \lambda^2\left[\Gamma\left(1+\frac{2}{\beta}\right) - \Gamma^2\left(1+\frac{1}{\beta}\right)\right] \tag{6.5}$$

$$S(g) = \lambda^3\left[\Gamma\left(1+\frac{3}{\beta}\right) - 3\Gamma\left(1+\frac{2}{\beta}\right)\Gamma\left(1+\frac{1}{\beta}\right) + 2\Gamma^3\left(1+\frac{1}{\beta}\right)\right]\frac{1}{\sigma^2(g)} \tag{6.6}$$

where the Gamma function is defined as

$$\Gamma(x) = \begin{cases} \displaystyle\int_0^\infty e^{-t}t^{x-1}dt \ (for \ x>0) \\[2mm] \displaystyle\lim_{n\to\infty}\frac{n!n^{x-1}}{x(x+1)(x+2)...(x+n-1)} \ (for \ any \ x\in\Re) \end{cases} \tag{6.7}$$

Finally, the reliability index is obtained as

$$R = \exp\left[-\left(\frac{\bar{x}}{\lambda}\right)^\beta\right] \tag{6.8}$$

The application of this type of analysis to a simple two–component composite beam is shown in [179], for instance. From the computational point of view it should be underlined that the mathematical packages for symbolic computation are very useful in inversion of the Gamma function and in obtaining a direct numerical solution of the equations system presented above.

The methodology shown above and applied for homogeneous media can be used for simulation of the composite materials as well. Having proposed a general algorithm for usage of the limit function g, the corresponding various limit functions adequate to composite materials are summarised below. The most simplified and natural formulation of the limit function is a difference between allowable and computed values of the structural state function or functions.

All limit state functions proposed and used for composites can be divided basically into three different groups. The most generalised functions, independent from the composite components type, and even from homogeneity or heterogeneity of a medium and fracture character as well as physical mechanisms of the whole process, can be classified into the first group. The functions included in the second one obey a precise definition of material fracture mechanism in terms of elastoplastic behaviour, crack formation and its propagation into the composite during the whole process. The last group is characterised by the presence of the failure function in the limit function and is therefore usually oriented to the specific groups and types of composite materials.

The most general relations are maximum stress and strain laws formulated in terms of longitudinal and transverse stresses and strain for both compression and tension as follows:

- maximum stress law:

$$g_X(X) = \begin{cases} \sigma_{L,t} - \sigma_X \left(\sigma_X \geq 0\right) \\ \sigma_{L,c} + \sigma_X \left(\sigma_X < 0\right) \end{cases}$$

$$g_y(X) = \begin{cases} \sigma_{L,t} - \sigma_y \left(\sigma_y \geq 0\right) \\ \sigma_{L,c} + \sigma_y \left(\sigma_y < 0\right) \end{cases}$$

$$g_y(X) = \sigma_{LT} - \left|\sigma_S\right|$$

(6.9)

- maximum strain law:

$$g_X(X) = \begin{cases} \varepsilon_{L,t} - \varepsilon_X \left(\varepsilon_X \geq 0\right) \\ \varepsilon_{L,c} + \varepsilon_X \left(\varepsilon_X < 0\right) \end{cases}$$

$$g_y(X) = \begin{cases} \varepsilon_{L,t} - \varepsilon_y \left(\varepsilon_y \geq 0\right) \\ \varepsilon_{L,c} + \varepsilon_y \left(\varepsilon_y < 0\right) \end{cases}$$

$$g_y(X) = \varepsilon_{LT} - \left|\varepsilon_S\right|$$

(6.10)

As can be seen, the limit functions are independent from of composite material type (fibre–reinforced or laminated) as well as from the character of its components (polymer–based, metal matrix, etc.). They originate from the mechanics of homogeneous media. However, brittle or ductile character of material damage is not taken into account in the analysis as well as the possibility of crack formation during the fatigue process. That is why more sophisticated criteria are proposed as, for instance, the one formulated as

$$g(X) = \frac{2X_1}{X_2 + X_3} - \frac{X_1\sqrt{\pi X_6}}{K_{1c}} \frac{8}{\pi_2} \sqrt{\ln \sec\left(\frac{\pi}{2} \frac{2X_1}{X_2 + X_3}\right)} \qquad (6.11)$$

where X_1 is loading stress, X_2 yield strength, X_3 tensile stress, X_4 fracture toughness, X_5 initial crack length, X_6 crack length and calculation of K_{1c} is presented by [91]. This limit state function allows us to combine brittle and ductile fracture type of the analysed material specimen, even in the elastoplastic range. However, as in previous formula, it is quite non-sensitive to the composite material type. Considering that, the limit state functions are combined with the failure stress or strain functions in the form of so−called quadratic polynomial failure criteria, for instance. The limit state functions proposed using such a criterion can be used for the unidirectional composite laminate in both stress and strain formulations:
− Hill−Chamis:

$$g(X) = 1 - \sigma_X^T F_{A,X} \sigma_X , \quad g(X) = 1 - \varepsilon_X^T G_{A,X} \varepsilon_X \qquad (6.12)$$

− Hoffman and Tsai−Wu [352]:

$$g(X) = 1 - \sigma_X^T F_{A,X} \sigma_X - F_{B,X}^T , \quad g(X) = 1 - \varepsilon_X^T G_{A,X} \varepsilon_X - G_{B,X}^T \qquad (6.13)$$

Starting from the equations describing the limit function g, its probabilistic moments are calculated using the formula proposed above, but in such a case the knowledge of failure function probabilistic moments is necessary. In this context, analogous to the previous considerations, the second order perturbation method can be applied to randomise any of the reliability criteria i.e. Tsai−Hill failure criterion.

6.2 Perturbation−based Reliability for Contact Problem

To illustrate the reliability analysis implementation, the stochastic perturbation reliability analysis of the linear elastic contact analysis is carried out for a composite reinforced with spherical particles. Since the solution for the deterministic problem is known and has been worked out analytically, the probabilistic analysis is made using the package MAPLE. The reliability limit function and probabilistic moments of the contact stress computations as well as some sensitivity numerical studies are carried out by the use of this program together with the visualisation of all computed functions. This methodology can be successfully applied for randomisation of all contact problem reliability studies, where contact stresses are described by the closed form equations. Otherwise, Stochastic Finite [88,162] or Boundary Element Method [46,51,185] computational implementations are to be made in order to get general approximate probabilistic solutions for the composite contact problems. Furthermore, the numerical approach to stochastic reliability, stochastic contact modelling and the relevant analytical computation aspects can be applied and explored in various areas of modern engineering, especially in the field of composite materials.

Let us consider the contact phenomenon between two linear elastic isotropic regions characterised by the Young moduli (e_1, e_2) and Poisson ratios (v_1, v_2). Let us assume that the regions have spherical shapes with radii R_1 and R_2, respectively, and that the contact is considered in a point denoted by C, as it is shown in Figure 6.1 below. The 3D view of the particle–reinforced composite plane cross–section is shown in Figure 6.2.

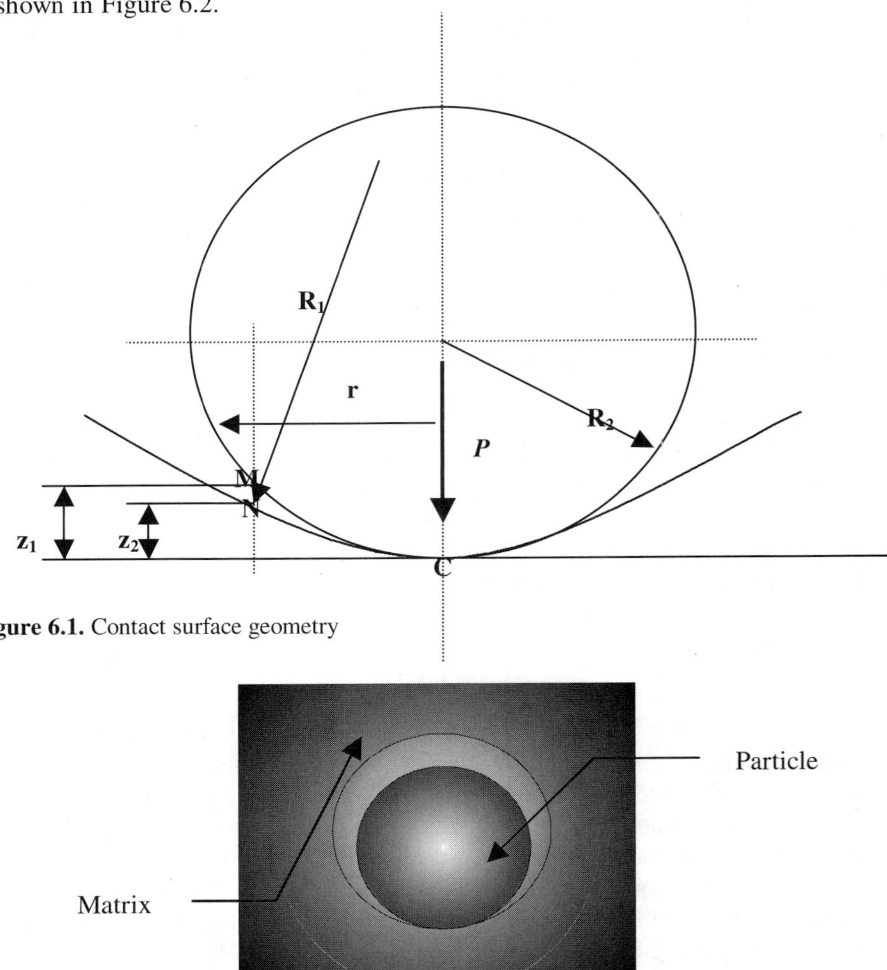

Figure 6.1. Contact surface geometry

Figure 6.2. 3D view of the particle–reinforced composite plane cross–section

Let us observe that the contact problem is axisymmetric with respect to the vertical axis introduced at the centre of the spherical particle and at the bottom of this sphere (Figure 6.2). It is assumed that there is no pressure between the composite constituents and therefore the contact appears at the point C only. The distance between the points on the contacting surfaces and the plane transverse to the vertical axis of both surfaces is assumed to be small and can be described as

$$z_2 - z_1 = \frac{r^2(R_1 - R_2)}{2R_1 R_2}$$
(6.14)

where r denotes the distance between the points M, N and the symmetry axis introduced at C. If the composite is loaded by the vertical force P acting along the vertical axis at the point C, then some local strains are induced in the neighbourhood of this point. They are a result of a contact phenomenon on a small circular surface (contact area). Assuming that the composite constituents radii R_1 and R_2 are sufficiently greater than the radius of the contact area, then the results of the Bussinesq problem of the linear elastic half–space loaded by the concentrated force can be adopted here. For this purpose let us denote by w_1 the vertical displacement induced by the local vertical strain of the point M belonging to the matrix; w_2 is the corresponding displacement of the point N in a vertical direction. Finally, assuming that the tangential plane in point C remains unmovable during a local compression, the close–up of the two points M and N can be expressed by some real η as [344]

$$\eta = \alpha - (w_1 + w_2) = \frac{r^2(R_1 - R_2)}{2R_1 R_2}$$
(6.15)

If M and N belong to the contact area, their displacements w_i for $i=1,2$ can be written as

$$w_i = \frac{1 - v_i^2}{\pi E_i} \iint q \, ds \, d\varphi$$
(6.16)

which follows the symmetry of the pressure intensity q and the corresponding local strains with respect to the vertical axis at the point C. Integration in this formula is carried out over the entire contact surface. Therefore

$$(k_1 + k_2) \iint q \, ds \, d\varphi = \alpha - \frac{r^2(R_1 - R_2)}{2R_1 R_2}$$
(6.17)

Now, we are looking for such an expression for q to fulfil the above equation. It can be obtained for the pressure distribution on the contact surface represented by the coordinates of the hemisphere with the radius a constructed on a contact surface. If q_0 is taken as the pressure at the point C, then one can show that

$$\int q\,ds = \frac{q_0}{a} A \tag{6.18}$$

where

$$A = \frac{\pi}{2}\left(a^2 - r^2 \sin^2 \varphi\right) \tag{6.19}$$

which gives

$$\frac{\pi(k_1 + k_2)q_0}{a} \int_0^{\frac{\pi}{2}}\left(a^2 - r^2 \sin^2 \varphi\right)d\varphi = \alpha - \frac{r^2(R_1 - R_2)}{2R_1 R_2} \tag{6.20}$$

Finally, the parameters a and α can be determined for this problem as

$$\left[\begin{array}{l} a = \sqrt[3]{\dfrac{3\pi}{4}\dfrac{P(k_1 + k_2)R_1 R_2}{R_1 - R_2}} \\[3ex] \alpha = \sqrt[3]{\dfrac{9\pi^2}{16}\dfrac{P^2(k_1 + k_2)^2(R_1 - R_2)}{R_1 R_2}} \end{array}\right. \tag{6.21}$$

which gives maximal pressure on the contact surface equal to

$$q = \frac{3P}{2\pi a^2} \tag{6.22}$$

Then, the normal stress can be defined as

$$\begin{aligned} \sigma_z &= -\int_0^a 3qr\,dr z^3 \left(r^2 + z^2\right)^{-5/2} \\[1ex] &= qz^3\left|\left(r^2 + z^2\right)^{-3/2}\right|_0^a = q\left[-1 + \frac{z^3}{\left(a^2 + z^2\right)}\right] \end{aligned} \tag{6.23}$$

Let us note that the shear stresses are equal to 0, which result from the spherical symmetry of the reinforcing particle. However in the case of ellipsoidal reinforcement the shear stresses differ from 0.

The main purpose of further analysis is to determine the probabilistic characteristics of maximal contact stresses as well as contact surface geometrical parameters. Since the spherical particle surrounding the matrix is considered, let us assume that the difference $R_1-R_2=\varepsilon$ is smaller than R_2. This parameter is treated as input design parameter in further sensitivity analysis. The characteristics mentioned above are necessary in the final stochastic reliability computations and, considering the complexity of the equations describing reliability parameters, the stochastic second order perturbation method is proposed. The second order perturbation follows a traditional approach in this area (and the lack of convergence studies with respect to the Taylor expansion order). The third probabilistic moment method reflects the need of unsymmetric random variables modelling. Adopting the same notation as before (see Chapter 1) the skewness parameter $S(u_i)$ is calculated by

$$S(u_i) = \frac{1}{\sigma^3(u_i)} \int_{-\infty}^{+\infty} (u_i - E[u_i])^3 p_R(b) db \tag{6.24}$$

In further applications, the Weibull distribution is used with the probability density function defined as

$$p_R = \left\{ \begin{array}{l} \dfrac{\beta}{\lambda} \left(\dfrac{x - \bar{x}}{\lambda} \right)^{\beta-1} \exp\left[-\left(\dfrac{x - \bar{x}}{\lambda} \right)^{\beta} \right]; x > \bar{x} \\ \\ 0, x < \bar{x} \end{array} \right. \tag{6.25}$$

where β is the Weibull shape parameter, λ denotes the scale parameter and $\bar{x}$ is the location parameter, which indicates the smallest value of the random variable $\mathbf{x}$ for which the probability density function is positive. Considering this definition, the Weibull PDF is used for general mechanical applications, where many random variables must be nonnegative (Young modulus and some geometrical parameters, for instance) and especially in composite failure and fatigue modelling. Let us note that if discrete representation of a random variable $b(\mathbf{x};\theta)$ is used, then statistical estimators may be applied to approximate any order probabilistic moments of this variable.

Starting from probabilistic moments and the stochastic perturbation methodology presented above, we compute the first three probabilistic orders of the vertical stresses $E[\sigma_z(x;\omega)]$, $Var(\sigma_z(x;\omega))$ and $S(\sigma_z(x;\omega))$ as

$$E[\sigma_z] = \sigma_z^0 + \frac{1}{2} \sum_{i=1}^{n} \left(\frac{\partial^2 \sigma_z}{\partial b_i^2} \right) \sigma^2(b_i) \tag{6.26}$$

$$\sigma^2(\sigma_z) = \sigma_z^2 + \sum_{i=1}^{n} \left[\left(\frac{\partial \sigma_z}{\partial b_i} \right)^2 + \sigma_z \frac{\partial^2 \sigma_z}{\partial b_i^2} \right] \sigma^2(b_i)$$

$$+ \sum_{i=1}^{n} \left(\frac{\partial \sigma_z}{\partial b_i} \frac{\partial^2 \sigma_z}{\partial b_i^2} \right) S(b_i) \sigma^3(b_i) - E^2[\sigma_z] \tag{6.27}$$

and

$$S(\sigma_z) = \left\{ \sigma_z^3 + \frac{3}{2}\sum_{i=1}^{n}\left[2\sigma_z\left(\frac{\partial\sigma_z}{\partial b_i}\right)^2 + \sigma_z^2\frac{\partial^2\sigma_z}{\partial b_i^2}\right]\sigma^2(b_i) \right.$$

$$+ \left(\sum_{i=1}^{n}\left[\left(\frac{\partial\sigma_z}{\partial b_i}\right)^3 + 3\sigma_z\frac{\partial\sigma_z}{\partial b_i}\frac{\partial^2\sigma_z}{\partial b_i^2}\right]S(b_i)\sigma^3(b_i)\right)\frac{1}{\sigma^3(\sigma_z)} \qquad (6.28)$$

$$\left. - E^3[\sigma_z] - 3E[\sigma_z]\sigma^2(\sigma_z) \right\}\frac{1}{\sigma^3(\sigma_z)}$$

Having computed the first three probabilistic moments of contact stresses (expected values, standard deviations and skewness coefficients), the random field of the limit function $g(z;\omega)$ is to be proposed. Usually, it can be introduced as a difference between allowable and actual stresses $\sigma_z(z;\omega)$ induced in the composite as

$$g(z;\omega) = \sigma_{all}(\omega) - \sigma_z(z;\omega) \qquad (6.29)$$

Let us underline that allowable stresses are most frequently analysed as random variables in the interior of statistically homogeneous materials, whereas actual stresses are random fields. That is why the computational analysis presented later is carried out for the specific value of the vertical coordinate z. The random variable of allowable stresses $\sigma_{all}(\omega)$ is specified by the use of the first three probabilistic moments $E[\sigma_{all}(\omega)]$, $Var(\sigma_{all}(\omega))$ and $S(\sigma_{all}(\omega))$. Then, the corresponding probabilistic characteristics of the limit function are calculated as

$$E[g] = g^0 + \frac{1}{2}\sum_{i=1}^{n}\left(\frac{\partial^2 g}{\partial b_i^2}\right)\sigma^2(b_i) \qquad (6.30)$$

$$\sigma^2(g) = \left(g^0\right)^2 + \sum_{i=1}^{n}\left[\left(\frac{\partial g}{\partial b_i}\right)^2 + g^0\frac{\partial^2 g}{\partial b_i^2}\right]\sigma^2(b_i) \qquad (6.31)$$

$$+ \sum_{i=1}^{n}\left(\frac{\partial g}{\partial b_i}\frac{\partial^2 g}{\partial b_i^2}\right)S(b_i)\sigma^3(b_i) - E^2[g]$$

as well as

$$S(g) = \left\{ \left(g^0\right)^3 + \frac{3}{2}\sum_{i=1}^{n}\left[2g^0\left(\frac{\partial g}{\partial b_i}\right)^2 + \left(g^0\right)^2\frac{\partial^2 g}{\partial b_i^2}\right]\sigma^2(b_i) \right.$$

$$+ \left(\sum_{i=1}^{n}\left[\left(\frac{\partial g}{\partial b_i}\right)^3 + 3g^0\frac{\partial g}{\partial b_i}\frac{\partial^2 g}{\partial b_i^2}\right]S(b_i)\sigma^3(b_i)\right)\frac{1}{\sigma^3(g)} \tag{6.32}$$

$$\left. - E^3[g] - 3E[g]\sigma^2(g) \right\}\frac{1}{\sigma^3(g)}$$

Inserting the limit state function g from (6.29) into (6.30)–(6.32) and assuming that the random variable of allowable stresses and the random field of actual stresses are uncorrelated it is obtained that

$$E[g] = \sigma_{all}^0 - \sigma_z^0 - \frac{1}{2}\sum_{i=1}^{n}\left(\frac{\partial^2\sigma_z}{\partial b_i^2}\right)\sigma^2(b_i) \tag{6.33}$$

$$\sigma^2(g) = \left(\sigma_{all}^0 - \sigma_z^0\right)^2 + \sum_{i=1}^{n}\left[\left(\frac{\partial\sigma_z}{\partial b_i}\right)^2 - \left(\sigma_{all}^0 - \sigma_z^0\right)\frac{\partial^2\sigma_z}{\partial b_i^2}\right]\sigma^2(b_i)$$

$$+ \sum_{i=1}^{n}\left(\frac{\partial\sigma_z}{\partial b_i}\frac{\partial^2\sigma_z}{\partial b_i^2}\right)S(b_i)\sigma^3(b_i) - E^2\left[\sigma_{all} - \sigma_z\right] \tag{6.34}$$

and, finally

$$S(g) = \left\{ \left(\sigma_{all}^0 - \sigma_z^0\right)^3 \right.$$

$$+ \frac{3}{2}\sum_{i=1}^{n}\left[2\left(\sigma_{all}^0 - \sigma_z^0\right)\left(\frac{\partial\sigma_z}{\partial b_i}\right)^2 + \left(\sigma_{all}^0 - \sigma_z^0\right)^2\frac{\partial^2\sigma_z}{\partial b_i^2}\right]\sigma^2(b_i)$$

$$+ \sum_{i=1}^{n}\left[\left(\frac{\partial g}{\partial b_i}\right)^3 + 3g^0\frac{\partial g}{\partial b_i}\frac{\partial^2 g}{\partial b_i^2}\right]S(b_i)\sigma^3(b_i) \tag{6.35}$$

$$\left. - E^3\left[\sigma_{all}^0 - \sigma_z^0\right] - 3E\left[\sigma_{all}^0 - \sigma_z^0\right]\sigma^2\left(\sigma_{all}^0 - \sigma_z^0\right) \right\}\frac{1}{\sigma^3\left(\sigma_{all}^0 - \sigma_z^0\right)}$$

Comparing the second order second moment (SOSM) approach with the second order third moment (SOTM) approach, it is seen that the expected values are described by exactly the same equation, while standard deviations (or variances) have some extra components connected with the skewness of analysed PDF; the third order parameter of the output PDF is taken into account in the SOTM−based analysis [282].

Computational experiments are conducted by the use of the symbolic computation system MAPLE, where the stochastic second order perturbation method in W–SOTM reliability analysis of the contact problem has been implemented. The entire analysis is divided into three groups of essentially various numerical examples: (1) deterministic analysis and sensitivity study of a contact problem with respect to the vertical spatial coordinate, (2) stochastic second order perturbation based numerical modelling by randomising most of the input problem parameters and (3) stochastic numerical modelling according to the Weibull second order third moment approach.

Deterministic analysis (Figure 6.3) and the sensitivity of contact stresses in a two-component composite with spherical particles is verified with respect to the vertical spatial coordinate. The following data are adopted for the computational analysis: e_2=2.0E9, v_1=0.3, v_2=0.2, R_2=0.18, P=10.0E5, α=e_1/e_2=2.0~8.0, β=R_1/R_2=1.001−1.01.

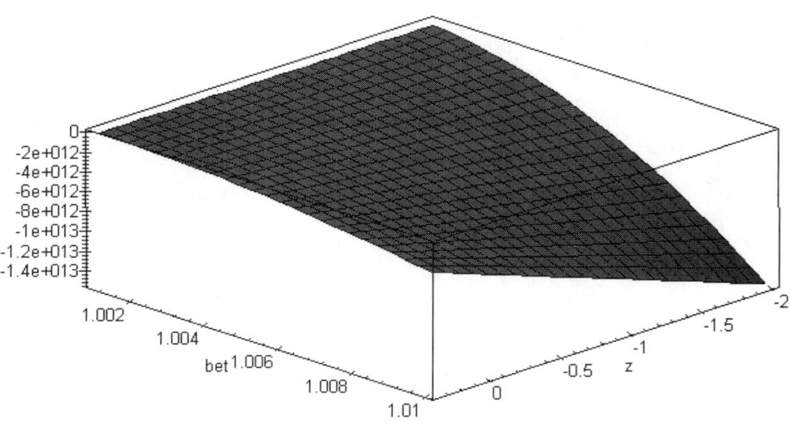

Figure 6.3. Contact stresses for the spherical particle–reinforced composite

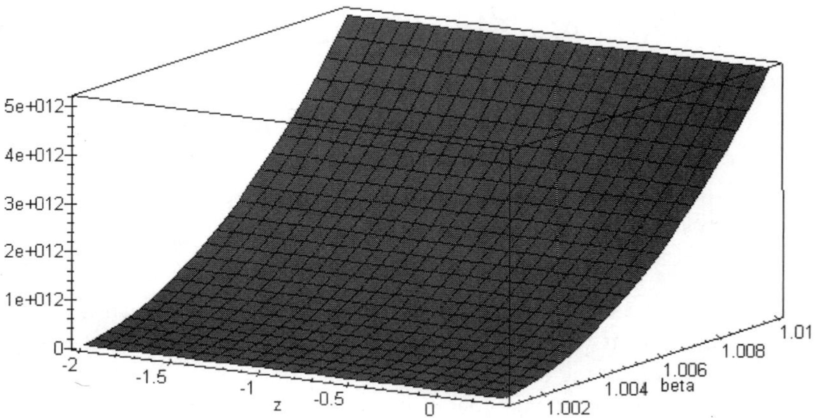

Figure 6.4. Sensitivity of contact stresses to vertical spatial coordinate 'z'

Computational analysis of vertical contact stresses and their sensitivity gradients ($d\sigma_z/dz$) with respect to the spatial coordinate is presented in Figure 6.4. The reinforcement ratio (α) and the radii ratio (β) are marked on the vertical axes in Figures 6.3 and 6.4. The compressive contact stresses are most sensitive to the parameter β for its value tending to 0 (matrix perfectly surrounds the matrix) and for the parameter α tending to 1 (Young modulus of the reinforcing particle tends to the matrix Young modulus). One of the main benefits of the MAPLE computations, i.e. visualisation of the stress variations and their sensitivity gradients, can be studied in these figures.

All the input parameters of the analysed contact problem are treated as random variables: Young moduli and Poisson ratios of the composite components as well as their radii. Deterministically calculated vertical stresses in the contact area are compared below with expected values, standard deviations and probabilistic envelope values of the vertical contact stresses for $z=0.1$. The standard deviations of the variables are taken in the range of 10% of the corresponding expectations; all variables are assumed to be uncorrelated.

The computed deterministic contact stresses are shown in Figures 6.5, 6.9, 6.13 for the particle centre ($z=0.09$) and for the matrix ($z=-0.5$ and $z=-2.0$, respectively). The expected values of contact stresses are shown in Figures 6.6, 6.10 and 6.14 for the same points, the standard deviations are shown in Figures 6.7, 6.11 and 6.15, while the probabilistic envelopes for these stresses surfaces are presented in Figures 6.8, 6.12 and 6.16. The vertical contact stress parameters are marked on the vertical axes; horizontal axes define the reinforcement ratio of the composite (α) and the ratio between particle and the surrounding matrix radii (β). All the surfaces shown in these figures have the same character and the variability with respect to the input parameters α and β, apart from the standard deviations plots.

Analysing Figures 6.5–6.6, 6.9–6.10 and 6.13–6.14, it is seen that the expected values of contact stress surfaces are quite close to those obtained in the corresponding deterministic analyses. Essential differences are observed between Figures 6.5–6.8, 6.9–6.12 and 6.13–6.16, where probabilistic envelopes of these stresses are shown. These envelopes are determined for a particular $\mathbf{x}$ on the basis of the results presented in Figs. 6.6–6.7, 6.10–6.11 and 6.14–6.15 as

$$Env(f(x); x) = E[f(x); x] - 3sig(f(x); x) \qquad (6.36)$$

Let us note that (6.36) is frequently used in the Stochastic Finite Element computations and stochastic fatigue analysis. The values of probabilistic envelope surfaces are significantly smaller than the corresponding values obtained from deterministic analysis, which means that stochastic perturbation based computational analysis more restrictive than the classical model as well as the corresponding expected values. All the surfaces combined in the probability envelope show that vertical stresses tend to 0 for the reinforcement ratio and matrix radius tending to a spherical particle radius. Comparing all deterministic and stochastic results, it is clear that the contact stresses are most sensitive to the vertical spatial coordinate.

Analysing Figures 6.5–6.8, 6.10–6.12 and 6.14–6.16 in terms of the contact stress variations with respect to the composite reinforcement ratio, it is observed that the greatest sensitivity appears for $\alpha \to 1$, which means that the greatest variations of examined probabilistic stresses are obtained for the homogeneous contact problem. Further numerical sensitivity analysis with respect to the Poisson ratio interrelations of both composite components is necessary.

The computational study on structural reliability, proposed in the theoretical considerations on structural reliability, is the main subject of the next example. The set of input data together with their probabilistic characteristics is given in Table 6.1 for the same composite contact problem as before. The Weibull probability density function (PDF) of the limit function is determined together with its up to third order probabilistic moments (cf. Table 6.1) obtained by a symbolic computational solution of the nonlinear equations system (6.30)–(6.32). The PDF of a limit function is presented in Figure 6.17 – probabilistic vertical stresses are shown on the horizontal axis, while the probability on the vertical axis.

First, it can be seen that even for a relatively small input coefficient of variation of input parameters (not greater than 0.1), the randomness level of the output function is about 18% of the relevant expected value. That is why the proposed third order approach is more accurate for the analysed contact problem. Furthermore, we observe that even for input skewnesses equal to 0, the corresponding third order probabilistic characteristics differ from 0, which reflects the differences in algebraic combinations of lower order characteristics. In further analysis it is necessary to verify the sensitivity (both in deterministic and stochastic context) of output Weibull PDF probabilistic moments with respect to all input mechanical parameters and their random characteristics. At the same time, the cross−correlation function of contact stresses can be symbolically computed using the program MAPLE.

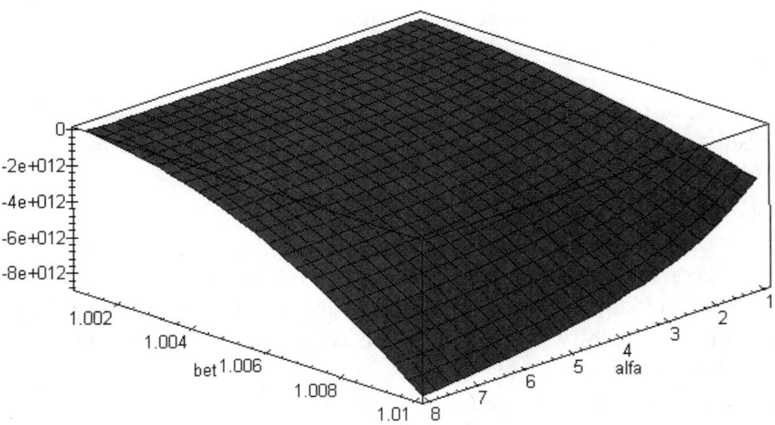

Figure 6.5. Deterministic contact stresses (z=0.018)

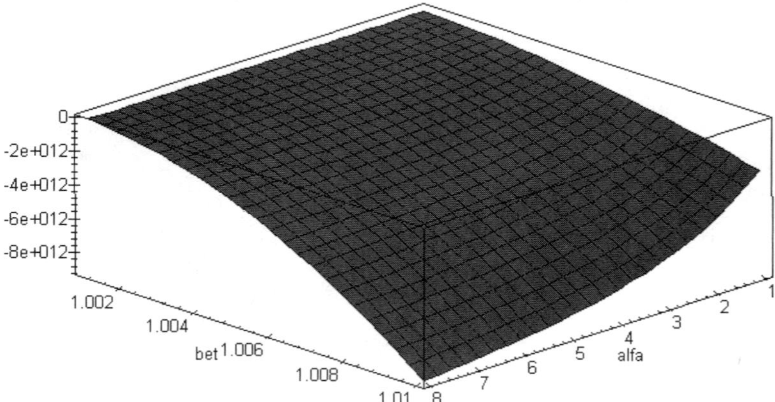

Figure 6.6. Expected values of contact stresses (z=0.018)

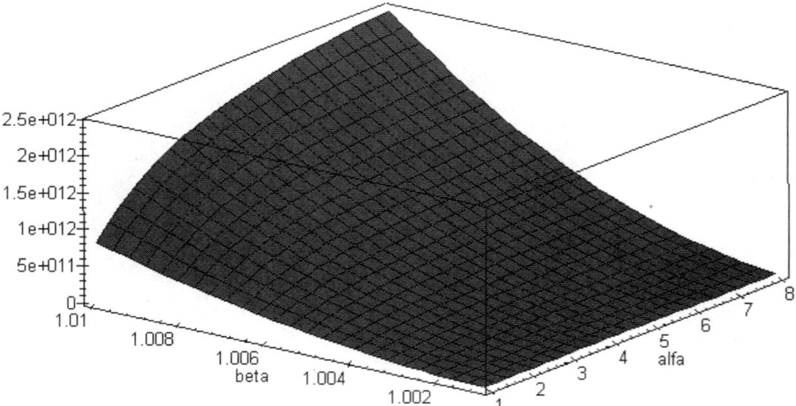

Figure 6.7. Standard deviations of contact stresses (z=0.018)

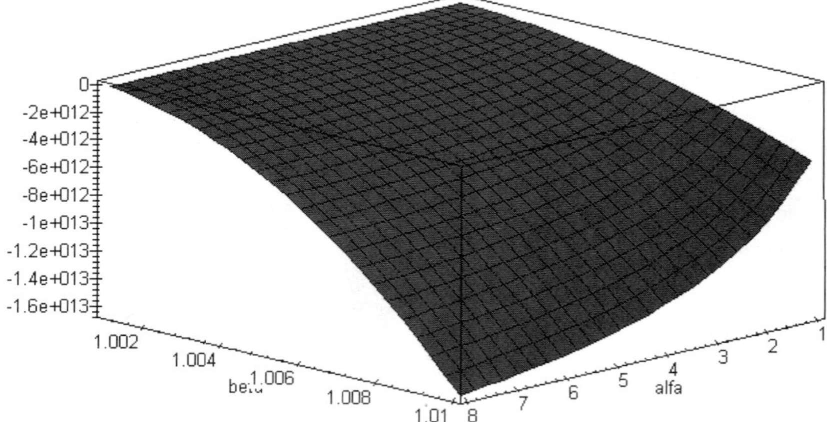

Figure 6.8. Probabilistic envelope of contact stresses (z=0.018)

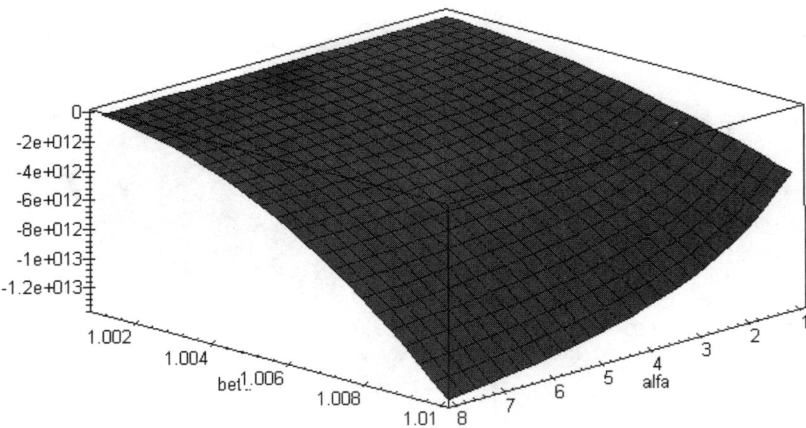

Figure 6.9. Deterministic contact stresses (z=-0.5)

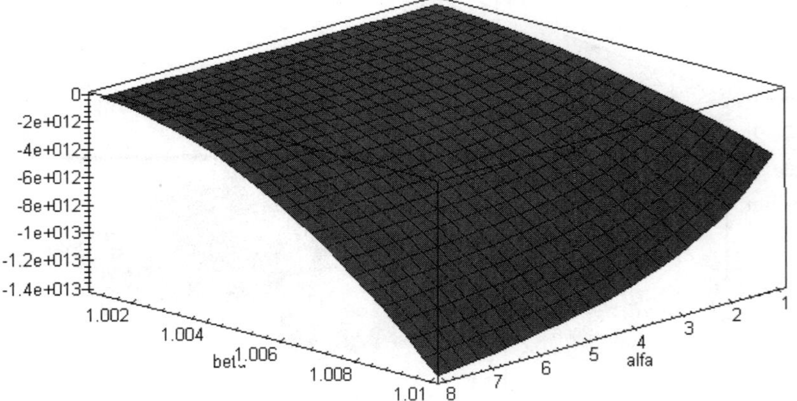

Figure 6.10. Expected values of contact stresses (z=-0.5)

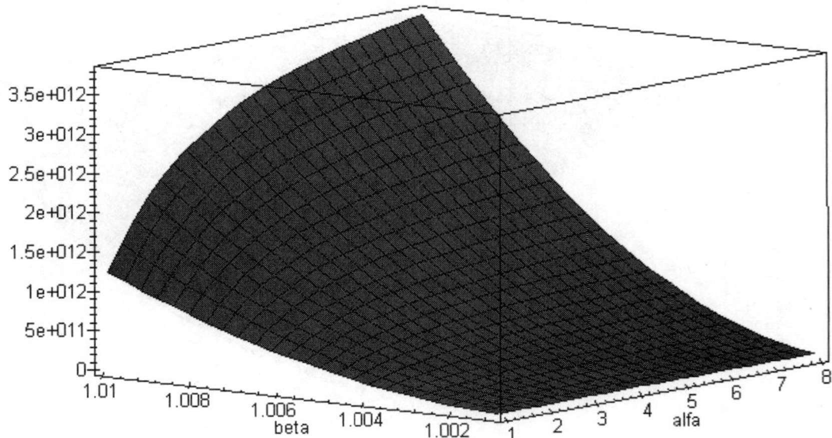

Figure 6.11. Standard deviations of contact stresses (z=-0.5)

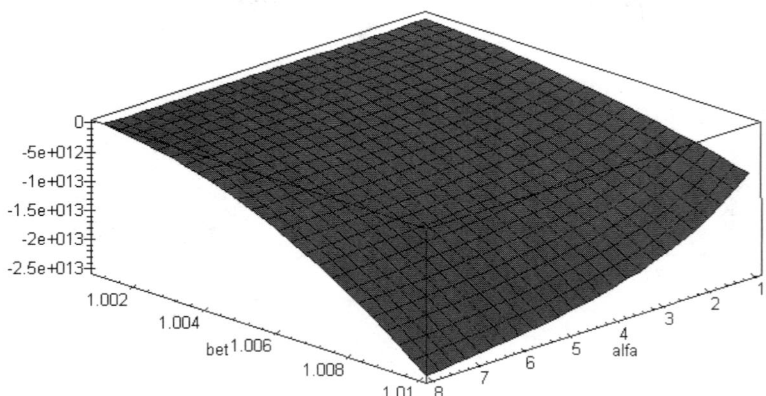

Figure 6.12. Probabilistic envelope of contact stresses (z=-0.5)

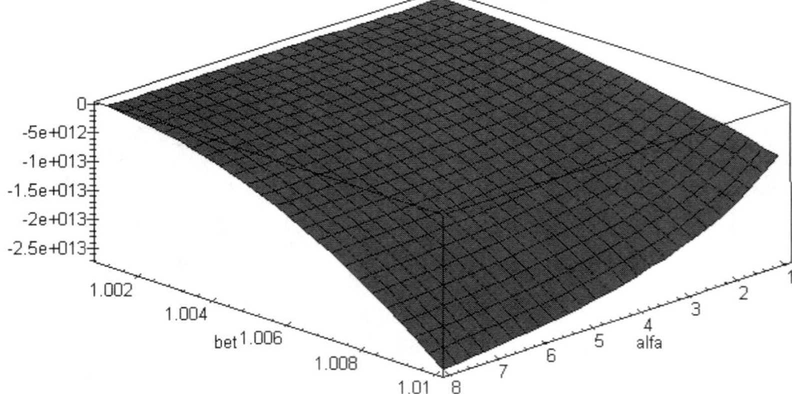

Figure 6.13. Deterministic contact stresses (z=-2.0)

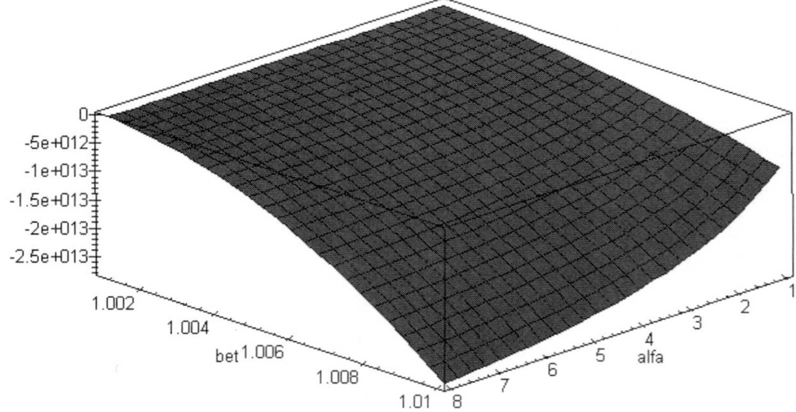

Figure 6.14. Expected values of contact stresses (z=-2.0)

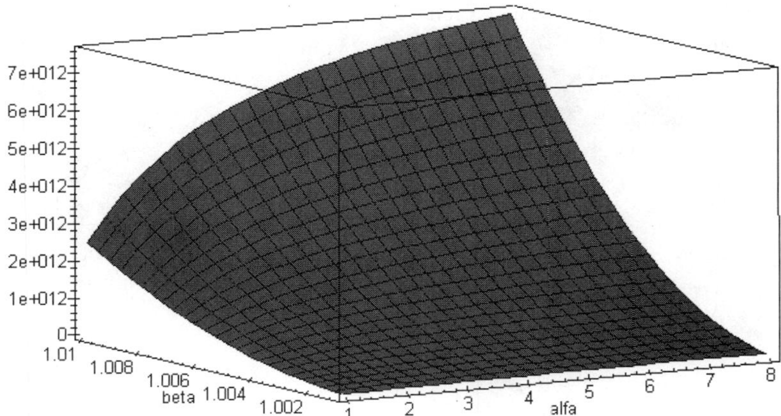

Figure 6.15. Standard deviations of contact stresses (z=-2.0)

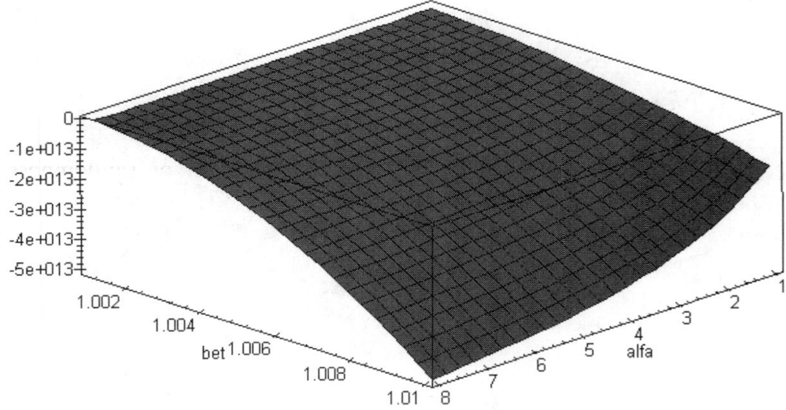

Figure 6.16. Probabilistic envelope of contact stresses (z=-2.0)

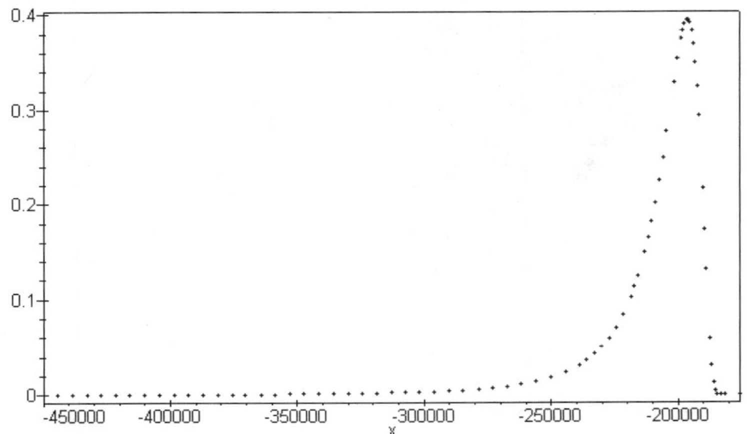

Figure 6.17. Equivalent Weibull distribution for the limit function

Table 6.1. Probabilistic input data for reliability index computations

Parameter	Value
E_2	2.0E6
v_1	0.3
v_2	0.2
R_2	1.8
P	5.0E2
Z	-0.018
$\sigma(e_2)$	0.2E6
$S(e_2)$	0.0
$\sigma(R_2)$	0.018
$S(R_2)$	0.0
σ_{all}	-4.0E5
α	10.0
β	1.01
$E[g]$	-211378.33
$\sigma(g)$	38213.61838 (α=0.18)
$S(g)$	5.158577

The analysis presented above reflects various sources of randomness and stochasticity in contact problems of the spherical particle reinforced composites. In comparison to the second order second probabilistic moment approach, third order probabilistic moments of both input and output parameters are analysed. It is demonstrated that even for skewnesses of the inputs equal to 0, the output third order probabilistic moments in reliability studies slightly differ from 0. It results from the main idea of the SOTM approach and from the interrelations between lower order probabilistic characteristics. Further, it is observed that deterministic values of the state functions are quite close to the computed expected values. They are considerably greater and well approximated by their probabilistic envelopes, which confirms the usefulness of these envelopes in various stochastic numerical experiments.

The most interesting extension of this study would be introducing: (1) the randomness of non-spherical contact surface (ellipsoidal one) and, next, (2) more realistic incremental Stochastic Finite or Boundary Element Method (SFEM or SBEM, respectively) of nonlinear geometry of the contacting surface. Next, the application of a computational W–SOTM reliability study in various numerical analyses of composites would be interesting, too. Neglecting relatively simple character of the deterministic contact problem, the geometrical sensitivity of the contact stress values is decisive for this analysis, both in deterministic and stochastic cases. Considering the above, one can have a conclusion that the stochastic second order perturbation analysis in a conjunction with mathematical symbolic computations is a very powerful stochastic computational tool. However, the limitations on the input randomness level typical for such an analysis must be fulfilled [208].

6.3. Stochastic Model of Degradation Process

Let us consider an engineering system $\Omega \in \Re^3$ under stochastic degradation processes (SDP) with n uncorrelated components $D(\mathbf{x};\omega;t) = D_i(\mathbf{x};\omega;t)$, $i=1,...,n$ where $t \in [0,\infty)$. It is assumed that for every $\tau \in [0,\infty)$ the components $D_i(\mathbf{x};\omega;\tau)$ are Gaussian random variables, i.e. they are uniquely defined by their first two probabilistic moments: the expected values $E[D_i(\mathbf{x};\omega;\tau)]$ and the variances $Var(D_i(\mathbf{x};\omega;\tau))$. Due to the uncorrelation assumption, the covariance matrix for any $\tau \in [0,\infty)$ between the SDP components, it yields [176]

$$Cov(D_i(\mathbf{x};\omega;\tau); D_j(\mathbf{x};\omega;\tau)) = 0; \quad i,j=1,...,n, \quad i \neq j \tag{6.37}$$

Moreover, because of the lack of respective experimental results, we assume that there are no time correlations between the SDP components

$$Cov(D_i(\mathbf{x};\omega;\tau^{(1)}); D_j(\mathbf{x};\omega;\tau^{(2)})) = 0; \quad i,j=1,...,n \tag{6.38}$$

However, in contrast to the above, the spatial correlation of particular components have non−zero values

$$Cov(D_i(\mathbf{x}^{(1)};\omega;\tau); D_i(\mathbf{x}^{(2)};\omega;\tau)) \neq 0 \tag{6.39}$$

and are computed by use of the statistical estimation methods. Let us note that, from an engineering point of view, every $D_i(\mathbf{x};\omega;t)$ for $i=1,...,n$ represents some material (elastic characteristics or yield stress) or geometrical properties (section area, element thickness) of the system Ω under considerations.

Further, let us assume that all SDP components are statistically measured (obtained in the experimental way) in the moments $t_1,t_2,...,t_m$, for some $m \in N$. On the basis of a measured M series of these components, the basic statistical parameters are estimated by use of the following formulae:

- the expected values estimator:

$$E[D_i(\mathbf{x};\omega;t_k)] = \frac{1}{M}\sum_{i=1}^{M} D_i^{(j)}(\mathbf{x};\omega;t_k) \tag{6.40}$$

where $D_i^{(j)}(\mathbf{x};\omega;t_k)$ denotes the jth measurement of the ith SDP component in the moment t_k;

- unbiased variance estimator of the ith SDP component in the moment t_k:

$$Var(D_i(\mathbf{x};\omega;t_k)) = \frac{1}{M-1} \sum_{j=1}^{M} \left\{ D_i^{(j)}(\mathbf{x};\omega;t_k) - E[D_i(\mathbf{x};\omega;t_k)] \right\}^2 \qquad (6.41)$$

- standard deviation of the ith SDP component in the moment t_k :

$$\sigma(D_i(\mathbf{x};\omega;t_k)) = \sqrt{Var(D_i(\mathbf{x};\omega;t_k))} \qquad (6.42)$$

- coefficient of variation of the ith SDP component in the moment t_k :

$$\alpha(D_i(\mathbf{x};\omega;t_k)) = \sqrt{\frac{Var(D_i(\mathbf{x};\omega;t_k))}{E^2[D_i(\mathbf{x};\omega;t_k)]}} \qquad (6.43)$$

- covariance matrix estimator of the ith SDP component in the moment t_k :

$$\begin{aligned} &Cov\left(D_i\left(\mathbf{x}^{(1)};\omega;t_k\right), D_i\left(\mathbf{x}^{(2)};\omega;t_k\right)\right) \\ &= \frac{1}{M-1} \sum_{j=1}^{M} \left(D_i^{(j)}\left(\mathbf{x}^{(1)};\omega;t_k\right) - E\left[D_i^{(j)}\left(\mathbf{x}^{(1)};\omega;t_k\right)\right]\right) \\ &\times \left(D_i^{(j)}\left(\mathbf{x}^{(2)};\omega;t_k\right) - E\left[D_i^{(j)}\left(\mathbf{x}^{(2)};\omega;t_k\right)\right]\right) \end{aligned} \qquad (6.44)$$

Next, on the basis of all statistical estimators of the SDP components $D_i(\mathbf{x};\omega;t)$ computed for the moments $t_1, t_2, \ldots, t_m$, let us introduce the polynomial approximation of the respective probabilistic moments. This approximation is shown for the example of the expected values and the variances:

$$E[D_i(\mathbf{x};\omega;t)] = \sum_{i=1}^{p} A_p \cdot t^p \ ; \ p \le k \qquad (6.45)$$

$$Var(D_i(\mathbf{x};\omega;t_k)) = \sum_{i=1}^{q} B_q \cdot t^q \ ; \ q \le k \qquad (6.46)$$

where the coefficients A_p, B_q depend on estimated values of the probabilistic moments approximated in the moments $t_1, t_2, \ldots, t_m$. Thus, on the basis of discrete values of these moments, their continuous time functions are obtained. It should be underlined that (6.45) and (6.46) enable us generally to provide an extrapolation of the expected values and variances which is the basis of the approach proposed.

Finally, let us introduce the following time continuous functions, being stochastic upper $U^{(i)}(x;\tau)$ and lower $L^{(i)}(x;\tau)$ bounds for every SDP components $D_i(\mathbf{x};\omega;t)$ in the form

$$\forall_{\substack{D_i(x;\omega;\tau)\,U^{(i)}(x;\tau) \\ L^{(i)}(x;\tau)}} \exists \; P\big(D_i(x;\omega;\tau)L^{(i)} \le D_i(x;\omega;\tau) \le U^{'(i)}\big) \cong 1 \tag{6.47}$$

To obtain these bounds for some $\tau \in [0,\infty)$, the well–known following bounds for Gaussian variables are used

$$U^{(i)}(x;\tau) = E[D_i(x;\omega;\tau)] + 3\sqrt{Var(D_i(x;\omega;\tau))} \tag{6.48}$$

$$L^{(i)}(x;\tau) = E[D_i(x;\omega;\tau)] - 3\sqrt{Var(D_i(x;\omega;\tau))} \tag{6.49}$$

It should be noted that the interval $\big[L^{(i)}(x;\tau), U^{(i)}(x;\tau)\big]$ can be contracted by decreasing the coefficient multiplied by the standard deviations of $D_i(\mathbf{x};\omega;\tau)$ in (6.48) and (6.49). However the probability value specified in (6.47) will decrease respectively as a result.

As was stated above, the main purpose of our analysis is to make a prognosis of the stochastic reliability and failure time and/or to compute the safety interval for the respective design parameters of the engineering system Ω considered. Taking this into account, there are two kinds of boundary conditions: the *1*st kind, of stress (load capacity conditions) and the 2nd kind, of displacement type (service conditions). Finally, the following inequalities are to be verified simultaneously to find out the time prognosis of the engineering structural safety:

$$\begin{cases} U\big[\sigma_{max}(x;\omega;t)\big] \le L\big[\sigma_{all}(x;\omega;t)\big] \\ U\big[\mathbf{u}_{max}(x;\omega;t)\big] \le L\big[\mathbf{u}_{all}(x;\omega;t)\big] \end{cases} \tag{6.50}$$

where u_{max} and σ_{max} are maximal values of displacements and stresses, while quantities indexed with 'all' are allowable values. Solving the set of inequalities (6.50) iteratively with given time increment Δt, the failure time t_f can be found as such a value, for which one of these inequalities does not hold as the first one.

It should be noted that these inequalities are based on the comparison of the upper bounds of the maximal stresses and displacement stochastic processes and the lower bounds of the allowable stresses and displacement stochastic processes. Moreover, the lower bounds from the right sides of the system (6.50) can be derived on the basis of the given SDP components $D_i(\mathbf{x};\omega;\tau)$ or given explicitly as deterministic values being an effect of simplified engineering calculations. On the other hand, the probabilistic moments of the maximal stresses and displacements can be evaluated by the collocation of the simulation technique or stochastic perturbation method with analytical solutions of the given problem or various numerical methods. Finally, let us note that the methodology presented can be efficiently used in conjunction with stochastic fatigue and fracture theories [89,377] and can extend the existing probabilistic strength models [142].

7 Multiresolutional Wavelet Analysis

7.1 Introduction

Multiscale analysis based on wavelet analysis, being a very modern and extensively developed numerical technique in signal theory [147,148,380], even in probabilistic context [289], introduces the capability to analyse the composite systems with multiple geometrical scales, which is very realistic for most engineering composites (the scales of microdefects, interface, reinforcement and the entire structure). Nowadays, this technique is employed in porous materials modelling [104], general FEM and BEM solutions for boundary problems [119], in vibration analysis [235] as well as in crack detection and impact damages [293,331,343], for instance. Figure 7.1 below presents the MATLAB illustration of the signal that can be interpreted as the information about the variability of heterogeneous medium physical properties in time (and/or in space). It is seen how such a signal can be decomposed using discrete wavelet transforms on the partially homogeneous parameters on different levels [169,170]. After such a decomposition, the traditional or wavelet–based discrete numerical methods can be applied for computational physical modelling.

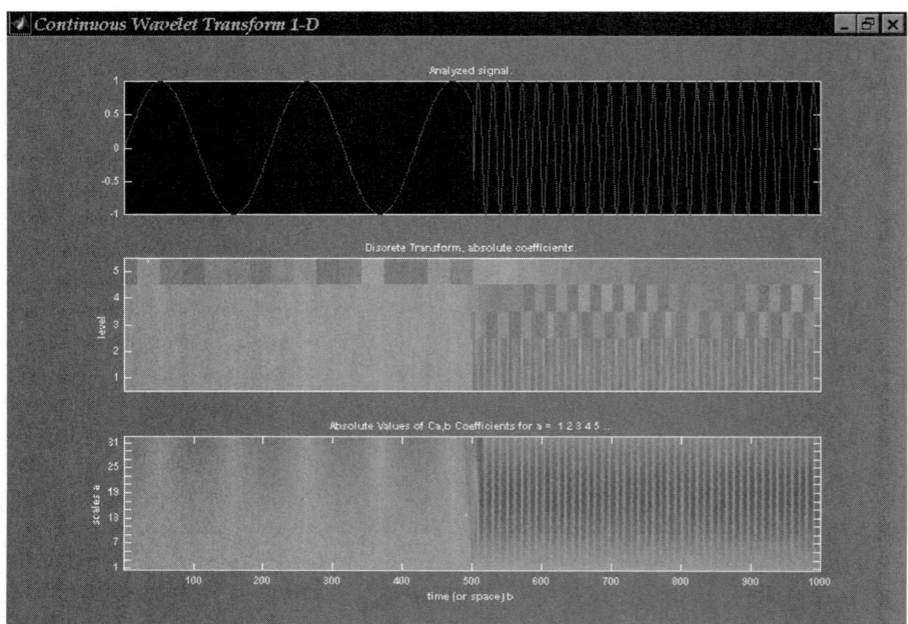

Figure 7.1. Discrete and continuous wavelet signal transform

The homogenisation method is still the most efficient way of computational modelling of composite systems. Usually it is assumed that there exists some scale relation between composite components and the entire system – two scales are introduced that are related by a scale parameter being some small real value tending most frequently to 0. An essential disadvantage of all these techniques is the impossibility of sensitivity analysis of composite homogenised characteristics with respect to geometrical scales relations.

Wavelet analysis became very popular in the area of composite materials modelling because of their multiscale and stochastic nature. The most interesting issue is composite global behaviour, which is more important than the multiphysical phenomena appearing at different levels of their complicated multiscale structure. That is why it is necessary to build an efficient mathematical and numerical multiresolutional algorithm to analyse composite materials and structures.

As is known, two essentially different ways are proposed to achieve this goal. First, the composite can be analysed directly using the wavelet decomposition–based FEM approach where the multiresolutional analysis can recover the material properties of any component at practically any geometrical level. The method leads to an exponential increase of the total number of degrees of freedom in the model – each new decomposition level increases this number.

Alternatively, a multiscale homogenisation algorithm can be applied to determine effective material parameters of the entire composite and next, to carry out the classical FEM or other related method–based computations. The basic difference between these two approaches is that the wavelet decomposition and construction algorithms are incorporated into the matrix FEM computations in the first method. The second method is based on the determination of the effective material parameters and Finite Element analysis of the equivalent homogeneous system, where the dimensions of the original heterogeneous and homogenised problems are almost the same. An analogous two methodologies had been known before the wavelet analysis was incorporated in engineering computations. However the homogenisation method assumptions dealing with the interrelations between macro– and microscales were essentially less realistic.

Considering the above, the aim of this chapter is to demonstrate the use of the wavelet–based homogenisation method in comparison with its preceding classical formulations. Effective material parameters of a periodic composite beam are determined symbolically in MAPLE and next, the temporal and spatial variability of thermal responses of homogenised systems are determined numerically and compared with the real structure behaviour. It is assumed here that material properties are temperature–independent, which should be extended next to the thermal–dependent behaviour. As is verified by the computational experiments, all homogenisation methods (classical and multiresolutional) give a satisfactory approximation of real heat transfer phenomena in the multiscale heterogeneous structure. The approach should be verified next for other types of composites as well as various physical and structural problems in both a deterministic and stochastic context. Separate studies should be carried out for the computer implementation of wavelet analysis in the Finite Element Method programs and comparison with the multiscale algorithm.

Further, we demonstrate the application of the wavelet–based homogenisation method in comparison with its preceding classical formulation. Effective material parameters of the periodic composite beam are determined symbolically in MAPLE and next, the structural responses of the linear elastic homogenised systems are determined numerically and compared with the real structure vibrations. The eigenproblems for various combinations of the effective parameters are computed thanks to the specially adopted Finite Element Method computer code to determine the most efficient homogenisation method for the periodic multiscale composite. It is done for two–, three– and five–bay free supported periodic composite beams having their applications in the aerospace industry as well as in the modelling of bridge vibrations, for instance. As is verified by the computational experiments, the homogenisation methods (classical and multiresolutional) give a satisfactory approximation of the periodic composite beam eigenfrequencies. The approach should be verified next for other types of structures as well as for other structural problems in both deterministic and probabilistic context.

Wavelet analysis is an especially promising tool in the domain of composite materials. It enables: (1) constructing the multiscale heterogeneous structures using particular wavelets which has to perfectly reflect the manufacturing process, for instance, and (2) multidimensional decomposition of the spatial distribution of composite materials and physical properties by the use of the wavelets of various types defined in different scales (heat conductivity or Young modulus along the heterogeneous specimen). The first opportunity corresponds to the analysis of experimental results (image analysis of composite morphology), while the second reflects the theoretical and computational analysis.

Let us notice that the wavelet analysis introduces new meaning for the term *composite*. In the view of the analysis below we can distinguish homogeneous materials from composites using the following definition: *the composite material and/or structure is such a heterogeneous continuum in which material or physical properties are related in macro– and microscales by at least a single wavelet transform.* This definition extends traditional, rather engineering approach to composites where laminated or fibre–reinforced structures were considered (partially constant character of material characteristics) to those media with sinusoidal variability in one direction of these properties at least (see Figures 7.2–7.7 below). Figure 7.2 shows the spatial variability of the Young modulus using the following wavelet function [188]:

$$e(x) = e_0 + \sin\left(\frac{\pi x}{l}\right) + 0.1\sin\left(\frac{\pi 10^2 x}{l}\right) + 0.1\sin\left(\frac{\pi 10^4 x}{l}\right), \quad l=10.$$

The next figures present the contributions of various scales to the macroscale elastic characteristic of the entire composite structure.

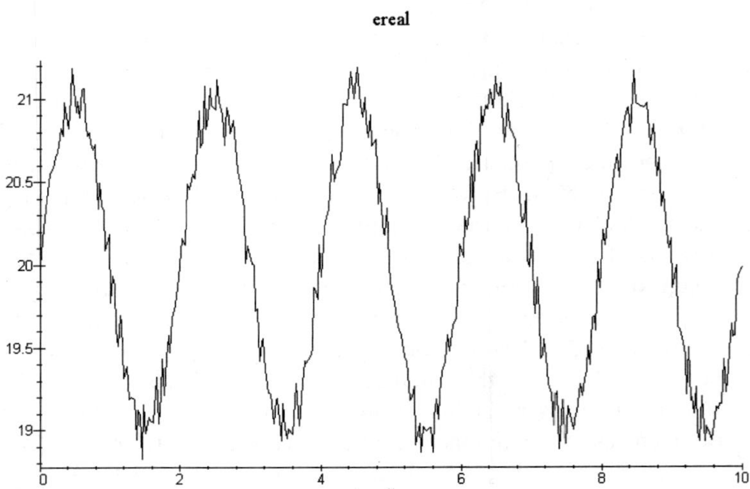

Figure 7.2. Distribution of the Young modulus in the real composite

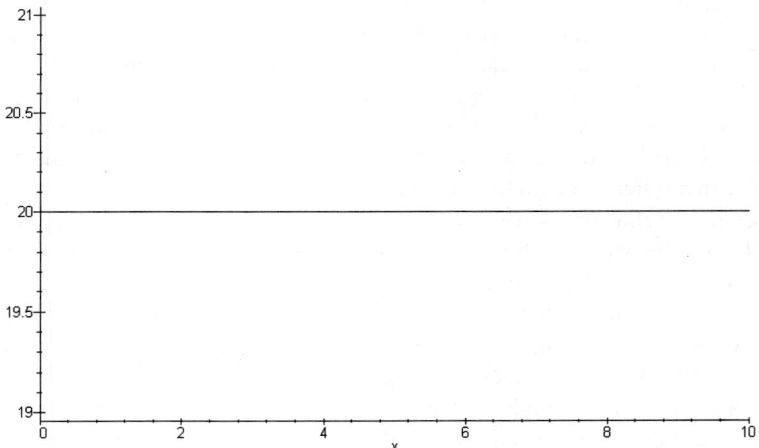

Figure 7.3. Zeroth order wavelet approximation of Young modulus in zeroth scale

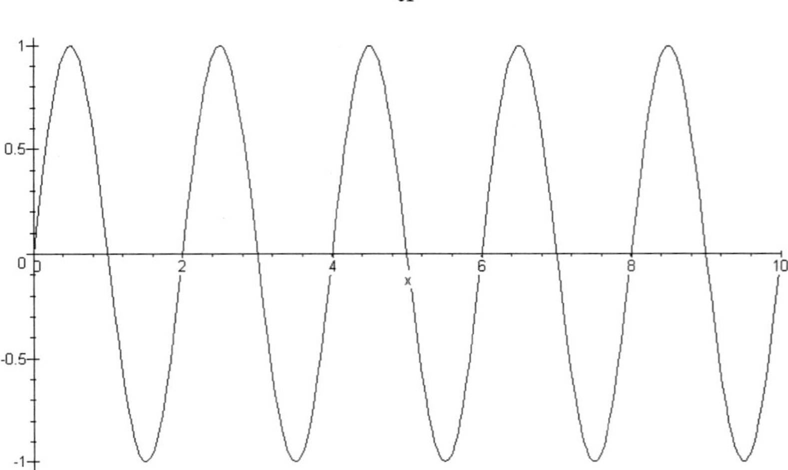

Figure 7.4. First order wavelet approximation of Young modulus in zeroth scale

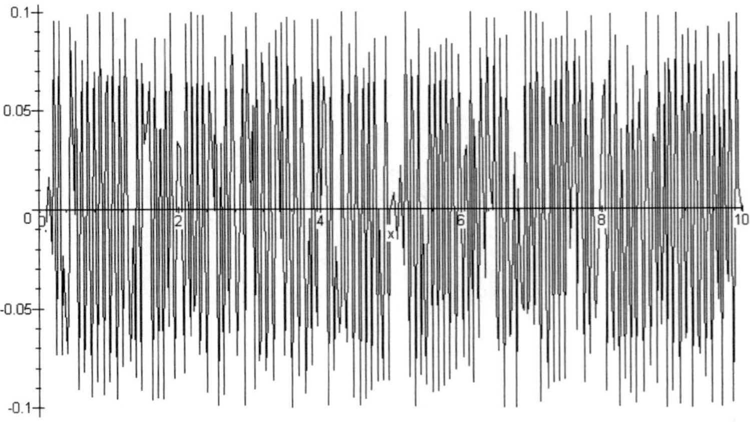

Figure 7.5. Second order wavelet approximation of Young modulus in zeroth scale

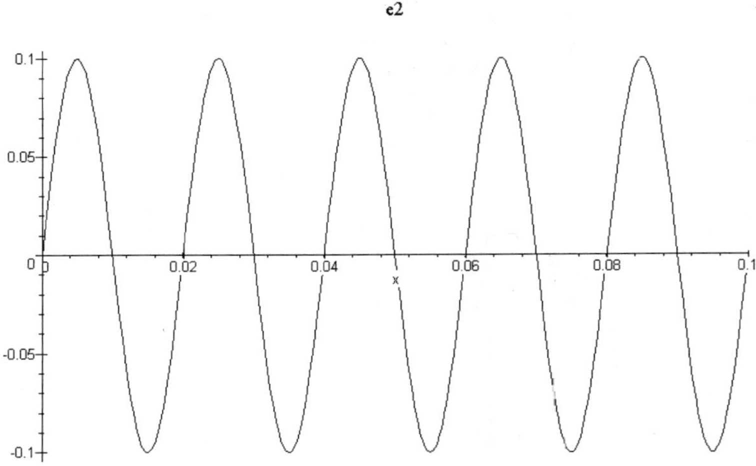

Figure 7.6. Second order wavelet approximation of Young modulus in first scale

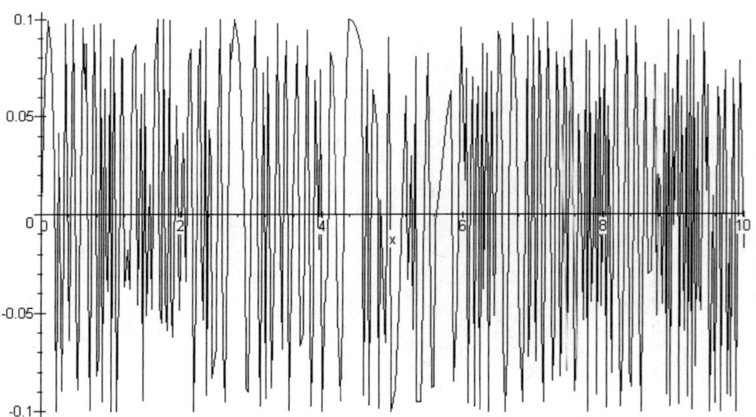

Figure 7.7. Third order wavelet approximation of Young modulus in first scale

As is shown in the next figures (Figures 7.8 and 7.9), using some special combinations of the basic wavelets (Haar, Mexican hat, Gabor, Morlet, Daubechies and/or sinusoidal waves [323]), the spatial variability of Young modulus for the two component composite with and without some interphase can be computationally simulated using a theoretical description of the spatial distribution of this modulus and the symbolic computation package MAPLE, for instance. For illustration of the problem we consider the Representative Volume Element (RVE) of a two−component composite with the following elastic characteristics: e_1=209E9 and e_2=209E8 with the RVE length l=1.0 and equal volume fractions of both components. The following wavelet function is proposed to achieve the multiscale character of Young modulus spatial variability in the RVE without the interface defects (Figure 7.8):

$$e(x) = h(x) + 0.2 \times 10^{10} \sin\left(5 \times 10^{1} x\right) + 0.2 \times 10^{10} \sin\left(5 \times 10^{4} x\right)$$

for $h(x)$ being the Haar wavelet function. It can be noticed that, thanks to the multiscale character of the choosen functions, the picture of composite Young modulus shows the randomness on its microscale. However the character of the spatial variability of this modulus is still deterministic. Furthermore, we can illustrate much more complicated and sometimes more realistic effects in composites – the RVE can be almost damaged at the interface and, according to ageing and fatigue processes, the spatial distribution of elastic properties can be far from constant along the heterogeneity main axis. It is approximated by a combination of Haar, some sinusoidal and the so–called Mexican hat wavelets as

$$e(x) = h(x) + 0.2 \times 10^{10} \sin\left(5 \times 10^{1} x\right) + 0.2 \times 10^{10} \sin\left(5 \times 10^{4} x\right) + 0.6 \times 10^{10}$$
$$- 0.76597 \times 10^{11} \times \exp\left(-8.00 \times x^{2}\right) \times \left(16.0 \times x^{2} - 1\right)$$

The algebraic structure of this wavelet is a little complicated: however (1) it illustrates very well the capabilities of the wavelet–based approximation of mechanical and physical properties of the real composites, (2) it can be used together with structural image analysis tools for the relevant analyses of composites and (3) it enables direct symbolic homogenisation of such media.

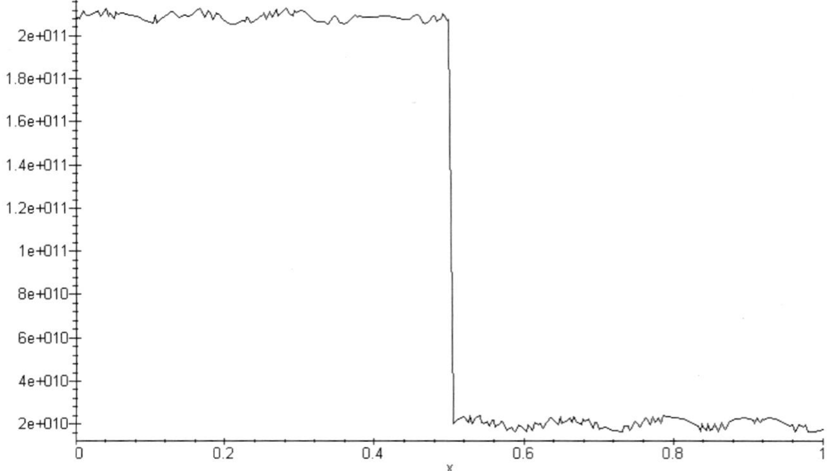

Figure 7.8. Wavelet approximation of elastic properties of two–component composite

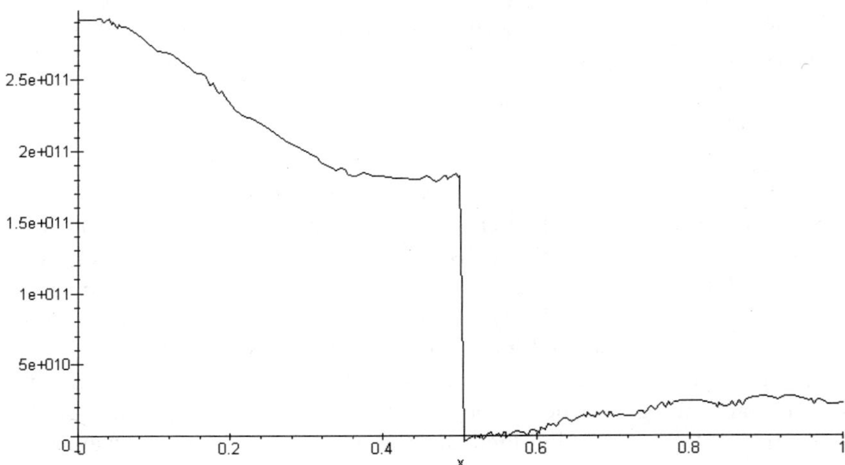

Figure 7.9. Wavelet approximation of the elastic properties of two–component composite with interface defects

As far as this composite is unidirectional, some classical homogenisation closed–form equations can be used to construct the equivalent medium using the relevant differential equilibrium equations directly. In this case it does not matter whether deterministic or probabilistic distribution of material coefficients are given – the PDF symbolic integration can be carried out using a computer. Fortunately, the structural sensitivity analysis may be performed with respect to the variabilities of material properties in quite different scales of the composite; it can be carried out analogously to the considerations presented in [167].

The situation complicates significantly in the case of planar distribution of material tensors, where the cell problems are to be solved by wavelet decomposition and construction to determine the effective behaviour of the entire composite. However, it is mathematically proved in this chapter, that when the structure is heterogeneous in many scales, the effective elastic modulus differs from that obtained for the corresponding classical two–scale and two–component composite beam.

Another disadvantage of the wavelet–based analysis of composite materials is the assumption of a very arbitrary character that the physical model and the accompanying equations of thermodynamical equilibrium have exactly the same form in each scale of the considered medium which follows purely mathematical nature of the wavelet transform. It eliminates the opportunity of the physical transition from the particle scale through chemical interface reactions in various composites to the global scale of the entire engineering structure. It reflects the intuitive feeling that the transition between the corresponding medium scale must strongly depend on the physical scale we are operating on.

7.2 Multiscale Reduction and Homogenisation

Therefore, a multiresolutional homogenisation method is proposed for numerical analysis together with various stochastic computational techniques, which makes it possible to determine probabilistic characteristics of various multiscale composites. Considering the fact that the multiresolutional method makes it possible to determine the effective physical characteristics in a closed form, the stochastic second order perturbation approach is applied to analyse the multiscale randomness of the entire composite in the most general form.

Let us consider the following differential equilibrium equation to distinguish the differences between a classical asymptotic approach and multiresolutional scheme:

$$-\frac{d}{dx}\left(e(x)\frac{d}{dx}u(x)\right)=f(x)\,;\qquad x\in[0,1] \tag{7.1}$$

where $e(x)$, defining material properties of the heterogeneous medium, varies arbitrarily on many scales (macro, meso and micro, etc.). The unit interval denotes here the Representative Volume Element (RVE), also called the periodicity cell.

The classical result obtained through the asymptotic homogenisation theory is given by (2.71) for deterministic composites exhibiting two separate geometrical scales linked by the scale parameter ε – this is the weakest point of this approach. Sometimes ε is treated as a positive real number tending to 0 (practically an infinite number of the RVEs in the composite) and, alternatively, some small positive parameter. As it can be demonstrated, the essential differences are observed in these two models. Now, this parameter is treated as some real functions introduced as the wavelet function relating two or more separate geometrical scales of the composite.

In contrast to the classical approach to the homogenisation problem, the multiresolution approach uses the algebraic transformation between scales provided by the multiresolution analysis to solve for the fine–scale behaviour and explicitly eliminate it from the equation. This approach has the advantage that the coefficients may vary on arbitrarily many scales. The chain of subspaces

$$...\subset V_2\subset V_1\subset V_0\subset V_{-1}\subset V_{-2}\subset... \tag{7.2}$$

defines the hierarchy of scales that the multiresolution scheme uses. This chain of subspaces is defined in such a way that the space V_j is "finer" than the space V_{j+1} in the sense that (1) all of V_{j+1} is contained in V_j, and (2) the component of V_j which is not in V_{j+1} consists of functions which resolve features on a scale finer than any function in V_{j+1} may resolve. The difference between successive spaces in this chain is captured by the so–called wavelet space W_{j+1}, defined to the orthogonal complement of V_{j+1} in V_j. An orthogonal basis for the wavelet space W_{j+1} is constructed which has vanishing moments, i.e. the basis elements are L^2–orthogonal to low–degree polynomials. The existence of orthogonal wavelet bases with vanishing moments distinguishes the multiresolution approach from typical multi–scale discretisations provided by finite–element or hierarchical bases. If we are considering a multiresolution analysis defined on a bounded domain, then the hierarchy of scales defined above has the coarsest scale (which is called V_0), and we write instead

$$V_0 \subset V_{-1} \subset V_{-2} \subset ... \tag{7.3}$$

Let us review the multiresolution strategy for the reduction and homogenisation of linear problems. Let us consider to this purpose a bounded linear operator $S_j : V_j \rightarrow V_j$. Since V_j is spanned by translations of the function $\phi(2^j x - k)$, we know that the operator S_j may be written in the form of a matrix. If the multiresolution analysis is defined on a bounded domain, then this matrix is finite; otherwise it is an infinite matrix, which we consider as an operator on L^2. Let us consider the equation

$$S_j x = f \tag{7.4}$$

The decomposition $V_j = V_{j+1} \oplus W_{j+1}$ allows us to split the operator S_j into four pieces (recall that W_{j+1} is called the wavelet space and is the "detail" or fine–scale component of V_j) and write

$$\begin{pmatrix} A_{S_j} & B_{S_j} \\ C_{S_j} & T_{S_j} \end{pmatrix} \begin{pmatrix} d_x \\ s_x \end{pmatrix} = \begin{pmatrix} d_f \\ s_f \end{pmatrix} \tag{7.5}$$

where we have

$$\begin{aligned}
A_{S_j} &: W_{j+1} \rightarrow W_{j+1} \\
B_{S_j} &: V_{j+1} \rightarrow W_{j+1} \\
C_{S_j} &: W_{j+1} \rightarrow V_{j+1} \\
T_{S_j} &: V_{j+1} \rightarrow V_{j+1}
\end{aligned} \tag{7.6}$$

and $d_x, d_y \in W_{j+1}$, $s_x, s_y \in V_{j+1}$ are the L^2–orthogonal projections of x and f onto the W_{j+1} and V_{j+1} spaces. The projection s_x is thus the coarse–scale component of the solution x and d_x is its fine–scale component. Formally eliminating d_x from (7.5) by substituting $d_x = A_{S_j}^{-1}\left(d_f - B_{S_j} s_x\right)$ yields

$$\left(T_{S_j} - C_{S_j} A_{S_j}^{-1} B_{S_j}\right) s_x = s_f - C_{S_j} A_{S_j}^{-1} d_f \tag{7.7}$$

This equation is called a reduced equation, while the operator

$$R_{S_j} = T_{S_j} - C_{S_j} A_{S_j}^{-1} B_{S_j} \tag{7.8}$$

is a one step reduction of the operator S_j also known as the Schur complement of the block matrix $\begin{pmatrix} A_{S_j} & B_{S_j} \\ C_{S_j} & T_{S_j} \end{pmatrix}$.

Note that the solution s_x of the reduced equation is exactly $P_{j+1}x$, where P_{j+1} is the projection onto V_{j+1} and x is the solution of (7.4). Note that the reduced equation is not the same as the averaged equation, which is given by

$$T_{S_j} \tilde{s}_x = s_f \tag{7.9}$$

Once we have obtained the reduced equation, it may formally be reduced again to produce an equation on V_{j+2} and the solution of this equation is just the V_{j+2} component of the solution for (7.4). Likewise, we may reduce these equations recursively n times (assuming that, if the multiresolution analysis is on a bounded domain, then $j+n \leq 0$) to produce an equation on V_{j+n}, the solution of which is the projection of the solution of (7.4) on V_{j+n}.

We note that in the finite–dimensional case, if we are considering a multiresolution analysis defined on a domain in $\mathbf{R}$, the reduced equation (7.5) has half as many unknowns as the original equation (7.4). If the domain is in $\mathbf{R}^2$, then the reduced equations have one–fourth as many unknowns as the original equation. Reduction, therefore, preserves the coarse–scale behaviour of solutions while reducing the number of unknowns.

In order to iterate the reduction step over many scales, we need to preserve the form of the equation as a way of deriving a recurrence relation. In (7.4) and (7.5), both S_j and R_{S_j} are matrices, and thus the procedure may be repeated. However, identification of the matrix structure is usually not sufficient. In particular, even though the matrix $\mathbf{A}$ for ODEs and PDEs is sparse, the component $A_{S_j}^{-1}$ term may become dense, changing the equation from a local one to a global one. It is important to know under what circumstances the local nature of the differential operator may be (approximately) preserved. Furthermore, if the equation is of the form of

$$-\nabla\big(e(x)\nabla u(x)\big)= f(x) \tag{7.10}$$

or some other variable–coefficient differential equation, we should verify if the reduction procedure preserves this form, so that we may find effective coefficients of the equation on the coarse scale. This process is the basic goal of homogenisation techniques, and it extracts information from the reduced equation based on the form of the original equation. Thus, within the multiresolution approach, reduction and homogenisation are closely related but have different goals: homogenisation attempts to find effective equations and their coefficients on the coarse scale, whereas reduction merely finds a coarse–scale version of a given system of equations.

The multiresolutional (MRA) homogenisation procedure is applied to the systems of ODEs, which may be written in the form

$$Bx+q+\lambda = K\big(Ax+p\big) \tag{7.11}$$

In particular, we consider equations of the form

$$\big(I+B(t)\big)x(t)+q(t)+\lambda = \int_0^t \big(A(s)x(s)+p(s)\big)ds , \ t\in (0,1) \tag{7.12}$$

on $L^2(0,1)$, where $B(t)$ and $A(t)$ are n x n matrix–valued functions, $p(t)$ and $q(t)$ are vector forcing terms, and $x(t)$ is the solution vector. As a differential equation this is written as

$$\frac{d}{dt}\big((I+B(t))x(t)+q(t)\big)= A(t)x(t)+p(t) \tag{7.13}$$

with the initial conditions $\big(I+B(0)\big)x(0) = -q(0)-\lambda$. On V_j, $j<0$, the projection of (7.11) is written as

$$B_jx_j +q_j +\lambda = K_j\big(A_jx_j +p_j\big) \tag{7.14}$$

or

$$S_jx_j = f_j \tag{7.15}$$

where

$$S_j = B_j - K_jA_j , \ f_j = K_jp_j -q_j -\lambda , \ x_j = P_jx_j \tag{7.16}$$

After a single reduction, our goal is to have an equation on V_{j+1} of the form

$$B_{j+1}^{(j)} x_{j+1}^{(j)} + q_{j+1}^{(j)} + \lambda = K_{j+1}\left(A_{j+1}^{(j)} x_{j+1}^{(j)} + p_{j+1}^{(j)}\right) \tag{7.17}$$

where $x_{j+1}^{(j)} = P_{j+1} x_j$, $B_{j+1}^{(j)} = P_{j+1} x_j$, etc. We use the notation $B_l^{(j)}$ to indicate that the equation is first projected to a scale V_j, and then the reduction procedure is applied l-j times to obtain an equation on V_l. This notation therefore indicates that (7.17) was obtained by a single reduction of the same form of equation on V_j one time to produce an equation on the coarser scale V_{j+1}.

It allows one to establish a recurrence relation for $k=j,j+1,\ldots,0$ between the operators and forcing terms $B_k^{(j)}, x_k^{(j)}, p_k^{(j)}, q_k^{(j)}$ on V_{k+1}. It turns out that this task of finding the recurrence relations is simplified significantly if one uses a multiresolution analysis whose basis functions have non−overlapping support. We use the Haar basis, but a multiwavelet basis may be used if higher order elements are necessary.

In the Haar basis, the operators B_j, A_j and K_j derived from equations of the form of (7.14) have a simple form. Each of these is in an $(N_j n)$ x $(N_j n)$ matrix, where $N_j=2^j$ is the number of unknowns on the scale V_j and n denotes the number of equations in the original system. Furthermore, B_j and A_j are both block-diagonal matrices. The diagonal blocks of B_j and A_j are n x n matrices. There are therefore N_j diagonal blocks, each of which is an n x n matrix. For B_j and A_j we denote their ith diagonal blocks by $\left(B_j\right)_i$ and $\left(A_j\right)_i$. The matrices are given by the Haar coefficients of the n x n matrix−valued functions $B(x)$ and $A(x)$ on the scale V_j. It can be written that

$$B_j = diag\left\{I + \left(B_j\right)_i\right\}_{i=0}^{i=2^j-1} \tag{7.18}$$

and

$$A_j = diag\left\{\left(A_j\right)_i\right\}_{i=0}^{i=2^j-1} \tag{7.19}$$

where

$$K_j = \delta_j \begin{pmatrix} \frac{1}{2}I & 0 & 0 & \cdots & 0 \\ I & \frac{1}{2}I & 0 & \cdots & 0 \\ I & I & \frac{1}{2}I & & \cdots \\ \cdots & & & \cdots & \\ I & \cdots & & I & \frac{1}{2}I \end{pmatrix} \tag{7.20}$$

where $\delta_j = 2^{-j}$, I is the $n \times n$ identity matrix, and $\left(B_j\right)_i$ and $\left(A_j\right)_i$ are the ith Haar coefficients on scale V_j of the $n \times n$ matrix–value functions $B(x)$ as well as $A(x)$. For (7.17), the recursion relations are given by

$$\left(A_{k+1}^{(j)}\right)_i = \left(S_A\right)_i - \left(D_A\right)_i F^{-1}\left(\left(D_B\right)_i + \frac{\delta_k}{2}\left(S_A\right)_i\right) \tag{7.21}$$

$$\left(B_{k+1}^{(j)}\right)_i = \left(S_B\right)_i - \frac{\delta_k}{2}\left(D_A\right)_i - \left(\left(D_B\right)_i - \frac{\delta_k}{2}\left(S_A\right)_i\right)F^{-1}\left(\left(D_B\right)_i + \frac{\delta_k}{2}\left(S_A\right)_i\right) \tag{7.22}$$

$$\left(p_{k+1}^{(j)}\right)_i = \left(S_p\right)_i - \frac{\delta_k}{2}\left(D_A\right)_i F^{-1}\left(\left(D_q\right)_i + \left(S_p\right)_i\right) \tag{7.23}$$

$$\left(q_{k+1}^{(j)}\right)_i = \left(S_q\right)_i - \frac{\delta_k}{2}\left(D_p\right)_i - \frac{\delta_k}{2}\left(\left(D_B\right)_i - \left(S_A\right)_i\right)F^{-1}\left(\left(D_q\right)_i - \left(S_p\right)_i\right) \tag{7.24}$$

where

$$\left(S_A\right)_i = \tfrac{1}{2}\left(\left(A_k^{(j)}\right)_{2i} + \left(A_k^{(j)}\right)_{2i+1}\right), \quad \left(D_A\right)_i = \tfrac{1}{2}\left(\left(A_k^{(j)}\right)_{2i} - \left(A_k^{(j)}\right)_{2i+1}\right) \tag{7.25}$$

$$\left(S_B\right)_i = \tfrac{1}{2}\left(\left(B_k^{(j)}\right)_{2i} + \left(B_k^{(j)}\right)_{2i+1}\right), \quad \left(D_B\right)_i = \tfrac{1}{2}\left(\left(B_k^{(j)}\right)_{2i} - \left(B_k^{(j)}\right)_{2i+1}\right) \tag{7.26}$$

$$\left(S_p\right)_i = \tfrac{1}{\sqrt{2}}\left(\left(p_k^{(j)}\right)_{2i} + \left(p_k^{(j)}\right)_{2i+1}\right), \quad \left(D_p\right)_i = \tfrac{1}{\sqrt{2}}\left(\left(p_k^{(j)}\right)_{2i} - \left(p_k^{(j)}\right)_{2i+1}\right) \tag{7.27}$$

$$\left(S_q\right)_i = \tfrac{1}{\sqrt{2}}\left(\left(q_k^{(j)}\right)_{2i} + \left(q_k^{(j)}\right)_{2i+1}\right), \quad \left(D_q\right)_i = \tfrac{1}{\sqrt{2}}\left(\left(q_k^{(j)}\right)_{2i} - \left(q_k^{(j)}\right)_{2i+1}\right) \tag{7.28}$$

and, finally,

$$F = I + \left(S_B\right)_i + \frac{\delta_k}{2}\left(D_A\right)_i \tag{7.29}$$

Note that the recurrence relations are local and can be carried out over many scales as needed (assuming the existence of F^{-1} at each scale). Starting with (7.17) on V_{-j} and, reducing j times, yields on V_0

$$B_0^{(j)} x_0^{(j)} + q_0^{(j)} + \lambda = K_0\left(A_0^{(j)} x_0^{(j)} + p_0^{(j)}\right) \tag{7.30}$$

where to compute $B_0^{(j)}, A_0^{(j)}, p_0^{(j)}, q_0^{(j)}$ we use the recurrence relations j times.

Multiresolutional homogenisation is formulated as follows. First, we consider the limit of (7.30) as $j \to -\infty$, therefore

$$B_0^{(-\infty)} x_0^{(-\infty)} + q_0^{(-\infty)} + \lambda = K_0\left(A_0^{(-\infty)} x_0^{(-\infty)} + p_0^{(-\infty)}\right) \tag{7.31}$$

It is employed to eliminate infinite number of fine scales from the original equation. The matrices $B_0^{(-\infty)}, A_0^{(-\infty)}$ are called the reduced coefficients of (7.14). Then, we look for the operators and forcing terms $B^h(t), A^h(t), p^h(t), q^h(t)$ with certain desired qualities (e.g. constant values) such that the equation

$$\left(I + B^h(t)\right)x(t) + q^h(t) + \lambda = \int_0^t \left(A^h(s)x(s) + p^h(s)\right)ds \ , \ t \in (0,1) \tag{7.32}$$

subjected to the same reduction and limit procedure as (7.12), yields on V_0 the same equation as in (7.31). For (7.12) we usually require that B^h, A^h, p^h, q^h be constant. The result of homogenisation in this case is summarised as follows:

Theorem
Given (7.12), if the limits, which determine the matrices $B_0^{(-\infty)}$ and $A_0^{(-\infty)}$ exist, then there exist constant matrices B^h, A^h and forcing terms p^h, q^h, such that the reduced coefficients and forcing terms of (7.32) are given by $B_0^{(-\infty)}, A_0^{(-\infty)}, q_0^{(-\infty)}, p_0^{(-\infty)}$. The homogenised coefficients B^h, A^h and forcing terms p^h, q^h are defined by

$$A^h = A_0^{(-\infty)} \tag{7.33}$$
$$B^h = A^h \tilde{A}^{-1} - I \tag{7.34}$$
$$p^h = p_0^{(-\infty)} \tag{7.35}$$

$$q^h = q_0^{(-\infty)} + \left(I - \tfrac{1}{2}\tilde{A} - \tilde{A}\left(\exp(\tilde{A} - I)\right)^{-1} A^h\right)^{-1} p^h \tag{7.36}$$

where

$$\tilde{A} = \log\left(I + \left(I + B_0^{(-\infty)} - \tfrac{1}{2} A^h\right)^{-1} A^h\right) \tag{7.37}$$

Proof
It is observed that for the constant coefficients the recurrence relations (7.21) and (7.22) simplify to

$$A_{k+1}^h = A_k^h \qquad (7.38)$$

$$B_{k+1}^h = B_k^h + \frac{\delta_k^2}{4} A_k^h \left(I + B_k^h\right)^{-1} A_k^h \qquad (7.39)$$

Likewise, the recurrence relations for the forcing terms simplify to

$$p_{k+1}^h = p_k^h \qquad (7.40)$$

$$q_{k+1}^h = q_k^h - \frac{\delta_k}{2} A_k^h \left(I + B_k^h\right)^{-1} p_k^h \qquad (7.41)$$

Since the term A^h is unchanged by reduction, it is clear that $A^h = A_0^{(-\infty)}$. Similarly, p^h is unchanged by reduction, so $p^h = p_0^{(-\infty)}$. The situation for B^h and q^h is more complicated. We solve for them analytically using the solution of (7.32). Consider the case $p_0^{(-\infty)} = 0$. Clearly, then, it is the case that $q^h = q_0^{(-\infty)}$. The solution of (7.32) is therefore given by

$$x(t) = -\exp\left(\tilde{A}t\right)\tilde{q} \qquad (7.42)$$

where $\tilde{A} = \left(I + B^h\right)^{-1} A^h$, $\tilde{q} = \left(I + B^h\right)^{-1}\left(q^h + \lambda\right)$. The average of this solution must also solve (7.31) since it is the equation for the average value of the solution by definition. The average value of $x(t)$ in (7.32) on the interval $[0,1]$ is given by

$$\langle x \rangle = \left(-\int_0^1 \exp\left(\tilde{A}t\right)dt\right)\tilde{q} = \left(I - \exp\left(\tilde{A}\right)\right)\tilde{A}^{-1}\tilde{q} \qquad (7.43)$$

The solution to (7.31) is given by

$$x_0^{(-\infty)} = -\left(I + B_0^{(-\infty)} - \tfrac{1}{2}A_0^{(-\infty)}\right)^{-1}\left(q_0^{(-\infty)} + \lambda\right) \qquad (7.44)$$

The right hand sides of (7.43) and (7.44) are demonstrated to be equal for all λ; setting $\lambda=0$ and solving for B^h yields the solution given in the statement of the proposition. The case when $p_0^{(-\infty)} \neq 0$ proceeds similarly.

Solutions of (7.32) have the same "average" or coarse–scale behaviour as solutions of (7.12). The main point is that this homogenisation procedure allows for coefficients to vary on arbitrarily many intermediate scales, which is in contrast to the classical homogenisation examples, which did not allow for intermediate scales.

As formulated above, the multiresolution approach to homogenisation requires the computation of $A_0^{(-\infty)}$ and $B_0^{(-\infty)}$, i.e. a limit over infinitely many scales. The typical practice is to compute successive $A_0^{(-J)}$ and $B_0^{(-J)}$ terms until finer approximations vary by less than some specified tolerance, and use these matrices as approximations to $A_0^{(-\infty)}$ and $B_0^{(-\infty)}$.

Besides establishing the general framework for multiresolution reduction and homogenisation, it is observed that for systems of linear ordinary differential equations, using the Haar basis (or a multiwavelet basis) provides a technical advantage. Since the functions of the Haar basis on a fixed scale do not have overlapping supports, the recurrence relations for the operators and forcing terms in the equation may be written as local relations and solved explicitly. Thus, for ODEs, an explicit local reduction and homogenisation procedure is possible.

Let us consider for illustration (7.1) with initial conditions at $x=0$. It may be rewritten as the coupled first−order system

$$\begin{cases} \dfrac{d}{dx} v(x) = -f(x) \\[2mm] \dfrac{d}{dx} u(x) = e(x)^{-1} v(x) \end{cases} \tag{7.45}$$

By writing in an integral form one can obtain

$$\begin{pmatrix} u(x) \\ v(x) \end{pmatrix} - \begin{pmatrix} u(0) \\ v(0) \end{pmatrix} = \int_0^x \left(\begin{pmatrix} 0 & e(t)^{-1} \\ 0 & 0 \end{pmatrix} \begin{pmatrix} u(t) \\ v(t) \end{pmatrix} + \begin{pmatrix} 0 \\ -f(t) \end{pmatrix} \right) dt \tag{7.46}$$

Thus, in the notation of (7.11), $B(x)=0$, $A(x) = \begin{pmatrix} 0 & e(x)^{-1} \\ 0 & 0 \end{pmatrix}$, $\lambda = \begin{pmatrix} u(0) \\ v(0) \end{pmatrix}$ and $q(t)=0$ as well as $p(t) = \begin{pmatrix} 0 \\ -f(t) \end{pmatrix}$.

Using the reduction procedure in the Haar basis for a system of linear differential equations, the goal is to find constants B^h, A^h, p^h, q^h such that

$$\left(I + B^h\right) \begin{pmatrix} u(x) \\ v(x) \end{pmatrix} + q^h + \lambda = \int_0^x \left(A^h \begin{pmatrix} u(t) \\ v(t) \end{pmatrix} + p^h \right) dt \tag{7.47}$$

after reduction to the scale V_0 will be the same as (7.46) reduced to that scale. This is accomplished by solving the recursion relations between the operators in the reduced equations explicitly, element−by−element in each matrix. This is possible to do because of the non−overlapping supports of the Haar basis functions on a fixed scale. The result for the first two coefficients is

$$B^h = \begin{pmatrix} 0 & 0 \\ 0 & 0 \end{pmatrix}, A^h = \begin{pmatrix} 0 & M_1 - 2M_2 \\ 0 & 0 \end{pmatrix} \tag{7.48}$$

where

$$M_1 = \int_0^1 \frac{1}{e(t)} dt, \quad M_2 = \int_0^1 \frac{t - \frac{1}{2}}{e(t)} dt \tag{7.49}$$

Similar expressions for p^h and q^h can be found. Note that we have $p^h = q^h = 0$ if $f(x)=0$ identically. Furthermore, in general B^h, A^h do not depend on p and q. As a first–order system of ordinary differential equations, the homogenised equation yields

$$\begin{cases} \dfrac{d}{dx} v(x) = f^h(x) \\ \dfrac{d}{dx} u(x) = (M_1 - 2M_2) v(x) \end{cases} \tag{7.50}$$

what is somewhat different from the classical result. This difference results from the fact that the multiresolution homogenisation procedure allows the coefficients $e(x)$ to vary on arbitrarily many scales, whereas the classical approach presented before allows only for coefficients of the form $e(x/\varepsilon)$. In the multiresolution context this amounts to restricting the coefficients to an asymptotically fine scale. Let us apply the same limit in the preceding section to the coefficients appearing in the multiresolution approach. We start with the coefficients of the form $e(x/\varepsilon)$.

Applying this homogenisation scheme to the elliptic equation with these coefficients yields two terms, $M_1(\varepsilon)$ and $M_2(\varepsilon)$. If we take the limit as $\varepsilon \to 0$, it is found that

$$\lim_{\varepsilon \to 0} M_1(\varepsilon) = M_1 \tag{7.51}$$

and

$$\lim_{\varepsilon \to 0} M_2(\varepsilon) = 0 \tag{7.52}$$

Thus, the factor M_2 is present in the multiresolution context but does not appear in the classical approach, and it is zero when the limit found in the classical method is applied to the result of the multiresolution methodology.

Let us note that this formula of the homogenised parameter $e^{(eff)}$ introduces new, closer bounds on the wavelet function defining material parameters than is done by classical formulation: integrals $\int_0^L \frac{dx}{e(x)}$ and $\int_0^L \frac{xdx}{e(x)}$ must be of real values and $e(x)$ must be positive defined to assure homogenisability of the problem. The counter−example is the family of sinusoidal wavelets of the form $e(x) = e_0 + \alpha \sin\left(\frac{\pi x}{L}\right)$, where $e_0, \alpha, L \in \Re$. Taking for example $L=10$, $e_0=20$ and $\alpha=0.1$, MAPLE symbolic integration returns $\int_0^L \frac{xdx}{e(x)} = -2.99242 - 19.07199i$. The classical small parameter homogenisation method should be applied in that case; otherwise another wavelet decomposition of the real composite is to be performed.

7.3 Multiscale Homogenisation for the Wave Propagation Equation

For illustration, let us consider the following ordinary differential equation (ODE) corresponding to unidirectional acoustic wave propagation in a multiscale medium with uniaxial distribution of nonhomogeneities [71,188]:

$$\frac{d}{dx}u(x) = i\omega M(x)u(x); \quad x \in [0,1] \tag{7.53}$$

where physical coefficients $M(x)$ for both composite layers are defined by

$$M(x) = \begin{cases} M_0, \ 0 \le x < \frac{1}{2} \\ M_1, \ \frac{1}{2} \le x \le 1 \end{cases} \tag{7.54}$$

These equations are solved using the methods typical for a deterministic problem and are derived for equal volume ratios of both layers. Otherwise, they should be complemented with the ratios c_1 and c_2. The corresponding homogenised equation can be rewritten for the deterministic system as

$$\frac{d}{dx}u(x) = K^{(eff)}u(x) \tag{7.55}$$

It can be demonstrated [71] that the homogenised coefficient $K^{(eff)}$ is equal to

$$K^{(eff)} = \log\left(I + \left(I + B_0^{(-\infty)} - \tfrac{1}{2}A_0^{(-\infty)}\right)^{-1} A_0^{(-\infty)}\right) \tag{7.56}$$

for

$$B_0^{(-\infty)} = S_B' - \tfrac{1}{4} D_A' - \left(D_B' - \tfrac{1}{4} S_A'\right) F^{-1} \left(D_B' + \tfrac{1}{4} S_A'\right) \tag{7.57}$$

and

$$A_0^{(-\infty)} = S_A' - D_A' F^{-1} \left(D_B' + \tfrac{1}{4} S_A'\right) \tag{7.58}$$

The right hand side coefficients denote

$$S_A' = \tfrac{1}{2} \left(\left(A_1^{(-\infty)}\right)_0 + \left(A_1^{(-\infty)}\right)_1\right), \quad D_A' = \tfrac{1}{2} \left(\left(A_1^{(-\infty)}\right)_0 - \left(A_1^{(-\infty)}\right)_1\right) \tag{7.59}$$

$$S_B' = \tfrac{1}{2} \left(\left(B_1^{(-\infty)}\right)_0 + \left(B_1^{(-\infty)}\right)_1\right), \quad D_B' = \tfrac{1}{2} \left(\left(B_1^{(-\infty)}\right)_0 - \left(B_1^{(-\infty)}\right)_1\right) \tag{7.60}$$

$$F = I + S_B' + \tfrac{1}{4} D_A' \tag{7.61}$$

where

$$A_j^{(-\infty)} = \lim_{J \to \infty} A_j^{(1-J)}, \quad B_j^{(-\infty)} = \lim_{J \to \infty} B_j^{(1-J)} \tag{7.62}$$

After some algebra it is found that

$$\left(A_1^{(-\infty)}\right)_j = A_j, \quad \left(B_1^{(-\infty)}\right)_j = \frac{A_j}{2} \left(\left(\exp\left(\frac{A_j}{2}\right) - I\right)^{-1} + \frac{1}{2} I\right) - I, \quad j=1,2 \tag{7.63}$$

where the following extension is used:

$$\exp\left(\frac{A_j}{2}\right) = \exp\left(\frac{i\omega}{2} M_j\right)$$

$$= I + \frac{i\omega}{2} M_j - \frac{\omega^2}{8} M_j^2 + \frac{i\omega^3}{48} M_j^3 + \ldots + O(\omega^n) \tag{7.64}$$

Taking into account that the coefficients B and A in (7.63) represent physical properties of the composite components with the total number of various scales tending to infinity, it is possible to determine an analogous definition of the homogenised coefficient for a composite with some finite scale number. Furthermore, using the stochastic second order perturbation second probabilistic moment methodology, it is relatively easy to determine the first two probabilistic moments of the homogenised coefficient defined by (7.63).

Further, real and imaginary parts of $K^{(eff)}$ are computed according to (7.56)–(7.58). The following data are adopted: $M_0=10.0$, $M_1=1.0$ with $c_1=c_2=0.5$, where the parameter $\omega\to0$ (cf. Figures 7.10 and 7.11) and $\omega\to\infty$ (see Figs. 7.12 and 7.13). As can be observed, the real and imaginary parts tend to 0 in both cases, which finally gives $K^{(eff)}\to0$, too. Further, such a combination of input parameters results in a minimum of the $K^{(eff)}$ real part for $\omega\approx1.15$. On the other hand, the singularity of $\mathrm{Im}(K^{(eff)})$ is obtained with $\omega\approx0.75$.

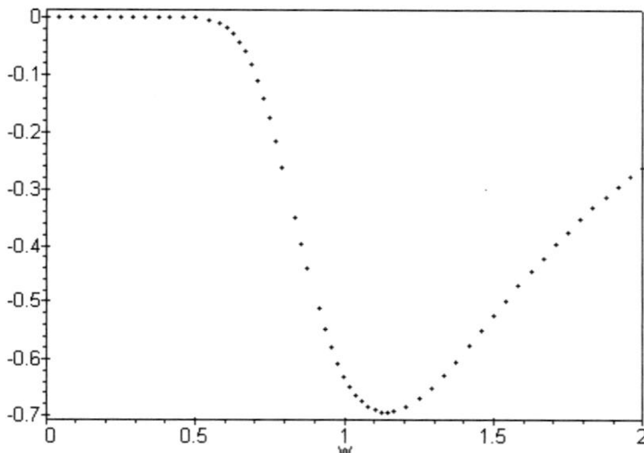

Figure 7.10. Real part of $K^{(eff)}$ near 0

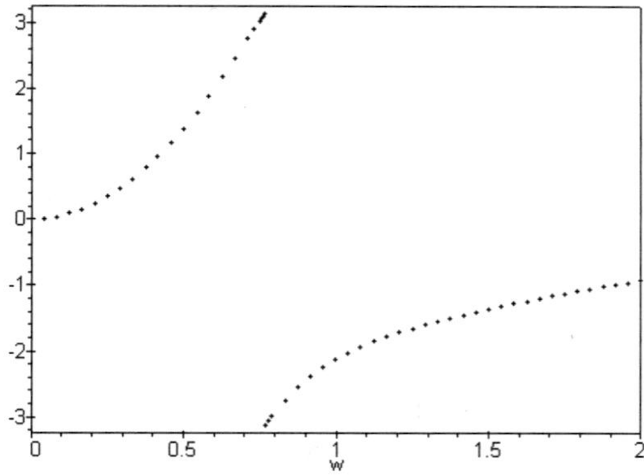

Figure 7.11. Imaginary part of $K^{(eff)}$ near 0

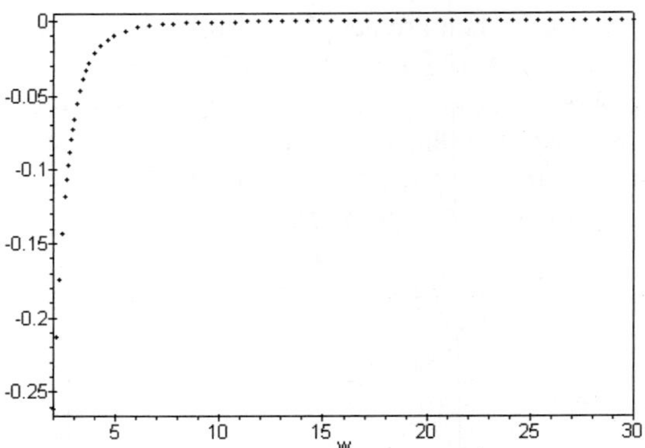

Figure 7.12. Real part of $K^{(eff)}$ in ω domain

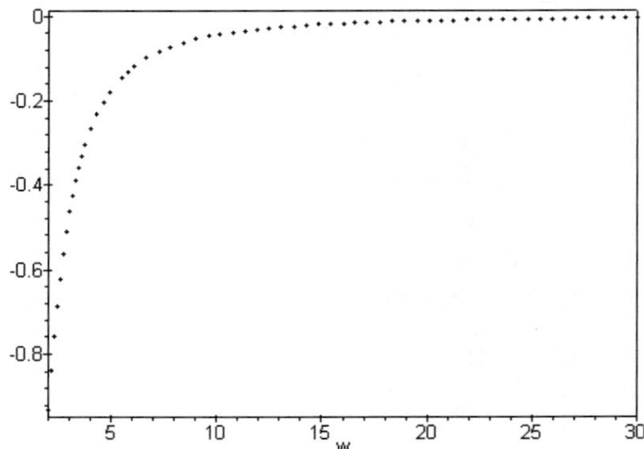

Figure 7.13. Imaginary part of $K^{(eff)}$ in ω domain

Next, the effective parameter in its real and imaginary part is determined as a function of the ω value and the ratio relating material parameters of the composite components $M_0=2-20$. The results are presented in Figures 7.14 and 7.15 below. As can be compared with Figures 7.12 and 7.13, the material parameter interrelation influences significantly the effective parameters in the same range as the ω values. Analogous limiting values in real and imaginary parts of the homogenised parameter as well as imaginary part singularities are noticed as second order functions of both design parameters of the study.

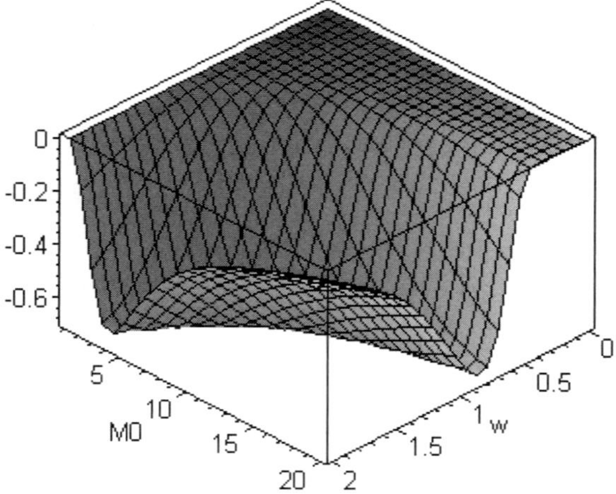

Figure 7.14. Real part of $K^{(eff)}$

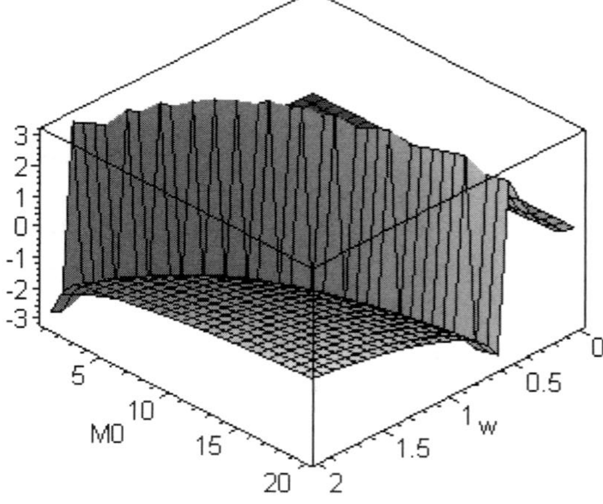

Figure 7.15. Imaginary part of $K^{(eff)}$

Probabilistic moments of real and imaginary surfaces are expected in the probabilistic case. However a more important problem (from the physical point of view) is to determine the relations for homogenised coefficients in terms of volume fractions of the layers as well as to extend the homogenisation method to the heterogeneous multiscale media with a more general periodic geometry of the RVE. The entire methodology can be adopted with minor changes to computational analysis of the wave propagation in random media [26], where material properties are defined using a combination of harmonic functions with random coefficients.

7.4 Introduction to Multiresolutional FEM Implementation

Let us consider the following boundary value problem for a homogeneous medium:

$$-e\nabla^2 u + \gamma u = f , \quad x \in \Omega \tag{7.65}$$

with

$$u = 0 , \quad x \in \Gamma_u \subset \partial\Omega \tag{7.66}$$

The variational formulation of this problem for the multiscale medium for $k=1,\ldots,n$, indexing its various scales is obtained at the scale k as

$$\int_\Omega e\nabla u_k \nabla \varphi_k d\Omega + \int_\Omega \gamma u_k \varphi_k d\Omega = \int_\Gamma f\varphi_k d\Gamma , \quad x \in \Omega \tag{7.67}$$

Solution of the problem must be found recursively by using some transformation between neighbouring medium scales. That is why the following nonsingular $n \times n$ wavelet transform matrix W_k is introduced:

$$\mathbf{W}_k = \mathbf{T}_k \begin{bmatrix} \mathbf{T}_{k-1} & 0 \\ 0 & \mathbf{I}_{k-1} \end{bmatrix} \tag{7.68}$$

where I_k is an identity matrix and

$$\mathbf{\psi}_k = \mathbf{W}_k^T \mathbf{\varphi}_k \tag{7.69}$$

T_k is a two-scale transform such that

$$\begin{Bmatrix} \mathbf{\varphi}_{k-1} \\ \mathbf{\psi}_k \end{Bmatrix} = \mathbf{T}_k^T \mathbf{\varphi}_k \tag{7.70}$$

with

$$\psi_k^j = \varphi_k^{2j-1} , \quad j=1,\ldots,N_k \tag{7.71}$$

where N_k denotes the total number of the FEM nodal points at the scale k. Let us illustrate the wavelet–based FEM idea using the example of a 1D linear two–node finite element. The classical shape functions are defined as [78]

$$\mathbf{N}^T = \left\{ \begin{matrix} N_1 \\ N_2 \end{matrix} \right\} = \left\{ \begin{matrix} \frac{1}{2}(1-\xi) \\ \frac{1}{2}(1+\xi) \end{matrix} \right\} \tag{7.72}$$

where N_1 is valid for $\xi=-1$ and N_2 – for $\xi=1$. The scale effect is introduced on the finite element level by inserting new extra degrees of freedom at each new scale. Then, the scale 1 corresponds to one multiscale DOF per the original finite element, scale 2 to two multiscale DOFs, etc., which may be characterised as [66]

$$\psi_k(\xi) = \psi_k\left(2^{k-1}(1+\xi) - 2j - 1\right) \tag{7.73}$$

and

$$\begin{cases} 2^{2-k}\,j - 1 \le \xi \le 2^{2-k}\,j + 2^{1-k} - 1 \\ 2^{2-k}\,j + 2^{1-k} - 1 \le \xi \le 2^{2-k}\,j + 2^{2-k} - 1 \end{cases} \tag{7.74}$$

where k defines the actual scale, while j characterises the translates in the finite element parametric space. Thus, the reconstruction algorithm starts from the original solution for the original mesh and next, introduction of the new scales is made using the reconstruction

$$\mathbf{u}_k^{2+2^{k-1}+j} = \sum_{i=1}^{N_{old}} \mathbf{N}_i \mathbf{u}_0^i + \sum_{i=1}^{N_{new}} \mathbf{\psi}_k^{2+2^{k-1}+j} \Delta \mathbf{u}_k^{2+2^{k-1}+j} \tag{7.75}$$

The wavelet algorithm for stiffness matrix reconstruction starts at scale 0 with the stiffness matrix

$$\mathbf{K}_0 = \frac{e}{h} \begin{bmatrix} 1 & -1 \\ -1 & 1 \end{bmatrix} \tag{7.76}$$

where h is the node spacing parameter. Then, the diagonal components of the stiffness matrix for any $k>0$ are equal to

$$K_k^{2+2^{K-1}+J} = \frac{2^{k+1} e}{h} \tag{7.77}$$

It should be underlined that the FEM so modified reflects perfectly the needs of computational modelling of multiscale media. When the homogenisation based modelling is performed, then the effective stiffness matrix is introduced as

$$\mathbf{K}_0^{(eff)} = \frac{e^{(eff)}}{h}\begin{bmatrix} 1 & -1 \\ -1 & 1 \end{bmatrix}$$

(7.78)

and in practice there is no need for a wavelet decomposition of this matrix. We observe that the projection algorithm can be applied for such $n \in N$ that ensure a sufficient mesh zoom on the smallest scale details in the composite microstructure.

The effectiveness of this approach can be illustrated with the following projection on the wavelet space for the function $f(t) = \cos(2\pi t)$ where $t \in [0,1]$ performed by use of the symbolic computation package MAPLE [70,182]. It is done for $n=2,\ldots,7$ and is presented correspondingly in Figures 7.16−7.21 below with computational performance indices collected in Table 7.1 (valid for a COMPAQ 475 MHz). As is observed, the increasing projection order decisively increases the computational time of wavelet decomposition of a multiscale phenomenon necessary for the FEM approach.

Table 7.1. Computational symbolic projection of cosine wavelets

Projection order 'n'	Finite elements number	Computational time [sec]	Memory[MB]
2	4	3.9	2.00
3	8	8.0	2.31
4	16	11.1	2.69
5	32	23.9	3.37
6	64	48.6	4.62
7	128	132.1	7.12

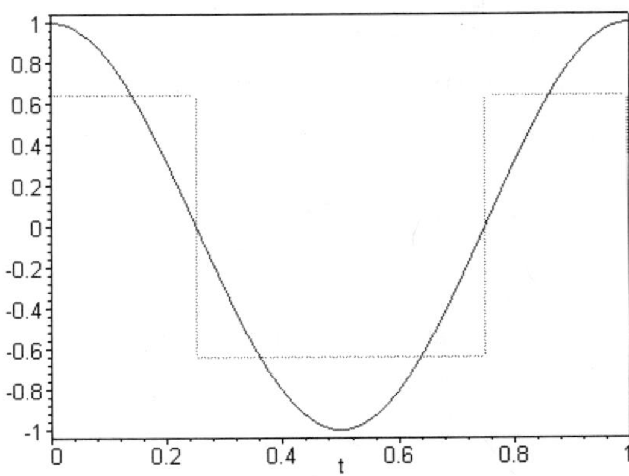

Figure 7.16. Wavelet projection for $n=2$

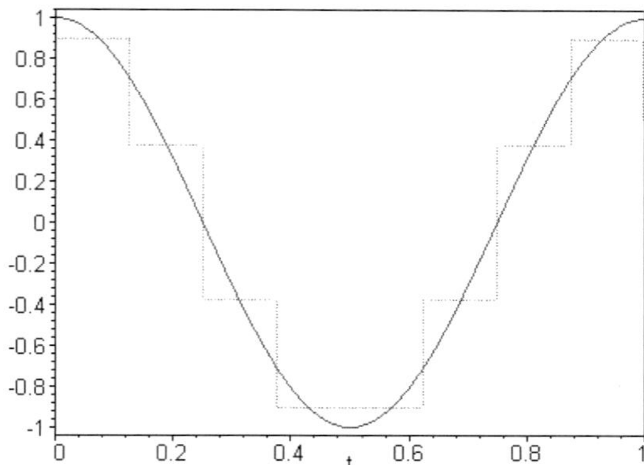

Figure 7.17. Wavelet projection for $n=3$

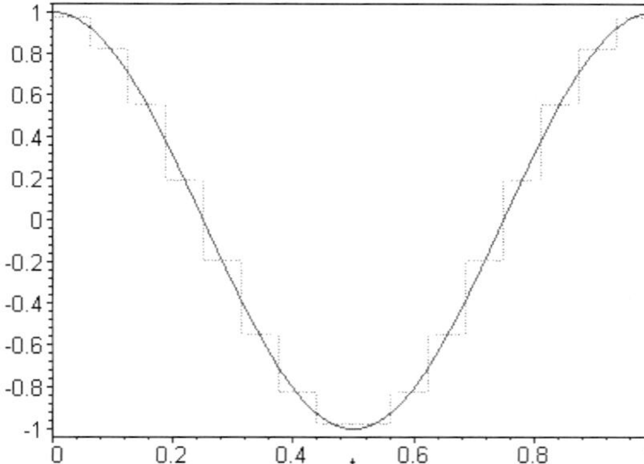

Figure 7.18. Wavelet projection for $n=4$

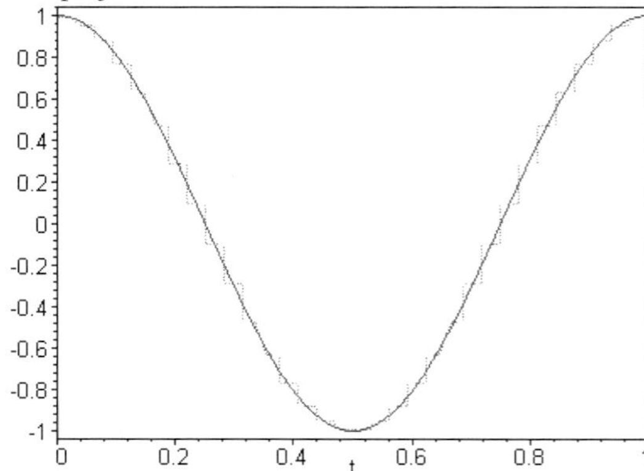

Figure 7.19. Wavelet projection for $n=5$

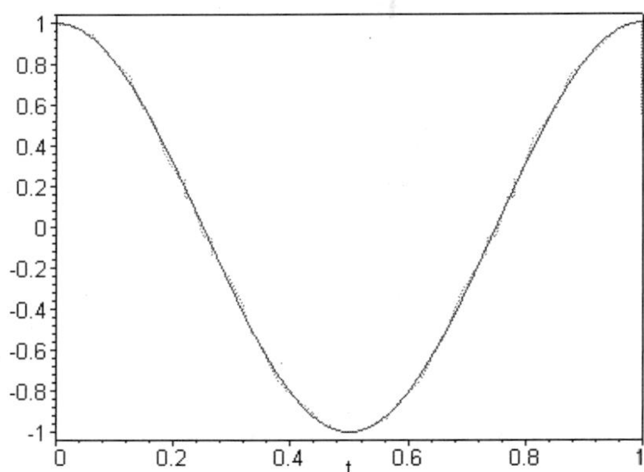

Figure 7.20. Wavelet projection for $n=6$

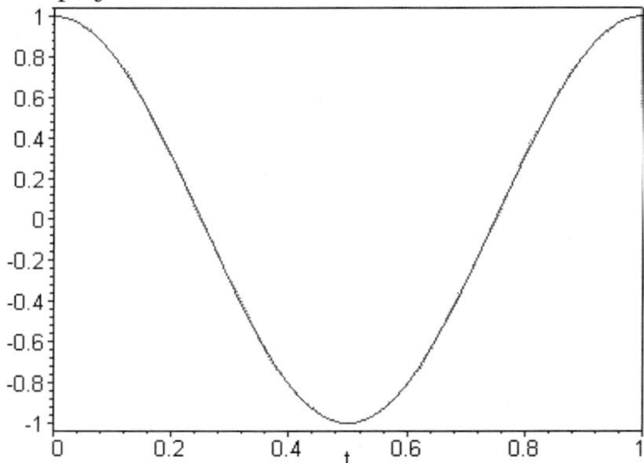

Figure 7.21. Wavelet projection for $n=7$

Computational experiments are performed using the system MAPLE and the additional implementation of the multiresolution homogenisation analysis. Basic computations are carried out with respect to interrelations between physical constants of both layers as well as the expansion order. Furthermore, deterministic and stochastic sensitivities of complex effective parameters (real and imaginary parts) are computed with respect to the first probabilistic moments of input physical parameters of composite layers. Finally, let us observe that a homogenised system, both in terms of deterministic or stochastic effective coefficients, can be analysed numerically using a classical Finite Element Method (FEM), for instance, or by application of various stochastic numerical methods (simulation, perturbation-based or spectral). A homogenisation–based numerical approach will considerably speed up the process of computational modelling of composites and, in the case of very complicated multiscale heterogeneous media, it can be the only available method.

7.5 Free Vibrations Analysis

The main idea of homogenisation problem solution now is a separate calculation of the effective elastic modulus and spatial averaging of the mass density, where the first part only needs multiresolutional approach [189]. The alternative wavelet–based methodology is presented in [328,329], for a plate wave propagation in [152], whereas some classical unidirectional examples are contained in [330]. Let us consider the following differential equilibrium equation:

$$-\frac{d}{dx}\left(e(x)I(x)\frac{d}{dx}u(x)\right)=M(x); \quad x\in[0,1] \tag{7.79}$$

where $e(x)$, defining material properties of the heterogeneous medium, varies arbitrarily on many scales together with the inertia momentum $I(x)$. A multiresolutional homogenisation starts now from the following decomposition of the equilibrium equation:

$$\begin{cases} \dfrac{d}{dx}v(x) = -M(x) \\ \dfrac{d}{dx}u(x) = \dfrac{v(x)}{e(x)I(x)} \end{cases} \tag{7.80}$$

to determine the homogenised coefficient $e^{(eff)}$ constant over the interval $x\in[0,1]$, which takes the integral form

$$\begin{pmatrix} u(x) \\ v(x) \end{pmatrix} - \begin{pmatrix} u(0) \\ v(0) \end{pmatrix} = \int_0^x \left(\begin{pmatrix} 0 & e(t)^{-1}I(t)^{-1} \\ 0 & 0 \end{pmatrix} \begin{pmatrix} u(t) \\ v(t) \end{pmatrix} + \begin{pmatrix} 0 \\ -M(t) \end{pmatrix} \right) dt \tag{7.81}$$

On the other hand, the reduction algorithm between multiple scales of the composite consists in determination of such effective tensors $B^{(eff)}$, $A^{(eff)}$, $p^{(eff)}$ and $q^{(eff)}$, such that

$$\left(I+B^{(eff)}\right)\begin{pmatrix} u(x) \\ v(x) \end{pmatrix} + q^{(eff)} + \lambda = \int_0^x \left(A^{(eff)}\begin{pmatrix} u(t) \\ v(t) \end{pmatrix} + p^{(eff)} \right) dt \tag{7.82}$$

It can be shown that

$$B^{(eff)} = \begin{pmatrix} 0 & 0 \\ 0 & 0 \end{pmatrix}; \; A^{(eff)} = \begin{pmatrix} 0 & C_1 - 2C_2 \\ 0 & 0 \end{pmatrix} \tag{7.83}$$

where

$$C_1 = \int_0^1 \frac{dt}{e(t)I(t)}; \, C_2 = \int_0^1 \frac{\left(t - \frac{1}{2}\right)dt}{e(t)I(t)} \tag{7.84}$$

Furthermore, for $f(x)=0$ there holds $p^{(\mathit{eff})} = q^{(\mathit{eff})} = 0$, while, in a general case, $B^{(\mathit{eff})}$ and $A^{(\mathit{eff})}$ do not depend on p and q. Finally, the homogenised ODEs are obtained as

$$\begin{cases} \dfrac{d}{dx}v(x) = f^{(\mathit{eff})} \\ \dfrac{d}{dx}u(x) = \left(C_1 - 2C_2\right)v(x) \end{cases} \tag{7.85}$$

which is essentially different to the classical result of the asymptotic homogenisation shown previously. Effective mass density of a composite can be derived by a spatial averaging method, which is completely independent from the space configuration and periodicity conditions of a composite structure. The relation is used for classical and wavelet–based homogenisation approaches as well. Finally, the following variational equation is proposed to achieve the dynamic equilibrium for the linear elastic system [208]:

$$\int_\Omega \rho \ddot{u}_i \delta u_i \, d\Omega + \int_\Omega C_{ijkl} \varepsilon_{ij} \delta \varepsilon_{kl} \, d\Omega = \int_\Omega \rho f_i \delta u_i \, d\Omega + \int_{\partial\Omega_\sigma} \hat{t}_i \delta u_i \, d(\partial\Omega) \tag{7.86}$$

where u_i represents displacements of the system Ω with elastic properties and mass density defined by the elasticity tensor $C_{ijkl}(x)$ and the function $\rho=\rho(\mathbf{x})$; the vector $\hat{t}_i$ denotes the stress boundary conditions defined on $\partial\Omega_\sigma \subset \partial\Omega$.

An analogous equation rewritten for the homogenised heterogeneous medium has the following form:

$$\int_\Omega \rho^{(\mathit{eff})} \ddot{u}_i \delta u_i \, d\Omega + \int_\Omega C_{ijkl}^{(\mathit{eff})} \varepsilon_{ij} \delta \varepsilon_{kl} \, d\Omega = \int_\Omega \rho^{(\mathit{eff})} f_i \delta u_i \, d\Omega + \int_{\partial\Omega_\sigma} \hat{t}_i \delta u_i \, d(\partial\Omega) \tag{7.87}$$

where all material properties of the real system are replaced with the effective parameters. Let us introduce a discrete representation of the function u_i by the following vector of the generalised coordinates for the needs of the Finite Element Method implementation:

$$u_i(x) = \phi_{i\alpha}(x) q_\alpha = \left[\sum_{e=1}^E \phi_{i\alpha}^{(e)}(x)\right] q_\alpha \tag{7.88}$$

which gives us for the strain tensor components

$$\varepsilon_{ij}(x) = \frac{1}{2}\left[\phi_{i\alpha,j}(x) + \phi_{j\alpha,i}(x)\right]q_\alpha = B_{ij\alpha}(x)q_\alpha \tag{7.89}$$

The matrix description for stiffness, mass, damping components as well as the RHS vector is proposed as

$$K_{\alpha\beta} = \int_\Omega C_{ijkl} B_{ij\alpha} B_{kl\beta} d\Omega \,, \quad K_{ijkl}^{(eff)} = \int_\Omega C_{ijkl}^{(eff)} B_{ij\alpha} B_{kl\beta} d\Omega \tag{7.90}$$

$$M_{\alpha\beta} = \int_\Omega \rho\phi_{i\alpha}\phi_{i\beta} d\Omega \,, \quad M_{\alpha\beta}^{(eff)} = \int_\Omega \rho^{(eff)}\phi_{i\alpha}\phi_{i\beta} d\Omega \tag{7.91}$$

$$Q_\alpha = \int_\Omega \rho f_i \phi_{i\alpha} d\Omega + \int_{\partial\Omega_\sigma} \hat{t}_i \phi_{i\alpha} d(\partial\Omega) \tag{7.92}$$

$$Q_\alpha^{(eff)} = \int_\Omega \rho^{(eff)} f_i \phi_{i\alpha} d\Omega + \int_{\partial\Omega_\sigma} \hat{t}_i \phi_{i\alpha} d(\partial\Omega)$$

Usually, it is assumed that the damping matrix can be decomposed into the part having the nature of body forces with the proportionality coefficient c_M and the rest composes the viscous stresses multiplied by the quantity c_K, so that

$$C_{\alpha\beta} = c_M M_{\alpha\beta} + c_K K_{\alpha\beta} \,, \quad C_{\alpha\beta}^{(eff)} = c_M M_{\alpha\beta}^{(eff)} + c_K K_{\alpha\beta}^{(eff)} \tag{7.93}$$

After such a discretisation of all the state functions and structural parameters in (7.86) and (7.87), the following matrix equation for real heterogeneous system is obtained:

$$M_{\alpha\beta}\ddot{q}_\beta + C_{\alpha\beta}\dot{q}_\beta + K_{\alpha\beta}q_\beta = Q_\alpha \tag{7.94}$$

Therefore, the equivalent homogenised dynamic equilibrium equation to be solved for the deterministic problem has the form

$$M_{\alpha\beta}^{(eff)}\ddot{\bar{q}}_\beta + C_{\alpha\beta}^{(eff)}\dot{\bar{q}}_\beta + K_{\alpha\beta}^{(eff)}\bar{q}_\beta = Q_\alpha^{(eff)} \tag{7.95}$$

where the barred unknowns represent the response of the homogenised system. The RHS vector is equal to 0, so the homogenised operators are to be computed for the LHS components only in the case of free vibrations. The eigenvalues and eigenvectors for the undamped systems are determined from the following matrix equations:

$$\left(K_{\alpha\beta} - \omega_{(\alpha)}M_{\alpha\beta}\right)\Phi_{\beta\gamma} = 0; \quad \left(K_{\alpha\beta}^{(eff)} - \overline{\omega}_{(\alpha)}M_{\alpha\beta}^{(eff)}\right)\overline{\Phi}_{\beta\gamma} = 0 \tag{7.96}$$

which are implemented and applied below to compare homogenised and real composites.

Numerical analysis illustrating presented ideas is carried out in two separate steps. First, homogenised characteristics of a periodic composite determined thanks to different homogenisation models are obtained by the use of the MAPLE symbolic computation. Then, the FEM analysis of the free vibration problems is made for the simply supported two-, three- and five-bay periodic beams, made of the original and homogenised composites, having applications in aerospace and other engineering structures subjected to vibrations [189]. The periodicity is observed in macroscale (equal length of each bay) as well as in microstructure – each bay is obtained by reproduction of the identical RVE whose elastic modulus is defined by some wavelet function.

The formulae presented above are implemented in the program MAPLE together with the spatial averaging method in order to compare the homogenised modulus computed by various ways (spatial averaging, classical and multiresolutional) for the same composite. Figure 7.22 illustrates the variability of this modulus along the RVE, where the function $e(x)$ is subtracted from the following Haar and Mexican hat wavelets:

$$h(x) = \begin{cases} 20.0E9; \ 0 \le x \le 0.5 \\ 2.0E9 \ 0.5 < x \le 1 \end{cases} \tag{7.97}$$

$$m(x) = 2 + \frac{1}{\sqrt{2\pi}\sigma^3} \frac{x^2}{\sigma^2 - 1} \exp\left(\frac{-x^2}{2\sigma^2}\right), \ \sigma=-0.4 \tag{7.98}$$

as

$$e(x) = 10.0h(x) + 2.0E9\ m(x) \tag{7.99}$$

Mass density of the composite is adopted as the wavelet of similar nature

$$\tilde{h}(x) = \begin{cases} 200; \ 0 \le x \le 0.5 \\ 20; \ 0.5 < x \le 1 \end{cases} \tag{7.100}$$

with

$$\rho(x) = 0.5\ \tilde{h}(x) + 0.5\ m(x) \tag{7.101}$$

which is displayed in Figure 7.23.

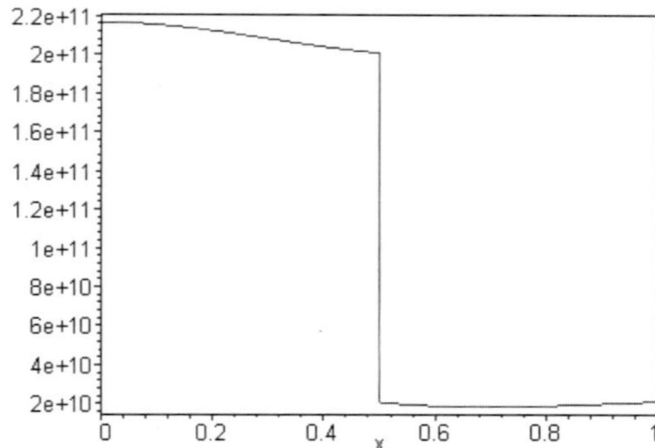

Figure 7.22. Wavelet-based definition of elastic modulus in the RVE

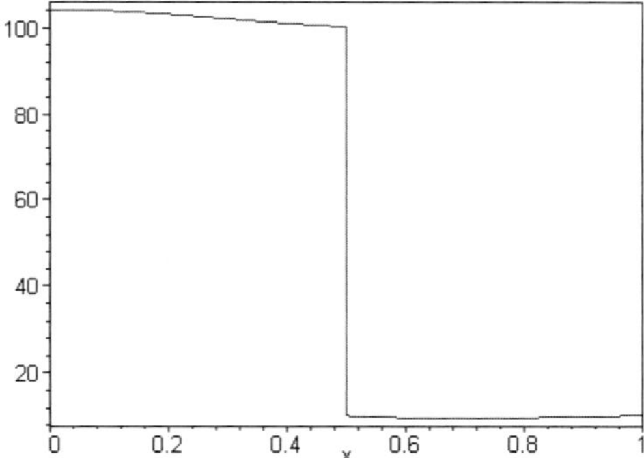

Figure 7.23. Wavelet-based definition of mass density in the RVE

The final form of these functions is established on the basis of the mathematical conditions for homogenisability analysed before as well as to obtain the final variability of composite properties similar to the traditional multi-component structures. Let us note that classical definition of periodic composite material properties contained the piecewise constant Haar basis only.

The following homogenised material properties are obtained from this input:

$$\rho^{(eff)} = 56.137 \, , e^{(av)} = 114.548 \, E9 > \quad e^{(eff,wav)} = 60.217 \, E9 > \quad e^{(eff)} = 35.437 \, E9 \, ,$$

which means that for this particular example, the highest value is obtained for the spatial averaging method, then – for the wavelet approach at least – for classical homogenisation method based on the small parameter assumption. The effectiveness of such homogenisation results is verified in the next section by comparison of the eigenvalues and the eigenfunctions of some periodic composite beams being homogenised with its real material distribution.

The free vibration problems for two–, three– and five–bay periodic beams are solved using the classical and homogenisation–based Finite Element Method implementation [13,387]. The unitary inertia momentum is taken in all computational cases, ten periodicity cells compose each bay, while material properties inserted in the numerical model are calculated from (a) spatial averaging, (b) the classical homogenisation method and (c) the multiresolutional scheme proposed above and compared against the real structure response. The results of eigenproblem solutions are presented as the first 10 eigenvalue variations for the beams in Figures 7.24, 7.26 and 7.28 together with the maximum deflections of these beams in Figures 7.25, 7.27 and 7.29 – the resulting values are marked on the vertical axes, while the number of the eigenvalue being computed is on the horizontal axes. The particular solutions for 1st, 2nd, 3rd and lower next eigenvalues are connected with the continuous lines to better illustrate interrelations between the results obtained in various homogenisation approaches related to the real composite model.

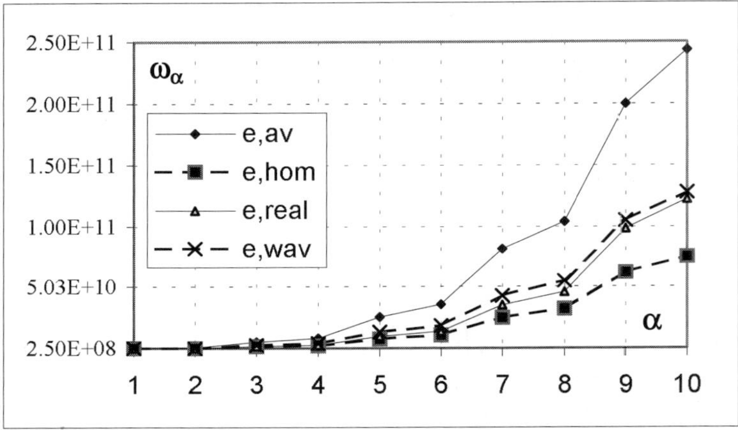

Figure 7.24. Eigenvalues progress for various two−bay composite structures

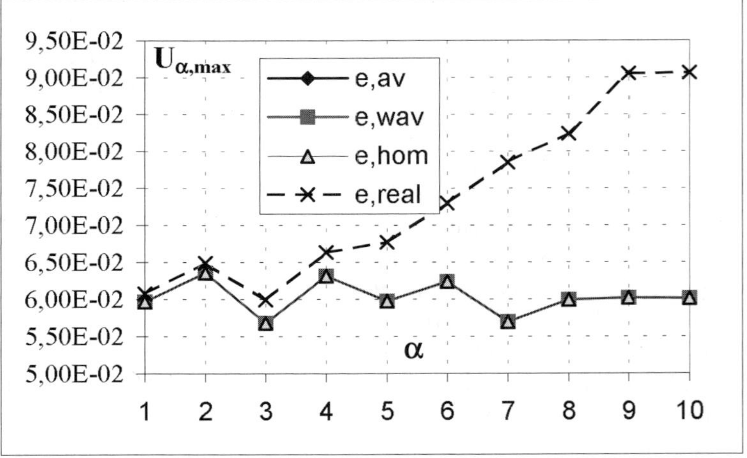

Figure 7.25. Maximum deflections for the eigenproblems of two−bay composite structures

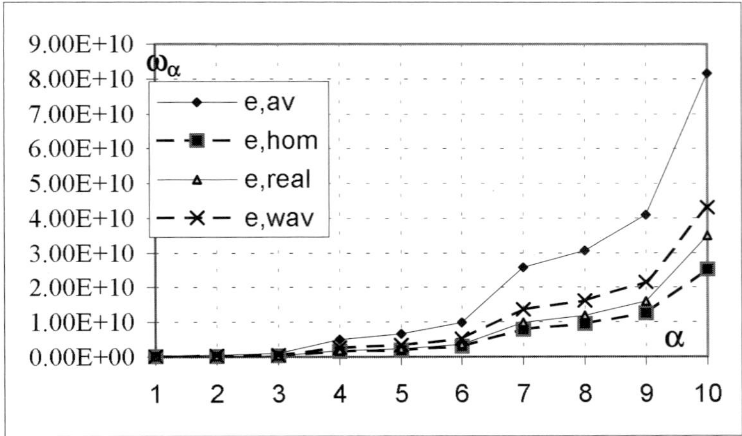

Figure 7.26. Eigenvalues progress for various three−bay composite structures

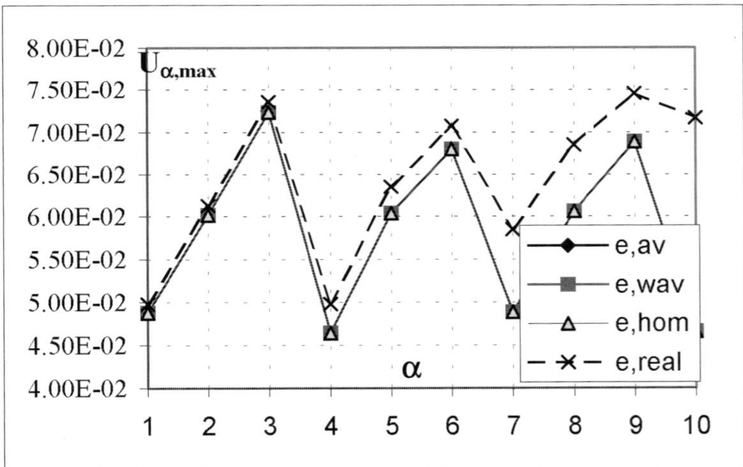

Figure 7.27. Maximum deflections for the eigenproblems of three−bay composite structures

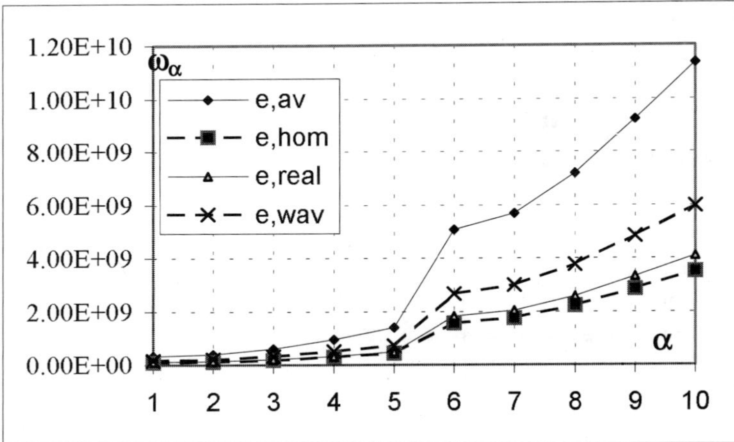

Figure 7.28. Eigenvalues progress for various five–bay composite structures

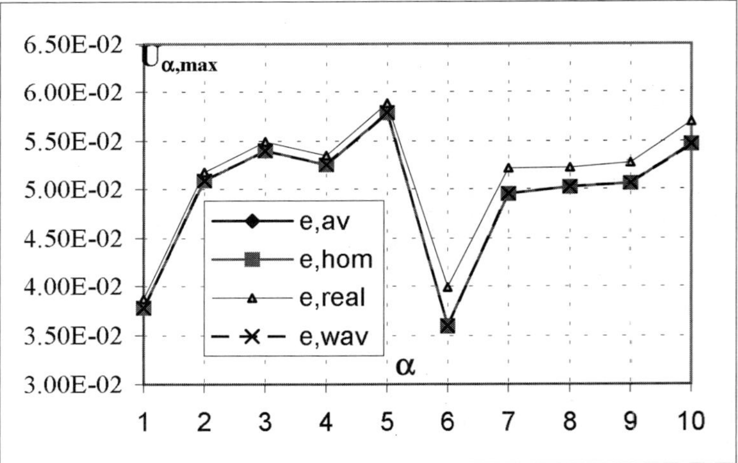

Figure 7.29. Maximum deflections for the eigenproblems of five–bay composite structures

As can be observed, the eigenvalues obtained for various homogenisation models approximate the values computed for the real composite with different accuracies, and the maximum deflections are the same. The weakest efficiency in eigenvalue modelling is detected in the case of a spatially averaged composite – the difference in relation to the real structure results increases together with the eigenvalue number. Wavelet–based and classical homogenisation methods give more accurate results – the first method is better for smaller numbers of bays (and the RVEs along the beam) see Figure 7.24, whereas the classical homogenisation approach is recommended in the case of increasing number of the bays and the RVEs, cf. Figures 7.26 and 7.28. The justification of this observation comes from the fact that the wavelet function appears to be of less importance for the increasing number of periodicity cells in the structure. Another interesting result is that the efficiency of the approximation of the maximum deflections for a multibay periodic composite beam by the deflections encountered for homogenised systems increases together with an increase of the total number of bays. The agreement between the eigenvalues for the real and homogenised systems will allow usage of the stochastic spectral finite element techniques [261], where the random process expansions are based on the relevant eigenvalues.

Finally, let us note that further extensions of this model on vibration analysis of fibre-reinforced composites [60] using 2D wavelets are possible. An application of wavelet technique is justified by the fact that the spatial distribution of the constituents in the composite specimen is recently a subject of digital image analysis [341]. On the other hand, chaotic behaviour of real and homogenised composites [199] may be studied in the above context.

7.6 Multiscale Heat Transfer Analysis

The idea of transient heat transfer homogenisation, i.e. calculation of the effective material parameters, consists in separate spatial averaging of the volumetric heat capacity and the solution (analytical or numerical) of the heat conduction homogenisation problem [15,165,166,195]. As is illustrated below, the final form of the effective heat conductivity coefficient varies with the composite model, whereas a composite with piecewise constant properties and/or defined by some wavelet functions can have the same homogenised volumetric heat capacity. That is why first the heat conduction equation for a 1D periodic composite is homogenised and the effective heat capacity and mass density are determined by a spatial averaging approach. The multiresolutional homogenisation method starts from the following decomposition of heat conduction equation [23,55] as follows:

$$\begin{cases} \dfrac{d}{dx} v(x) = -Q(x) \\ \dfrac{d}{dx} T(x) = \dfrac{v(x)}{k(x)} \end{cases} \qquad (7.102)$$

The main goal is to determine the homogenised coefficient $k^{(eff)}$ being constant over the interval $x \in [0,1]$. Therefore, the equation system (7.102) can be rewritten as

$$\begin{pmatrix} T(x) \\ v(x) \end{pmatrix} - \begin{pmatrix} T(0) \\ v(0) \end{pmatrix} = \int_0^x \left(\begin{pmatrix} 0 & k(t)^{-1} \\ 0 & 0 \end{pmatrix} \begin{pmatrix} T(t) \\ v(t) \end{pmatrix} + \begin{pmatrix} 0 \\ -Q(t) \end{pmatrix} \right) dt \qquad (7.103)$$

On the other hand, the reduction algorithm between multiple scales of the composite consists in the determination of such effective operators $B^{(eff)}$, $A^{(eff)}$, $p^{(eff)}$, $q^{(eff)}$, that

$$\left(I + B^{(eff)}\right)\binom{T(x)}{v(x)} + q^{(eff)} + \lambda = \int_0^x \left(A^{(eff)}\binom{T(t)}{v(t)} + p^{(eff)} \right) dt \tag{7.104}$$

It can be shown that

$$B^{(eff)} = \begin{pmatrix} 0 & 0 \\ 0 & 0 \end{pmatrix}; \quad A^{(eff)} = \begin{pmatrix} 0 & k_1 - 2k_2 \\ 0 & 0 \end{pmatrix} \tag{7.105}$$

where

$$k_1 = \int_0^1 \frac{dt}{k(t)}; \quad k_2 = \int_0^1 \frac{\left(t - \frac{1}{2}\right)dt}{k(t)} \tag{7.106}$$

Furthermore, for $Q(x)=0$ there holds $p^{(eff)} = q^{(eff)} = 0$ (in a general case, $B^{(eff)}$ and $A^{(eff)}$ do not depend on p and q). Finally, the system of two homogenised ordinary differential equations are obtained as

$$\begin{cases} \dfrac{d}{dx} v(x) = q^{(eff)} \\[2mm] \dfrac{d}{dx} T(x) = \left(k_1 - 2k_2\right)v(x) \end{cases} \tag{7.107}$$

which is essentially different than the classical result of the asymptotic homogenisation shown previously. Let us observe that in the case of the heat conductivity variability in two separate scales $k = k\left(x, \dfrac{x}{\varepsilon}\right)$ the multiresolutional scheme reduces to the classical macro–micro methodology where the following limits are demonstrated:

$$\lim_{\varepsilon \to 0} k_1(\varepsilon) = k_1 \quad \text{and} \quad \lim_{\varepsilon \to 0} k_2(\varepsilon) = 0 \tag{7.108}$$

Finally, the effective volumetric heat capacity of a composite is determined by the spatial averaging method, which relation does not depend either on the space configuration or on the periodicity conditions of a composite structure, and is used for both classical and multiresolutional homogenisation approaches.

Using traditional FEM discretisation of the temperature field and its gradients by the nodal temperatures vector θ_α [7,21,213,283]

$$T(y) = H_\alpha(y)\theta_\alpha \; ; \quad \alpha=1,...,N \tag{7.109}$$

$$T_{,y}(y) = H_{\delta,y}(y)\theta_\delta \; ; \quad \delta=1,...,N \tag{7.110}$$

the following transient problems are solved:
- averaged material properties

$$C_{\delta\beta}^{(av)}\dot{\theta}'_\beta + K_{\delta\beta}^{(av)}\theta'_\beta = P_\delta^{(av)}, \quad \delta,\beta=1,2,...,N \;, \tag{7.111}$$

- asymptotically homogenised material properties

$$C_{\delta\beta}^{(eff)}\dot{\theta}''_\beta + K_{\delta\beta}^{(eff)}\theta''_\beta = P_\delta^{(eff)}, \quad \delta,\beta=1,2,...,N \;, \tag{7.112}$$

- for multiresolutionally homogenised material properties in the system

$$C_{\delta\beta}^{(eff)w}\dot{\theta}'''_\beta + K_{\delta\beta}^{(eff)w}\theta'''_\beta = P_\delta^{(eff)w}, \quad \delta,\beta=1,2,...,N \;. \tag{7.113}$$

Numerical analysis illustrating the ideas presented is carried out in two separate steps. First, homogenised characteristics of a periodic composite obtained through different homogenisation models are determined by the use of MAPLE symbolic computations. This numerical approach is used also to verify input parameter variability of the homogenised characteristics as well as design sensitivities of these characteristics with respect to the contrast parameter (interrelation between the heat conductivities of the composite components) and the interface location along the RVE length (g). Next, the FEM analysis of transient heat transfer is made to discuss the differences between temperature and heat flux histories resulting from various homogenisation models contrasted with the real system. An alternative way to model multiscale transient heat transfer phenomena in composites is to expand the classical FEM methodology using a wavelet based both space and time adaptive numerical methods, as it was discussed in [17], for instance; the other aspects of this problem have been studied in [40].

The formulae for effective heat conductivity are implemented in the program MAPLE together with the spatial averaging method in order to compare the homogenised modulus computed by various ways for the same composite. Figure 7.30 illustrates the variability of this modulus along the RVE, where the function $k(x)$ is subtracted from the following Haar basis and Mexican hat wavelet:

$$h(x) = \begin{cases} k_1 \; ; \; 0 \le x \le 0.5 \\ k_2 \; ; \; 0.5 < x \le 1 \end{cases} \tag{7.114}$$

$$m(x) = 2 + \frac{1}{\sqrt{2\pi}\sigma^3}\frac{x^2}{\sigma^2-1}\exp\left(\frac{-x^2}{2\sigma^2}\right), \quad \sigma=-0.5 \tag{7.115}$$

as

$$k(\,x\,) = h(\,x\,) + 0.001\,m(\,x\,) \tag{7.116}$$

Further, volumetric heat capacity of the composite is adopted as the wavelet of a similar form

$$\tilde{h}(x) = \begin{cases} \rho_1 c_1;\ 0 \le x \le 0.5 \\ \rho_2 c_2;\ 0.5 < x \le 1 \end{cases} \tag{7.117}$$

with

$$\rho(\,x\,)c(\,x\,) = \tilde{h}(\,x\,) + 10^3\,m(\,x\,) \tag{7.118}$$

which is demonstrated in Figure 7.31.

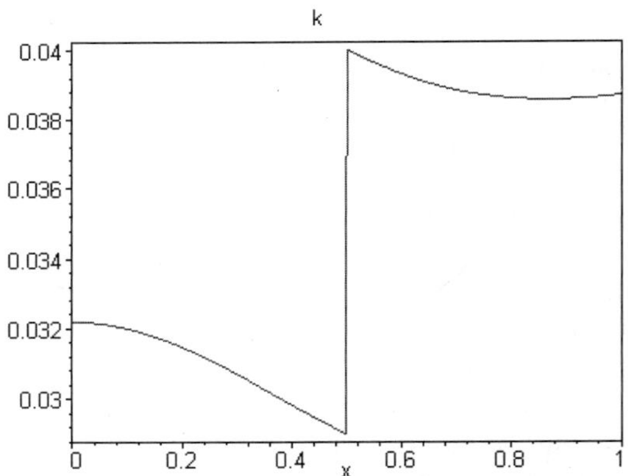

Figure 7.30. Wavelet-based definition of heat conductivity coefficient in RVE

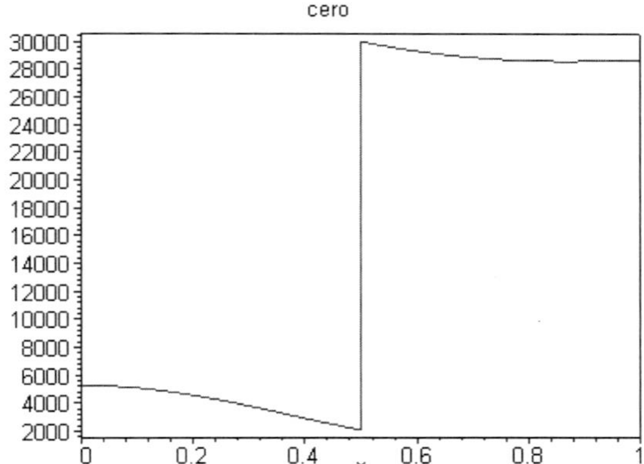

Figure 7.31. Wavelet-based definition of the volumetric heat capacity in RVE

The final form of these functions is established on the basis of the mathematical conditions for homogenisability analysed before as well as to obtain the final variability of composite properties similar to the traditional multi-component structures. Let us note that the classical definition of periodic composite material properties contained the piecewise constant Haar basis only.

Symbolic computations of the MAPLE system were used next to perform the comparison between the spatial averaging, classical and multiresolutional homogenisation scheme for various values of the composite constituents contrast and the interface position g. The results of the analysis are demonstrated in Figures 7.32, 7.33 and 7.34, respectively. However it could be expected, the results of spatial averaging are globally the greatest for the entire variability ranges of the design parameters, while the interrelation between the classical and wavelet-based methods differ on the input parameter values.

The separate, very interesting numerical problem would be to determine the intersection of the surfaces plotted in Figures 7.33 and 7.34. It can be interpreted as the curve equivalent to such pairs of the contrast and interface location in the RVE for which both multiresolutional and classical homogenisation methods can result in the same effective quantity. Let us note that the problem is independent from physical interpretation of homogenised characteristics).

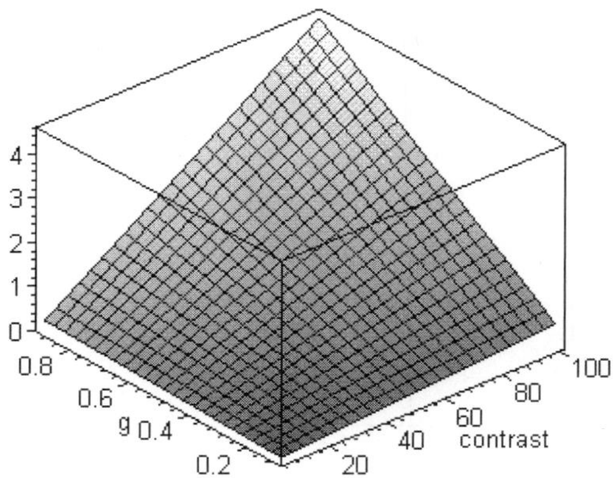

Figure 7.32. Parameter variability of $k^{(av)}$

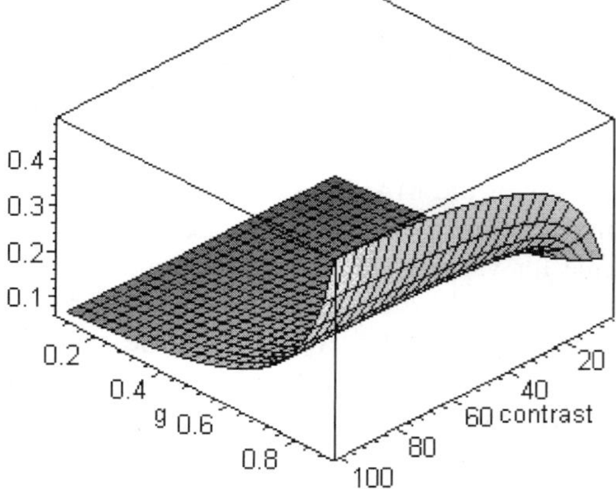

Figure 7.33. Parameter variability of $k^{(eff)}$

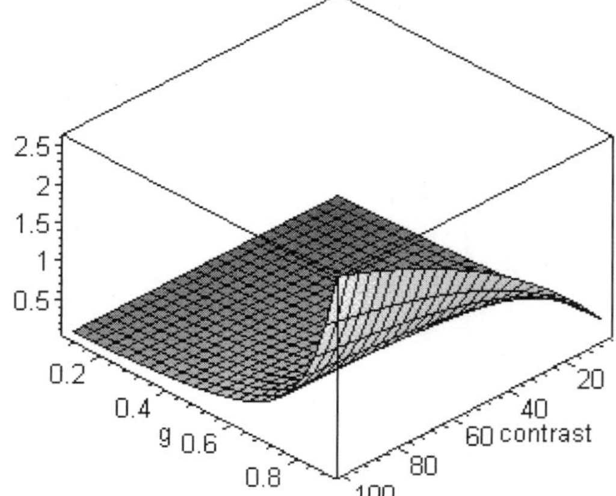

Figure 7.34. Parameter variability of $k^{(eff)w}$

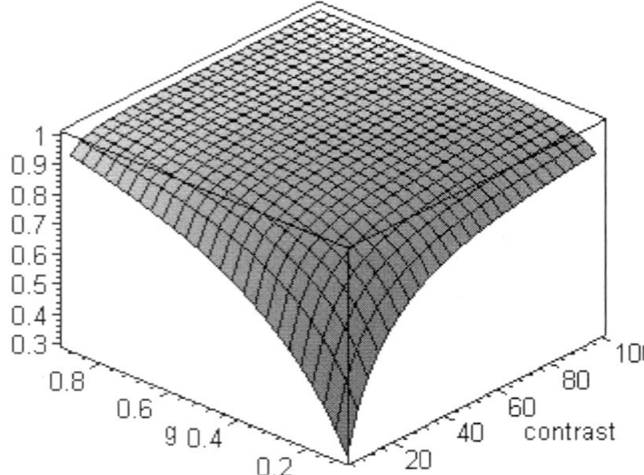

Figure 7.35. Sensitivity of $k^{(av)}$ wrt contrast parameter

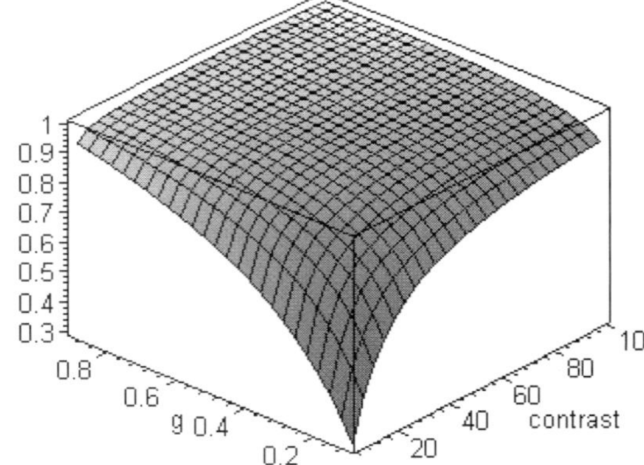

Figure 7.36. Sensitivity of $k^{(av)}$ wrt the interface location

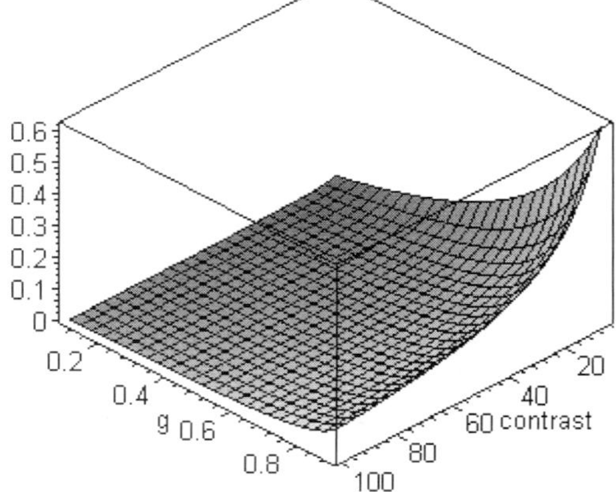

Figure 7.37. Sensitivity of $k^{(eff)}$ coefficient wrt components contrast

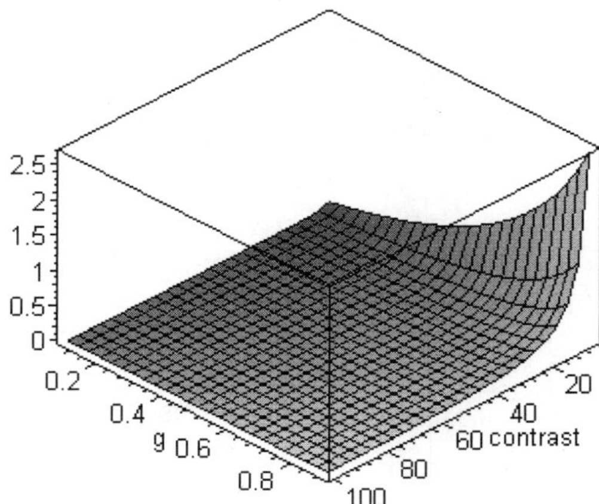

Figure 7.38. Sensitivity of $k^{(eff)}$ wrt interface location

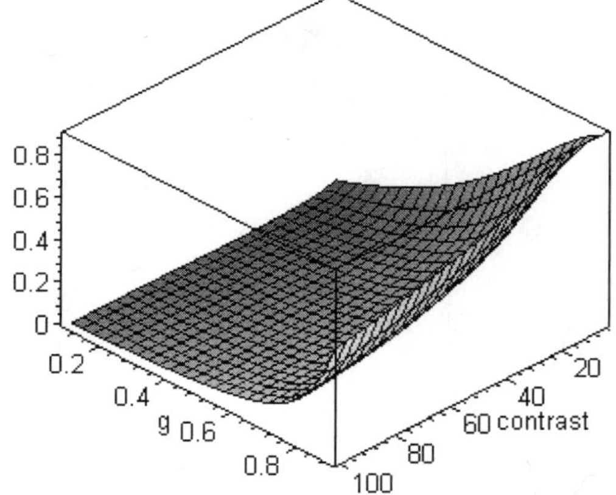

Figure 7.39. Parameter sensitivity of $k^{(eff)w}$ wrt contrast parameter

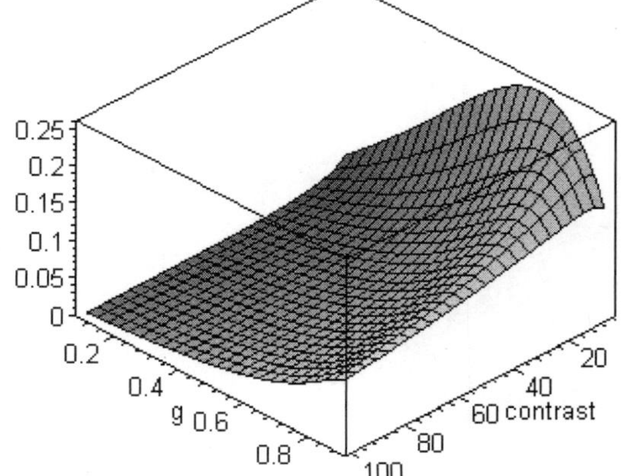

Figure 7.40. Parameter sensitivity of $k^{(eff)w}$ wrt interface location

Partial derivatives of the averaged, asymptotically and multiresolutionally homogenised heat conductivity are normalised using the factor h/k where h denotes the contrast or the parameter g, while $k \equiv \{k^{(av)}, k^{(eff)}, k^{(eff)w}\}$. The results of symbolic computations are presented in Figures 7.35–7.40 and it is clear that the spatial averaging method results in the composite with an extremely different parameter sensitivity in comparison to the other homogenisation models (both quantitatively and qualitatively). Sensitivity gradients for asymptotic and multiresolutional homogenisations have very analogous surfaces – the only differences are observed for higher values of the design parameters. The numerical results obtained can be effectively used in the optimisation of composite materials according to the methodology based on the homogenisation approach. Moreover, they can be applied to the homogenisation of random composites where first and second order parameter sensitivities are necessary to determine the first two probabilistic moments of the effective parameter in the second order perturbation approach at least.

The transient heat transfer phenomenon in a two–layer unidirectional composite structure has been modelled using the commercial Finite Element Method program ANSYS [2]. The division of the periodicity cell with unit length $L=1.0$ m into two components with equal lengths and 1000 of 4–noded isoparametric heat transfer finite elements PLANE55 (500 elements for each material) is schematically shown in Figure 7.41. Constant temperature $T=0$ is applied at the left boundary and the unit heat flux Q at the right edge, whereas initial temperatures along the composite are taken as equal to 0. Material properties used in numerical analysis are calculated for (a) real composite structure – test no 1, (b) spatially averaged composite – test no 2, (c) classical homogenisation method – test no 3, and (d) multiresolutional homogenisation scheme proposed now – test no 4. Input material data for particular computational tests are collected in Table 7.2 below.

Table 7.2. Material data for the FEM analysis

Computational test number	k [W/m°C]	c [J/kg°C]
1	0.031 / 0.0385	4000 / 29000
2	0.0349	16465.20
3	0.0345	16465.20
4	0.0328	16465.20

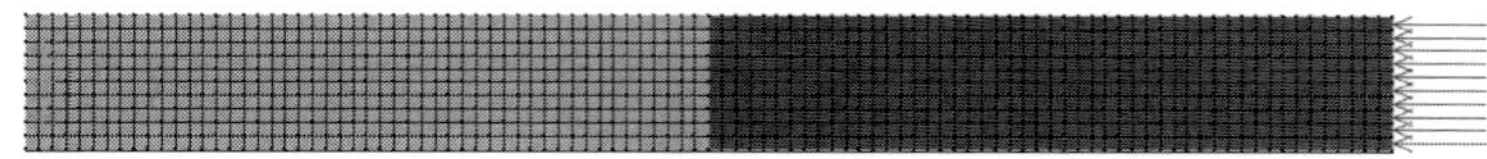

Figure 7.41. Finite Element mesh for the composite structure

The results for the steady–state analysis are shown in Figures 7.42–7.45 in the form of a spatial temperature distribution and the analogous heat flux distribution along the composite; their error approximations are computed and visualised also.

Considering the nonstationary character of the transient heat transfer, the temperature distributions for various moments of the heating process are collected in Figures 7.46–7.53. Analysing the temperatures fields along the composite structure it can be observed that the best agreement with the structural behaviour is obtained for the test related to the multiresolutionally homogenised composite. The classical homogenisation method gives more accurate results in the neighbourhood of the heated surface only. In the case of temperature gradients it can be concluded that the wavelet–based homogenisation approach gives the highest averages temperature gradient and greater than the classical method and spatial averaging, respectively. It is important considering reliability analysis based on the homogenisation methods; this gradient is however a few percent smaller than the maximum gradient for the real composite.

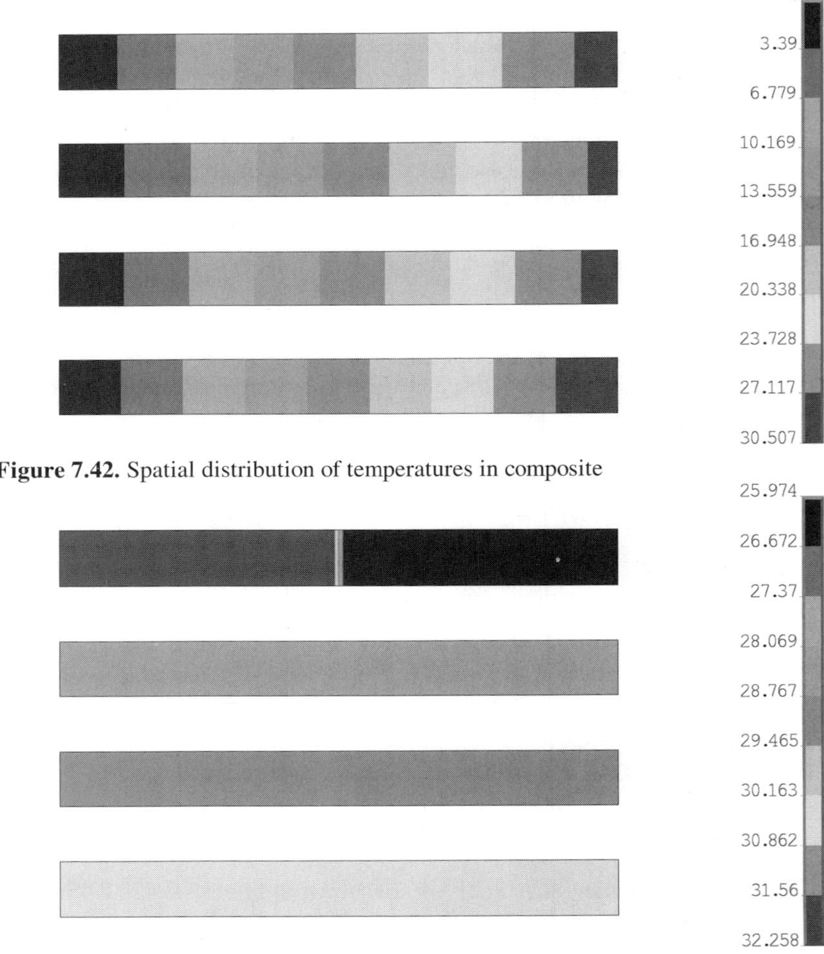

Figure 7.42. Spatial distribution of temperatures in composite

Figure 7.43. Temperature gradients along the composite

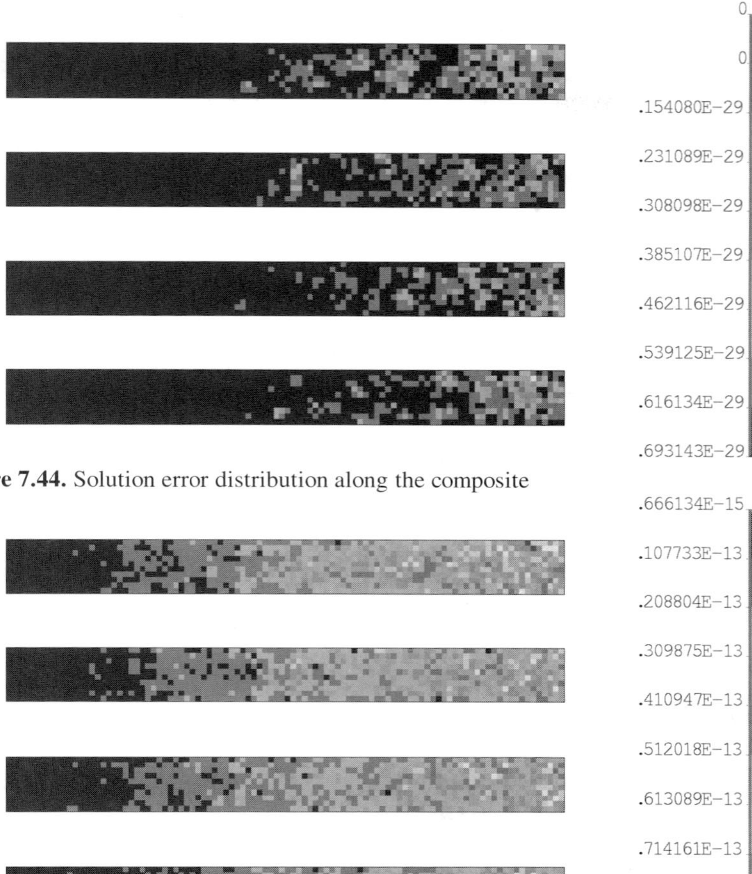

Figure 7.44. Solution error distribution along the composite

Figure 7.45. Temperature gradient error along the composite

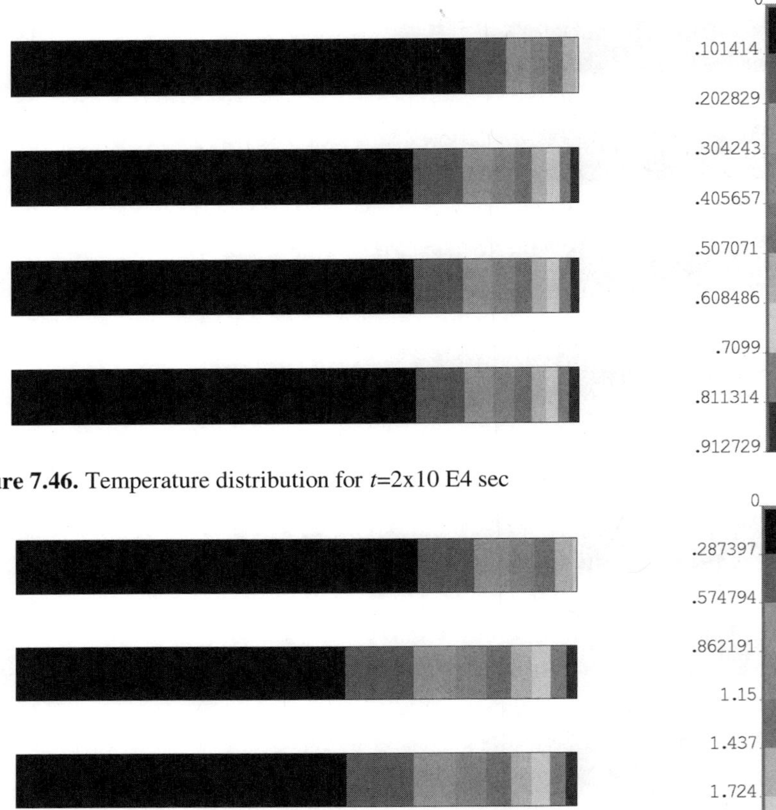

Figure 7.46. Temperature distribution for t=2x10 E4 sec

Figure 7.47. Temperature distribution for t=4x10 E4 sec

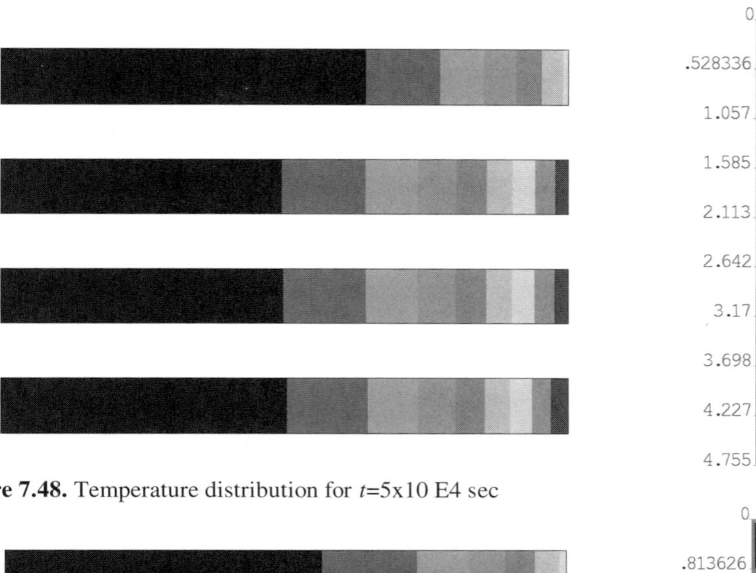

Figure 7.48. Temperature distribution for t=5x10 E4 sec

Figure 7.49. Temperature distribution for t= 8x10 E4 sec

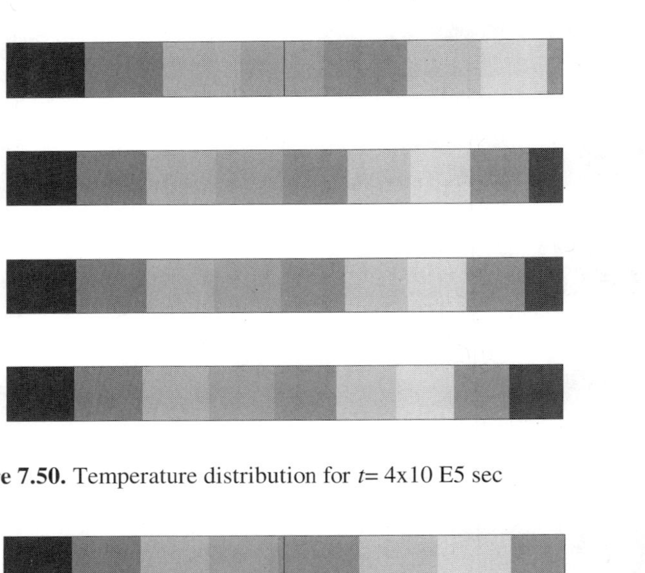

Figure 7.50. Temperature distribution for $t= 4 \times 10$ E5 sec

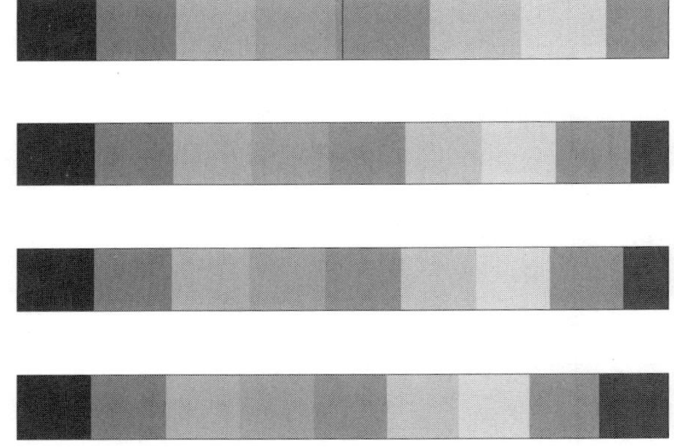

Figure 7.51. Temperature distribution for $t=6 \times 10$ E5 sec

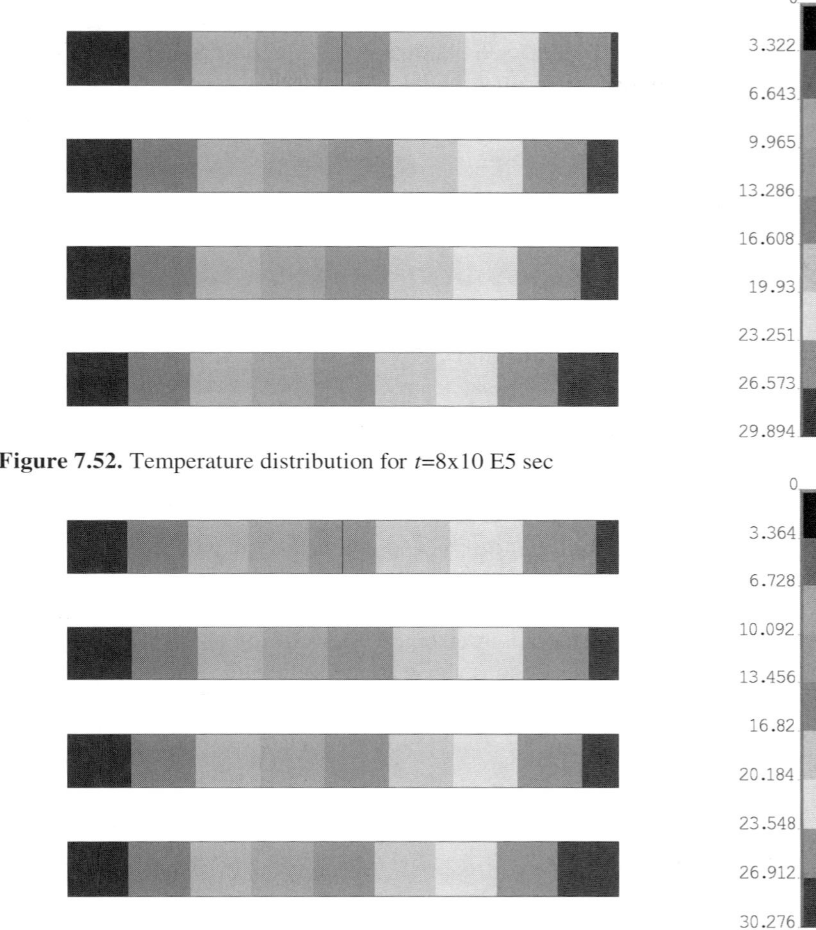

Figure 7.52. Temperature distribution for t=8x10 E5 sec

Figure 7.53. Temperature distribution for t=1x10 E6 sec

The temperature solution error related to the real composite behaviour numerical tests is best approximated by the error computed for the structure homogenised by the wavelet−based methodology also − it shows analogous spatial distribution and maximum values, although spatial distribution is analogous in all cases as well. In further analysis the results obtained should be contrasted with the implementation of the wavelet decomposition of initial material properties in the Finite Element Method program.

Finally, transient behaviour of the composite is analysed numerically and presented for various time moments of the heating process in Figures 7.46−7.53. The real composite is heated at the boundary relevant to the material with higher volumetric heat capacity and the contrast between heat capacities is very high. That is why the heating process in the real composite is very slow − significantly slower than takes place in all homogenised models (Figure 7.53 corresponds to almost a steady state for comparison). The opposite relation can be noticed in the case of inverted materials in the analysed laminate. Neglecting temperature scale differences between the real and effective models, the best approximation for the original structure behaviour is done by the spatially averaged system.

7.7 Stochastic Perturbation–based Approach to the Wavelet Decomposition

Let us consider a multiresolutional wavelet–based algorithm and its application in the solution of the linear algebraic equations system [334] being a basis for various discrete numerical techniques [206]. There holds

$$\mathbf{K}\mathbf{q} = \mathbf{f} \qquad (7.119)$$

where the matrix K is positive definite and represents the behaviour of some linear engineering system, q is a discretised vector of the engineering system response resulting from the excitation expressed by a vector f. Further, let us assume for the needs of the algorithm applicability, that matrix $\mathbf{K}$ is of the size $2^n \mathrm{x} 2^n$ and let us introduce the Haar transform for the vector $\mathbf{q}$ in the following way:

$$\mathbf{s}^{(k)} = \frac{1}{\sqrt{2}}\left(\mathbf{q}^{(2k-1)} + \mathbf{q}^{(2k)}\right) \qquad (7.120)$$

$$\mathbf{d}^{(k)} = \frac{1}{\sqrt{2}}\left(\mathbf{q}^{(2k-1)} - \mathbf{q}^{(2k)}\right) \qquad (7.121)$$

with $k=1,...,2^{n-1}$. Let us observe that $\mathbf{s}^{(k)}$ are introduced to scale averages of the vector $\mathbf{q}$ values in the neighbouring points while $\mathbf{d}^{(k)}$ is to scale their differences. Let us introduce the matrix M_n such that

$$\mathbf{M} = M_n = \frac{1}{\sqrt{2}}\begin{bmatrix} 1 & 1 & 0 & 0 & & & \cdots \\ 0 & 0 & 1 & 1 & 0 & 0 & \cdots \\ & & & & & & \cdots \\ 1 & -1 & 0 & 0 & & & \cdots \\ 0 & 0 & 1 & -1 & 0 & 0 & \cdots \\ & & & & & & \cdots \\ \cdots & \cdots & \cdots & \cdots & \cdots & \cdots & \cdots \end{bmatrix} \qquad (7.122)$$

having dimensions $2^n \mathrm{x} 2^n$ and such that

$$M_n^T M_n = M_n M_n^T = I \qquad (7.123)$$

whose top half is denoted by L_n, while the bottom one is H_n. Then, the orthogonality gives

$$M_n^T M_n = H_n^T H_n + L_n^T L_n = I \qquad (7.124)$$

and

$$H_n^T H_n = I \, , \; L_n^T L_n = I \qquad (7.125)$$

where

$$\mathbf{Lq = s}, \quad \mathbf{Hq = d} \qquad (7.126)$$

Let us rewrite (7.119) in the form of a pair of equations with unknown $\mathbf{s}$ and $\mathbf{d}$ as follows:

$$\mathbf{LKq} = \left(\mathbf{LKL}^T\right)\mathbf{Lq} + \left(\mathbf{LKH}^T\right)\mathbf{Hq} = \mathbf{Lf} \qquad (7.127)$$

Similarly, there holds

$$\mathbf{HKq} = \left(\mathbf{HKL}^T\right)\mathbf{Lq} + \left(\mathbf{HKH}^T\right)\mathbf{Hq} = \mathbf{Hf} \qquad (7.128)$$

Denoting further by

$$\mathbf{LKL}^T = \mathbf{T}, \mathbf{LKH}^T = \mathbf{C} \qquad (7.129)$$

and

$$\mathbf{HKL}^T = \mathbf{B}, \mathbf{HKH}^T = \mathbf{A} \qquad (7.130)$$

as well as

$$\mathbf{Lf} = \mathbf{f}_s, \mathbf{Hf} = \mathbf{f}_d \qquad (7.131)$$

we obtain (7.131) as

$$\begin{cases} \mathbf{Ts + Cd = f}_s \\ \mathbf{Bs + Ad = f}_d \end{cases} \qquad (7.132)$$

Assuming that $\mathbf{A}$ is invertible, the unknown $\mathbf{d}$ can be eliminated from (7.132) to get a reduced system of equations, and finally to calculate $\mathbf{s}$. Therefore

$$\mathbf{d} = -\mathbf{A}^{-1}\mathbf{B}\mathbf{s} + \mathbf{A}^{-1}\mathbf{f}_d \tag{7.133}$$

and, by substitution of (7.133) into (7.132) it is obtained that

$$\left(\mathbf{T} - \mathbf{C}\mathbf{A}^{-1}\mathbf{B}\right)\mathbf{s} = \mathbf{f}_s + \mathbf{C}\mathbf{A}^{-1}\mathbf{f}_d \tag{7.134}$$

The procedure of transformation of (7.133) is called a reduction step – the total number of unknowns is reduced here two times. Let us introduce the following recursions:

$$\begin{cases} \mathbf{K}_0 = \mathbf{K}, \mathbf{f}_0 = \mathbf{f} \\ \mathbf{K}_1 = \mathbf{T}_0 - \mathbf{C}_0\mathbf{A}_0^{-1}\mathbf{B}_0, \mathbf{f}_1 = \mathbf{L}\mathbf{f}_0 + \mathbf{C}_0\mathbf{A}_0^{-1}\mathbf{H}_0\mathbf{f}_0 \end{cases} \tag{7.135}$$

Since that, we obtain

$$\mathbf{K}_1\mathbf{s}_1 = \mathbf{f}_1 \tag{7.136}$$

which is similar to the starting equations where the unknown is given as $\mathbf{s}_1 = \mathbf{L}\mathbf{q}$. The process shown above can be repeated up to n times according to the following recursion:

$$\begin{cases} \mathbf{K}_{j+1} = \mathbf{T}_j - \mathbf{C}_j\mathbf{A}_j^{-1}\mathbf{B}_j \\ \mathbf{f}_{j+1} = \mathbf{L}_j\mathbf{f}_j + \mathbf{C}_j\mathbf{A}_j^{-1}\mathbf{H}_j\mathbf{f}_j \end{cases} \tag{7.137}$$

where

$$\begin{cases} \mathbf{T}_j = \mathbf{L}_j\mathbf{K}_j\mathbf{L}_j^T \\ \mathbf{B}_j = \mathbf{H}_j\mathbf{K}_j\mathbf{L}_j^T \\ \mathbf{C}_j = \mathbf{L}_j\mathbf{K}_j\mathbf{H}_j^T \\ \mathbf{A}_j = \mathbf{H}_j\mathbf{K}_j\mathbf{H}_j^T \end{cases} \tag{7.138}$$

It is seen that considering the dimensions of the initial stiffness matrix in the form of $2n \times 2n$, then repeating the reduction scheme n times, the resulting equation has the single scalar unknown where the general unknown reconstruction scheme is given by the formula:

$$\mathbf{q}^{(2k-1)} = \frac{1}{\sqrt{2}}\left(\mathbf{s}^{(k)} + \mathbf{d}^{(k)}\right)$$
(7.139)

$$\mathbf{q}^{(2k)} = \frac{1}{\sqrt{2}}\left(\mathbf{s}^{(k)} - \mathbf{d}^{(k)}\right)$$
(7.140)

An analogous situation takes place when $\mathbf{K}$ is a stochastic linear operator describing the behaviour of some engineering system, $\mathbf{f}$ is a random external excitation, while $\mathbf{q}$ represents the random response of the system. Then, using second order perturbation theory, (7.119) can be expanded as follows:

$$\begin{cases} \mathbf{K}^0\mathbf{q}^0 = \mathbf{f}^0 \\ \mathbf{K}^0\mathbf{q}^{,r} = \mathbf{f}^{,r} - \mathbf{K}^{,r}\mathbf{q}^0 \\ \mathbf{K}^0\mathbf{q}^{(2)} = \left(\mathbf{f}^{,rs} - 2\mathbf{K}^{,r}\mathbf{q}^{,s} - \mathbf{K}^{,rs}\mathbf{q}^0\right)Cov\left(b^r, b^s\right) \end{cases}$$
(7.141)

where the first two probabilistic moments of the response are calculated as

$$E[\mathbf{q}] = \mathbf{q}^0 + \tfrac{1}{2}\mathbf{q}^{,rs}Cov\left(b^r, b^s\right)$$
(7.142)

and

$$Cov\left(\mathbf{q}^r, \mathbf{q}^s\right) = \mathbf{q}^{,r}\mathbf{q}^{,s}Cov\left(b^r, b^s\right)$$
(7.143)

Applying analogous assumptions as previously, i.e.

$$O\left(\mathbf{K}^0\right) = O\left(\mathbf{K}^{,r}\right) = O\left(\mathbf{K}^{,rs}\right) = 2^n \; ; \; n \in N$$
(7.144)

we decompose mth order displacement vectors $\mathbf{q}^{(m)}$ as

$$\mathbf{s}_k^{(m)} = \frac{1}{\sqrt{2}}\left(\mathbf{q}_{2k-1}^{(m)} + \mathbf{q}_{2k}^{(m)}\right)$$
(7.145)

$$\mathbf{d}_k^{(m)} = \frac{1}{\sqrt{2}}\left(\mathbf{q}_{2k-1}^{(m)} - \mathbf{q}_{2k}^{(m)}\right)$$
(7.146)

Since the fact, that the matrix M_n consists of the real numbers only, it is defined in exactly the same way. Then, the decomposition of $\mathbf{q}^{(m)}$ into the vectors $\mathbf{s}^{(m)}$ and $\mathbf{d}^{(m)}$ is introduced as

$$\mathbf{Lq}^{(m)} = \mathbf{s}^{(m)}$$
(7.147)

and

$$\mathbf{Hq}^{(m)} = \mathbf{d}^{(m)} \tag{7.148}$$

Therefore, full multiresolutional decomposition of up to second order equilibrium equations is carried out as

$$\mathbf{LK}^{(m)}\mathbf{q}^{(m)} = \left(\mathbf{LK}^{(m)}\mathbf{L}^T\right)\mathbf{Lq}^{(m)} + \left(\mathbf{LK}^{(m)}\mathbf{H}^T\right)\mathbf{Hq}^{(m)} = \mathbf{Lf}^{(m)} \tag{7.149}$$

and

$$\mathbf{HK}^{(m)}\mathbf{q}^{(m)} = \left(\mathbf{HK}^{(m)}\mathbf{L}^T\right)\mathbf{Lq}^{(m)} + \left(\mathbf{HK}^{(m)}\mathbf{H}^T\right)\mathbf{Hq}^{(m)} = \mathbf{Hf}^{(m)} \tag{7.150}$$

Denoting further by

$$\mathbf{LK}^{(m)}\mathbf{L}^T = \mathbf{T}^{(m)}, \mathbf{LK}^{(m)}\mathbf{H}^T = \mathbf{C}^{(m)} \tag{7.151}$$

and

$$\mathbf{HK}^{(m)}\mathbf{L}^T = \mathbf{B}^{(m)}, \mathbf{HK}^{(m)}\mathbf{H}^T = \mathbf{A}^{(m)} \tag{7.152}$$

there holds

$$\mathbf{Lf}^{(m)} = \mathbf{f}_s^{(m)}, \mathbf{Hf}^{(m)} = \mathbf{f}_d^{(m)} \tag{7.153}$$

Finally, the reduction equations are obtained as follows:
- zeroth order equations:

$$\begin{cases} \mathbf{T}^0\mathbf{s}^0 + \mathbf{C}^0\mathbf{d}^0 = \mathbf{f}_s^0 \\ \mathbf{B}^0\mathbf{s}^0 + \mathbf{A}^0\mathbf{d}^0 = \mathbf{f}_d^0 \end{cases} \tag{7.154}$$

- first order equations:

$$\begin{cases} \mathbf{T}^{,r}\mathbf{s}^0 + \mathbf{T}^0\mathbf{s}^{,r} + \mathbf{C}^{,r}\mathbf{d}^0 + \mathbf{C}^0\mathbf{d}^{,r} = \mathbf{f}_s^{,r} \\ \mathbf{B}^{,r}\mathbf{s}^0 + \mathbf{B}^0\mathbf{s}^{,r} + \mathbf{A}^{,r}\mathbf{d}^0 + \mathbf{A}^0\mathbf{d}^{,r} = \mathbf{f}_d^{,r} \end{cases} \tag{7.155}$$

- second order equations:

$$\begin{cases} \mathbf{T}^{,rs}\mathbf{s}^0 + 2\mathbf{T}^{,r}\mathbf{s}^{,s} + \mathbf{T}^0\mathbf{s}^{,rs} + \mathbf{C}^{,rs}\mathbf{d}^0 + 2\mathbf{C}^{,r}\mathbf{d}^{,s} + \mathbf{C}^0\mathbf{d}^{,rs} = \mathbf{f}_s^{,rs} \\ \mathbf{B}^{,rs}\mathbf{s}^0 + 2\mathbf{B}^{,r}\mathbf{s}^{,s} + \mathbf{B}^0\mathbf{s}^{,rs} + \mathbf{A}^{,rs}\mathbf{d}^0 + 2\mathbf{A}^{,r}\mathbf{d}^{,s} + \mathbf{A}^0\mathbf{d}^{,rs} = \mathbf{f}_d^{,rs} \end{cases} \tag{7.156}$$

Since that, we derive the reduction equations for the mth order of a vector $\mathbf{d}$ [182]. The process is much more complicated than in the deterministic case. The three up to the second order equation systems are obtained as follows:

-      zeroth order equations:

$$\mathbf{d}^0 = \left(\mathbf{A}^{-1}\right)^0\left(-\mathbf{B}^0\mathbf{s}^0 + \mathbf{f}_d^0\right) \tag{7.157}$$

-      first order equations:

$$\mathbf{d}^{,r} = \left\{\left(\mathbf{A}^{-1}\right)^{,r}\left(-\mathbf{B}^0\mathbf{s}^0 + \mathbf{f}_d^0\right) + \left(\mathbf{A}^{-1}\right)^0\left(-\mathbf{B}^{,r}\mathbf{s}^0 - \mathbf{B}^0\mathbf{s}^{,r} + \mathbf{f}_d^{,r}\right)\right\} \tag{7.158}$$

-      second order equations:

$$\mathbf{d}^{(2)} = \left\{\left(\mathbf{A}^{-1}\right)^{,rs}\left(-\mathbf{B}^0\mathbf{s}^0 + \mathbf{f}_d^0\right) + 2\left(\mathbf{A}^{-1}\right)^{,r}\left(-\mathbf{B}^0\mathbf{s}^0 + \mathbf{f}_d^0\right)^{,s}\right.$$
$$\left. + \left(\mathbf{A}^{-1}\right)^0\left(-\mathbf{B}^{,rs}\mathbf{s}^0 - \mathbf{B}^0\mathbf{s}^{,rs} + \mathbf{f}_d^{,rs}\right)\right\} Cov\left(b^r, b^s\right) \tag{7.159}$$

Then, the reduced equation has the following form:

-      zeroth order

$$\left(\mathbf{T}^0 - \mathbf{C}^0\left(\mathbf{A}^{-1}\right)^0\mathbf{B}^0\right)\mathbf{s}^0 = \mathbf{f}_s^0 - \mathbf{C}^0\left(\mathbf{A}^{-1}\right)^0\mathbf{f}_d^0 \tag{7.160}$$

-      first order

$$\left(\mathbf{T}^{,r} - \mathbf{C}^{,r}\left(\mathbf{A}^{-1}\right)^0\mathbf{B}^0 - \mathbf{C}^0\left(\mathbf{A}^{-1}\right)^{,r}\mathbf{B}^0 - \mathbf{C}^0\left(\mathbf{A}^{-1}\right)^0\mathbf{B}^{,r}\right)\mathbf{s}^0$$
$$+ \left(\mathbf{T}^0 - \mathbf{C}^0\left(\mathbf{A}^{-1}\right)^0\mathbf{B}^0\right)\mathbf{s}^{,r} \tag{7.161}$$
$$= \mathbf{f}_s^{,r} - \mathbf{C}^{,r}\left(\mathbf{A}^{-1}\right)^0\mathbf{f}_d^0 - \mathbf{C}^0\left(\mathbf{A}^{-1}\right)^{,r}\mathbf{f}_d^0 - \mathbf{C}^0\left(\mathbf{A}^{-1}\right)^0\mathbf{f}_d^{,r}$$

-      second order

$$\left(\mathbf{T}^{,rs} - \mathbf{C}^{,rs}\left(\mathbf{A}^{-1}\right)^{0}\mathbf{B}^{0} - \mathbf{C}^{0}\left(\mathbf{A}^{-1}\right)^{,rs}\mathbf{B}^{0} - \mathbf{C}^{0}\left(\mathbf{A}^{-1}\right)^{0}\mathbf{B}^{,rs} - 2\mathbf{C}^{,r}\left(\mathbf{A}^{-1}\right)^{,s}\mathbf{B}^{0} \right)\mathbf{s}^{0}$$

$$+ \left(-2\mathbf{C}^{0}\left(\mathbf{A}^{-1}\right)^{,r}\mathbf{B}^{,s} - 2\mathbf{C}^{,r}\left(\mathbf{A}^{-1}\right)^{0}\mathbf{B}^{,s} \right)\mathbf{s}^{0} + \left(\mathbf{T}^{0} - \mathbf{C}^{0}\left(\mathbf{A}^{-1}\right)^{0}\mathbf{B}^{0} \right)\mathbf{s}^{,rs}$$

$$+ 2\left(\mathbf{T}^{,r} - \mathbf{C}^{,r}\left(\mathbf{A}^{-1}\right)^{0}\mathbf{B}^{0} - \mathbf{C}^{0}\left(\mathbf{A}^{-1}\right)^{,r}\mathbf{B}^{0} - \mathbf{C}^{0}\left(\mathbf{A}^{-1}\right)^{0}\mathbf{B}^{,r} \right)\mathbf{s}^{,s} \qquad (7.162)$$

$$= \mathbf{f}_{s}^{,rs} - \mathbf{C}^{,rs}\left(\mathbf{A}^{-1}\right)^{0}\mathbf{f}_{d}^{0} - \mathbf{C}^{0}\left(\mathbf{A}^{-1}\right)^{,rs}\mathbf{f}_{d}^{0} - \mathbf{C}^{0}\left(\mathbf{A}^{-1}\right)^{0}\mathbf{f}_{d}^{,rs}$$

$$- 2\left(\mathbf{C}^{,r}\left(\mathbf{A}^{-1}\right)^{,s}\mathbf{f}_{d}^{0} + \mathbf{C}^{0}\left(\mathbf{A}^{-1}\right)^{,r}\mathbf{f}_{d}^{,s} + \mathbf{C}^{,r}\left(\mathbf{A}^{-1}\right)^{0}\mathbf{f}_{d}^{,s} \right)$$

Since that, the first recursive step for mth order stochastic equations is obtained as

$$\begin{cases} \mathbf{K}_{0}^{(m)} = \mathbf{K}^{(m)}, \mathbf{f}_{0}^{(m)} = \mathbf{f}^{(m)} \\ \mathbf{K}_{1}^{(m)} = \mathbf{T}_{0}^{(m)} - \left(\mathbf{C}_{0}\mathbf{A}_{0}^{-1}\mathbf{B}_{0}\right)^{(m)}, \mathbf{f}_{1}^{(m)} = \mathbf{L}_{0}^{(m)}\mathbf{f}_{0}^{(m)} + \left(\mathbf{C}_{0}\mathbf{A}^{-1}\mathbf{H}\mathbf{f}_{0}\right)^{(m)} \end{cases} \qquad (7.163)$$

As a result, the triple up to the second order equations for $s_{1}^{(m)}$ are rewritten as follows:

$$\mathbf{K}_{1}^{0}\mathbf{s}_{1}^{0} = \mathbf{f}_{1}^{0} \qquad (7.164)$$

$$\mathbf{K}_{1}^{,r}\mathbf{s}_{1}^{0} + \mathbf{K}_{1}^{0}\mathbf{s}_{1}^{,r} = \mathbf{f}_{1}^{,r} \qquad (7.165)$$

$$\mathbf{K}_{1}^{,rs}\mathbf{s}_{1}^{0} + 2\mathbf{K}_{1}^{,r}\mathbf{s}_{1}^{,s} + \mathbf{K}_{1}^{0}\mathbf{s}_{1}^{,rs} = \mathbf{f}_{1}^{,rs} \qquad (7.166)$$

where $s_{1}^{(m)} = Lq^{(m)}$. Repeating this process up to n times, which is possible considering initial dimensions of the matrix $\mathbf{K}$, it is obtained for mth order

$$\begin{cases} \mathbf{K}_{j+1}^{(m)} = \mathbf{T}_{j}^{(m)} - \left(\mathbf{C}_{j}\mathbf{A}_{j}^{-1}\mathbf{B}_{j}\right)^{(m)} \\ \mathbf{f}_{j+1}^{(m)} = \mathbf{L}_{n-j}^{(m)}\mathbf{f}_{j} + \mathbf{L}_{j}\mathbf{f}^{(m)} + \left(\mathbf{C}_{j}\mathbf{A}_{j}^{-1}\mathbf{H}_{n-j}\mathbf{f}_{j}\right)^{(m)} \end{cases} \qquad (7.167)$$

with

$$\begin{cases} \mathbf{T}_{j}^{(m)} = \mathbf{L}_{n-j}\mathbf{K}_{j}^{(m)}\mathbf{L}_{n-j}^{T} \\ \mathbf{B}_{j}^{(m)} = \mathbf{H}_{n-j}\mathbf{K}_{j}^{(m)}\mathbf{L}_{n-j}^{T} \\ \mathbf{C}_{j}^{(m)} = \mathbf{L}_{n-j}\mathbf{K}_{j}^{(m)}\mathbf{H}_{n-j}^{T} \\ \mathbf{A}_{j}^{(m)} = \mathbf{H}_{n-j}\mathbf{K}_{j}^{(m)}\mathbf{H}_{n-j}^{T} \end{cases} \qquad (7.168)$$

while the reconstruction scheme for the mth order solution vector is given by the following formula:

$$\mathbf{q}_{2k-1}^{(m)} = \frac{1}{\sqrt{2}}\left(\mathbf{s}_k^{(m)} + \mathbf{d}_k^{(m)}\right) \tag{7.169}$$

$$\mathbf{q}_{2k}^{(m)} = \frac{1}{\sqrt{2}}\left(\mathbf{s}_k^{(m)} - \mathbf{d}_k^{(m)}\right) \tag{7.170}$$

Let us consider for illustration the following transformation of the random variables:

$$Y = X^p \cos \omega t \tag{7.171}$$

where $X \in (\Omega, \sigma, P)$, $p \in Z$ and $\omega, t \in \Re$. Therefore, the first two probabilistic moments of Y can be calculated as

$$E[Y] = Y^0 + \frac{1}{2}\frac{\partial^2 Y}{\partial X^2}Var(X) \tag{7.172}$$

$$Var(Y) = \left(\frac{\partial Y}{\partial X}\right)^2 Var(X) \tag{7.173}$$

according to the presented second order perturbation technique. It is obtained by the classical differentiation calculus that

$$\frac{\partial Y}{\partial X} = pX^{p-1} \cos \omega t \tag{7.174}$$

and

$$\frac{\partial^2 Y}{\partial X^2} = p(p-1)X^{p-2} \cos \omega t \tag{7.175}$$

The following iterative formula can be proposed for the nth perturbation approach:

$$\frac{\partial^k Y}{\partial X^k} = \left[\prod_{l=0}^{k}(p-l)\right]X^{p-(l+1)} \cos \omega t \tag{7.176}$$

Therefore, the expected values are determined

$$E[Y] = E^p[X]\cos \omega t + \frac{1}{2}p(p-1)X^{p-2} \cos \omega t Var(X) \tag{7.177}$$

and variances as

$$Var(Y) = \left(pX^{p-1}\right)^2 \cos^2 \omega t\, Var(X) \tag{7.178}$$

in the second order perturbation approach. The visualisation of all wavelet functions and their approximations are presented below using the symbolic computation package MAPLE [182]. The following function is used $f(t) = \dfrac{1}{l(\omega)}\cos(2\pi t)$, $t \in [0,1]$ where $l(\omega)$ belongs to the additional random space with the expected value $E[l]=10$ and the variance equal to $Var(l)=4$; p=−1. The wavelet projection are shown for $n=3,\ldots,6$ in case of the expected values − in Figures 7.54−7.57 and the wavelet approximations for the variance for $n=4,5,6$ are shown in Figures 7.58−7.60.

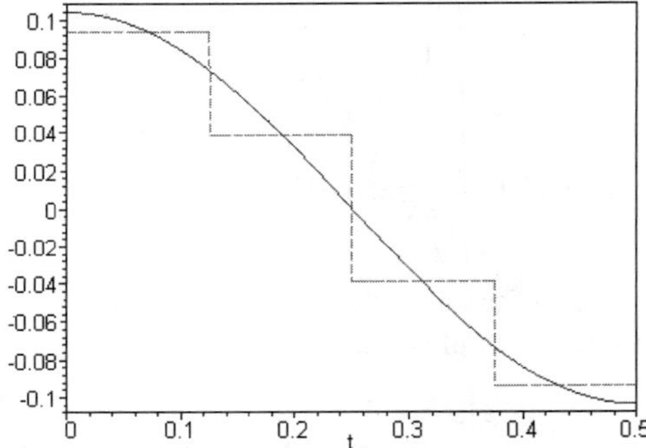

Figure 7.54. Wavelet projection of expected values for $n=3$

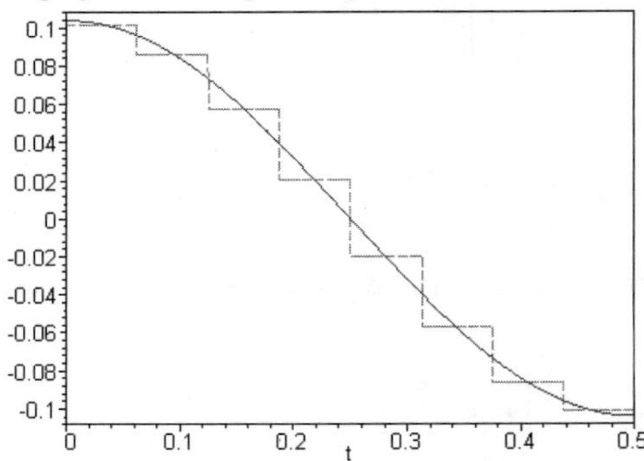

Figure 7.55. Wavelet projection of expected values for $n=4$

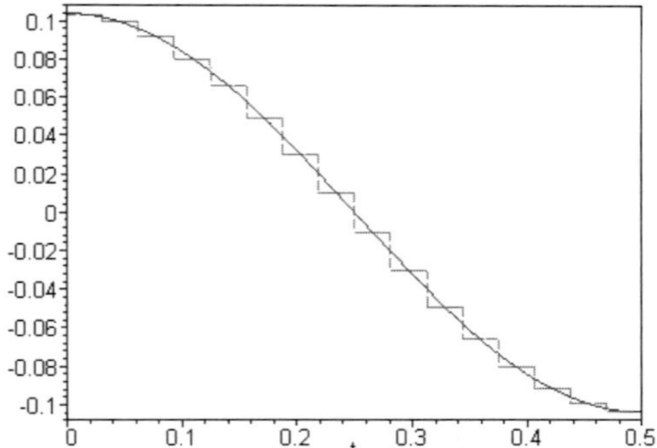

Figure 7.56. Wavelet projection of expected values for $n=5$

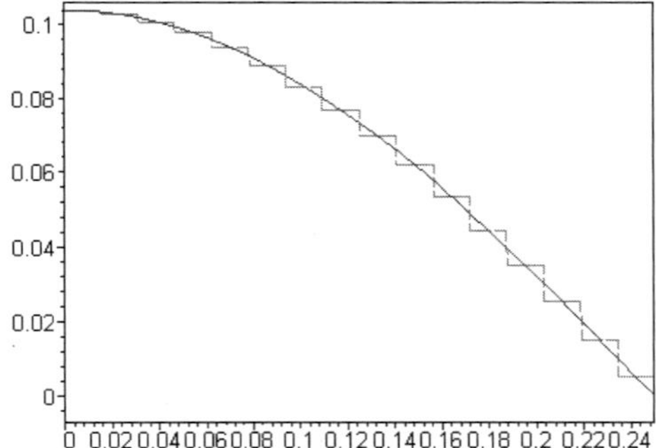

Figure 7.57. Wavelet projection of expected values for $n=6$

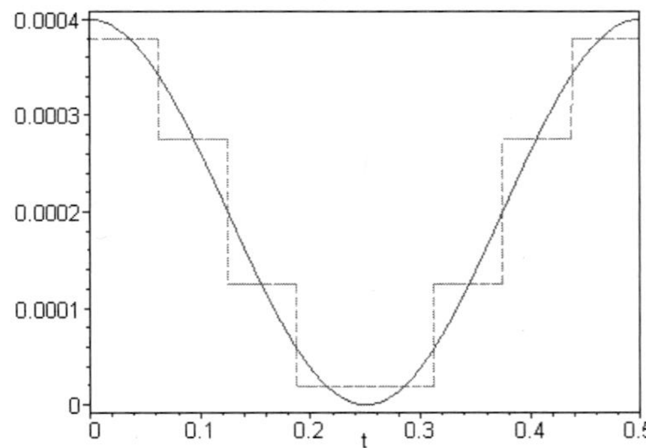

Figure 7.58. Wavelet projection of variances for $n=4$

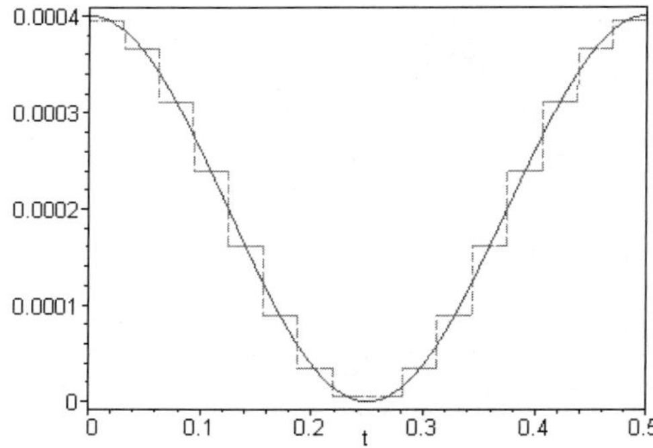

Figure 7.59. Wavelet projection of variances for $n=5$

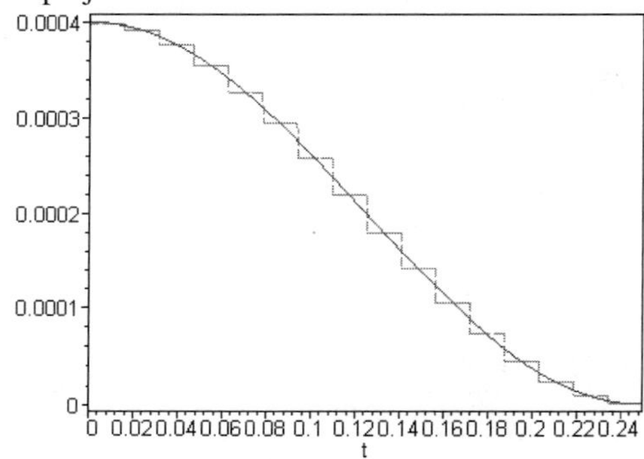

Figure 7.62. Wavelet projection of variances for $n=6$

The expected values and their wavelet projections are greater than the corresponding deterministic values of $f(t)$ computed for Var(l)=0. Since the expectations and their deterministic origins are very similar, the convergence of analysed projections is quite the same – for $n=6$ the approximation error on the interval [0,1] in practice can be neglected. The situation changes in the case of variances where projection of the 6th order is not quite smooth; for $n=2$ cannot be accepted at all because of the constant function resulting from the wavelet projection.

As is documented in Table 7.3, the total computational cost by means of the consumed time and memory necessary to obtain wavelet projection increases nonlinearly together with this projection order. Taking into account that the time of the linear equation system solution shows the same tendency, the very exact solution of (7.120) with 7th and even higher order wavelet projection needs more powerful computers. The last column of the computer test shows that the approximation of variances needs essentially more time and memory than the analogous projection of zeroth order moments (deterministic values) and the expectations (first moments). It should be documented by the relevant numerical tests, if the computational symbolic projection cost increases together with the order of the probabilistic moment being projected onto the same wavelet family.

Table 7.3. Computational cost of wavelet projection (for COMPAQ 475 MHz)

Projection order	q total dimension	f(t) secs/MB	E[f(t)]	Var[f(t)]
2	4	7.4/1.94	8.4/1.94	8.9/2.06
3	8	10.1/2.19	10.9/2.19	11.0/2.31
4	16	14.1/2.62	14.9/2.62	17.3/2.69
5	32	20.7/3.25	25.7/3.31	31.8/3.44
6	64	53.0/4.56	53.7/4.56	70.5/4.94
7	128	131.0/7.06	130.7/7.06	185.4/7.75
8	256	395.3/12.2	360.8/12.2	593.2/13.50

7.8 Concluding Remarks

As was demonstrated above, the wavelet–based multiresolutional computational techniques can be very efficient, considering the capability of heterogeneity analysis on extremely different geometrical scales in the same time. Such phenomena appear frequently in engineering composites – at the interface between the components, on microscale connected with the periodicity cell, for a window on mesoscale for a couple of reinforcing fibres or particles as well as for the macroscale connected with the global composite structure. As can be observed, the wavelet-based numerical methods (especially the Finite Element Method) can be successfully used even for the heterogeneous media with random or stochastic microstructure thanks to implementation of a randomisation method (simple algebra, PDF integration, Monte Carlo simulation, stochastic perturbation or even spectral analyses).

The homogenisation method discussed in this chapter enables us to apply an alternative approach, where the effective material parameters (or its probabilistic moments) are determined first and then the entire composite is analysed using traditional computational techniques. Wavelet-based multiresolutional approach to the homogenisation problem should, however, be formulated to introduce the components characteristics on many scales into the final effective structural parameters. As was demonstrated in the mathematical considerations, homogenised properties in multiscale analysis and classical macro–micro passage are essentially different, even in a deterministic formulation, which was observed previously in three scale Monte Carlo simulation based homogenisation studies for the fibre–reinforced composites [191,197].

Finally, let us note that due to the character of the homogenised 1D elastostatic problem, computational studies on effective coefficient probabilistic behaviour can be applied without any further modifications in the heat conduction problem of a composite with exactly the same multiscale internal structure as well as for any linear field problem with random coefficients defined by their first two probabilistic moments. The real and imaginary parts of the effective coefficient for the wave equation can be used in acoustic wave propagation in random media. It is observed that for wave propagation, homogenised coefficients strongly depend on the same range on angular velocity and the interrelation of material properties of the layered medium components.

The most important result of the homogenisation–based Finite Element modelling of the periodic composite beams is that replacing the real composite behaviour is very well approximated by the homogenised model response. For a smaller number of bays in the periodic structure, wavelet–based homogenisation gives more accurate results, while the classical approach is more efficient for the increasing number of bays. Maximum deflections of the analysed beams are approximated by all the models with the same precision, which increases for increasing number of bays in the whole structure.

The wavelet–based multiresolutional homogenisation method introduces new opportunities to calculate effective parameters for the composites with material properties given in various scales by some wavelet functions. This method is more attractive from the mathematical point of view. However it is characterised by new, closer bounds on the homogenisability of composite structures, but it eliminates all formal problems resulting from the assumption of small parameter existence between macro– and microscales. Now, practically any number of various scales can be considered in composite materials and structures, which is important in all these cases, where material properties are obtained through signal detection and its analysis. Finally, obtaining satisfactory agreement between the real and homogenised structures enables the application of this method to the forced vibrations of deterministic systems as well as the use of dynamical systems with stochastic parameters.

The second order perturbation wavelet projection gives complicated formulae for approximation of the original functions or matrices, which enables fast wavelet–based discretisation of random variables and/or fields. It is necessary to recall the algebraic restrictions on the first two probabilistic moments of the input to achieve the coefficient of variation to be essentially smaller than 0.15.

However it is documented by the above numerical examples that the wavelet projection of the expected value and its deterministic origin have almost the same character – the same order of approximation is necessary to achieve the same convergence and error level. Wavelet projection of variance (and higher order probabilistic characteristics) needs greater precision, especially for smaller values of the projection order n. Let us note that analogous projection for random functions or operators defined in two– or three–dimensional spaces can be done by the use of Daubechies wavelets in a similar manner to that presented here.

Symbolic computations package MAPLE [61,70] (as well as other numerical tools of this class) is very efficient in wavelet projections of various discrete and/or continuous functions because the efficiency of the projection (and its averaged error) can be recognised graphically in specially adopted plots. Otherwise, a special purpose numerical error routine should be implemented and applied.

The most important result of the homogenisation−based Finite Element modelling of the periodic composite beams is that the real composite behaviour is very well approximated by the homogenised model response. The multiresolutional homogenisation technique giving a more accurate approximation of the real structure behaviour is decisively more complicated in numerical implementation because of the necessity of applying the combined symbolic−FEM approach. A wavelet−based space−time decomposition should be applied in computational modelling of the transient heat transfer problems in heterogeneous media.

Furthermore, mathematical and numerical studies should be conducted to increase nonstationary heat transfer modelling in unidirectional composites by the application of the homogenisation method. In the case of small contrast between heat capacities of the constituents, the method proposed was verified as effective; the situation changes when the value of contrast parameter increases dramatically.

8 Appendix

8.1 Procedure of MCCEFF Input File Preparation

The instructions described below deal with the preparing of input data file to the MCCEFF analysis in the case there is no need to use the mesh generator.

1. Heading line (12a4) general information
2. General information about the problem homogenised (6i5)

Column	Variable	Description
1–5	NUMNP	Total number of nodal point in the structure discretised
6–10	NELTYP	Total number of finite element groups (=1)
11–15	LL	Total number of load cases (=3)
36–40	KEQB	Total number of non-zero degrees of freedom in the main matrix
66–70	MK	Total number of random trials
71–75	NBN	Total number of nodal points of the interfaces

General comments:

A. NELTYP variable is provided due to the original POLSAP code to extend in the next version the MCCEFF code with the analysis of the engineering structures homogenised (e.g. fibre-reinforced plates and shells). However due to its constant value it may have been omitted.

B. LL variable is provided taking into account that in the next versions of the program the rest of the effective tensor components will be computed (in the 3D homogenisation problem). There are three different components of the elasticity tensor homogenised for the plane strain problems being solved by the program.

C. KEQB parameter should be modified (default value is equal to 0) if the program MCCEFF in the process of main stiffness matrix formation or solution of the fundamental algebraic equations system stops running. The value of the parameter is to be taken from the interval [0,NEQB], where NEQB is the total number of the degrees of freedom of the composite cell. The probability of the successful computations increases with decreasing KEQB parameter.

3. Nodal points data (7i5,4d10.0,3i2)

Column	Variable	Description
2–5	N	Nodal point number
7–10	IX(N,1)	
11–15	IX(N,2)	
16–20	IX(N,3)	Displacement boundary conditions codes
21–25	IX(N,4)	=0, free degree of freedom
26–30	IX(N,5)	=1, fixed degree of freedom
31–35	IX(N,6)	
36–45	X(N)	X coordinate
46–55	Y(N)	Y coordinate
56–65	Z(N)	Z coordinate
66–70	K(N)	Nodal point generation code
71–72	M1	Number of the internal region
73–74	M2	Number of the external region
75–77	M3	The interface end code (=1)

General comments:

A. Nodal point numbering has to be continuous and to start from number 1, which should denote the centre of the fibre (considering stress boundary conditions computations).

B. Interface nodal points numbering has to be provided in the anticlockwise system and the distances between any two points must be equal.

C. The structure being discretised should be placed in the YZ plane; the X coordinate will be used in the next version for the analysis of the 3D composite problems.

D. The regions of the different materials should have increasing number starting from the central component (fibre in two-component composites) and continuous to the external boundary of the cell.

E. In the case of half or quarter of the periodicity cell analysis the M3 parameter should be used to underline the ends of the interface being cutted.

4. General finite elements data (3i5)

Column	Variable	Description
1–5	='3'	Plane strain code
6–10	NUMEL	Total number of finite elements
11–15	NUMMAT	Total number of composite components

5. Material data
5.1. General data (2i5,2d10.0)

Column	Description
1–5	Material number
6–10	Total number of different temperatures
11–20	Gravity loading
21–30	Mass density

General comment:

The total number of the materials used should be greater than 10.

5.2. Detailed data two lines for any different temperature (8d10.0/3d10.0)

Column	Description
1–10	Temperature
11–20	Elasticity modulus E_n
21–30	Elasticity modulus E_s
31–40	Elasticity modulus E_t
41–50	Poisson coefficient ν_n
51–60	Poisson coefficient ν_s
61–70	Poisson coefficient ν_t
71–80	Shear modulus G
1–10	Coefficient of thermal expansion α_n
11–20	Coefficient of thermal expansion α_s
21–30	Coefficient of thermal expansion α_t

6. Probabilistic parameters (2d10.0/2i5,4d10.0)

Column	Description
1–10	Variance of Young modulus
11–20	Variance of Poisson ratio
1–5	Averaged material type: = 1, material without defects = 2, material with interface defects = 3, material with volume defects
6–10	Structural defects type: = 1, circle = 2, triangle = 3, rectangle = 4, hexagon
11–20	Expected value of the geometrical parameter
21–30	Variance of the geometrical parameter
31–40	Expected value of defects total number
41–50	Variance of defects total number

7. Finite elements description (7i5)

Column	Description
1–5	Finite element number
6–10	I node number
11–15	J node number
16–20	K node number
21–25	L node number
26–30	Material number
56–60	Finite elements generation code: =0 (default) - the lack of generation =1, generation

8.2 Input Data for ABAQUS Reinforced Concrete Plate Analysis

Example input data file for the ABAQUS [1] analysis of the steel reinforced concrete plate analysed in the book. The comment lines are indicated by '**' to enable the user to gain a better understanding of model computed. The lines indicated by '*' or without any indication are program execution lines.

```
*heading
corner supported shell
**all material and geometrical parameters are defined in US units
*node
1,0.,0.
7,18.,0.
61,0.,18.
67,18.,18.
** node numbers, their coordinates
*ngen,nset=y-sym
1,7
** node generation on the 'y-sym' boundary with numbers from 1 to 7
*ngen,nset=x-sym
1,61,10
** node generation on the 'x-sym' boundary with numbers from 1 to 61 with
increment equal to 10
*ngen,nset=lx2
61,67
** node generation on the 'lx2' boundary with numbers from 61 to 67
*ngen,nset=ly2
7,67,10
** nodes generation on the 'ly2' boundary with numbers from 7 to 67 with
increment equal to 10
*nset,nset=one
1,
**definition of the nodes set called 'one'
*nfill
x-sym,ly2,6,1
**generation ('filling') of the nodes contained in the internal of the rectangular
given by parallel boundaries 'y-sym', 'x-sym', 'lx2' and 'ly2'
*element,type=s8r,elset=slab
**element type and element set definition
1,1,3,23,21,2,13,22,11
**master element definition: corner and midpoint nodes in anti-clockwise order
*elgen,elset=slab
**element generation for the element set 'slab'
```

```
1,3,2,1,3,20,3
```
**master element number, number of elements to be defined in the first
**row generated including the master element, increment in node numbers
**of corresponding nodes from element to element in row (default is 1),
**increment in element numbers in row (default is 1), numbers of rows to
**be defined (default is 1), increment in node numbers of corresponding
**nodes from row to row, increment in element numbers of corresponding
**elements from row to row, numbers of layers to be defined (defined is
**1), increment in node numbers of corresponding nodes from layer to
**layer, increment in element numbers of corresponding elements from
**layer to layer
*shell section,elset=slab,material=a1
```
1.75,9
```
**shell thickness, total number of the integration points through its thickness
*material,name=a1
**concrete material parameters definition
*elastic
```
4.15e6,.15
```
**Young modulus, Poisson ratio
*concrete
```
3000.,0.
5500.,.0015
```
**absolute value of compressive stress, absolute value of plastic strain (the
**first stress-strain point must be at zero plastic strain and defines the
**initial yield point)
*failure ratios
```
1.16 , .0836
```
**ratio of the ultimate biaxial compressive stress to the uniaxial
**compressive ultimate stress (default is 1.16), absolute value of the ratio
**of uniaxial tensile stress at failure to the uniaxial compressive stress at
**failure (default is 0.09), the ratio of the magnitude of a principal
**component of plastic strain at ultimate stress in biaxial compression to
**the plastic strain at ultimate stress in uniaxial compression, the ratio of
**the tensile principal stress value at cracking, in plane stress, when the
**other non-zero principal stress component is at the ultimate compressive
**stress value, to the tensile cracking stress under uniaxial tension.
*tension stiffening
**definition of retained tensile stress normal to a crack is a function of the
**deformation in the direction of the normal to the crack

1.,0.

0.,2.e-3

**fraction of remaining stress to stress at cracking, absolute value of the

**direct strain minus the direct strain at cracking

*rebar,element=shell,material=slabmt,geometry=isoparametric,name=yy

**definition of the rebars, reinforced element type, material name, rebar

**geometry type (isoparametric or skew), name of the rebars group

slab,.014875,1.,-.435,4

**definition of rebars geometry, cross-sectional area of each rebar, spacing

**of the rebars in the plane of the shell, position of the rebars in the shell

**direction, edge number to which rebar are similar [4]

*rebar,element=shell,material=slabmt,geometry=isoparametric,name=xx

slab,.014875,1.,-.435,1

*material,name=slabmt

*elastic

29.e6

**Young modulus, Poisson ratio is default

*plastic

50.e3

**Yield stress value

*boundary

**displacement boundary condition definition

y-sym,ysymm

x-sym,xsymm

**symmetry conditions on x-sym, y-sym boundaries

67,3

*restart,write,frequency=999

**option RESTART controls the writing to and reading of the restart file,

**which is used by the postprocessor; the option will create a restart

**file after each increment at which the increment number is exactly

**divisible by N, and at the end of each step of the analysis, regardless of

**the value of N at that time

*step,inc=30

**option STEP must begin each step definition, parameter INC is equal to

**the maximum number of increments in a step (upper bound, the default

**value is 10)

*static,riks

**this option indicates that the step should be analysed as a static load

**step; the Riks method is chosen by the RIKS parameter

.05,1.,,,,1,3,-1.
**initial time increment, time period of the step, minimum time increment
**allowed, maximum increment allowed, maximum value of the load
**proportionality factor for the Riks method, node number at which the
**value is being monitored, degree of freedom being monitored, value of
**the total displacement (or rotation) at the node and degree of freedom
**which, if crossed during an increment, ends the step at that increment
*cload
**concentrated loading definition
1,3,-5000.
**node number, number of the corresponding degree of freedom, loading
**magnitude in the orientation given by the user by ordering nodes into
**shell elements
*el print,frequency=10
**option provided tabular printed output of element variables; parameter
**FREQUENCY is equal to the output frequency measured in the
**increments performed (if this option is omitted, very large printed output
**files will be produced by large models in multiple increment analysis!)
s
**all stress components
sinv
**all stress invariants (MISES,TRESC,PRESS-equivalent pressure stress,
**INV3-third stress invariant)
e
**all strain components
pe
** all plastic strain component
crack
**crack orientations in concrete
*el file,frequency=10
s
sinv
e
pe
crack
*node file,nset=one
**this option allows nodal variables to be written to the ABAQUS results
**file (no nodal variables will be written to the results file unless this
**option is used!)

u
*end step
**FURTHER COMMENTS
**it is possible to provide SHEAR RETENTION parameter to describe the
**reduction of the shear modulus as a function of the tensile strain across the
**crack; if this parameter is omitted (should be placed after TENSION
**STIFFENING lines), it is assumed that the shear retention behaviour depends
**only on temperature. EXPANSION parameter may be used to introduce thermal
**volume change effects in the concrete. The NLGEOM parameter may be
**included in the STEP option when the large strains and rotations associated with
**failure of concrete are observed.

8.3 MAPLE Script for Computations of the Homogenised Heat Conductivity Coefficients

Mexican hat - Haar basis plot of the input heat conductivity

```
> restart; sig:=-.5; k1:=0.1; k2:=0.4;
> kmexh:=2+1/(sqrt(2*Pi)*sig^3)*((exp(-x^2/(2*sig^2)))*(x^2/sig^2-1));
> khaar:=piecewise(x<=0.5,k1,x>=0.5,k2);
> k:=khaar+0.005*kmexh;
> plot(k,x=0..1,title='k');
```

Mexican hat - Haar basis heat conductivity contrast variability

```
> restart; sig:=-.5; k1:=contrast*k2; k2:=0.05;
> kmexh:=2+1/(sqrt(2*Pi)*sig^3)*((exp(-x^2/(2*sig^2)))*(x^2/sig^2-1));
> khaar:=piecewise(x<=0.5,k1,x>=0.5,k2);
> k:=khaar+0.005*kmexh;
> plot3d(k,x=0..1,contrast=5..100,title='k');
```

Spatial averaging of the Mexican hat - Haar type heat conducitivity

```
> restart; sig:=-.5; k1:=0.1; k2:=0.4;
> kmexh:=2+1/(sqrt(2*Pi)*sig^3)*((exp(-x^2/(2*sig^2)))*(x^2/sig^2-1));
> khaar:=piecewise(x<=0.5,k1,x>=0.5,k2);
> k:=khaar+0.005*kmexh;
> plot(k,x=0..1,title='k');
> kav:=evalf(int(k,x=0..1));
```

Classical homogenisation of the Mexican hat - Haar basis type heat conductivity

```
> restart; sig:=-.5; k1:=0.1; k2:=0.4;
```

```
> kmexh:=2+1/(sqrt(2*Pi)*sig^3)*((exp(-x^2/(2*sig^2)))*(x^2/sig^2-1));
> khaar:=piecewise(x<=0.5,k1,x>=0.5,k2);
> k:=khaar+0.005*kmexh;
> kk:=1/k: kkk:=int(kk,x=0..1): khom:=evalf(1/kkk);
```

Wavelet homogenisation of the Mexican hat - Haar basis type heat conductivity

```
> restart; sig:=-.5; k1:=0.1; k2:=0.4;
> kmexh:=2+1/(sqrt(2*Pi)*sig^3)*((exp(-x^2/(2*sig^2)))*(x^2/sig^2-1));
> khaar:=piecewise(x<=0.5,k1,x>=0.5,k2);
> k:=khaar+0.005*kmexh;
> kk:=1/k; kkk:=x/k; kkkk:=1/(2*k);
> kkk1:=evalf(int(kk,x=0..1)); kkk2:=evalf(int(kkk,x=0..1));
kkk3:=evalf(int(kkkk,x=0..1));
> khomwav:=evalf(1/(kkk1-2*kkk2+2*kkk3));
```

Parametric variability of the spatial average of the Mexican hat - Haar type heat conductivity with respect to the contrast and volumetric ratio

```
> restart; sig:=-.5; k1:=contrast*k2; k2:=0.05;
> kmexh:=2+1/(sqrt(2*Pi)*sig^3)*((exp(-x^2/(2*sig^2)))*(x^2/sig^2-1));
> khaar:=piecewise(x<=g,k1,x>=g,k2);
> k:=khaar+0.005*kmexh;
> kav:=evalf(int(k,x=0..1));
> plot3d(kav,contrast=5..100,g=0.1..0.9,title='kav');
```

Parametric sensitivity of the Mexican hat - Haar basis effective heat conductivity with respect to the contrast and volumetric ratio

```
> restart; sig:=-.5; k1:=contrast*k2; k2:=0.05;
> kmexh:=2+1/(sqrt(2*Pi)*sig^3)*((exp(-x^2/(2*sig^2)))*(x^2/sig^2-1));
> khaar:=piecewise(x<=g,k1,x>=g,k2);
> k:=khaar+0.005*kmexh;
> kav:=evalf(int(k,x=0..1));
> dkavdcontrast:=diff(kav,contrast)*contrast/kav;
> plot3d(dkavdcontrast,contrast=5..100,g=0.1..0.9,title='dkavdcontrast');
> dkavdg:=diff(kav,g)*g/kav;
> plot3d(dkavdg,contrast=5..100,g=0.1..0.9,title='dkavdg');
```

Parametric variability of the classically homogenised the Mexican hat - Haar basis heat conductivity with respect to the contrast and volumetric ratio

```
> restart; sig:=-.5; k1:=contrast*k2; k2:=0.05;
> kmexh:=2+1/(sqrt(2*Pi)*sig^3)*((exp(-x^2/(2*sig^2)))*(x^2/sig^2-1));
> khaar:=piecewise(x<g,k1,x>g,k2);
> k:=khaar+0.005*kmexh;
```

```
> kk:=1/k: kkk:=int(kk,x=0..1): khom:=eval(1/kkk);
> plot3d(khom,contrast=5..100,g=0.1..0.9,title='khom');
```

Parametric sensitivity of the Mexican hat - Haar basis effective heat conductivity with respect to the contrast and volumetric ratio

```
> restart; sig:=-.5; k1:=contrast*k2; k2:=0.05;
> kmexh:=2+1/(sqrt(2*Pi)*sig^3)*((exp(-x^2/(2*sig^2)))*(x^2/sig^2-1));
> khaar:=piecewise(x<g,k1,x>g,k2);
> k:=khaar+0.005*kmexh;
> kk:=1/k: kkk:=int(kk,x=0..1): khom:=eval(1/kkk);
> dkhomdcontrast:=diff(khom,contrast)*contrast/khom;
> plot3d(dkhomdcontrast,contrast=5..100,g=0.1..0.9, title='dkhomdcontrast');
> dkhomdg:=diff(khom,g)*g/khom;
> plot3d(dkhomdg,contrast=5..100,g=0.1..0.9,title='dkhomdg');
```

Parametric variability of the multiresolutional homogenisation of the Mexican hat - Haar basis heat conductivity with respect to the contrast and volumetric ratio

```
> restart; sig:=-.5; k1:=contrast*k2; k2:=0.05;
> kmexh:=2+1/(sqrt(2*Pi)*sig^3)*((exp(-x^2/(2*sig^2)))*(x^2/sig^2-1));
> khaar:=piecewise(x<=g,k1,x>=g,k2);
> k:=khaar+0.005*kmexh;
> kk:=1/k; kkk:=x/k; kkkk:=1/(2*k);
> kkk1:=eval(int(kk,x=0..1)); kkk2:=eval(int(kkk,x=0..1));
kkk3:=eval(int(kkkk,x=0..1));
> khomwav:=eval(1/(kkk1-2*kkk2+2*kkk3));
> plot3d(khomwav,contrast=5..100,g=0.1..0.9,title='khomwav');
```

Parametric sensitivity of the Mexican hat - Haar basis homogenised heat conductivity with respect to the contrast and volumetric ratio

```
> restart; sig:=-.5; k1:=contrast*k2; k2:=0.05;
> kmexh:=2+1/(sqrt(2*Pi)*sig^3)*((exp(-x^2/(2*sig^2)))*(x^2/sig^2-1));
> khaar:=piecewise(x<=g,k1,x>=g,k2);
> k:=khaar+0.005*kmexh;
> kk:=1/k; kkk:=x/k; kkkk:=1/(2*k);
> kkk1:=eval(int(kk,x=0..1)); kkk2:=eval(int(kkk,x=0..1));
kkk3:=eval(int(kkkk,x=0..1));
> khomwav:=eval(1/(kkk1-2*kkk2+2*kkk3));
> dkhomwavdcontrast:=diff(khomwav,contrast)*contrast/khomwav;
> plot3d(dkhomwavdcontrast,contrast=5..100,g=0.1..0.9,title='dkhomwavdcontrast');
> dkhomwavdg:=diff(khomwav,g)*g/khomwav;
> plot3d(dkhomwavdg,contrast=5..100,g=0.1..0.9,title='dkhomwavdg');
```

References

1. ABAQUS, v. 5.8, (1999) User's Manual. Hibbitt, Karlsson & Sorensen, Pawtucket.
2. ANSYS v.5.5. User's Manual (1999), Swanson, Houston, PA.
3. Abdelal GF, Caceres A, Barbero EJ, (2002) A micro-mechanics damage approach for fatigue of composite materials. Comp. Struct. 56:413-422.
4. Aboudi J, Elastoplasticity theory for composite materials, (1986) Solid Mech. Arch. 11:141-183.
5. Achenbach JD, Zhu H, (1989) Effect of interfacial zone on mechanical behavior and failure of fiber-reinforced composites. J. Mech. Phys. Sol. 37(3):381-393.
6. Adrianov IV et al, (1999) Homogenization procedure and Pade approximants for effective heat conductivity of composite materials with cylindrical inclusions having square cross-section. Proc. Roy. Soc. London A 455:3401-3413.
7. Akin JE (1982) Application and the implementation of finite element methods, Academic Press, New York
8. Algam M, Bennett RM, Zureick AH, (2002) Three-parameter vs. two-parameter Weibull distribution for pultruded composite material properties. Comp. Struct. 58:497-503.
9. Alzebdeh K et al., (1998) Fracture of random matrix-inclusion composites: scale effects and statistics. Int. J. Sol. Struct. 35(19):2537-2566.
10. Amaniampong G, Burgoyne CJ, (1996) Monte-Carlo simulations of the time dependent failure of bundles of parallel fibres. Eur. J. Mech. A/Solids 15(2):243-266.
11. Anthoine A, (1995) Derivation of the in-plane elastic characteristics of masonry through homogenization theory. Int. J. Sol. Struct. 32(2):137-163.
12. Aravas N, (1987) On the numerical integration of a class of pressure-dependent plasticity models. Int. J. Num. Meth. Engrg. 24:1395-1416.
13. Argyris J, Mlejnek HP (1991) Dynamics of Structures, North Holland
14. Augusti G, Baratta A, Casciati F (1984) Probabilistic Methods in Structural Engineering, Chapman and Hall
15. Auriault JL, (1983) Effective macroscopic description for heat conduction in periodic composites. Int. J. Heat and Mass Transfer 26:861-869.
16. Babuska I, (1976) Homogenization and its application. Mathematical and computational problems. In: Numerical solution of partial differential equations III, Academic Press, pp. 89-116.
17. Bacry E, Mallat S, Papanicolaou G, (1990) A wavelet based space-time adaptive numerical method for partial differential equations. Rep. AFOSR-90-0040, New York
18. Bahei-El-Din YA, (1996) Finite element analysis of viscoplastic composite materials and structures. Int. J. Comp. Mat. Struct. 3:1-28.

19.   Barbero E, Fernandez-Saez J, Navarro C, (2000) Statistical analysis of the mechanical properties of composite materials. Comp. Part B: Engineering 31:375-381.

20.   Bast CC, Boyce L, (1995) Probabilistic Material Strength Degradation Model for Inconel 718 Components Subjected to High Temperature, High-Cycle and Low-Cycle Mechanical Fatigue, Creep and Thermal Fatigue Effects. NASA Rep. 198246, Univ. of Texas, San Antonio

21.   Bathe KJ (1996) Finite Element Procedures. Prentice Hall, New York

22.   Bazant ZP (1985) Mechanics of fracture and progressive cracking in concrete structures. In: Sih GC, DiTommaso P (eds.) Fracture Mechanics of Concrete: Structural Application and Numerical Calculation, pp 1-94.

23.   Becker M (1986) Heat Transfer. A Modern Approach. Plenum Press, New York

24.   Beckert W, Lauke B, (1996) Finite element calculation of energy release rate for single-fibre pull-out test. Comput. Mat. Sci. 5:1-11.

25.   Beltzer AI (1990) Variational and Finite Element Methods. A Symbolic Computation Approach. Springer-Verlag

26.   Belyaev AK, Ziegler F, (1998) Uniaxial waves in randomly heterogeneous elastic media. Prob. Engrg. Mech. 13(1):27-38.

27.   Benabou L, Benseddiq N, Naït-Abdelaziz M, Comparative analysis of damage at interfaces of composites. Comp. Part B: Engrg. 33:215-224.

28.   Ben-Amoz M, (1976) A dynamic theory for composite materials. J. Appl. Math. Phys. ZAMP 27:83-99.

29.   Bendat JS, Piersol AG (1971) Random Data: Analysis and Measurement Procedures. Wiley

30.   Bensoussan A, Lions JL, Papanicolaou G (1978) Asymptotic Analysis for Periodic Structures. North-Holland, Amsterdam

31.   Benveniste Y, (1987) A new approach to the Mori-Tanaka's theory in composite materials. Mech. Mat. 6:147-157.

32.   Beran MJ, (1974) Application of statistical theories for the determination of thermal, electrical and magnetic properties of heterogeneous materials. In: Broutman LJ et al. (eds.) Mechanics of Composite Materials. Academic Press, pp.209-249.

33.   Besterfield G, (1988) Probabilistic finite elements for fracture mechanics. AIAA J.: 1635-1639.

34.   Beyerlein IJ, Landis CM, (1999) Shear-lag model for failure simulations of unidirectional fiber composites including matrix stiffness. Mech. Mat. 31:331-350.

35.   Beyerlein IJ, Phoenix SL, (1997) Statistics of fracture for an elastic notched composite lamina containing Weibull fibers. Engrg. Fract. Mech. 57(2/3):241-265, 267-299.

36.   Binder K, Heermann DW, (1997) Monte Carlo Simulation in Statistical Physics. An Introduction. Springer

37.   Bolotin VV (1999) Fatigue of composite materials. Wiley

38. Bolzon G, Ghilotti D, Maier G, (2002) Strength of periodic elastic-brittle composites evaluated through homogenization and parameter identification. Eur. J. Mech. A/Sol. 21:355-378.

39. Boswell MT et al., (1991) The Art of Computer Generation of Random Variables. In: C.R. Rao (ed.) Handbook of Statistics, 9: Computational Statistics. Elsevier, pp 662-721.

40. Bourgeois C, Nicaise S, (2001) Prewavelet analysis of the heat equation. Numer. Math. 87:407-434.

41. Boutin C, Auriault JL, (1993) Acoustics of a bubbly fluid at large bubble concentration. Eur. J. Mech. B/Fluids 12(3):367-399.

42. Boyce L, Chamis CC, (1988) Probabilistic constitutive relationships for cyclic material strength models. AIAA J.:1299-1306.

43. Braides A, Defrancheschi A (1998) Homogenization of Multiple Integrals. Oxford Sci. Publ., Clarendon Press, Oxford

44. Brandt S, (1999) Data Analysis, Statistical and Computational Methods for Scientists and Engineers, Springer Verlag

45. Brandt A, Jendo S, Marks W, (1984) Probabilistic approach to reliability-based optimum structural design. Engrg. Trans. 32(1):57-74.

46. Breitung K, Casciati F, Faravelli L, A stochastic boundary element model for soil properties. Proc. of IUTAM/IABEM BEM Symp., Kraków 1999.

47. Brown HC, Chamis CC, (1991) Computational characterization of high temperature composites via METCAN. AMD-vol. 118, Mechanics of Composites at Elevated and Cryogenic Temperature, pp. 133-144.

48. Bucinell RB, (1998) Development of stochastic free edge delamination model for laminated composite materials subjected to constant amplitude fatigue loading. J. Comp. Mat. 32(12):1138-1155.

49. Buczkowski R, Kleiber M, (1999) A stochastic model of rough surfaces for finite element contact analysis. Comput. Meth. Appl. Mech. Engrg. 169: 43-59.

50. de Buhan P, Taliercio A, (1991) A homogenization approach to the yield strength of composite materials. Eur. J. Mech. A/Solids 10(2):129-154.

51. Burczyński T (1995) Boundary Element Method in Mechanics (in Polish). WNT, Warsaw

52. Byström J, (2003) Influence of the inclusions distribution on the effective properties of the heterogeneous media. Comp. Part B: Engrg. 34:587-592.

53. Caprino G, D'Amore A, Facciolo F, (1998) Fatigue sensitivity of random glass fibre reinforced plastics. J. Comp. Mat. 32(12):1203-1220.

54. Carlsson LA, Gillespie Jr JW, (1989): Mode-II interlaminar fracture of composites. In: Friedrich K (ed.) Application of fracture mechanics to composite materials. Elsevier, pp 113- 154.

55. Carslaw HS, Jeager JC (1959) Conduction of Heat in Solids. Oxford Univ. Press, London

56. Casciati F, Faravelli L, (1995) Non-linear Stochastic Dynamics by Equivalent Linearization. Proc. 2nd Int. Workshop on Stochastic Methods in Structural Mechanics, Pavia.

57. Castañeda PP, (2002) Second-order homogenization estimates for nonlinear composites incorporating field fluctuations: I – Theory, J. Mech. Phys. Solids 50:737-757.

58. Castañeda PP, (2002) Second-order homogenization estimates for nonlinear composites incorporating field fluctuations: II – Applications, J. Mech. Phys. Solids 50:759-782.

59. Chaboche JL et al., (1997) Numerical analysis of composite systems by using interphase/interface models. Comp. Mech. 20:3-11.

60. Chandra R, Singh SP, Gupta K, (2003) A study of damping in fiber-reinforced composites. J. Sound Vibr. 262:475-496.

61. Char BW et al. (1992) First Leaves: A Tutorial Introduction to Maple V, Springer-Verlag.

62. Chiang KT et al., (1995) Materials characterization of silicon carbide reinforced titanium (Ti/SCS-6) metal matrix composites: part II. Theoretical modeling of fatigue behavior. Metall. Mat. Trans. A 26A:3249-3255.

63. Choi CK, Noh HC, (1996) Stochastic finite element analysis of plate structures by weighted integral method. Struct. Engrg. Mech. 4(6):703-715.

64. Choi DK, Nomura S, (1992) Application of symbolic computation to two-dimensional elasticity. Comput. & Struct. 43(4):645-649.

65. Christensen RM (1979) Mechanics of Composite Materials. Wiley-Interscience

66. Christon MA, Roach DW, (2000) The numerical performance of wavelets for PDEs: the multi-scale finite element. Comp. Mech. 25:230-244.

67. Chung PW, Namburu RR, (2003) On a formulation for a multiscale atomistic-continuum homogenization method. Int. J. Sol. Struct. 40:2563-2588.

68. Collatz L (1966) The Numerical Treatment of Differential Equations. 3rd Edition, Springer-Verlag

69. Corigliano A, Mariani S, (2002) Identification of a constitutive model for the simulation of time-dependent interlaminar debonding processes in composites. Comput. Meth. Appl. Mech. Engrg. 191:1861-1894.

70. Cornil JM, Testud P (2001) An Introduction to Maple V. Springer-Verlag. Berlin-Heidelberg-New York

71. Coult NA (1997) A Multiresolutional Strategy for Homogenization of Partial Differential Equations. Ph.D. Thesis, Colorado Univ.

72. Crisfield MA (1991) Non-linear Finite Element Analysis of Solids and Structures. Wiley

73. Cruz ME, Patera AT, (1995) A parallel Monte-Carlo finite element procedure for the analysis of multicomponent media. Int. J. Num. Meth. Engrg. 38:1087-1121.

74. Dai LH, Huang GJ, (2001) An incremental micromechanical scheme for nonlinear particulate composites. Int. J. Mech. Sci. 43:1179-1193.

75. Dems K, Mróz Z, (1993) On shape sensitivity approaches in the numerical analysis of structures. Struct. Optimiz. 6:86-93.

76. Dems K, Mróz Z, (1987) Variational approach to sensitivity analysis in thermoelasticity. J. Thermal Stresses 10:283-306.

77. Desai CS, Nagaraj K, (1988) Modeling for cyclic normal and shear behavior of interfaces. ASCE J. Engrg. Mech. 114(7):1198-1217.

78. Dhatt G, Touzot G, (1984) The Finite Element Method Displayed. John Wiley & Sons

79. Di Sciuva M, Lomario D, (2003) A comparison between Monte Carlo and FORMs in calculating the reliability of a composite structure. Comp. Struct. 59:155-162.

80. Ditlevsen O, Sobczyk K, (1986) Random fatigue crack growth with retardation. Engrg. Fract. Mech. 24(6):861-878.

81. Dolbow J, Belytschko T, (1999) Numerical integration of the Galerkin weak form in meshfree methods. Comput. Mech. 23:219-230.

82. Donahue RJ, Clark HM, Atammi P, Kumble R, McEvily AJ, (1972) Crack Opening Displacement and the Rate of Fatigue Crack Growth. Int. J. Fract. Mech. 8:209-219.

83. Drugan J, (2003) Two exact micromechanics-based nonlocal constitutive equations for random linear elastic composite materials. J. Mech. Phys. Solids 51:1745-1772.

84. Duvaut G et al., (2000) Optimization of fiber reinforced composites. Comp. Struct. 48:83-89.

85. Dvorak GJ, (2000) Composite materials: inelastic behavior, damage, fatigue and fracture. Int. J. Sol Struct. 37:155-170.

86. Dvorak GJ, Bahei-El-Din YA, Wafa AM, (1994) Implementation of the transformation field analysis for inelastic composites materials, Comput. Mech. 14: 201-228.

87. Elishakoff I et al. (1992) Random Vibrations and Reliability of Composite Structures. Technomic Press

88. Elishakoff I, Ren YJ, Shinozuka M, (1995) Improved finite element method for stochastic problems, Chaos, Solitons and Fractals 5(5):833-846.

89. Ermer DS, Notohardjono BD, (1984) Fatigue failure identification by time series model. Engrg. Fract. Mech. 20(5/6):705-718.

90. Evans G, Blackledge J, Yardley P (2000) Numerical Methods for Partial Differential Equations. Springer-Verlag London

91. Feddersen CE, (1970) Evaluation and prediction of the residual strength of center cracked tension panels. ASTM STP 486, pp. 50-78.

92. Feng XQ, (2001) On estimation methods for effective moduli of microcracked solids. Arch. Appl. Mech. 71:537-548.

93. Figiel Ł, Kamiński M, (2003) Computational fatigue crack growth analysis in layered composites. In: Bathe KJ (ed.) Proc. 2nd MIT Conf. Comput. Fluid & Solid Mech., Elsevier, pp. 258-260.

94. Figiel Ł, Kamiński M, (2001) Effective elastoplastic properties of the periodic composites. Comput. Mat. Sci. 22(3-4):221-239.

95. Figiel Ł, Kamiński M, (2003) Mechanical and thermal fatigue delamination of curved layered composites. Comput. & Struct. 81(18-19):1865-1873.

96. Figiel Ł, Kamiński M, (2001) Mechanical and thermal fatigue of curved composite beams. In: Topping BHV (ed.) Proc. 8th Int. Conf. Civil Struct. Engrg. Comput. Techn., Civil-Comp Press, Stirling, UK.

97. Figiel Ł, Kamiński M, (2003) Numerical analysis of damage evolution in composite pipe joints under cyclic static axial tension. Engrg. Trans. 51(4): 469-485.

98. Fish J (edr.), (1999) Computational advances in modeling composites and heterogeneous materials. Comput. Meth. Appl. Mech. Engrg. 172(1-4).

99. Fish J, Ghouali A, (2001) Multiscale analytical sensitivity analysis for composite materials. Int. J. Num. Meth. Engrg. 50:1501-1520.

100. Fong JT, (1982) What is fatigue damage?. Reifsnider KL (ed.) Damage in Composite Materials. ASTM STP 775, pp 243-266.

101. Forman RG, Kearney VE, Engle RM, (1967) Numerical Analyses of Crack Propagation in Cyclic-Loaded Structures. J. Basic Engrg. 89:459-464.

102. Frangopol DM, Recek S, (2003) Reliability of fiber-reinforced composite laminate plates. Prob. Engrg. Mech. 18:119-137.

103. Frank PM (1978) Introduction to System Sensitivity Theory. Academic Press

104. Frantziskonis G, (2002) Wavelet-based analysis of multiscale phenomena: application to material porosity and identification of dominant scales. Prob. Engrg. Mech. 17:349-357.

105. Frantziskonis G et al., (1997) Heterogeneous solids - Part I: analytical and numerical 1-D results on boundary effects. Eur. J. Mech. A/Solids 16(3):409-423.

106. Furmański P, (1997) Heat conduction in composites: homogenization and macroscopic behaviour. Appl. Mech. Rev. 50(6):327-355.

107. Galvanetto U, Pellegrino C, Schrefler BA, (1998) Plane stress plasticity in periodic composites. Comput. Mat. Sci. 624:1-11.

108. Gałka A, Gambin B, Telega JJ, (1998) Variational bounds on the effective moduli of anisotropic piezoelectric composites. Arch. Mech. 50(4):675-689.

109. Gamstedt EK, (2000) Effects of debonding and fiber strength distribution on fatigue-damage propagation in carbon fiber-reinforced epoxy. J. Appl. Polymer Sci. 76:457-474.

110. Gamstedt EK, (2000) Fatigue in composite laminates – A qualitative link from micromechanisms to fatigue life performance. In: Cardon AH et al., Recent developments in durability analysis of composite systems, pp. 87-100.

111. Gander W, Hrebicek J, (1995) Solving Problems in Scientific Computing Using MAPLE and MATLAB. Springer-Verlag Berlin-Heidelberg, 2nd Edition

112. Geindreau C, Auriault JL, (1999) Investigation of the viscoplastic behavior of alloys in the semi-solid state by homogenization. Mech. Mat. 31:535-551.

113. Ghanem RG, Spanos PD, (1997) Spectral techniques for stochastic finite elements. Arch. Comput. Meth. Engrg. 4(1):63-100.

114. Ghanem RG, Spanos PD (1991) Stochastic Finite Elements: A Spectral Approach. Springer-Verlag

115. Ghanem RG, Spanos PD, (1992) Spectral stochastic finite-element formulation for reliability analysis. J. Engrg. Mech. ASCE 117(10):2351-2372.

116. Ghosh S, Lee K, Moorthy S, (1995) Multiple scale analysis of heterogeneous elastic structures using homogenization theory and Voronoi cell finite element method. Int. J. Sol. Struct. 32(1): 27-62.

117. Golanski D, Terada K, Kikuchi N, (1997) Macro and micro scale modeling of thermal residual stresses in metal matrix composite surface layers by the homogenization method. Comput. Mech. 19:188-202.

118. Gonzalez C, Llorca J, (2000) A self-consistent approach to the elasto-plastic behaviour of two-phase materials including damage. J. Mech. Phys. Solids 48:675-692.

119. González P, Cabalerio JC, Pena TF, (2002) Parallel iterative solvers involving fast wavelet transforms for the solution of BEM systems. Adv. Engrg. Software 33:417-426.

120. Graham LL, Gurley K, Masters F, (2003) Non-Gaussian simulation of local material properties based on a moving-window technique. Prob. Engrg. Mech. 18:223-234.

121. Grigoriu M, (1999) Stochastic mechanics. Int. J. Sol. Struct. 37:197-214.

122. Guinovart-Diaz R et al., (2001) Closed-form expressions for the effective coefficients of fibre-reinforced composite with transversely isotropic constituents. I: elastic and hexagonal geometry. J. Mech. Phys. Solids 49:1445-1462.

123. Guinovart-Diaz R et al., (2001) Closed-form expressions for the effective coefficients of fibre-reinforced composite with transversely isotropic constituents. II: piezoelectric and hexagonal geometry. J. Mech. Phys. Solids 49:1463-1479.

124. Gutierrez MA, de Borst R, (1999) Numerical analysis of localization using a viscoplastic regularization: influence of stochastic materials defects. Int. J. Num. Meth. Engrg. 44:1823-1841.

125. Hammersley JM, Handscomb DC, (1964) Monte Carlo methods. Wiley

126. Hammond DA et al., (1993) Cavitation erosion performance of fiber reinforced composites. J. Comp. Mat. 27(16):1522-1544.

127. Hansen U, (1999) Damage development in woven fabric composites during tension-tension fatigue. J. Comp. Mat. 33(7):614-639.

128. Harper CA (ed.), (1996) Handbook of plastics, elastomers and composites. McGraw-Hill.

129. Hashin Z, (1983) Analysis of Composite Materials - A Survey. Appl. Mech. 50:481-503.

130. Hashin Z, (1983) Statistical cumulative damage theory for fatigue life prediction. J. Appl. Mech. 50:571-579.

131. Hashin Z, (2002) Thin interphase/imperfect interface in elasticity with application to coated fiber composites. J. Mech. Phys. Solids 50:2509-2537.

132. Hasofer AM, Ditlevsen O, Tarp-Johansen NJ, Positive random fields for modeling materials stiffness and compliance (in press).

133. Hassani B, Hinton E (1999) Homogenization and Structural Topology Optimization. Theory, Practice and Software. Springer-Verlag, London

134. Haug EJ, Choi KK, Komkov V (1986) Design Sensitivity Analysis of Structural Systems. Series Math. Sci. Eng., Academic Press

135. Hien TD, Kleiber M, (1997) Stochastic finite element modelling in linear transient heat transfer. Comput. Meth. Appl. Mech. Engrg. 144:111-124.

136. Hill R, (1967) The essential structure of constitutive laws for metal composites and polycryctals. J. Mech. Phys. Sol. 15:79-95.

137. Hobson PD, (1982) The Formation of a Crack Growth Equation for Short Cracks. Fatigue Fract. Engrg. Mat. Struct. 5:323-327.

138. Hori M, Nemat-Nasser S, (1998) Universal bounds for effective piezoelectric moduli. Mech. Mat. 30:1-19.

139. Hurtado JE, Barbat AH, (1998) Monte-Carlo techniques in computational stochastic mechanics. Arch. Comput. Meth. Engrg. 5(1):3-30.

140. Hutchinson JW et al., (1987) Crack paralleling an interface between dissimilar materials. J. Appl. Mech. 54:828-832.

141. Huyse L, Maes MA, (2001) Random field modeling of elastic properties using homogenization. J. Engrg. Mech. ASCE 127(1):27-36.

142. Hwang TK, Hong CS, Kim CG, (2003) Probabilistic deformation and strength prediction for a filament wound pressure vessels. Comp. Part B: Engrg. 34:481-497.

143. Hwang W, Han KS, (1986) Cumulative damage models and multi-stress fatigue life prediction. J. Comp. Mat. 20:125-153.

144. Hwang W, Han KS, (1986) Fatigue of composites - fatigue modulus concept and life prediction. J. Comp. Mat. 20:154-165.

145. Ichikawa Y, Kawamura K, Fujii N, (2003) Molecular simulation and multiscale homogenization analysis for microinhomogeneous clay materials. Engrg. Comput. 20(5/6):559-582.

146. Ismar H, Reinert U, (1997) Modelling and simulation of the macromechanical nonlinear behaviour of fibre-reinforced ceramics on the basis of a micromechanical-statistical material description. Acta Mech. 120:47-60.

147. Jameson L, (1998) A wavelet-optimized, very high order adaptive grid and order numerical method. SIAM J. Sci. Comput. 19(6):1980-2013.

148. Jameson L et al., (1998) Wavelet-based numerical methods. Int. J. Comput. Fluid Dyn. 10:267-280.

149. Janicki A, (1996) Numerical and Statistical Approximation of Stochastic Differential Equations with Non-Gaussian Measures. Technical University of Wrocław

150. Jankovic I, Fiori A, Dagan G, (2003) Effective conductivity of an isotropic heterogeneous medium of lognormal conductivity distribution. Multiscale Model. Simul. 1(1):40-56.

151. Jasiuk I, Chen J, Thorpe MF, (1994) Elastic moduli of two-dimensional materials with polygonal and elliptical holes, Appl. Mech. Rev. ASME 47(1):18-28.

152. Jeong H, Jang YS, (2000) Wavelet analysis of plate wave propagation in composite laminates. Comp. Struct. 49:443-450.

153. Jeong HK, Shenoi RA, (1998) Reliability analysis of midplane symmetric laminated plates using direct simulation method. Comp. Struct. 43:1-13.

154. Jeulin D, Ostoja-Starzewski M, (2001) Mechanics of Random and Multiscale Microstructures. CISM Courses and Lectures, No. 430, Springer Wien New York, Udine

155. Jędrysiak J, (1998) On dynamics of thin plates with a periodic structure. Engrg. Trans. 46(1):73-87.

156. Jiang M, Ostoja-Starzewski M, Jasiuk I, (2001) Scale-dependent bounds on effective elastoplastic response of random composites. J. Mech. Phys. Solids 49:655-673.

157. Johnson SW et al. (1998) Application of fracture mechanics to the durability of bonded composite joints. Report DOT/FAA/AR-97/56, OAR, Washington D.C.

158. Kachanov LM (1986) Introduction to Damage Mechanics. Kluwer Academic Publishers

159. Kalamkarov AL, Kolpakov AG (1997) Analysis, Design and Optimization of Composite Structures. Wiley

160. Kamiński M, (1999) Computational engineering homogenization of steel reinforced concrete plates. Adv. Compos. Letters 8(5):213-218.

161. Kamiński M, (2002) Current computational issues in composite materials modeling. In: Obrębski JB (edr.), Lightweight structures in civil engineering. pp. 755-766.

162. Kamiński M, (2000) Homogenization of 1D elastostatics by the stochastic second order approach. Mech. Res. Comm. 27(3):273-280.

163. Kamiński M, (2000) Homogenized properties of n-components composites, Int. J. Engng. Sci. 38(4):405-427.

164. Kamiński M, (2001) Homogenization method in stochastic finite element analysis of some 1D composite structures. In: Topping BHV (ed.) Proc. 8th Int. Conf. Civil Struct. Engrg. Comput. Techn., Civil-Comp Press, Stirling, UK.

165. Kamiński M, (2003) Homogenization of transient heat transfer problems for some composite materials. Int. J. Engrg. Sci. 41(1):1-29.

166. Kamiński M, (2003) Homogenization technique for transient heat transfer in unidirectional composites. Comm. Num. Meth. Engrg. 19(7):503-512.

167. Kamiński M, (2001) Material sensitivity analysis in homogenization of the linear elastic composites. Arch. Appl. Mech. 71(10):679-694.

168. Kamiński M, (2003) Material sensitivity studies for homogenized superconducting composites. In: Topping BHV (ed.) Proc. 9th Int. Conf. Civil Struct. Engrg. Comput. Techn., Civil-Comp Press, Stirling, UK

169. Kamiński M, (2002) Multiresolutional homogenization technique in transient heat transfer for unidirectional composites. In: Topping BHV, Bittnar Z. (edrs.) Proc. 3rd Int. Conf. Engrg. Comput. Techn., Civil-Comp Press, Stirling, UK.

170. Kamiński M, (2001) Multiresolutional wavelet-based homogenization of random composites. In: Proc. Second ECCOMAS Conference, Kraków.

171. Kamiński M, (1999) Monte-Carlo simulation of effective conductivity for fiber composites. Int. Comm. Heat and Mass Transfer 26(6):801-810.

172. Kamiński M, (2002) On probabilistic fatigue models for composite materials. Int. J. Fatigue 24:477-495.

173. Kamiński M, (2001) On sensitivity of effective elastic moduli for fibre-reinforced composites. Comm. Num. Meth. Engrg. 17(2):127-135.

174. Kamiński M, (1998) Probabilistic bounds on effective elastic moduli for the superconducting coils. Comput. Mat. Sci. 11:252-260.

175. Kamiński M, (1998) Probabilistic effective heat conductivity of fiber composites, Int. J. Mech. & Mech. Engng. 2:175-207.

176. Kamiński M, (1998) Probabilistic reliability analysis of lightweight structures. In: Obrębski JB (ed.) Lightweight Structures in Civil Engineering, pp 206-214.

177. Kamiński M, (2003) Sensitivity analysis of homogenized characteristics of some elastic composites. Comput. Meth. Appl. Mech. Engrg. 192(16-18):1973-2005.

178. Kamiński M, (2003) Some notes on numerical convergence of the stochastic perturbation method. In: Sloot PMA et al. (edrs.) Computational Science – ICCS 2003, vol. I, Lecture Notes in Comput. Sci. No 2657, pp. 521-530.

179. Kamiński M, (2000) Stochastic computational mechanics of composite materials. In: de Wilde WP, Blain WR, Brebbia CA (edrs.), Advances in Composite Materials & Structures VII. WIT Press, pp 219-228.

180. Kamiński M, (2001) Stochastic finite element method homogenization of heat conduction problem in fiber composites. Int. J. Struct. Engrg. Mech. 11(4):373-392.

181. Kamiński M, (2002) Stochastic perturbation approach to engineering structure vibrations by the Finite Difference Method. J. Sound & Vibr. 251(4):651-670.

182. Kamiński M, (2004) Stochastic perturbation approach to wavelet-based multiresolutional analysis. Num. Linear Algebra Engrg. Appl. 11(4):355-370.

183. Kamiński M, (2002) Stochastic problem of fiber-reinforced composite with interface defects. Engrg. Comput. 19(7):854-868.

184. Kamiński M, (2001) Stochastic problem of the viscous incompressible fluid flow with heat transfer. ZAMM 81(12): 827-837.

185. Kamiński M, (1999) Stochastic second-order perturbation BEM formulation. Engrg. Anal. Bound. Elem. 2:123-130.

186. Kamiński M, (2001) Stochastic second order perturbation approach to the stress-based finite element method. Int. J. Sol. Struct. 38(21):3831-3852.

187. Kamiński M, (2001) The stochastic second order perturbation technique in the Finite Difference Method. Comm. Num. Meth. Engrg. 17(9):613-622.

188. Kamiński M, (2003) Wavelet-based homogenization of unidirectional multiscale composites. Comput. Mat. Sci. 27(4):446-460.

189. Kamiński M, (2002) Wavelet-based finite element elastodynamic analysis of composite beams. In: H.A. Mang, F.G. Rammerstorfer, J. Eberhardsteiner (edrs.) Proc. 5th World Congress on Computational Mechanics WCCM V, Vienna.

190. Kamiński M, Hien TD, (1999) Stochastic finite element analysis of transient heat transfer in composite materials with interface defects. Arch. Appl. Mech. 51(3-4):269-288.

191. Kamiński M, Kleiber M, (2000) Numerical homogenization of n-component composites with stochastic interface defects. Int. J. Num. Meth. Engrg. 47:1001-1025.

192. Kamiński M, Kleiber M, (2000) Perturbation based stochastic finite element method for homogenization of two-component elastic composites. Comput. & Struct. 78(6):811-826.

193. Kamiński M, Kleiber M, (1996) Stochastic finite element method in random non-homogeneous media. In: J.A. Desideri et al., (edrs.) Numerical Methods in Engineering'96, Wiley, pp 35-41.

194. Kamiński M, Kleiber M, (1996) Stochastic structural interface defects in composite materials. Int. J. Sol. Struct. 33(20-22):3035-3056.

195. Kamiński M, Pawlik M, (2002) Homogenization of transient heat transfer problems for some composite materials. In: Topping BHV, Bittnar Z. (edrs.) Proc. 3rd Int. Conf. Engrg. Comput. Techn., Civil-Comp Press, Stirling, UK.

196. Kamiński M, Pawlik M, (2003) Transient heat transfer in layered composites with random geometry. In: Bathe KJ (ed.) Proc. 2nd MIT Conf. Comput. Fluid & Solid Mech., Elsevier, pp. 360-363.

197. Kamiński M, Schrefler BA, (2000) Probabilistic effective characteristics of cables for superconducting coils. Comput. Meth. Appl. Mech. Engrg. 188(1-3):1-16.

198. Kamiński M, Spanos PD, (2001) Stochastic second order perturbation approach to finite difference method in vibration analysis. Int. J. Appl. Mech. Engrg. 6(2):369-393.

199. Kapitaniak T, (2000) Chaos for Engineers. Theory, Applications and Control, Springer

200. Karpur P, Matikas TE, Pagano NJ (edrs.), (1995) Fiber-matrix interface. Int. J. Sol. Struct. 5(6).

201. Kennedy CA, Lennox WC, (2001) Moment operations on random variables, with applications for probabilistic analysis. Prob. Engrg. Mech. 16:253-259.

202. Khoroshun LP, (2000) Mathematical models and methods of the mechanics of stochastic composites. Int. J. Appl. Mech. 36(10):1284-1316.

203. Kim JK, Mai YW (1998) Engineering Interfaces in Fiber Reinforced Composites. Elsevier

204. Kingman JFC, (1993) Poisson Processes. Clarendon Press, Oxford

205. Kinloch AJ, (1994) The service life of adhesive joints. In: Adhesion and adhesives. Science and Technology. Chapman & Hall, pp 339-404.

206. Kleiber M (ed.) (1998) Handbook of Computational Solid Mechanics. Springer-Verlag

207. Kleiber M et al. (1997) Parameter Sensitivity in Nonlinear Mechanics. Wiley

208. Kleiber M, Hien TD (1992) The Stochastic Finite Element Method. Wiley

209. Kleiber M, Siemaszko A, Stocki R, (1999) Interactive stability-oriented reliability-based design optimization. Comput. Meth. Appl. Mech. Engrg. 168:243-253.

210. Kleiber M, Woźniak Cz (1991) Nonlinear Mechanics of Structures. PWN/Kluwer

211. Knox EM, Cowling MJ, Hashim SA, (2000) Fatigue performance of adhesively bonded connections in GRE pipes. Int. J. Fatigue 22:513-519.

212. Kolling S, Gross D, (2001) Simulation of mictrostructural evolution in materials with misfitting precipitates. Prob. Engrg. Mech. 16:313-322.

213. Krishnamoorthy CS (1994) Finite Element Analysis, 2nd Edition, McGraw-Hill

214. Kushnevsky V, Morachkovsky O, Altenbach H, (1998) Identification of effective properties of particle reinforced composite materials. Comput. Mech. 22:317-325.

215. Kutt TV, Bieniek MP, (1988) Cumulative damage and fatigue life prediction. AIAA J. 26(2):213-219.

216. Kwon YW, Kim C, (1998) Micromechanical model for thermal analysis of particulate and fibrous composites. J. Thermal Stresses 21:21-39.

217. Landis CM, Beyerlein IJ, McMeeking RM, (2000) Micromechanical simulation of the failure of fiber reinforced composites. J. Mech. Phys. Solids 48:621-648.

218. Larsson H, Bernard J., (1978) Fracture of longitudinally cracked ductile tubes. Int. J. Press Vessels & Piping 6:223-243.

219. Lauke B, Fu SY, (1999) Strength anisotropy of misaligned short-fibre-reinforced polymers. Comp. Sci. Techn. 59:99-708.

220. Lee HK, Simunovic S, (2000) Modelling of progressive damage in aligned and randomly oriented discontinuous fiber polymer matrix composites. Comp. Part B: Engrg. 31:77-86.

221. Lefik M, Schrefler BA, (1992) Homogenized material coefficients for 3D elastic analysis of superconducting coils. In: Ladeveze P, Zienkiewicz OC, edrs. New Advances in Computational Structural Mechanics. Elsevier

222. Leguillon D, Lacroix C, Matrin E, (2000) Interface debonding ahead of a primary crack. J. Mech. Phys. Solids 48:2137-2161.

223. Lene F, (1986) Damage constitutive relations for composite materials. Engrg Fract. Mech. 25(5/6):713-728.

224. Lene F, Leguillon D, (1982) Homogenized constitutive law for a partially cohesive composite material. Int. J. Sol. Struct. 18(5):443-458.

225. Leo PH, Hu J, (1995) A continuum description of partially coherent interfaces. Continuum Mech. Thermodyn. 7:39-56.

226. Lesne PM, Allio N, Valle R, (1995) Combined effects of the fibre distribution and of the fibre matrix or interphase matrix transverse modulus ratio on the possible fracture modes of unidirectional composites submitted to a transverse loading. Acta Metall. Mater. 43(12):4247-4266, 1995.

227. Lewiński T, Telega JJ (2000) Plates, Laminates and Shells. Asymptotic Analysis and Homogenization. Ser. Adv. Math. Appl. Sci. 52, World Sci. Publ, Singapore.

228. Li JY, (1999) On micromechanics approximation for the effective thermoelastic moduli of multi-phase composite materials. Mech. Mat. 31:149-159.

229. Liaw PK et al., (1995) Materials characterization of silicon carbide reinforced titanium (Ti/SCS-6) metal matrix composites: part I. Tensile and fatigue behavior. Metall. Mat. Trans. A 26A:3225-3247.

230. Lim LG, Dunne FPE, (1996) The effect of volume fraction on reinforcement on the elastic-viscoplastic response of metal-matrix composites. Int. J. Sol. Struct. 38(1):19-39.

231. Lin SC, (2000) Reliability predictions of laminated composite plates with random system parameters. Prob. Engrg. Mech. 15:327-338.

232. Lin S, Garmestani H, (2000) Statistical continuum mechanics analysis of an elastic two-isotropic-phase composite material. Comp. Part B: Engrg. 31:39-46.

233. Lin YK, (1967) Probabilistic Theory of Structural Dynamics. McGraw-Hill, New York

234. Lin YK, Yang JN, (1982) A stochastic theory of fatigue crack propagation. AIAA J. 23(1):117-124.

235. Liu B, (2003) Adaptive harmonic wavelet transform with applications in vibration analysis. J. Sound Vibr. 262:45-64.

236. Liu WK et al., (2000) Multi-scale methods. Int. J. Num. Meth. Engrg. 47:1343-1361.

237. Liu WK, Belytschko T, Mani A, (1986) Probabilistic finite elements for nonlinear structural dynamics. Comput. Meth. Appl. Mech. Engrg. 56:61-81.

238. Liu ML, Yu J, (2003) Finite element modeling of delamination by layerwise shell element allowing for interlaminar displacements. Comp. Sci. Techn. 63:512-529.

239. Luciano R, Willis JR, (2001) Non-local constitutive response of a random laminate subjected to configuration-dependent body force. J. Mech. Phys. Solids 49:431-444.

240. Lund E, (1998) Shape optimization using Weibull statistics of brittle failure. Struct. Optimiz. 15:208-214.

241. Lutes LD (1987) An Introduction to Stochastic Structural Fatigue. Rice University Press

242. Lüth H (2001) Solid Surfaces, Interfaces and Thin Films. 4th Edition. Springer-Verlag Berlin Heidelberg

243. Ma F, Wong FS, Caughey TK, (1983) On the Monte-Carlo methodology for cumulative damage. Comput. & Struct. 17(2):177-181.

244. Madsen HO, (1984) Bayesian fatigue life prediction. Probabilistic Methods in the Mechanics of Solids and Structures. Eddwertz S, Lind NC, (edrs.) Proc. IUTAM Symp., Stockholm, pp 395-406.

245. Madsen B, Lilholt H, (2003) Physical and mechanical properties of unidirectional plant fibre composites – an evaluation of the influence of porosity. Comp. Sci. Techn. 63:1265-1272.

246. Mao H, Mahadevan S, (2002) Fatigue damage modeling of composite materials. Comp. Struct. 58:405-410.

247. McEvily AJ, Groeger J, (1977) On the Threshold for Fatigue Crack Growth. Proc. 4th Int. Conference on Fracture vol. II, Canada, pp 1293-1298.

248. Meguid SA, Kalamkarov AL, (1994) Asymptotic homogenization of elastic composite materials with a regular structure. Int. J. Sol. Struct. 31(3): 303-316.

249. Meirovitch L (1975) Elements of Vibration Analysis. McGraw & Hill

250. Michel JC et al., (1999) Effective properties of composite materials with periodic microstructures: a computational approach, Comput. Meth. Appl. Mech. Engrg. 172(1-4):109-144.

251. Mignolet MP, Mallick K, (1995) Random inelastic behavior of composite materials with local load sharing. Prob. Engrg. Mech. 10:83-93.

252. Miloh T, Benveniste Y, (1999) On the effective conductivity of composites with ellipsoidal inhomogeneities and highly conducting interfaces. Proc. Roy. Soc. Lond. A 455:2687-2706.

253. Milton GW (2002) The Theory of Composites. Cambridge University Press

254. Mischke CR (1986) Probabilistic views of the Palmgren-Miner damage rule. ASME 86-WA/E-23

255. Mital SK, Murthy PLN, Chamis CC, (1993) Interfacial microfracture in high temperature metal matrix composites. J. Comp. Mat. 27(17):1678-1694.

256. Moro T, El Hami A, El Moudni A, (2002) Reliability analysis of a mechanical contact between deformable solids. Prob. Engrg. Mech. 17:227-232.

257. Morrow JD (1986) The effect of selected subcycle sequences in fatigue loading histories. Random fatigue life prediction. ASME PVP 72, pp 43-60.

258. Moser B, (2003) The influence of non-linear elasticity on the determination of Weibull parameters using the fibre bundle test. Composites: Part A 34:907-912.

259. Mura T (1982) Micromechanics of Defects in Solids. Sijthoff and Noordhoff

260. Müller WH, Schmauder S, (1993) Interface stresses in fiber-reinforced materials with regular fiber arrangements. Comp. Struct. 24:1-21.

261. Nag A, (2003) A spectral finite element with embedded delamination for modeling of wave scattering in composite beams. Comp. Sci. Techn. 63:2187-2200.

262. Nakayasu H, Maekawa Z, Rackwitz R, (1989) Reliability - oriented materials design of composite materials. Proc. 5th Conf. Struct. Safety & Reliability, pp 2095-2098.

263. Nayfeh AH (1973) Perturbation Methods. Wiley

264. Nelli-Silva EC, Ono Fonseca JS, Kikuchi N, (1997) Optimal design of piezoelectric microstructures. Comput. Mech. 19:397-410.

265. Nemat-Nasser S, Hori M (1993) Micromechanics: Overall Properties of Heterogeneous Material. North-Holland, Amsterdam

266. Nomura S, (1987) Effective medium approach to matrix-inclusion type composite materials. ASME Trans. 54:880-883.

267. Nomura S, Ball DL, (1994) Stiffness reduction due to multiple microcracks in transverse isotropic media. Engrg. Fract. Mech. 48(5):649-653.

268. Noor AK (edr.), (1990) Symbolic Computations and Their Impact on Mechanics. ASME PVP-205.

269. Noor AK, Shah RS, (1993) Effective thermoelastic and thermal properties of unidirectional fiber-reinforced composites and their sensitivity coefficients, Int. J. Comp. Struct. 26:7-23.

270. Ochiai S et al., (2003) Modeling of residual stress-induced stress-strain behavior of unidirectional brittle fiber/brittle matrix composite with weak interface. Comp. Sci. Techn. 63:1027-1040.

271. Oden JT (1972) Finite Elements of Nonlinear Continua. McGraw-Hill

272. Oden JT et al., (2003) Research directions in computational mechanics. Comput. Meth. Appl. Mech. Engrg. 192:913-922.

273. Onkar AK, Yadav D, (2003) Non-linear response statistics of composite laminates with random material properties under random loading. Comp. Struct. 60:375-383.

274. Ostoja-Starzewski M, (1999) Microstructural disorder, mesoscale finite elements and macroscopic response. Proc. Roy. Soc. Lond. A 455:3189-3199.

275. Ostoja-Starzewski M, Jasiuk I, (1994) Micromechanics of random media. Proc. 1st Joint ASCE-ASME-SES Symp. "Micromechanics of Random Media" Appl. Mech. Rev. 47(1/2)

276. Owen DRJ, Hinton E (1980) Finite Elements in Plasticity Theory and Practice. Pineridge Press

277. Paepegem Van W, Degrieck J, De Baets P, (2001) Finite element approach for modeling fatigue damage in fibre-reinforced composite materials. Comp. Part B: Engrg. 32:575-588.

278. Pankov AA, (1999) Solution of the stochastic boundary-value problem of elasticity theory for composites with disordered structures in the correlative approximation of the method of quasi-periodic components. Mech. Comp. Mat. 35(4):315-324.

279. Paris F, Correa E, Cañas J, (2003) Micromechanical view of failure of the matrix in fibrous composite materials. Comp. Sci. Techn. 63:1041-1052.

280. Paris PC, Erdogan F (1962). A critical analysis of crack propagation laws. ASME 62-WA-234.

281. Pedersen P (edr.), (1993) Optimal Design with Advanced Materials. Elsevier

282. Peng XQ et al., (1998) A stochastic finite element method for fatigue reliability analysis of gear teeth subjected to bending. Comput. Mech. 21:253-261.

283. Pepper DW, Heinrich JC (1992) The Finite Element Method. Series in Computational and Physical Processes in Mechanics and Thermal Sciences. Hemisphere

284. Peric D, Owen DRJ, (1992) Computational model for 3-D contact problems with friction based on the penalty method. Int. J. Num. Meth. Engrg. 35:1289-1309.

285. Pham DC, (1999) Bounds for the elastic shear moduli of isotropic and quasi-symmetric composites and exact modulus of the coated sphere assemblage. Math. Mech. Sol. 4:57-69.

286. Philips LN (ed.), (1989) Design with Advanced Composite Materials. Springer-Verlag

287. Philippidis TP, Lekou DJ, Kalogiannakis GA, (2000) On the stochastic nature of thermomechanical properties in glass reinforced polyester laminates. Comp. Struct. 49:293-301.

288. Phoenix SL, (2000) Modeling the statistical lifetime of glass fiber/polymer matrix composites in tension. Comp. Struct. 48:19-29.

289. Phoon KK et al., (2002) Implementation of Karhunen-Loeve expansion for simulating using a wavelet-Galerkin scheme. Prob. Engrg. Mech. 17:293-303.

290. Pridle EK, (1976) High-cycle fatigue crack propagation under random and constant amplitude loading. Int. J. Press Vessels & Piping 4:89-117.

291. Pochiraju KV, Lau ACW, Wang ASD, (1994) A local-global matching method for the single fiber pullout problem with perfectly bonded interface. Comput. Mech. 14(1):84-99.

292. Qian W, Sun CT, (1998) A frictional interfacial crack under combined shear and compression. Comp. Sci. Techn. 58:1753-1761.

293. Quek ST et al., (2001) Sensitivity analysis of crack detection in beams by wavelet technique. Int. J. Mech. Sci. 43:2899-2910.

294. Rao HS et al., (1997) A model of heat transfer in brake pads by mathematical homogenization. Sci. Engrg. Comp. Mat. 6(4):219-224.

295. Rahul-Kumar P et al., (2000) Interfacial failures in a compressive shear strength test of glass/polymer laminates. Int. J. Sol. Struct. 37:7281-7305.

296. Ratwani MM, Kan HP, (1981) Compression fatigue analysis of fiber composites. AIAA J Aircraft 18(6).

297. Reddy JN (1984) Energy and Variational Methods in Applied Mechanics. Wiley

298. Reifsnider KL (edr.), (1991) Fatigue of Composite Materials. Elsevier

299. Reifsnider KL, Stinchcomb WW, (1986) A critical-element model of the residual strength and life of fatigue - loaded composite coupons. In: Hahn HT. (edr.), Composite Materials: Fatigue and Fracture, ASTM STP 907, Philadelphia, pp 298-313.

300. Renaudin P et al., (1997) Heterogeneous solids - Part II: numerical 2-D results on boundary and other relevant phenomena. Eur. J. Mech. A/Solids 16(3):425-443.

301. Rice JR et al., (1990) Mechanics and thermodynamics of brittle interfacial failure in bimaterial systems. In: Ruhle M et al. (edrs.), Metal-Ceramic Interfaces. Pergamon Press, Oxford, pp 269-294.
302. Roberts R, Kibler JJ, (1971) Some aspects of fatigue crack propagation. Engrg. Fract. Mech. 2:243-260.
303. Rocha PA, Cruz ME, (2001) Computation of the effective conductivity of unidirectional fibrous composites with an interfacial thermal resistance. Numer. Heat Transfer Part A 39:179-203.
304. Roh YS, Xi Y, (2000) A general formulation for transition probabilities of Markov model and the application to fracture of composite materials. Prob. Engrg. Mech. 15:241-250.
305. Rosen BW, Hashin Z, (1970) Effective thermal expansion coefficients and specific heats of composite materials. Int. J. Engrg. Sci. 8:157-173.
306. Sab K, (1992) On the homogenization and the simulation of random materials. Eur. J. Mech. A/Solids. 11:585-607.
307. Saleeb AF et al., (2003) An anisotropic viscoelastoplastic model for composites – sensitivity analysis and parameter estimation. Comp. Part B: Engrg. 34:21-39.
308. Sanchez-Palencia E (1980) Non-homogeneous Media and Vibration Theory. Lecture Not. Phys. No. 127, Springer-Verlag
309. Sanchez-Palencia E, Zaoui A, (1987) Homogenization Techniques for Composite Media. Lect. Notes Phys. No. 272, Springer-Verlag
310. Savic V, Tuttle ME, Zabinsky ZB, (2001) Optimization of composite I-sections using fiber angles as design variables. Comp. Struct. 53:265-277.
311. Schapery RA, (1968) Thermal expansion coefficients of composite materials based on energy principles. J. Comp. Mat. 2(3):380-404.
312. Schellekens JCJ (1990) Computational Strategies for Composite Structures, TU Delft
313. Schuëller GI, (1997) A state-of-the-art report on computational stochastic mechanics. Prob. Engrg. Mech. 12(4):197-321.
314. Selvadurai APS, Boulon MJ, (1995) Mechanics of Geomaterial Interfaces. Stud. Appl. Mech. 42. Elsevier
315. Sevostianov I, Kachanov M, (2002) Explicit cross-property correlations for anisotropic two-phase composite materials. J. Mech. Phys. Solids. 50: 253-282.
316. Sevostianov I, Verijenko V, Kachanov M, (2002) Cross-property correlations for short-fiber reinforced composites with damage and their experimental verification. Comp. Part B: Engrg. 33:205-213.
317. Shiao MC, Chamis CC, (1999) Probabilistic evaluation of fuselage-type composite structures. Prob. Engrg. Mech. 14:179-187.
318. Shih CF, (1991) Cracks on bimaterial interfaces: elasticity and plasticity aspects. Mat. Sci. Engrg. A143:77-90.
319. Shinozuka M, Deodatis G, (1988) Stochastic process models for earthquake ground motion. Prob. Engrg. Mech. 3(3):114-123.

320. Sigmund O, Materials with prescribed constitutive parameters: an inverse homogenization problem. Int. J. Sol. Struct. 31(17): 2313-2329, 1994.
321. Simkins DC, Li S, (2003) Effective bending stiffness for plates with microcracks. Arch. Appl. Mech. 73:282-309.
322. Simo JC et al., (1985) A perturbed Lagrangian formulation for the finite element solution of contact problems. Comput. Meth. Appl. Mech. Engrg. 51:163-180.
323. Simonovski I, Boltežar M, (2003) The norms and variances of the Gabor, Morlet and general harmonic wavelet functions. J. Sound. Vibr. 264:545-557.
324. Singh BN, Yadav D, Iyengar NGR, (2001) Natural frequencies of composite plates with random material properties using higher-order shear deformation theory. Int. J. Mech. Sci. 43:2193-2214.
325. Van der Sluys O, Schreurs PJG, Meijer HEH, (1999) Effective properties of a viscoplastic constitutive model obtained by homogenization. Mech. Mat. 31:743-759.
326. Sobczyk K, (1991) Stochastic Differential Equations with Applications to Physics and Engineering. Kluwer Academic Publishers
327. Sobczyk K, Spencer BF (1992) Random Fatigue: from Data to Theory. Academic Press
328. Steinberg BZ, McCoy JJ, Mirotznik M, (2000) A multiresolution approach to homogenization and effective modal analysis of complex boundary value problems. SIAM J. Appl. Math. 60(3):939-966.
329. Steinberg BZ, McCoy JJ, (1999) A multiresolution homogenization of modal analysis with application to layered media. Math. Comput. Simul. 50:393-417.
330. Strzelecki T, (1996) Mechanics of Heterogeneous Media. Homogenization Theory (in Polish). Dolnośląskie Wydawnictwo Edukacyjne, Wrocław
331. Sung DU, Kim CG, Hong CS, (2002) Monitoring of impact damages in composite laminates using wavelet transform. Comp. Part B: Engrg. 33:35-43.
332. Suquet P, Continuum Micromechanics. CISM Courses and Lectures No 377. Springer Wien New York.
333. Sutherland LS, Guedes Soares C, (1997) Review of probabilistic models of the strength of composite materials. Reliability Engrg & System Safety 56:183-196.
334. Sweldens W (1994) The Construction and Application of Wavelets in Numerical Analysis. Ph.D. Thesis, Columbia
335. Tabiei A, Sun J, (1999) Statistical aspects of strength size effect of laminated composite materials. Comp. Struct. 46:209-216.
336. Talreja R, (1986) Composite laminates with cracks. Engrg. Fract. Mech. 25(5/6):751-762.
337. Talreja S (edr.), (1994) Damage Mechanics of Composite Materials, Elsevier
338. Tanov R, Tabiei A, (2000) A note on finite element implementation of sandwich shell homogenization. Int. J. Num. Meth. Engrg. 48:467-473.

339. Tartar L, (1992) On mathematical tools for studying partial differential equations of continuum physics: H-measures and Young measures. In: Butazzo G et al., New Developments in Partial Differential Equations and Applications to Mathematical Physics, Plenum Press, pp. 201-217.

340. Terada K, Kikuchi N, (2001) A class of general algorithms for multi-scale analyses of heterogeneous media, Comput. Meth. Appl. Mech. Engrg. 190:5427-5464.

341. Terada K, Miura T, Kikuchi N, (1997) Digital image-based modeling applied to the homogenization analysis of composite materials. Comput. Mech. 20:331-346.

342. Thierauf G, (1995) Optimal topologies of structures. Homogenization, pseudo-elastic approximation and the bubble-method. Engrg. Comput. 13(1):86-102.

343. Tian J, Li Z, Su X, (2003) Crack detection in beams by wavelet analysis of transient flexural waves. J. Sound Vibr. 261:715-727.

344. Timoshenko S, Goodier SN (1951) Theory of Elasticity. McGraw-Hill, Inc.

345. Todinov MT, (2002) Distribution of properties from sampling inhomogeneous materials by line transects. Prob. Engrg. Mech. 17:131-141.

346. Todinov MT, (2003) Statistics of inhomogeneous media formed by nucleation and growth. Prob. Engrg. Mech. 18:139-149.

347. Torquato S, (2000) Modeling of physical properties of composite materials. Int. J. Sol. Struct. 37:411-422.

348. Trykozko A, Zijl W, (2002) Complementary finite element methods applied to the numerical homogenization of 3D absolute permeability. Comm. Num. Meth. Engrg. 18:31-41.

349. Tryon RG, Cruse TA, (1997) Probabilistic mesomechanical fatigue crack nucleation model. ASME J. Engrg. Mat. Techn. 119:65-70.

350. Tsai GG, Doyle JF, Sun CT, (1987) Frequency effects on the fatigue life and damage of graphite/epoxy composites. J. Comp. Mat. 21:2-13.

351. Tsai SW, Hahn HT (1980) Introduction to Composite Materials.

352. Tsai SW, Wu EM, (1971). A general theory of strength for anisotropic materials. J. Comp. Mat. 5:58-80.

353. Tsukrov I, Kachanov M, (2000) Effective moduli of an anisotropic material with elliptical holes of arbitrary orientational distribution. Int. J. Sol. Struct. 37:5919-5941.

354. Tsurui A et al., (1989) Time variant structural reliability analysis using diffusive crack growth models. Engrg. Fract. Mech. 34(1):153-167.

355. Tveergard V (1990) Failure by ductile cavity growth at a metal/ceramic interface. DCAMM Rep. No. 406

356. Van den Nieuwenhof B, Coyette JP, (2003) Modal approaches for the stochastic finite element analysis of structures with material and geometrical uncertainties. Comput. Meth. Appl. Mech. Engrg. 192:3705-3729.

357. Vanmarcke E (1983) Random fields. Analysis and Synthesis. MIT Press

358. Wang W, Jasiuk I, (1998) Effective elastic constants of particulate composites with inhomogeneous interphases. J. Comp. Mat. 32(15):1391-1424.
359. Wang X, Sun JQ, (2003) Random fatigue of a higher order sandwich beam with parameter uncertainties. J. Sound Vibr. 260:349-356.
360. Waite SR, (1990) Use of embedded optical fiber for significant fatigue damage detection in composite materials. Composites 21(3):225-231.
361. Wang P, (1992) Homogenization for inhomogeneous elastic body containing gas bubbles. Appl. Anal. 44:1-20.
362. Weihe S, Kröplin B, (1993) Micromechanical simulation of the initiation of a delamination. Proceedings of MECAMAT'93 Seminar: Micromechanics of Materials, France, pp 485-498.
363. Wheeler L, Luc C, (1999) On conditions at an interface between two materials in three-dimensional space. Math. Mech. Sol. 4:183-200.
364. Whitworth HA, (1987) Modeling stiffness reduction of graphite/epoxy composite laminates. J. Comp. Mat. 21:362-372.
365. Wiener N, (1958) Nonlinear Problems in Random Theory. MIT Press
366. Wilczyński AP, Klasztorny M, (2000) Determination of complex compliances of fibrous polymeric composites. J. Comp. Mat. 34(1):2-26.
367. Williams JG, (1987) Fracture Mechanics of Polymers. Ellis Horwood, Inc.
368. Więckowski Z (1999) Finite Element Method Application in Some Nonlinear Problems of Mechanics (in Polish). Sci. Bull. T.U. Łódź No. 815
369. Wolf JP, Song C, (2001) The scaled boundary finite-element method – a fundamental solution-less boundary-element method. Comput. Meth. Appl. Mech. Engrg. 190:5551-5568.
370. Woźniak Cz, (2001) Mechanics of Elastic Plates and Shells. PWN, Warszawa
371. Wriggers P, Zavarise G, (1996) On the application of augmented Lagrangian techniques for nonlinear constitutive laws in contact interfaces. Comm. Appl. Num. Meth., 9:815-824.
372. Wu WF, Cheng HC, Kang CK, (2000) Random field formulation of composite laminates. Comp. Struct. 49:87-93.
373. Wu WF, Ni CC, (2003) A study of stochastic fatigue crack growth modeling through experimental data. Prob. Engrg. Mech. 18:107-118.
374. Xiao F, Hui CY, (1994) A boundary element method for calculating the K field for cracks along a bimaterial interface. Comput. Mech. 15:58-78.
375. Yadav D, Verma N, (1998) Free vibration of composite circular cylindrical shells with random material parameters. Part I: general theory. Comp. Struct. 41:331-338.
376. Yang C, (2000) Design and analysis of composite pipe joints under tensile loading, J. Comp. Mat. 34(4):332-349.
377. Yao JTP. et al., (1986) Stochastic fatigue, fracture and damage analysis. Struct Safety 3:231-267.
378. Yeh HL, Yeh HY, (2000) Elastic moduli of laminated composites revisited by using statistical analysis. Comp. Part B: Engrg. 31:57-64.

379. Yokobori T, (1979) A critical evaluation of mathematical equations for fatigue crack growth with special reference to ferrite grain size and monotonic yield strength dependence. In: Fong JT, (edr.) Fatigue Mechanisms, Proc. ASTM-NBS-NSF Symp, ASTM STP 675, pp 683-706.

380. Young RK (1993) Wavelet Theory and its Applications. Kluwer

381. Zaidman M, Ponte-Castaneda P, (1995) Effective yield surfaces for anisotropic composite materials. In: Parker DF, England AH, (edrs.) IUTAM Symposium on Anisotropy, Inhomogeneity and Nonlinearity in Solid Mechanics. Kluwer, pp 415-422.

382. Zavarise G. et al., (1998) A method for solving contact problems. Int. J. Num. Meth. Engrg., 42(3):473-498.

383. Zeman J, Sejnoha M, (2001) Numerical evaluation of effective elastic properties of graphite fiber tow impregnated by polymer matrix. J. Mech. Phys. Solids 49:69-90.

384. Zhao YH, Weng GJ, (1996) Plasticity of a two-phase composite with partially debonded inclusions. Int. J. Plasticity 12(6):781-804.

385. Zheng XJ, Wang X, Zhou YH, (2000) Magnetoelastic analysis of non-circular superconducting partial torus. Int. J. Sol. Struct. 37:563-576.

386. Zieliński AP, Frey F, (2003) Nonlinear weighted residual approach: application to laminated beams. Comput. & Struct. 81:1087-1098.

387. Zienkiewicz OC, Taylor RL (2000) The Finite Element Method. 5th Edition. Butterworth-Heinemann

388. Zohdi TI, Wriggers P, (2001) A model for simulating the deterioration of structural-scale material responses of microheterogeneous solids. Comput. Meth. Appl. Mech. Engrg. 190(22-23):2803-2823.

389. Zweben C, Hahn HT, Chou TW (edrs.), (1989) Delaware Composites Design Encyclopedia, Lancaster, Technomic Publishing Co.

Index